T0297058

# Advances in Intelligent Systems and Computing

Volume 425

**Series editor**

Janusz Kacprzyk, Polish Academy of Sciences, Warsaw, Poland
e-mail: kacprzyk@ibspan.waw.pl

*About this Series*

The series "Advances in Intelligent Systems and Computing" contains publications on theory, applications, and design methods of Intelligent Systems and Intelligent Computing. Virtually all disciplines such as engineering, natural sciences, computer and information science, ICT, economics, business, e-commerce, environment, healthcare, life science are covered. The list of topics spans all the areas of modern intelligent systems and computing.

The publications within "Advances in Intelligent Systems and Computing" are primarily textbooks and proceedings of important conferences, symposia and congresses. They cover significant recent developments in the field, both of a foundational and applicable character. An important characteristic feature of the series is the short publication time and world-wide distribution. This permits a rapid and broad dissemination of research results.

*Advisory Board*

Chairman
Nikhil R. Pal, Indian Statistical Institute, Kolkata, India
e-mail: nikhil@isical.ac.in

Members
Rafael Bello, Universidad Central "Marta Abreu" de Las Villas, Santa Clara, Cuba
e-mail: rbellop@uclv.edu.cu
Emilio S. Corchado, University of Salamanca, Salamanca, Spain
e-mail: escorchado@usal.es
Hani Hagras, University of Essex, Colchester, UK
e-mail: hani@essex.ac.uk
László T. Kóczy, Széchenyi István University, Győr, Hungary
e-mail: koczy@sze.hu
Vladik Kreinovich, University of Texas at El Paso, El Paso, USA
e-mail: vladik@utep.edu
Chin-Teng Lin, National Chiao Tung University, Hsinchu, Taiwan
e-mail: ctlin@mail.nctu.edu.tw
Jie Lu, University of Technology, Sydney, Australia
e-mail: Jie.Lu@uts.edu.au
Patricia Melin, Tijuana Institute of Technology, Tijuana, Mexico
e-mail: epmelin@hafsamx.org
Nadia Nedjah, State University of Rio de Janeiro, Rio de Janeiro, Brazil
e-mail: nadia@eng.uerj.br
Ngoc Thanh Nguyen, Wroclaw University of Technology, Wroclaw, Poland
e-mail: Ngoc-Thanh.Nguyen@pwr.edu.pl
Jun Wang, The Chinese University of Hong Kong, Shatin, Hong Kong
e-mail: jwang@mae.cuhk.edu.hk

More information about this series at http://www.springer.com/series/11156

Sabu M. Thampi · Sanghamitra Bandyopadhyay
Sri Krishnan · Kuan-Ching Li
Sergey Mosin · Maode Ma
Editors

# Advances in Signal Processing and Intelligent Recognition Systems

Proceedings of Second International
Symposium on Signal Processing
and Intelligent Recognition Systems
(SIRS-2015), December 16–19, 2015,
Trivandrum, India

 Springer

*Editors*
Sabu M. Thampi
Indian Institute of Information Technology
   and Management – Kerala (IIITM-K)
Trivandrum, Kerala
India

Sanghamitra Bandyopadhyay
Machine Intelligence Unit
Indian Statistical Institute
Kolkata, West Bengal
India

Sri Krishnan
Department of Electrical and Computer
   Engineering
Ryerson University
Toronto, ON
Canada

Kuan-Ching Li
Department of Computer Science and
   Information Engineering
Providence University
Taichung
Taiwan

Sergey Mosin
Computer Engineering Department
Vladimir State University
Vladimir Region
Russia

Maode Ma
School of Electrical and Electronic
   Engineering
Nanyang Technological University
Singapore
Singapore

ISSN 2194-5357          ISSN 2194-5365   (electronic)
Advances in Intelligent Systems and Computing
ISBN 978-3-319-28656-3          ISBN 978-3-319-28658-7   (eBook)
DOI 10.1007/978-3-319-28658-7

Library of Congress Control Number: 2015958912

© Springer International Publishing Switzerland 2016
This work is subject to copyright. All rights are reserved by the Publisher, whether the whole or part
of the material is concerned, specifically the rights of translation, reprinting, reuse of illustrations,
recitation, broadcasting, reproduction on microfilms or in any other physical way, and transmission
or information storage and retrieval, electronic adaptation, computer software, or by similar or dissimilar
methodology now known or hereafter developed.
The use of general descriptive names, registered names, trademarks, service marks, etc. in this
publication does not imply, even in the absence of a specific statement, that such names are exempt from
the relevant protective laws and regulations and therefore free for general use.
The publisher, the authors and the editors are safe to assume that the advice and information in this
book are believed to be true and accurate at the date of publication. Neither the publisher nor the
authors or the editors give a warranty, express or implied, with respect to the material contained herein or
for any errors or omissions that may have been made.

Printed on acid-free paper

This Springer imprint is published by SpringerNature
The registered company is Springer International Publishing AG Switzerland

# Preface

This Edited Volume contains a selection of refereed and revised papers originally presented at the second International Symposium on Signal Processing and Intelligent Recognition Systems (SIRS-2015), December 16–19, 2015, Trivandrum, India. SIRS-2015 provided a forum for the sharing, exchange, presentation and discussion of original research results in both methodological issues and different application areas of signal processing and pattern recognition.

Credit for the quality of the symposium proceedings goes first and foremost to the authors. They contributed a great deal of effort and creativity to produce this work, and we are very thankful that they chose SIRS-2015 as the place to present it. All the authors who submitted papers, both accepted and rejected, are responsible for keeping the SIRS program vital. The program committee received 175 submissions. The committee had a very challenging task of choosing high quality submissions. Each paper was peer reviewed by at least three or more independent referees and the papers were selected based on the referee recommendations. The technical program of SIRS'15 comprises of 59 papers (41 regular papers and 18 short papers). This volume is organized into different topical sections. The papers offer stimulating insights into biometrics, digital watermarking, recognition systems, image and video processing, signal and speech processing, pattern recognition, machine learning and knowledge-based systems. Two workshops were co-located with the symposium: *workshop on Advances in Image Processing, Computer Vision, and Pattern Recognition (IWICP-2015)* and *workshop on Signal Processing for Wireless and Multimedia Communications (SPWMC'15).*

The success of such an event is mainly due to the hard work and dedication of a number of people and the collaboration of several institutions. We are grateful to the members of the program committee for reviewing and selecting papers in a very short period of time. Many thanks to all the Chairs and their involvement and support have added greatly to the quality of the symposium. We also wish to thank all the members of the Advisory Committee, whose work and commitment were invaluable. We would like to express our sincere gratitude to local organizing committees that has made this event a success. Our special thanks also to the keynote speakers and tutorial presenters for their effort in preparing the lectures. The EDAS conference system proved very helpful during the submission, review, and editing phases.

We wish to express our sincere thanks to Thomas Ditzinger, Senior Editor, Engineering/AppliedSciences Springer-Verlag and Janusz Kacprzyk, Series Editor for their help and cooperation.

Finally, we hope that you will find this edited book to be a valuable resource in your professional, research, and educational activities whether you are a student, academic, researcher, or a practicing professional.

Sabu M. Thampi
Sanghamitra Bandyopadhyay
Sri Krishnan
Kuan-Ching Li
Sergey Mosin
Maode Ma

# Organization

## Organized by

Indian Institute of Information Technology and Management-Kerala (IIITM-K), Trivandrum, India

http://www.iiitmk.ac.in

*in association with*

# Committee

## Advisory Committee

| | |
|---|---|
| Janusz Kacprzyk | Polish Academy of Sciences, Poland |
| Hideyuki Takagi | Kyushu University, Japan |
| Oge Marques | Florida Atlantic University (FAU) (Boca Raton, Florida), USA |
| Sankar Kumar Pal | Indian Statistical Institute, Kolkata, India |
| Sugata Sanyal | Tata Institute of Fundamental Research, India |
| Alexander Gelbukh | Instituto Politecnico Nacional, Mexico |
| Dharma P. Agrawal | University of Cincinnati, USA |
| Pramod K. Varshney | Syracuse University, USA |
| Sushmita Mitra | Indian Statistical Institute, Kolkata, India |
| Selwyn Piramuthu | University of Florida, USA |
| Mario Koeppen | Kyushu Institute of Technology, Japan |
| Suash Deb | INNS India Regional Chapter |
| Nallanathan Arumugam | King's College London, United Kingdom |
| Sri Krishnan | Ryerson University, Toronto, Canada |
| Salah Bournnane | Ecole Centrale Marseille, France |
| P. Nagabhushan | University of Mysore, India |
| Rajasree M.S | Director, IIITM-K, India |
| Elizabeth Sherly | IIITM-K, India |

## General Chair

| | |
|---|---|
| Sanghamitra Bandyopadhyay | Indian Statistical Institute, Kolkata, India |

## Organising Chair

| | |
|---|---|
| Sabu M. Thampi | IIITM-K, India |

## Program Chairs

| | |
|---|---|
| Kuan-Ching Li | Providence University, Taiwan |
| Sergey Mosin | Vladimir State University, Russia |
| Maode Ma | Nanyang Technological University, Singapore |

# TPC Members/Additional Reviewers

| | |
|---|---|
| Cesar Cardenas | Tecnologico de Monterrey - Campus Queretaro, Mexico |
| Marcelo Carvalho | University of Brasilia, Brazil |
| El-Sayed El-Alfy | King Fahd University of Petroleum and Minerals (KFUPM), Saudi Arabia |
| Rodolfo Oliveira | Nova University of Lisbon, Portugal |
| Aniruddha Bhattacharjya | Guru Nanak Institute of Technology (GNIT), India |
| Mohammed Khan | Indian Institute of Technology, Hyderabad, India |
| Ryszard Tadeusiewicz | AGH University of Science and Technology, Poland |
| Maaruf Ali | University of East London, United Kingdom |
| Biju Issac | Teesside University, Middlesbrough, United Kingdom |
| Felix Albu | Valahia University of Targoviste, Romania |
| Mihaela Albu | Politehnica University of Bucharest, Romania |
| Anna Antonyová | University of Prešov in Prešov, Slovakia |
| Tarek Bejaoui | University of Paris-Sud 11, France |
| Vikrant Bhateja | Shri Ramswaroop Memorial Group of Colleges, India |
| Igor Bisio | University of Genoa, Italy |
| Tufik Buzid | Algabel Algharbi University, Tripoli, Libya |
| Chinmay Chakraborty | Birla Institute of Technology, Mesra, India |
| Zhe Chen | Northeastern University, P.R. China |
| Wei-Yu Chiu | Yuan Ze University, Taiwan |
| Sung-Bae Cho | Yonsei University, Korea |
| Ashutosh Dubey | Trinity Institute of Technology and Research Bhopal, India |
| Mourad Fakhfakh | University of Sfax, Tunisia |
| Ponnambalam G | Monash University Sunway Campus, Malaysia |
| Manish Gupta | Hindustan Institute of Technology and Management, Agra, India |
| Saad Harous | UAE University, UAE |
| Sarangapani Jagannathan | Missouri University of Science and Technology, USA |
| Raveendranathan Kalathil Chellappan | LBS Institute of Technology for Women Poojappura, India |
| Abhishek Midya | National Institute of Technology, Silchar, India |
| Yusuke Nojima | Osaka Prefecture University, Japan |
| Radu-Emil Precup | Politehnica University of Timisoara, Romania |
| Priya Ranjan | Templecity Institute of Technology and Engineering, USA |

| Ramesh Rayudu | Victoria University of Wellington, New Zealand |
| Shubhajit Roy Chowdhury | School of Computing and Electrical Engineering, IIT Mandi, India |
| Ravi Subban | Pondicherry University, Pondicherry, India |
| Daisuke Umehara | Kyoto Institute of Technology, Japan |
| Gancho Vachkov | The University of the South Pacific (USP), Fiji |
| Jaap van de Beek | Luleå University of Technology, Sweden |
| Fanggang Wang | Beijing Jiaotong University, P.R. China |
| Boyang Zhou | State Grid Corporation of China, P.R. China |
| Brian Sadler | Army Research Laboratory, USA |
| Batu Krishna Chalise | Arraycomm, USA |
| Liau Eric | Intel Corporation, Germany |
| Sunil Kumar Kopparapu | Tata Consultancy Services, India |
| Erwin Daculan | University of San Carlos, Philippines |
| Pitoyo Hartono | Chukyo University, Japan |
| Wei-Chiang Hong | Oriental Institute of Technology, Taiwan |
| Radu Vasiu | Politehnica University of Timisoara, Romania |
| Paolo Crippa | Universita Politecnica delle Marche, Italy |
| Kuan-Chieh Huang | National Cheng Kung University, Taiwan |
| Sergey Mosin | Vladimir State University, Russia |
| Mohammadali Mohammadi | Shahrekord University, Iran |
| Mahfuzah Mustafa | Universiti Malaysia Pahang, Malaysia |
| Teddy Gunawan | International Islamic University Malaysia, Malaysia |
| Nor Hayati Saad | UiTM, Malaysia |
| Anas Abou El Kalam | UCA - ENSA/OSCARS Laboratory, Morocco |
| Siby Abraham | University of Mumbai, India |
| Amit Acharyya | IIT HYDERABAD, India |
| Ali Al-Sherbaz | The University of Northampton, United Kingdom |
| Angeliki Alexiou | University of Piraeus, Greece |
| Belal Amro | Hebron University, Palestine |
| Markos Anastasopoulos | University of Bristol, United Kingdom |
| Manjunath Aradhya | Sri Jayachamarajendra College of Engineering, India |
| Ognjen Arandjelovic | University of St Andrews, United Kingdom |
| Krishna Asawa | Jaypee Institute of Information Technology, India |
| Ouarda Assas | Universite of Msila, Algeria |
| Vahida Attar | College of Engineering Pune, India |
| Vinayak Bairagi | University of Pune, India |
| Valentina Balas | Aurel Vlaicu University of Arad, Romania |
| Faycal Bensaali | Qatar University, Qatar |
| Zoran Bojkovic | University of Belgrade, Serbia |
| Ivo Bukovsky | Czech Technical University in Prague, Czech Republic |

Joao Paulo Carvalho          Instituto Superior Tecnico - Technical University
                             of Lisbon, Portugal
Mehmet Celenk                Ohio University, USA
Nabendu Chaki                University of Calcutta, India
Jayasree Chakraborty         Mamorial Sloan Kettering Cancer Center, USA
Anjali Chandavale            University of Pune, India
Rama Seshagiri Rao           JNTU Hyderabad, India
  Channapragada
Amitava Chatterjee           Jadavpur University, India
Ouyang Chen-Sen              I-Shou University, Taiwan
Yordan Chervenkov            Naval Academy - Varna, Bulgaria
Silvana Costa                Instituto Federal de Educacao, Ciencia e
                             Tecnologia da Paraiba, Brazil
Pasquale Daponte             University of Sannio, Italy
Ashok Kumar Das              International Institute of Information Technology,
                             Hyderabad, India
Kenneth Dawson-Howe          Trinity College Dublin, Ireland
Tiago de Carvalho            Federal Rural University of Pernambuco, Brazil
Grzegorz Debita              Wroclaw University of Technology, Poland
Vani Devi                    IIST, India
Moussa Diaf                  Universite Mouloud Mammri, Algeria
Ibrahim El rube'             Taif University, Saudi Arabia
Vaibhav Gandhi               Middlesex University, United Kingdom
Rama Garimella               IIIT Hyderabad, India
Nithin George                IIT Gandhinagar, India
Sudhanshu Gonge              University of Pune, India
Steven Guan                  Xian Jiatong-Liverpool University, Australia
Xiaoning Guo                 Multimedia University, Malaysia
Maki Habib                   The American University in Cairo, Egypt
Bo Han                       Aalborg University, Denmark
Thomas Hanne                 University of Applied Sciences, Switzerland
Yong Hu                      The University of Hong Kong, Hong Kong
Hirotaka Inoue               National Institute of Technology, Kure College,
                             Japan
Marina Ivasic-Kos            University of Rijeka, Croatia
Frank Klawonn                Ostfalia University, Germany
Mario Köppen                 Kyushu Institute of Technology, Japan
Andrey Krylov                Lomonosov Moscow State University, Russia
Ajey Kumar                   Symbiosis Centre for Information Technology,
                             India
Zsofia Lendek                Technical University of Cluj-Napoca, Romania
Edwin Lughofer               University of Linz, Austria
Jayamohan M                  College of Applied Science, India
George Magoulas              Birkbeck College, University of London,
                             United Kingdom

| Noor Mahammad Sk | IIITDM Kancheepuram, India |
| M Manikandan | Anna University, India |
| Sapan Mankad | Nirma University, Ahmedabad, India |
| Joycee Mekie | IIT Gandhinagar, India |
| Varun Menon | S C M S School of Engineering and Technology, India |
| Deepak Mishra | IIST, India |
| Marek Miskowicz | AGH University of Science and Technology, Poland |
| Lahcène Mitiche | University of Djelfa, Algeria |
| Ravibabu Mulaveesala | Indian Institute of Technology Ropar, India |
| Sakthi Muthiah | LNMIIT, India |
| Jyothisha Nair | Amrita University, India |
| Ibrahim Nasir | Sebha University, Libya |
| Nizampatnam Neelima | Engineering, India |
| Uche Okonkwo | University of Pretoria, South Africa |
| Kalman Palagyi | University of Szeged, Hungary |
| Rosaura Palma-Orozco | Instituto Politecnico Nacional, Mexico |
| Hemprasad Patil | Visvesvaraya National Institute of Technology, India |
| Isabella Poggi | Roma Tre University, Italy |
| Rahul Pol | Pune University, India |
| M.V.N.K. Prasad | IDRBT, India |
| Padma Prasada | VTU Belgaum, India |
| V.B. Surya Prasath | University of Missouri-Columbia, USA |
| Hugo Proença | University of Beira Interior, Portugal |
| Grienggrai Rajchakit | Maejo University, Thailand |
| Ranjana Rajnish | Amity University, Lucknow, India |
| Alexandre Ramos | Federal University of Itajubá, Brazil |
| Ajita Rattani | Michigan State University, USA |
| Carlos Regis | IFPB, Brazil |
| Asharaf S | IIITMK, Trivandrum, India |
| Sachin Kumar S | Amrita Vishwa Vidyapeetham, India |
| Sumitra S | IIST, India |
| Beatriz Sainz | University of Valladolid, Spain |
| Ajit Samasgikar | VTU Belgaum, India |
| Andrews Samraj | Mahendra Engineering College, India |
| Luciano Sanchez | University of Oviedo, Spain |
| Valerio Scordamaglia | University of Reggio Calabria, Italy |
| Kandasamy Selvaradjou | Pondicherry Engineering College, India |
| Kaushal Shukla | Indian Institute of Technology, India |
| Patrick Siarry | University of Paris XII, France |
| Vladimir Spitsyn | Tomsk Polytechnic University, Russia |
| Mu-Chun Su | National Central University, Taiwan |
| Gorthi Subrahmanyam | IIST, India |

Roberto Tagliaferri          University of Salerno, Italy
Rohit Thanki                 C U Shah Unversity, India
Ciza Thomas                  College of Engineering Trivandrum, India
Shikha Tripathi              Amrita Vishwa Vidhyapeetham, India
Ralph Turner                 Eastern Kentucky University, USA
Jayaraman Valarmathi         Vellore Institute of Technology, India
Zita Vale                    Polytechnic Institute of Porto, Portugal
Michael Vrahatis             University of Patras, Greece
Haixin Wang                  Fort Valley State University, USA
Rolf Wurtz                   Ruhr-University of Bochum, Germany
Ales Zamuda                  University of Maribor, Slovenia
Hector Zenil                 Oxford University and Karolinska Institute,
                               United Kingdom
Ming-Yue Zhai                North China Electric Power University,
                               P.R. China
Shang-Ming Zhou              Swansea University, United Kingdom
Reyer Zwiggelaar             Aberystwyth University, United Kingdom
Arun Gopalakrishnan          Centre for Development of Advanced Computing,
                               India
V Satheesh Prabhu            Cdac, India
Sreeraman Rajan              Defence Research and Development
                               Canada-Ottawa, Canada
Sheeba Rani                  IIST Trivandrum, India
Anustup Choudhury            Sharp Laboratories of America, USA
Julius Eiweck                Alcatel-Lucent Austria, Austria
Chi-Keong Goh                Rolls-Royce Advanced Technology Centre,
                               Singapore
Rajeev Kumaraswamy           QuEST Global Engineering Services Pvt Ltd,
                               India
Kumar Padmanabh              Robert Bosch, India
Andrei Shin                  Samsung SDS Co., Ltd., Korea
Gustavo Fernández            AIT Austrian Institute of Technology, Austria
  Domínguez
Kwasi Opare                  MobileLink LAB, University of Electronic
                             Science and Technology, P.R. China

# Second International Workshop on Advances in Image Processing, Computer Vision, and Pattern Recognition (IWICP-2015)

## TPC Members/Additional Reviewers

| | |
|---|---|
| Ayan Mondal | Indian Institute of Technology, Kharagpur, India |
| Biju Issac | Teesside University, Middlesbrough, United Kingdom |
| Mohammad Faiz Liew Abdullah | Universiti Tun Hussein Onn Malaysia (UTHM), Malaysia |
| Anna Antonyová | University of Prešov in Prešov, Slovakia |
| Tarek Bejaoui | University of Paris-Sud 11, France |
| Ravi Subban | Pondicherry University, Pondicherry, India |
| Geert Verdoolaege | Ghent University, Belgium |
| Adib Chowdhury | University College of Technology Sarawak, Malaysia |
| Naveen Kolla | Geethanjali Institute of Science and Technology Nellore, India |
| Honglei Zhang | Tampere University of Technology, Finland |
| Moulay Akhloufi | Laval University, Canada |
| Vikrant Bhateja | Shri Ramswaroop Memorial Group of Professional Colleges, Lucknow (UP), India |
| Sung-Bae Cho | Yonsei University, Korea |
| Bijoy Ghosh | Texas Tech University, USA |
| Manish Gupta | Hindustan Institute of Technology and Management, Agra, India |
| Sunil Kumar Kopparapu | Tata Consultancy Services, India |
| Srimanta Mandal | Indian Institute of Technology Mandi, India |
| Badri Narayan Subudhi | Indian Statistical Institute, Kolkata, India |
| Davide Valeriani | University of Essex, United Kingdom |
| Abdul Halim Ali | Universiti Kuala Lumpur - International College, Malaysia |
| Burhan Gulbahar | Ozyegin University, Turkey |
| Thanikaiselvan V | VIT University, India |
| Paolo Crippa | Università Politecnica delle Marche, Italy |
| Kuan-Chieh Huang | National Cheng Kung University, Taiwan |
| Amita Kapoor | Shaheed Rajguru College of Applied Sciences for Women, India |
| Rakesh Manjappa | Indian Institute of Science, India |
| Ankit Chaudhary | Truman State University, USA |
| Prasant Sahu | IIT Bhubaneswar, India |
| Shikha Agrawal | Rajiv Gandhi Proudyogiki Vishwavidyalaya, Bhopal, India |

| | |
|---|---|
| Manjunath Aradhya | Sri Jayachamarajendra College of Engineering, India |
| Sreeparna Banerjee | West Bengal University of Technology, India |
| Krishna Battula | Jawaharlal Nehru Technological University Kakinada, India |
| Evgeny Belyaev | University of Oulu, Finland |
| Philip Branch | Swinburne University of Technology, Australia |
| Prabhakar C j | Kuvempu University, India |
| Azza Elaskary | Atomic Energy Authority, Egypt |
| Omar Farooq | Aligarh Muslim University, Aligarh, India |
| Gianluigi Ferrari | University of Parma, Italy |
| Katiganere Siddaramappa Hareesha | Manipal University, India |
| M. Udin Harun Al Rasyid | Politeknik Elektronika Negeri Surabaya (PENS) - Indonesia, Indonesia |
| Katerina Kabassi | TEI of the Ionian Islands, Greece |
| Maurice Khabbaz | Notre-Dame University, Lebanon |
| Yassine Khlifi | Umm Al-Qura University, KSA, Saudi Arabia |
| Sreeraj M | Cochin University of Science and Technology, India |
| Marek Miskowicz | AGH University of Science and Technology, Poland |
| Júlio Nievola | Pontifícia Universidade Catolica do Paraná – PUCPR, Brazil |
| Subhendu Pani | BPUT, India |
| Francesco Paolucci | Scuola Superiore Sant'Anna, Italy |
| M.V.N.K. Prasad | IDRBT, India |
| Kandarpa Sarma | Gauhati University, India |
| Sheikh Mohammed Shariful Islam | Center for Control of Chronic Diseases (CCCD), Bangladesh |
| Deepak Singh | Dayalbagh Educational Institute, India |
| Muthukumar Subramanyam | National Institute of Technology, Puducherry, India |
| Senthilkumar Thangavel | Amrita School of Engineering, India |
| Shikha Tripathi | Amrita School of Engineering, Amrita Vishwa Vidhyapeetham, India |
| Vani Vasudevan | Al Yamamah University, Saudi Arabia |
| Ujjwal Verma | Manipal University, India |
| Qiang Yang | Zhejiang University, P.R. China |
| Nuri Yilmazer | Texas A&M University-Kingsville, USA |
| Musheer Ahmad | Jamia Millia Islamia, New Delhi, India |
| Fayas Asharindavida | Taif University, Saudi Arabia |
| Bhaskar Belavadi | BGS Health & Education city, Uttarahalli Road, Bangalore, India |
| P. Pablo Garrido Abenza | Miguel Hernandez University, Spain |

| | |
|---|---|
| Rahul Gupta | Manipal Institute of Technology, India |
| Raza Hasan | Middle East College, Oman |
| Thiang Hwang Liong Hoat | Petra Christian University, Indonesia |
| Pavan Kumar C | VIT University, India |
| Anan Liu | Tianjin University, P.R. China |
| Changqing Luo | Mississippi State University, USA |
| K Mahantesh | SJBIT, India |
| Ibrahim Missaoui | National Engineering School of Tunis, Tunisia |
| Carlos Oliveira | IFRJ, Brazil |
| Prakornchai Phonrattanasak | North Eastern University, Thailand |
| Rajiv Singh | Banasthali University, India |
| Kapil Wankhade | G.H. Raisoni College of Engineering Nagpur, INDIA, India |
| Akash Yadav | Indian Institute of Technology, Patna, India |
| Suja P. | Amrita School of Engineering, Amrita Vishwa Vidyapeetham, India |
| Vimina R | Rajagiri College of Social Sciences, India |
| Siriporn Dachasilaruk | Naresuan University, Thailand |
| Amrita A Manjrekar | Shivaji University, India |
| Afaf Merazi | Djillali Liabès University of Sidi Bel-Abbès, Algeria |
| Sabri Abdelouahed | Faculty of Science Dhar el Mahraz Fes, Morocco |
| Anna Bartkowiak | University of Wroclaw, Poland |
| Salah Bourennane | Ecole Centrale Marseille, France |
| Deepak Choudhary | LPU, India |
| Simon Fong | University of Macau, Macao |
| Steven Guan | Xian Jiatong-Liverpool University, Australia |
| Alex James | Nazarbayev University, Kazakhstan |
| Agilandeeswari Loganathan | VIT University, India |
| Pascal Lorenz | University of Haute Alsace, France |
| Rosaura Palma-Orozco | Instituto Politécnico Nacional, Mexico |
| Grienggrai Rajchakit | Maejo University, Thailand |
| Abdelmadjid Recioui | Universitry of Boumerdes, Algeria |
| Patrick Siarry | University of Paris XII, France |
| Georgios Sirakoulis | Democritus University of Thrace, Greece |
| Elpida Tzafestas | University of Athens, Greece |
| Michael Vrahatis | University of Patras, Greece |
| Reyer Zwiggelaar | Aberystwyth University, United Kingdom |
| G Deka | Directorate General of Training, India |
| Shuaib Ahmed | Tata Research Development and Design Center, India |
| Julius Eiweck | Alcatel-Lucent Austria, Austria |

| | |
|---|---|
| Hrishikesh Sharma | Innovation Labs, Tata Consultancy Services Ltd., India |
| Andrei Shin | Samsung SDS Co., Ltd., Korea |
| Balaji Balasubramaniam | Tata Research Development and Design Centre (TRDDC), India |
| Roberto Herrera Lara | National Polytechnic School, Ecuador |
| Vishnu Pendyala | Santa Clara University, USA |
| Karthikeyan Ramasamy | TECLEVER SOLUTIONS PVT LTD, India |
| Tadrash Shah | Bank of America, USA |
| Muharrem Tümçakr | Aselsan Inc. Defense Systems Technologies Division, Turkey |
| Anil Yekkala | Freelancer in the area of Image and Video Processing, India |
| Hrudya P | Amrita Center for Cyber Security, India |
| Sharare Kiani Harchegani | Electronic Research Institute of Sharif University of Technology, Iran |
| Ali Abbas | Higher Institute for Applied Science and Technology, Syria |
| Michael Barros | Waterford Institute of Technology, Ireland |
| Ayoub Bouziane | University Sidi Mohamed Ben Abdellah, Morocco |
| Waleed Halboob | Universiti Putyra Malaysia, Malaysia |
| Saif Mahmood | Universiti Putra Malaysia, Malaysia |
| Jirapong Manit | University of Luebeck, Germany |
| Ajoy Mondal | Indian Statistical Institute, India |
| Abdul Quyoom | Central University Rajasthan, India |
| Manoj Sharma | University of Allahabad, India |
| Ashkan Tashk | Shiraz University of Technology, Iran |
| Savita Lothe | Dr Babasaheb Ambedkar Marathawada University, India |
| Xiaojing Zhang | North China Electric Power University, P.R. China |

## Workshop on Signal Processing for Wireless and Multimedia Communications (SPWMC'15)

### TPC Members/Additional Reviewers

| | |
|---|---|
| Kamran Arshad | University of Greenwich, United Kingdom |
| Antoine Bagula | University of the Western Cape, South Africa |
| Ramiro Barbosa | Institute of Engineering of Porto, Portugal |
| Bao Rong Chang | National University of Kaohsiung, Taiwan |
| Phan Cong-Vinh | NTT University, Vietnam |

| | |
|---|---|
| Floriano De Rango | University of Calabria, Italy |
| Salvatore Distefano | Politecnico di Milano, Italy |
| El-Sayed El-Alfy | King Fahd University of Petroleum and Minerals (KFUPM), Saudi Arabia |
| Mohamed El-Tarhuni | American University of Sharjah, UAE |
| Steven Guan | Xian Jiatong-Liverpool University, Australia |
| Vana Kalogeraki | Athens University of Economics and Business, Greece |
| George Karagiannidis | Aristotle University of Thessaloniki, Greece |
| Chandan Karmakar | The University of Melbourne, Australia |
| Pascal Lorenz | University of Haute Alsace, France |
| Júlio Nievola | Pontificia Universidade Catolica do Paraná – PUCPR, Brazil |
| Madan Pande | International Institute of Information Technology - Bangalore, India |
| Sherif Rashad | Florida Polytechnic University, USA |
| Antonio Ruiz-Martínez | University of Murcia, Spain |
| Jorge Sá Silva | University of Coimbra, Portugal |
| Maytham Safar | Kuwait University, Kuwait |
| Luciano Sanchez | University of Oviedo, Spain |
| Björn Schuller | Imperial College London, United Kingdom |
| Patrick Siarry | University of Paris XII, France |
| Toshio Tsuji | Hiroshima University, Japan |
| Elpida Tzafestas | University of Athens, Greece |
| Athanasios Vasilakos | National Technical University of Athens, Greece |
| Michael Vrahatis | University of Patras, Greece |
| Reyer Zwiggelaar | Aberystwyth University, United Kingdom |
| Ravi Kodali | National Institute of Technology, Warangal, India |
| E Hari Krishna | KU College of Engineering & Technology, India |
| Navin Kumar | Amrita University, India |
| Dnyanesh Mantri | Pune University, India |
| Pratima Patel | Autonomous, India |
| Vasanthi S | SRM University, India |
| Giridhar Mandyam | Qualcomm, USA |
| Kyriakos Manousakis | Applied Communication Sciences, USA |
| Stephane Senecal | Orange Labs, France |
| Pawan Bhandari | Intel, India |
| Eugénia Bernardino | Polytechnic Institute of Leiria, Portugal |

# Contents

## Pattern Recognition/Machine Learning/Knowledge-Based Systems

# Part I
# Biometrics/Digital Watermarking/Recognition Systems

# Emotion Recognition from Facial Expressions for 4D Videos Using Geometric Approach

V.P. Kalyan Kumar, P. Suja and Shikha Tripathi

**Abstract** Emotions are important to understand human behavior. Several modalities of emotion recognition are text, speech, facial expression or gesture. Emotion recognition through facial expressions from video play a vital role in human computer interaction where the facial feature movements that convey the emotion expressed need to be recognized quickly. In this work, we propose a novel method for the recognition of six basic emotions in 4D video sequences of BU-4DFE database using geometric based approach. We have selected key facial points out of the 83 feature points provided in the BU-4DFE database. A video expressing emotion has frames containing neutral, onset, apex and offset of that emotion. We have identified the apex frame from a video sequence automatically. The Euclidean distance between the feature points in apex and neutral frame is determined and their difference in corresponding neutral and the apex frame is calculated to form the feature vector. The feature vectors thus formed for all the emotions and subjects are given to Random Forests and Support Vector Machine (SVM) for classification. We have compared the accuracy obtained by the two classifiers. Our proposed method is simple, uses only two frames and yields good accuracy for BU-4DFE database. We have determined optimum number of key facial points that could provide better recognition rate using the computed distance vectors. Our proposed method gives better results compared with literature and can be applied for real time implementation using SVM classifier and kinesics in future.

**Keywords** Facial emotions · Key feature points · Apex frame · Euclidean distance · Random forests · Support Vector Machine

## 1 Introduction

Since few decades scientists are keen to improve the communication between human and computers. Human computer interaction has become indispensable as

V.P.K. Kumar(✉) · P. Suja · S. Tripathi
Amrita Robotic Research Centre, Amrita School of Engineering,
Amrita Vishwa Vidyapeetham, Bangalore 560035, India
e-mail: {kalyanvpkumar,shikha.eee}@gmail.com, suja_rajesh@yahoo.co.in

© Springer International Publishing Switzerland 2016
S.M. Thampi et al. (eds.), *Advances in Signal Processing and Intelligent Recognition Systems*,
Advances in Intelligent Systems and Computing 425,
DOI: 10.1007/978-3-319-28658-7_1

3

computing has become common in our daily life. To make an effective human computer interface, the interaction between them should be simple and easy as in human to human interaction. Emotions are fundamental to human beings and it plays a vital role in everyday life. The six basic emotions are anger, disgust, fear, happy, sad and surprise. Diverse applications include computer graphics, psychology, automatic driver fatigue detection, surveillance, etc. Emotion recognition consists of preprocessing, feature extraction and classification. In this work, we have used videos from BU-4DFE database, extracted neutral and apex frames automatically, extracted facial feature information that are contributing for recognizing emotions to form feature vector. They are given to classifiers for classification. We have implemented automatic retrieval of apex frame and with optimum number of feature points obtained encouraging accuracy. We have also determined optimum number of feature points as their selection plays an important role in recognizing emotions. Samples images with basic six emotions of BU-4DFE database are given in Fig. 1.

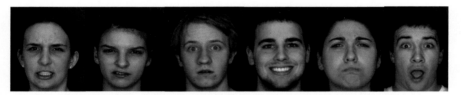

**Fig. 1** Six basic emotions from BU-4DFE Database (anger, disgust, fear, happy, sad and surprise).

The reminder of this paper is organized as follows. In section 2, background work is discussed. In section 3, we present our proposed method. Result and analysis are given in section 4. Future work and conclusion are discussed in the last section.

## 2    Background Work

Various techniques have been proposed for emotion recognition during the last few decades. Not much work has been done in this area with respect to dynamic apex detection and emotion recognition using 4D videos. For emotion recognition using videos, the two approaches that are widely used for feature extraction are appearance based and geometry based. In geometric feature based methods, the facial components or facial feature points are extracted to form a feature vector that represents the face geometry. In appearance based methods, image filters, such as Gabor wavelets, are applied to either whole-face or specific regions in a face image to extract a feature vector [1].To recognize the emotions accurately we need to calculate deformations occurred on the face by calculating the change in distance from eyes, mouth and nose. In the recent years, several emotion recognition algorithms have been proposed. In 2012 Sandbach [2] proposed a method that includes Iterative Closest Point (ICP), Free Form Deformation (FFD), vector projections and HMM classifier for recognizing emotions using BU-4DFE and also compared the results with a similar 2D system. In 2012 Songfan [3] developed an emotional avatar image concept based on Active Appearance Models (AAM), Local Binary Patterns (LBP) using FERA 2011 database. Chuan proposed a method using geometric based approach that uses Euclidean distance, PCA and SVM [4]. ASM model is used to localize the points automatically where the shape information has been extracted from the images and it is used to compute the distance parameters and finally classified emotions using SVM classification. Anwar developed a model using facial Points Localization Model PDM, where they have determined drop in recognition rate using general neutral model [5].

Ben in 2014 [6] proposed an approach using radial curves and Riemannian shape analysis. Chiranjeevi in 2015 [7] proposed a method to dynamically learn the neutral appearance at key emotion points using statistical texture model in a continuous video sequence. Peng in 2015 [8] proposed head motions in videos based on SIFT classification using SVM. Most of the existing methods discussed in literature deals with images (or image sequences) with large variation in appearance of facial expressions, identifying these expressions are a significant research challenge in various disciplines ranging from entertainment to medical applications and affective computing. Finding the optimum number of key points which gives maximum recognition rate using geometric based approach is indispensible. Literature suggests that it is difficult to recognize emotion when pose and head movements are seen in videos, occlusion of objects or person moving randomly in videos, ambiguity and uncertainty in face motion, etc.,.

The methods proposed in this area of research are complex involving high computation. Various challenges in recognizing emotions are all the subjects will not express the motion at same time, detecting the apex frame is a challenging task because as the emotion varies continuously in a video sequence and it is difficult to detect the apex of an emotion. The most important challenge is to determine optimum number of points that could provide maximum recognition rate. In this paper, we have proposed a simple approach using Euclidean distance through which the distance of feature points between two frames under consideration are calculated which then form the feature vector and classified in to six basic emotions. We have also determined optimum number of points that could provide better recognition using few frames. Our approach gives good accuracy for Support Vector Machine (SVM) and Random Forests.

## 3    Proposed Method

We have used BU-4DFE database [9] for implementation. It consists of 101 subjects expressing anger, disgust, happiness, fear, sadness, and surprise. Each expression sequence contains approximately 100 frames depending upon the duration of the video. 3D facial expressions are captured at a video rate (25 frames per second). 83 feature points are provided for every frame of the video sequence. The database comprises 606 3D facial expression sequences, with a total of approximately 60,600 frame models. Each 3D model of a 3D video sequence has the resolution of approximately 35,000 vertices. The texture video has a resolution of about

1040×1329 pixels per frame. The database consists of 58 female and 43 male subjects, with a variety of ethnic/racial ancestries, including Asian, Black, Hispanic/Latino, and White. We have used the videos of all six emotions expressed by 60subjects in our work. The steps involved in our proposed method are automatic peak detection, feature extraction, and classification which are explained in this section. The block diagram that describes the proposed method is shown in Fig. 2.

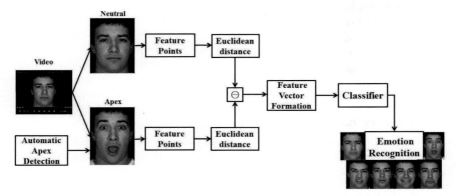

**Fig. 2** Proposed emotion recognition system for 4D videos

## 3.1    *Automatic Peak Detection*

In automatic peak detection, we extract the neutral and apex frames from each subject for each emotion. In BU-4DFE database the emotions posed by the various subjects has sequence of frames which starts with neutral expression and followed by apex of an expression and ends with neutral expression. In few subjects the frames start with apex expression and ends with neutral expression. We have developed a method which automatically identifies the apex expression from the sequence of frames. From the entire sequence of frames in the database for an emotion enacted by a subject, only few frames are considered and a method has been developed to automatically detect the apex frame by summing the Euclidean distances between the identified frames. We assume an integer variable 'n' which represents an interval between frames in a video sequence. The starting frame in the video sequence is numbered as 1 and frames of the order 1, 1n, 2n, 3n, etc., are selected till the last frame of the video sequence. 83 feature points are given in the database for each frame. For the selected frames, Euclidean distance is calculated for the corresponding 83 feature points between the frames 1 & 1n, 1 & 2n, 1 & 3n, 1 & 4n, 1 & 5n and so on. The Euclidean distance calculated between two frames, neutral and peak frames is given by (1).

$$W = \sqrt{(x_i - x_j)^2 + (y_i - x_j)^2 + (x_i - x_j)^2}, (i, j) \in [\![1, 83]\!] \qquad (1)$$

For example, if the number of frames in a video sequence is 65 and 'n' value is assumed as 5, then we select frames of the order 1,5,10,15,20,25,30........65. This means that significant movement of the features on the face can be identified for every 'n'[th] frame as there is only minor difference in every subsequent frame, and in our work we have selected 'n' value as 5 and only 14 frames are sufficient to recognize the apex of an emotion in a complete video sequence. The calculated distances between a set of frames is summed up. This is repeated for the video sequence of all selected 60 subjects and 6 emotions. The frame that has maximum value of sum implies that, it has more difference in terms of distance as well as change in movements of the features on the face and it is identified as the apex frame. In our investigation we have found that considering last frame as 65th frame and 5 as 'n' value will give better recognition rate for a video sequence. After identifying the neutral and apex, the next step is to extract the feature points for further processing which is explained in next section.

## 3.2 Feature Extraction

We have used geometric based method for feature extraction. 83 key facial points are marked on face for every frame and their coordinates are provided in BU-4DFE database. The key facial points will describe the geometric information about the features of the face like eyebrow, eye and mouth. Fig. 3 shows the 83 key facial points from which we have selected two different set of feature points. We have selected 25 and 8 key facial points out of 83 key facial points for both

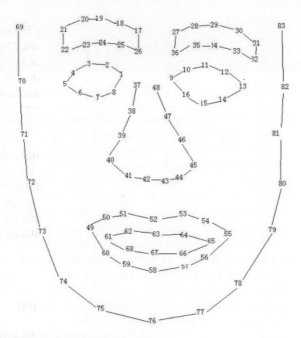

**Fig. 3** 83 key facial points or feature points

neutral and apex frames, although a different number of the points can be used ins-tead. After several experiments we have identified these key facial points on neutral and apex frame for a subject which is given in Fig. 4. When facial expression changes, the vertical displacement will be more and it is determined by the movement of eye, eyebrow & mouth and horizontal displacement is determined for corner points in mouth and eyebrows. The horizontal and vertical distance between the feature points $p_i$ & $p_j$ in one frame are represented as $(p_i, p_j)_x$ and $(p_i, p_j)_y$ respectively.

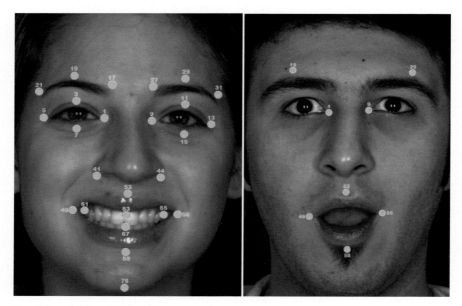

**Fig. 4** 25 and 8 key facial points

The horizontal and vertical distances given in table 1 are calculated using (2) and they are referred as Euclidean distances. To calculate the Euclidean distance in our proposed method for feature extraction, we make use of either 'x' or 'y' coordinate of the feature points based on horizontal or vertical distance as mentioned in table 1. For example $w_5$, $w_{13}$ and $w_{15}$ refers to horizontal distance which is calculated by using the 'x' coordinates of key facial points 17 & 27, 49 & 55, 61 & 65 and the remaining are vertical distances which are calculated by taking the 'y' coordinates of the fea-ture points as mentioned in table 1. After calculating the Euclidean distances using 22 distance vectors for neutral and apex frames individually, the difference between the corresponding Euclidean distance in neutral and apex frame are calculated by subtracting the Euclidean distances. They are then concatenated to form a feature vector which is given as $[dw_1, dw_2, dw_3, \ldots \ldots \ldots dw_{25}]$. To reduce the dimensionality and also for comparison purpose we have repeated the same procedure for 8 points [5] and a frame marked with 8 points is also given in Fig. 4. A drop in recognition rate is observed by using general neutral model instead of person specific model [5]. In our approach, we have used person specific model and Euclidean distance measure to form feature vector thus simplifying the process using 25 key facial points. Also only two frames, neutral and peak frame of each subject have been used for the formation of feature vector instead of the complete video sequence of frames. The Euclidean distance is calculated for the distances given in table 2, using (4) and the feature vector are formed which is given as $[dw_1, dw_2, dw_3 \ldots\ldots\ldots dw_8]$. This process is repeated for 60 subjects and 6 emotions.

**Table 1** Distance vectors and geometric features for 25 key facial points using BU-4DFE

| $w_i$ | Distance | Geometric Feature |
|---|---|---|
| $w_1$ | $(p_{17}, p_{21})_Y$ | Left Eyebrow |
| $w_2$ | $(p_{17}, p_{19})_Y$ | Left Eyebrow |
| $w_3$ | $(p_{27}, p_{31})_Y$ | Right Eyebrow |
| $w_4$ | $(p_{27}, p_{29})_Y$ | Right Eyebrow |
| $\boldsymbol{w_5}$ | $\boldsymbol{(p_{17}, p_{27})_X}$ | *Eyebrows* |
| $w_6$ | $(p_1, p_5)_Y$ | Left Eye |
| $w_7$ | $(p_3, p_7)_Y$ | Left Eye |
| $w_8$ | $(p_9, p_{13})_Y$ | Right Eye |
| $w_9$ | $(p_{11}, p_{15})_Y$ | Right Eye |
| $w_{10}$ | $(p_1, p_{17})_Y$ | Left Eye to Left Eyebrow |
| $w_{11}$ | $(p_9, p_{27})_Y$ | Right Eye to Right Eyebrow |
| $w_{12}$ | $(p_{41}, p_{44})_Y$ | Nose |
| $\boldsymbol{w_{13}}$ | $\boldsymbol{(p_{49}, p_{55})_X}$ | *Mouth* |
| $w_{14}$ | $(p_{52}, p_{58})_Y$ | Mouth |
| $\boldsymbol{w_{15}}$ | $\boldsymbol{(p_{61}, p_{65})_X}$ | *Mouth* |
| $w_{16}$ | $(p_{63}, p_{67})_Y$ | Mouth |
| $w_{17}$ | $(p_{41}, p_{49})_Y$ | Nose to Mouth |
| $w_{18}$ | $(p_{44}, p_{55})_Y$ | Nose to Mouth |
| $w_{19}$ | $(p_{41}, p_{58})_Y$ | Nose to Mouth |
| $w_{20}$ | $(p_{44}, p_{58})_Y$ | Nose to Mouth |
| $w_{21}$ | $(p_{52}, p_{76})_Y$ | Mouth to Chin |
| $w_{22}$ | $(p_{58}, p_{76})_Y$ | Mouth to Chin |

The feature vectors thus formed for 25 and 8 feature points are given in (3) & (5) respectively. So, the size of feature vector is 360 x 25 and 360 x 8 for 25 and 8 feature points respectively. After the formation of feature vector it is fed to classifiers for classifying into six basic emotions.

$$w_m = \sqrt{(p_i - p_j)^2}, \; m=1 \text{ to } 22, \; (i, j) \in [\![1,76]\!] \qquad (2)$$

$$F_e^s = [dw_1, dw_2, dw_3, \ldots\ldots\ldots dw_k]^T, e = 1 \text{ to } 6, s = 1 \text{ to } n \qquad (3)$$

$$w_n = \sqrt{(p_i - p_j)^2}, \, n = 1 \text{ to } 6, \, (i, j) \in [\![1,58]\!] \tag{4}$$

$$F_e^s = [dw_1, \, dw_2, \, dw_3, \ldots\ldots\ldots\ldots dw_k]^{\mathrm{T}}, e = 1 \text{ to } 6, s = 1 \text{ to } n \tag{5}$$

s: number of subjects, e: number of emotions

**Table 2** Distance vectors and geometric features for 8 key facial points using BU-4DFE

| $w_i$ | Distance | Geometric Feature |
|-------|----------|-------------------|
| $w_1$ | $(p_1, p_{19})_Y$ | Left Eye to Left Eyebrow |
| $w_2$ | $(p_9, p_{29})_Y$ | Right Eye to Right Eyebrow |
| $w_3$ | $(p_1, p_{52})_Y$ | Left Eye to Mouth |
| $w_4$ | $(p_9, p_{52})_Y$ | Right Eye to Mouth |
| $w_5$ | $(p_{49}, p_{55})_X$ | *Mouth* |
| $w_6$ | $(p_{52}, p_{58})_Y$ | Mouth |

## 3.3    *Classification Using Support Vector Machine*

We have used Multiclass SVM classifier for classifying six emotions. SVM is extensively used to solve the optimization and classification problems [10 - 12]. Range of kernel functions like Linear, Gaussian Radial Basis Function, Polynomial, Sigmoid etc., can be used for classification. Out of those kernels we found sigmoid kernel giving better results. The sigmoid kernel is given by (6)

$$\text{Sigmoid: } k(x, y) = tanh(\gamma x . y + c) \tag{6}$$

SVM perform classification by finding a decision surface that has a maximum distance to the closest point in the training set. Using the kernel function, decision surface is given by (7).

$$f(x) = \sum_{i=1}^{l} y_i a_i K(x, x_i) + b \tag{7}$$

We classified emotions using Linear, Polynomial, Gaussian RBF and Sigmoid kernel separately for different training and testing datasets. In our proposed method, it is observed that Sigmoid kernel yields better accuracy. The, comparison of results with literature, accuracy and classification time are given in table 3 to 6 in section 4.

## 3.4    *Classification Using Random Forests*

For the classification, we have also used the random forest algorithm. The algorithm was proposed by Breiman [13] and defined as a meta-learner comprised of many individual trees. It is designed to operate quickly over large datasets and more importantly to be diverse by using random samples to build each tree in the forest. A tree achieves highly nonlinear mappings by splitting the original problem into smaller ones, solvable with simple predictors.

Each node in the tree consists of a test, whose result directs a data sample toward the left or the right child. During training, the tests are chosen in order to group the training data in clusters where simple models achieve good predictions. Such models are stored at the leaves, computed from the annotated data, which reach each leaf during training. Once trained, a random forest is capable of classifying a new expression from an input feature vector by pruning it down each of the trees in the forest. Each tree gives a classification decision by voting for that class. Then, the forest chooses the classification that has maximum votes.

## 4     Results and Analysis

The performance of our system is evaluated on the videos of BU-4DFE database. 60 subjects with six emotions namely anger, disgust, fear, happy, sad and surprise per subject are considered. Out of the 360 feature vectors, samples are separated for training and testing. It is noticed that the system achieved better classification accuracy using the SVM classifier for 25 key facial points. Whereas, using random forests the accuracy obtained is low when compared with SVM for 8 key facial points. The accuracy obtained by random forests for 25 and 8 key facial points is comparatively lesser than SVM for most of the combinations of training and testing data sets. Also, 25 key facial points yields better accuracy than 8key facial points.

The accuracy of our proposed method is compared with literature in table 5. Accuracy for Proposed method using SVM classifier and Random forests for 40 training and 20 testing samples are given in table 3, where out of 360 samples 80% of the samples are given for training and 20% are given for testing. 100% accuracy is obtained for emotion 'happy' for 25 and 8 key facial points. 'Surprise' yield 100% for 25 key facial points. 'Disgust', 'fear', and 'sad' achieves 91.66% equally for 25 key facial points. 'Anger' achieves the lowest of all emotions for 25 key facial points. In case of 8 key facial points accuracy for 'fear' is the lowest and 'anger', 'disgust' and 'surprise' achieves equal accuracy. So, we can conclude that 25 key facial points yield better accuracy than 8 key facial points. The accuracy of two chosen classifiers is compared in table 4 and the result obtained by our proposed method is compared with literature in table 5. From table 5 it is observed that the accuracy obtained using 83 key facial points [14] using 3D mesh models obtained 93% where we have achieved the same accuracy using 25 key facial points. Also, if key facial points are less than 25, then the accuracy is comparatively less as seen in [12], [4] and [5] mentioned in table 5. So we can conclude that 25 key facial points is optimum. Our proposed method gives better accuracy using SVM classifier for BU-4DFE database compared to other approaches and also with a similar approach using CK database. SVM is faster than random forest classifier and the classification time is given in table 6.

**Table 3** Accuracy using SVM classifier and Random Forests for 25 & 8 key facial points (for basic emotions)

| Emotion | Accuracy (%) using SVM | | Accuracy (%) using Random Forests | |
|---|---|---|---|---|
| | **25 Points** | **8 Points** | **25 Points** | **8 Points** |
| Anger | 83.33 | 91.66 | 83.33 | 83.33 |
| Disgust | 91.66 | 91.66 | 91.66 | 75.00 |
| Fear | 91.66 | 58.33 | 66.66 | 91.66 |
| Happy | 100 | 100 | 100 | 75.00 |
| Sad | 91.66 | 83.33 | 100 | 66.66 |
| Surprise | 100 | 91.66 | 91.66 | 100 |
| **Overall** | *93.06* | *86.11* | *88.89* | *81.94* |

**Table 4** Comparison of accuracy using SVM and Random Forests for 25 and 8 key facial points

| No of key facial points and distance vectors | Classifier | Kernel | Accuracy (%) |
|---|---|---|---|
| 25, 22 | SVM | Sigmoid | *93.06* |
| | Random Forests | - | 88.89 |
| 8, 6 | SVM | Sigmoid | 86.11 |
| | Random Forests | - | 81.94 |

**Table 5** Comparison of results of our proposed method with literature

| Reference | Database | Algorithm | No of points | Accuracy (%) |
|---|---|---|---|---|
| [14] Drira | BU-4DFE | DVF, LDA | 83 | 93.00 |
| [12] Sohail | JAFFE<br>CK | Multi Detector Approach | 11 | 89.44<br>84.86 |
| [5] Saeed | CK<br>BU-4DFE | PDM | 8 | 89.00<br>83.89 |
| [4] Chuan Wan | CK | ASM, PCA | 20 | 80.10 |
| *Proposed* | *BU-4DFE* | *Euclidean Distance* | *25*<br>*8* | *93.06*<br>*86.11* |

**Table 6** Classification time for SVM and Random Forests using 25 and 8 key facial points

| No of key facial points | Classification time (Sec) | |
|---|---|---|
| | SVM | Random Forests |
| 25 Points | 2 - 3 | 13 - 17 |
| 8 Points | | |

## 5    Conclusion and Future Work

We proposed a novel method to recognize the basic six emotions from the video sequences of BU-4DFE database. The database consists of videos which are posed and expression varies from subject to subject, some subjects have the first frame of the video sequence that starts with neutral emotion and in few subjects the neutral emotion starts at a different frame. So, we have proposed automatic apex detection in order to detect the apex of the video sequence for any kind of posed video sequence. We have used 60 videos out of 101 subjects, 6 emotions from the BU-4DFE database. The proposed method automatically detects apex frame of a video sequence. In our proposed method we have used only two frames, neutral and peak frame of each subject and 25 or 8 key facial points for feature vector formation. This reduces the size of feature vector.

We have classified using random forests and SVM with different kernels. We have computed the classification time for SVM and random forests, where SVM classifier takes very less time compared with Random forests. We also performed the emotion recognition using 8 key facial points and found that the maximum accuracy is achieved at 86.11% and 83.88% for SVM and Random forests respectively. We have observed that 25 key facial points gives better accuracy than 8 key facial points and it is concluded to be the optimum number of key facial points for emotion recognition. In our experiments we have found that Sigmoid kernel gives better accuracy (93.06%) using 25 key facial points than random forests (88.89%). The accuracy obtained by the proposed method is better compared to the methods existing in literature. Our approach can be applied for real time implementation using SVM classifier and kinesics in future.

## References

1. Tian, Y., Kanade, T., Cohn, J.F.: Facial expression analysis. In: Anil, K.J., Stan, Z.L. (eds.) Handbook of Face Recognition, pp. 247–276. Springer, USA (2003)
2. Sandbach, G., Zafeiriou, S., Pantic, M., Rueckert, D.: Recognition of 3D facial expression dynamics. Image and Vision Computing **30**, 762–773 (2012)
3. Yang, S., Bhanu, B.: Understanding Discrete Facial Expressions in Video Using an Emotion Avatar Image. IEEE Transactions on Systems, Man, and Cybernetics – Part B: Cybernetics **42**(4), 980–992 (2012)

4. Wan, C., Tian, Y., Liu, S.: Facial expression recognition in video sequences. In: 10th World Congress on Intelligent Control and Automation, July 6–8, 2012, Beijing, China, pp. 4766–4770 (2012)
5. Saeed, A., Al-Hamadi, A., Niese, R.: The effectiveness of using geometrical features for facial expression recognition. In: 2013 IEEE International Conference on Cybernetics (CYBCONF), June 13–15, 2013, Lausanne, Switzerland, pp. 122–127 (2013)
6. Ben Amor, B., Drira, H., Berretti, S., Daoudi, M., Srivastava, A.: 4-D Facial Expression Recognition by Learning Geometric Deformations. IEEE Transactions on Cybernetics **44**(12), 2443–2457 (2014)
7. Chiranjeevi, P., Gopalakrishnan, V., Moogi, P.: Neutral Face Classification Using Personalized Appearance Models for Fast and Robust Emotion Detection. IEEE Transactions Image Processing **24**(9), 2701–2711 (2015)
8. Liu, P., Yin, L.: Spontaneous facial expression analysis based on temperature changes and head motions. In: 11th IEEE International Conference and Workshops on Automatic Face and Gesture Recognition (FG), vol. 1, pp. 1–6, May 4–8, 2015
9. Yin, L., Chen, X., Sun, Y., Worm, T., Reale, M.: A high-resolution 3D dynamic facial expression database. In: 8th IEEE International Conference on Automatic Face & Gesture Recognition, September 2008, Amsterdam, Netherlands, pp. 1–6 (2008)
10. Chang, C.-C., Lin, C.-J.: LIBSVM: A Library for Support Vector Machines. ACM Transactions on Intelligent Systems and Technology **2**, 27:1–27:27 (2011)
11. Hsu, W., Chang, C.-C., Lin, C.-J.: A practical guide to support vector classification (2010)
12. Sohail, A.S.M., Bhattacharya, P.: Classifying facial expressions using point-based analytic face model and support vector machines. In: IEEE International Conference on Systems, Man and Cybernetics, ISIC, October 2007, Montreal, Que, Canada, pp. 1008–1013 (2007)
13. Breiman, L.: Random forests. Mach. Learning **45**(1), 5–32 (2001)
14. Drira, H., Ben Amor, B., Daoudi, M., Srivastava, A.: 3D dynamic expression recognition based on a novel deformation vector field and random forest. In: 21st International Conference on Pattern Recognition (ICPR), November 11–15, 2012, pp. 1104–1107 (2012)

# Fraudulent Image Recognition
# Using Stable Inherent Feature

Deny Williams, G. Krishnalal and V.P. Jagathy Raj

**Abstract** To prove authenticity and originality of images, many techniques were recently released. This paper proposes a pixel based forgery detection technique for identifying forged images, which is an effective method for finding tampering in images. The proposed method detects splicing and copy-move forgery in images by locating the forged components in the input image. Splicing is the process of copying a component from an image and pasted to another image. Copy-move forgery is the process of copying a component from an image and pasted to another portion of the same image. To find the forged component in the input image, the noise variance remaining after denoising method and SURF features are used. In order to locate spliced component, image segmentation is done before finding the number of components in the image. For segmentation, segmentation based on combining spectral and texture features are used. To identify the number of components in the image, fuzzy c-means clustering is used. In order to locate copy-move forgery, SURF features are detected first and extracted for finding the similarity between keypoints. The experiment results show that the proposed method is very good at identifying whether an image is forged or not. The proposed method gives a high speed performance compared to the state-of-the-art methods. Results gained through the experiments on both manually edited images and visually realistic real images shows the effectiveness of the proposed method.

## 1 Introduction

As digital images are considered as evidences in court and in many legal matters, it becomes a necessity to prove the integrity and authenticity of digital images.

D. Williams(✉) · G. Krishnalal
Amal Jyothi College of Engineering, Kottayam, India
e-mail: denywilliams@cs.ajce.in, gkrishnalal@amaljyothi.ac.in

V.P.J. Raj
School of Management Studies, CUSAT, Kochi, India
e-mail: jagathy@cusat.ac.in

© Springer International Publishing Switzerland 2016
S.M. Thampi et al. (eds.), *Advances in Signal Processing and Intelligent Recognition Systems,*
Advances in Intelligent Systems and Computing 425,
DOI: 10.1007/978-3-319-28658-7_2

15

A lot of image manipulation software's and hardware's came into existence due to the advancement in the field of technology, which is a cause for losing trust in photographs. As the technology improves, the techniques employed in image forgery also increased. Many forgery detection methods came into existence till now, but no method performs at its best as almost all state-of-the-art methods presume low detection accuracy. Image forgery detection can be broadly classified into passive techniques and active techniques [8, 9]. Mainly there are two image authentication techniques under active method. They are digital watermarking [9, 10] and data hiding. Both active methods require prior information about the image. And also this technique needs cooperation from the image taker.

Passive techniques do not require active participation of an external agent. This technique uses already available artifacts (fingerprints) of image left by the camera for the authentication purpose. Passive methods provides many ways to detect forgery than active methods, thereby provides high accuracy in image authentication. Now people create fake digital images using high end manipulation software's for humor or for getting popularity. Images have been applied for malicious purposes in many real life situations.

For instance, O.J. Simpson is an actor and a retired football player, who was arrested for the murder of his wife and her friend during the year 1994. His mugshot was everywhere in the publication at that time. TIME magazine also published a mugshot of him as its cover, which makes him to appear darker by changing the photograph's color saturation. This becomes a big issue during that period.

## 2    Related Works

Several studies were conducted on this topic, and several image splicing and copy-move forgery detection methods were proposed and developed. Copy–move forgery detection techniques [1, 2] usually use a block-based scheme to detect image forgery. In copy-move forgery detection, an image is taken first and divided it into fixed-size over-lapping blocks and then features of the blocks are extracted and represented as feature vectors. In lexicographical order, the feature vectors are arranged to make almost similar blocks adjacent to each other. Finally, blocks with duplicate regions are pruned out using a similarity measure.

I-Cheng Chang et al. [3] proposed a novel forgery detection algorithm by detecting the inpainted regions from the tampered image. This forgery detection starts by searching suspicious region followed by forged region identification. Multi-region relation (MRR) is employed to identify forgery in images. Sergio Bravo-Solorio and Asoke K.Nandi proposed in [4] duplicated region detection mechanisms in geometrically manipulated images. This method is implemented by obtaining a 1-D descriptor with the help of overlapped block of pixels, which is invariant to geometrical manipulations by mapping the overlapped pixels to log-polar coordinates. The main problem faced by the method is in detecting duplicates in the areas of uniform luminance. Xunyu Pan and Siwei Lyu in [5] proposed a copy-move forgery detection method with the help of scale invariant feature transform (SIFT) keypoints. Random Sample Consensus (RANSAC) algorithm is employed to extract only robust or reliable SIFT keypoints from the extracted keypoints. The proposed method shows high performance on duplicated and distorted regions. W. Chen et al. in [6] adopt the method of obtaining information about wavelet sub-bands to find splicing in images. From the phase and magnitude function of the image, 120-D features are extracted to detect splicing. In the paper [7], W. Zhang et al., Markov features are used to identify splicing in images. Markov features are calculated separately for DWT and DCT. Another feature to prove authenticity in images is by using SPN (Sensor Pattern Noise) which is an imprint left by camera during the image acquisition time. By detecting SPN [11, 12, 13, 14] feature of the image, the source of an image can be determined. In this paper, we proposed a method to find splicing and copy-move forgery in images. This paper aims to evaluate the detection accuracy and computational time of proposed method by comparing it with state-of-the-art methods.

## 3   Proposed System

The system is developed to find tampering in images, mainly copy-move and splicing.

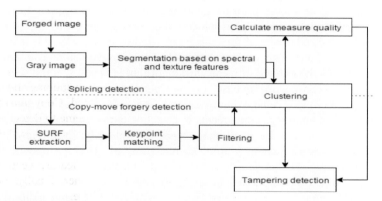

**Fig. 1** Proposed method

Fig.1 shows how the proposed method works. The blocks shown above dotted line indicates the methods employed for splicing detection and below indicates the methods used for copy-move forgery detection. As a preprocessing step, the input RGB image is converted to gray image in both detection methods. In both the methods, clustering is done for grouping the similar pixel value components. After clustering in splicing detection, it calculates the measure quality to detect tampering by determining the noise variance in each component. And in copy-move detection method, the tampering can be detected suddenly after clustering.

## 3.1    Image Splicing Detection

### 3.1.1    Segmentation and Clustering

This section explains mainly about the spliced component detection.The problem
can be defined in the following way. Given input image is divided into many segments based on spectral and texture features. Digital images vary one another with
the location and size of components and with different noise levels. The goal of
segmentation is to determine the identical regions and to localize them. It is
evident that the noise variance varies highly for each localized regions. So, each
image segments may have different noise variance. Before segmenting the input
image, the image undergoes denoising. Denoising is performed to remove noise
from the image for better segmentation process. Those intensity images that are
degraded by the constant power additive noise are taken to lowpass filters in
weiner 2-d filtering technique.

Wiener2 estimates statistics from the neighborhood of each pixel locally. Image
segmentation based on combining spectral and texture features [21] is employed
for estimating the components in an image. For obtaining enhanced spatial patterns, linear filters are used. Five types of filters are used in this work such as LoG
filter, sobel filter, guassian filter, laplacian filter and gabor filter. After the filtering
operation, the local histograms of all filtered input bands are concatenated. The set
of filters needs to be chosen prior to specifying the spectral histogram. Segmentation estimates combination weights, which indicates segment ownership of pixels.
To compute sub-band image of for each filter, window of size (W=45) is used. In
[15], the image components are obtained using the adaptive component detection
method, that uses the method of mean shift algorithm [17] and spectral segmentation with the k-means algorithm based on the eigenvectors of the matting laplacian
matrix [18]. And in [17], image segmentation based on local noise standard deviation is used, to obtain image components in detecting tampering. Fuzzy c-means is
used to group clusters in the given image based on similarity. Each image of size
m x n and the output obtained from segmentation are used as input to fuzzy
c-means clustering.  The grade of membership and the coordinates of each cluster
center are returned as output to clustering. 0 and 1 are used as the grades of
membership. 0 indicates no membership and 1 indicates full membership. Partial
membership is shown by the values between 0 and 1. After several iteration, exactly similar data points appear in a cluster. The clustering ends when the iteration
count reaches its maximum.

### 3.1.2    Calculating Measure Quality

For an authentic image, same pattern of noise can be seen over the entire image.
Adding extra noise may cause variation in the noise of the image. Therefore, the
detection of inconsistent noise levels in an image may signify tampering. Since no
duplicate region check is employed to find splicing, the variance of noise in image
is used to find splicing. To find the noise variance, the sensor pattern noise (SPN)

noise that left even after denoising is used. For denoising, weiner filter is used. The tampering can be obtained only if there is a large difference in the variance of two spliced component. After denoising, the remaining noise of each component is calculated using the equation:

$$Noise\_residual = Original\_image - Denoised\_image \quad (3.1)$$

If images of two cameras spliced together, the noise variance of the two spliced component will differ. Then the variance of the remaining noise in each image component is calculated based on noise residual. It is done by using the measure quality (MQ),

$$MQ = \frac{\max(noise\_var iance)}{\left(\frac{1}{n}\right)\sum_{k=1}^{n} noise\_var iance} \quad (3.2)$$

where max (noise_variance) is the maximum of noise variance. Then a threshold is set based on many trial and error basis by on many images. If MQ is greater than threshold, then we can say that the tested image is a tampered image otherwise it is not. LUM (Lower Upper Middle) filter [19] as used in [15] proposed by Wu-Chih Hu et al. and one level DWT [20] as used in [16] proposed by Mahdian et al. is employed for filtering operation. SPN is present in every digital image, which is an intrinsic characteristic of the sensor and is relatively stable. Sensor noise is extracted from the camera sensor which exists in the sensor during the sensor manufacturing process itself. Therefore, the variance of the remaining noise is used as the feature to obtain the tampering detection, where the variance of the remaining noise is a kind of SPN.

## 3.2   Copy-Move Forgery Detection

The copy-move forgery detection is done by using the method of speed up robust feature (SURF). As a first step, the RGB image is converted to the gray scale image. SURF, a robust local feature detector extracts the features of the image. It means extract interest point descriptors from the image. Descriptors are derived from pixel surrounding an interest point. They are needed to describe and match features specified by a single point location. Integral image, keypoint detection, orientation assignment and feature descriptor generation are the four main steps in feature extraction. After the feature extraction, these features are matched to find duplication by using nearest neighbor method. To reduce the probability of false matches, filtering is applied with the help of Euclidean distance. If the distance is less than a particular threshold, such pairs can be removed. The proposed method detects copy-move forgery and splicing simultaneously. If the tampering is due to splicing, then it uses noise variance method and if it is a copy-move forgery, the SURF method is used to detect tampering in images.

## 4    Experiment Results and Comparisons

In this section, we demonstrate how the forensic methods proposed in the previous sections are implemented. In the experiment, we randomly generated many set of forged images based on CASIA v1.0 dataset and also used many manually created photoshopped images. The algorithm is implemented in Matlab R2013a. CASIA v1.0 consists of 800 authentic and 920 spliced image blocks of size 384 x 256 pixels. Fig. 2 show some of the images used in the experiment.

(a)               (b)               (c)               (d)

**Fig. 2** Sample images used in the experiment

Detection accuracy ($D_a$) is used to illustrate the performance of proposed method. TP indicates the count of true positive pixels, FP shows the count of false positive pixels, TN gives the count of true negative pixels, and FN indicates total number of false negative pixels.

$$D_a = \frac{(TP + TN)}{(TP + TN + FP + FN)} \tag{4.1}$$

After calculating the Measure Quality (MQ), normalization is applied to the obtained variances.

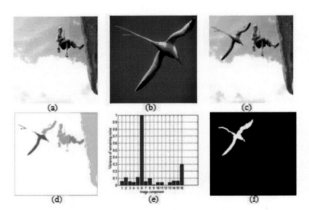

(a)               (b)               (c)

(d)               (e)               (f)

**Fig. 3** Results obtained using noise variance method

Fig. 3 shows the working system to detect splicing in images. Fig. 3 shows how splicing in images are detected. (a) Indicates the target image and (b) shows the source image. In (c), the eagle from the source image is copied and pasted to target image. (d) depicts the result obtained after the component detected through segmentation based on spectral and texture features and fuzzy c-means clustering. A bar graph obtained after calculating the noise variance of remaining noise and number of components in the image is plotted against y and x axis is indicated in (e). The spliced component eagle is detected in (f) which is the result of tampering detection. The total time taken for splicing detection alone is 20.14s.

## 4.1  Choosing of Denoising Filter

Table 1, shows image formats used to make composite images.

**Table 1** Image formats employed to make composite images

| Source Image | Target Image | Composite Image |
|---|---|---|
| Jpg | Tif | Tif |
| Bmp | Gif | Gif |
| Gif | Jpg | Jpg |
| Tif | Png | Bmp |
| Jpg | Jpg | Jpg |

LUM filter and Weiner filter is compared using the Peak Signal-to-Noise Ratio (PSNR) value. This denoising filter is applied before when an image is taken to segmentation and surf extraction. High PSNR value indicates that the image holds high quality even after denoising. Decrease in MSR value means that after denoising the noise ratio decreases. The Weiner filter gives high PSNR value compared to LUM filter in the experiment and analysis. Table 2 shows comparison of LUM and Weiner filter based on PSNR value of some images used in the work. From the figure, we can infer that by using Weiner filter, we get better PSNR value and the quality after denoising is better.

**Table 2** Comparison between LUM and Weiner filter

| Figures | LUM Filter | | Weiner Filter | |
|---|---|---|---|---|
| | PSNR | MSE | PSNR | MSE |
| Fig. 2(a) | 27.97 | 103.7 | 33.86 | 26.71 |
| Fig. 2(b) | 30.82 | 53.78 | 33.58 | 28.48 |
| Fig. 2(c) | 27.63 | 112.07 | 33.20 | 23.05 |
| Fig. 2(d) | 26.98 | 130.29 | 33.65 | 28.02 |

## 4.2    Component Detection

Compared to the method proposed in [15], the proposed method works better and show high tampering detection accuracy. In [15], Wu-Chih Hu et al. adopt the method of mean shift algorithm [17] and spectral segmentation with the $k$-means algorithm with eigenvectors of the matting Laplacian matrix [18] for adaptive component detection. In [16] Mahdian et al., used image segmentation based on local noise standard deviation. When using both methods, they have low detection accuracy compared to the proposed method. The method proposed by Mahdian et al. failed to find the tampered regions when the noise degradation is very small. Fig. 4 shows the output obtained after main shift segmentation and segmentation based on combining spectral and texture features.

**Fig. 4** a) Result of mean shift based segmentation, (b) Result of segmentation based on spectral and texture features.

From the output, we can see that the result obtained after segmentation based on spectral and texture features, show high level of segmentation compared to mean shift algorithm. In mean shift, the man, eagle and the rock indicate somewhat same color and also it segments the image into more three components, which is a disadvantage while segmenting. While looking at the segmentation results based on spectral and texture features, the man, eagle and rock are detected separately with different colors and the three components are segmented more accurately.

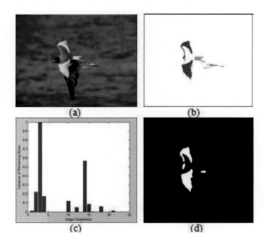

**Fig. 5** Tampering detection of image with birds

Fig. 5, shows the tampering detection of image with birds using proposed method, where Fig. 5 (a) represents the composite image with BMP format; Fig. 5 (b) shows the image components; Fig. 5 (c) specify the normalization of variances of remaining noise of image components; and Fig. 5 (d) gives the obtained result of tampering detection using the proposed method.

## 4.3  Copy-Move Forgery Detection

The time consumed for computing SURF and approximate nearest neighbor method is 3.550 seconds. The True Positive Rate and False Positive Rate are determined for the tested images and the performance for the existing method is compared with this work. Fig. 6 shows the detection and localization of copy-move forgery. In the copy-move forgery detection, if the matched feature is greater than index-pair "12", a copy-move forgery is detected in the image. In the figure, one giraffe is a copy of the other giraffe which is shown in Fig. 6 (a). In Fig. 6 (b), the detection and localization of the similar regions is shown.

(a)                                    (b)

**Fig. 6** Example showing detection of copy-move forgery

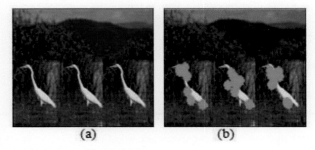

(a)                                    (b)

**Fig. 7** Multiple cloned component detection

This method can also detect multiple cloned components in an image. The same single bird is copied twice as shown in Fig. 7(a). The proposed method identifies the forged location correctly without taking much computation time as shown in Fig. 7(b).

## 4.4 After JPEG Compression

When the image is saved even after JPEG compression with quality 8, the proposed method is able to detect the tampering in images. And the experiment shows that if the forged mage is compressed with any quality, the splicing and copy-move forgery in images can be detected with high accuracy. As the compression quality decreases, the detection and localization of tampering also decreases. But the proposed method is able to find the tampering in images, even if the quality is low. Fig. 8. show the detection accuracy of image saved with quality "1".

**Fig. 8** (a) Image saved with quality "1", (b) Result of copy-move forgery detection

From Fig. 8, it is clear that when compared to Fig. 6, the number of matched keypoints decreased as the quality becomes low. The performance evaluation of the proposed method compared to previous methods using the selected CASIA v1.0 dataset is listed here in table 3.

**Table 3** Comparison of detection capability in CASIA v1.0 dataset

| Methods | TP | TN | FP | FN | Detection accuracy |
|---|---|---|---|---|---|
| NA-DJPG compression [23] | 150 | 471 | 329 | 310 | 49.29% |
| BPPM-based method [24] | 77 | 561 | 239 | 383 | 50.63% |
| Feature Inconsistency [15] | 127 | 643 | 157 | 333 | 61.11% |
| Proposed Method | 172 | 695 | 120 | 273 | 68.80% |

The proposed method has a detection accuracy of 68.80%, which is obtained using the equation (4.1). The proposed method is compared with many other methods based on JPEG- compression [23], EXIF-noise [22], SPN [11] and feature inconsistency [15]. From table 4, it is evident that the proposed method outperforms the methods of [22, 23, 11, 15], when comparing the tampering detection capability in CASIA v1.0 dataset. Table 4, shows the detection performance of the proposed method compared to many other methods using CASIA v1.0 dataset.

**Table 4** Comparison of detection capability in CASIA v1.0 dataset

| JPEG Compression | EXIF noise | SPN | Feature inconsistency | Proposed method |
|---|---|---|---|---|
| Effective | In-effective | In-effective | Effective | Effective |

## 5    Conclusion and Future Works

In this work, we presented a new pixel based forgery detection technique based on noise variance calculation and SURF feature. Compared to other state of the art methods, the proposed method shows high accuracy. During denoising, we observed that high PSNR value leads to high quality which helps in careful selection of denoising technique. In the future, more suitable image component detection methods can be devised to effectively improve the tampering detection accuracy of the proposed method. In case of any geometrical operations applied on images after copy-move, we could accurately detect copy-move forgery in the images by improving the proposed method.

## References

1. Lee, J.-C., Chang, C.-P., Chen, W.-K.: Detection of copy–move image forgery using histogram of orientated gradients. Inf. Sciences (2015)
2. Amerini, I., Ballan, L., Caldelli, R., et al.: Copy-move forgery detection and localization by means of robust clustering with J-linkage. Signal Processing: Image Communication **28**, 659–669 (2013)
3. Chang, I.-C., Yu, J.C., Chang, C.-C.: A forgery detection algorithm for exemplar-based inpainting images using multi-region relation. Image and Vision Computing **31**, 57–71 (2013)
4. Bravo-Solorio, S., Nandi, A.K.: Automated detection and localisation of duplicated regions affected by reflection, rotation and scaling in image forensics. Signal Process. **91**, 1759–1770 (2011)
5. Pan, X., Lyu, S.: Region duplication detection using image feature matching. IEEE Trans. on Inf. Forensics and Secur. **5**(4), 857–867 (2010)
6. Zhang, W., Cao, X., Qu, Y., et al.: Detecting and extracting the photo composites using planar homography and graph cut. IEEE Trans. Inf. Forensics Secur. **5**(3), 544–555 (2010)
7. He, Z., Lu, W., et al.: Digital image splicing detection based on markov features in dct and dwt domain. Pattern Recognit. **45**(2012), 4292–4299 (2012)
8. Birajdar, G.K., Mankar, V.H.: Digital image forgery detection using passive tech-niques: a survey. Digit. Investig. **10**, 226–245 (2013)
9. Liu, K.C.: Color image watermarking for tamper proofing and pattern-based recovery. IET Image Process. **6**(5), 445–454 (2012)

10. Han, Q., Han, L., et al.: Dual watermarking for image tamper detection and self-recovery. In: Proc. of the 9th Int. Conference on Intell. Inf. Hiding and Multimedia Signal Process., pp. 33–36 (2013)
11. Lukas, J., Fridrich, J., Goljan, M.: Digital camera identification from sensor pattern noise. IEEE Trans. on Inf. Forensics and Secur. **1**(2), 205–214 (2006)
12. Li, C.-T.: Source camera identification using enhanced sensor pattern noise. IEEE Trans. on Inf. Forensics and Secur. **5**(2), 280–287 (2010)
13. Kang, X., Li, Y., Qu, Z., Huang, J.: Enhancing source camera identification performance with a camera reference phase sensor pattern noise. IEEE Trans. on Inf. Forensics and Secur. **7**(2), 393–402 (2012)
14. Tomioka, Y., Ito, Y., Kitazawa, H.: Robust digital camera identification based on pairwise magnitude relations of clustered sensor pattern noise. Trans. on Inf. Forensics and Secur. **8**(12), 1986–1995 (2013)
15. Wu-Chih, H., Dai, J.-S., Jian, J.-S.: Effective composite image detection method based on feature inconsistency of image components. Digit. Signal Processing **39**, 50–62 (2015)
16. Mahdian, B., Saic, S.: Using noise inconsistencies for blind image forensics. Image Vis. Compu., Special Section: Computer Vision Methods for Ambient Intell. **10**, 1497–1503 (2009)
17. Comaniciu, D., Meer, P.: Mean shift:a robust approach toward feature space analysis. IEEE Trans. Pattern Anal. Mach. Intell. **24**(5), 603–619 (2002)
18. Levin, A., Rav-Acha, A., Lischinski, D.: Spectral matting. IEEE Trans. Pattern Anal. Mach. Intell. **30**(10), 1699–1712 (2008)
19. Hardie, R.C., Boncelet, C.G.: Lum filters:a class of rank-order-based filters for smoothing and sharpening. IEEE Trans. Signal Process. **41**(3), 1061–1076 (1993)
20. Mallat, S.G.: A theory for multiresolution signal decomposition: the wavelet representation. IEEE Trans. on Pattern Analysis and Machine Intell. **11**(7), 674–693 (1989)
21. Yuan, J., Wang, D., Li, R.: Remote sensing image segmentation by combining spectral and texture Features. IEEE Trans. on Geoscience and Remote Sensing **52**(1), 16–24 (2014)
22. Fan, J., Cao, H., Kot, A.C.: Estimating EXIF parameters based on noise features for image manipulation detection. IEEE Trans. Inf. Forensics Secur. **8**(4), 608–618 (2013)
23. Bianchi, T., Piva, A.: Image forgery localization via block-grained analysis of JPEG artifacts. IEEE Trans. Inf. Forensics Secur. **7**(3), 1003–1017 (2012)
24. Lin, Z., He, J., Tang, X., Tang, C.K.: Fast, automatic and fine-grained tampered JPEG image detection via DCT coefficient analysis. Pattern Recognit. **42**(11), 2492–2501 (2009)

# Kernel Visual Keyword Description for Object and Place Recognition

Abbas M. Ali and Tarik A. Rashid

**Abstract** The most important aspects in computer and mobile robotics are both visual object and place recognition; they have been used to tackle numerous applications via different techniques as established previously in the literature, however, combining the machine learning techniques for learning objects to obtain best possible recognition and as well as to obtain its image descriptors for describing the content of the image fully is considered as another vital way which can be used in computer vision. Thus, the ability of the system is to learn and describe the structural features of objects or places more effectively, which in turn; it leads to a correct recognition of objects. This paper introduces a method that uses Naive Base to combine the Kernel Principle Component (KPCA) features with HOG features from the visual scene. According to this approach, a set of SURF features and Histogram of Gradient (HOG) are extracted from a given image. The minimum Euclidean Distance between all SURF features is computed from the visual codebook which was constructed by K-means previously to be combined with HOG features. A classification method such as Support Vector Machine (SVM) was used for data analysis and the results indicate that KPCA with HOG method significantly outperforms bag of visual keyword (BOW) approach on Caltech-101 object dataset and IDOL visual place dataset.

**Keywords** SURF · K-means · BOW · KPCA · HOG

## 1 Introduction

The visual recognition of both objects and places has been concentrated in the area of computer vision and has attracted many researchers for last several years. It can be stated that in visual recognition of objects and places there are some features

A.M. Ali · T.A. Rashid(✉)
Software Engineering Department, College of Engineering,
Salahuddin University, Erbil, Kurdistan, Iraq
e-mail: tarik.rashid@su.edu.krd

© Springer International Publishing Switzerland 2016
S.M. Thampi et al. (eds.), *Advances in Signal Processing and Intelligent Recognition Systems,*
Advances in Intelligent Systems and Computing 425,
DOI: 10.1007/978-3-319-28658-7_3

27

which are informative, notable and very useful for fully describing the content of visual object and place, and they ultimately will determine the accuracy rate of the recognition system. Both SIFT and SURF as local feature descriptor techniques were used by many researchers to describe images via collections of local feature vectors. These descriptors have been broadly used for visual object and place recognition [1, 2].

It is worth noticing that these local descriptors in an image can present large representations of feature, which, might be difficult to be tackled via machine learning techniques. Thus, in [3, 4] the BOW representations has been suggested. It is noticed that BOW produced reliable performance which has influenced researchers to construct further relationships between different visual keywords to enhance some particular goals [5, 6, 7, 8].

In [9] the researchers used the histogram of the less organized order of visual keyword which is known as the hard bag-of-features (HBOF) or the Hard Assignment. It is indicated that when every single feature in the image is used to represent centroids of winning cluster or a keyword representation in the visual codebook, the other cluster centroids can be ignored and are not used for describing the image content. Research works in [10, 11, 12] produced several fresh and innovation models to enhance the hard bag of features of visual words. The models used soft assignment in which all cluster centroids are taken into account for better improvement in describing the contents of images.

The computation time was increased via minimising the features in such a way that can display invariance against various conditions on the sights, these conditions are translation, scaling, noise and illumination changes. Therefore, an automated learning process of low dimensional model from some training objects is generated via the concept of reducing features. In this regards, principle component analysis has been used to tackle various problems for applications in computer vision such as face and object recognition, detection, and tracking. Principle component analysis has also showed its success in robot localization [4].

Sch"olkopf introduced the concept of Kernel Principle Analysis as a generalisation of principle component analysis for the investigation purpose [13, 4], the idea was applied effectively to tackle problems in image processing, face recognition, image de-noising, texture classification and others. In [14], the linear discrimination analysis is used successfully for tackling the classification problems. Yet, this approach cannot tackle the nonlinear problems, as a consequence, it was expanded to kernel based approaches in Baudat [14] which is called GDA.

The contribution of this paper is to use Naïve Base technique with the Kernel Principle Component features and histogram of gradient features of the visual scene to improve the recognition accuracy. Based on this technique, a set of SURF features and histogram of gradient are extracted from a given image, and then, the minimum Euclidean Distance between all local features is computed from the visual codebook which was constructed by K-means approach. The kernel analysis is applied to the distance result and also HOG is extracted from the same image, then, these features are combined through Naive Base technique. SVM and KNN are used to analyse and classify the data.

## 2    Related Works

It was proven that KPCA is better than PCA in recognition performance of places for the purpose of localization; in addition, PCA is also used for the same purpose. The impact of illumination on the PCA and the use of principle analysis to filter invariant features without referencing to original image are shown in [14, 3, 15].

In [16] KPCA was used for Gabor features and SVM is used to classify objects, besides, the main purpose of using KPCA is to provide the scheme of nonlinearity for classifying objects. In [13, 17, 18] the use of the concept of incremental PCA and the investigation on issue of batch learning were introduced. This paper aims primarily at tackling the on-line learning approach for the robot landmarks in the hope of avoiding the redundant calculation of the PCA for all samples. In [19, 20], KPCA showed better performance than PCA, the study established an evaluation way among different vision based robot localisation approaches. Additionally, the study demonstrated that PCA needs extra computational power.

PCA implementations showed acceptable performance in applications of robot localization and face recognition. On the other hand, using PCA with a Gaussian distribution is appropriate for generating data. Nevertheless, it is obvious that the natural image distributions are extremely non-Gaussian [4, 17].

This paper aims at improving the accuracy recognition of PCA features in visual place to make better localization for the mobile robotics.

## 3    SURF and PCA

The content of images will mainly determine the characteristics of local features. These features or local features are discriminative, can straight forwardly be calculated and not influenced via the rotation or limited lighting changes of the content of the image.

Basically, the technique involves clustering features such as SURF via K-means clustering approach in the bag of visual keywords, which established good results in image scene recognition applications [8, 9, 21].

The approach K-Means clustering has been widely used to cluster features which are similar to bag of visual keywords [10]. HOBF algorithm is essentially used to relate the features to the cluster centroids which match up to the minimum Euclidian Distance and results in a feature vector that is used to label the specific feature. This is expressed in equation below [3].

$$\text{HBOF(w)} = \sum_{i=1}^{n} \begin{cases} 1 & \text{if } w = \arg\ \min_c (\text{dis}(c, r_i)) \\ 0 & \text{Otherwise} \end{cases} \qquad (1.1)$$

Where $n$ represents regions number in the image, $w$ is the visual word, *dist* is the Euclidean Distance between the feature vector $r$ and cluster centroid $c$. The Hard assignment for bags of visual keywords varies from the soft assignment arrangement. In [4, 5], it can be noticed that weights were used for each cluster centroid and the non-linear distribution is required for increasing the precision of the soft assignment scheme, as a consequence, KPCA is used for this reason. In this way, a non-linear distribution for the visual key points will be given which occasionally it provides more accurate results for object categorization. Alternatively, the required time to compute KPCA is luxurious due to the computational cost to calculate Eigen values for the whole data set. Evidently, the computation cost is augmented when the size of the data set gets increased. Accordingly, the clustering applied upon the SURF features taken out from the scene of the image so that to reduce the training samples, as it is stated that clustering is more effective than sparse KPCA.

In this research paper, the visual words are built through the computation of the minimum distance of the local features from cluster centres which are used as training samples for KPCA of each category.

# 4    SURF Features

In 2006 in [22], a new technique called Speeded Up Robust Features is proposed, it is considered as a robust image detector and descriptor which is applied to solve problems in computer vision such as object recognition. It is moderately the practice stages to extract the features stimulated by the SURF descriptor. The normal SURF type is quicker than SURF since it uses integral of images and its competing with SIFT in robustness.  SURF algorithm uses Hessian Matrix, it has three key stages: interest point extraction, repeatable angle computation and descriptor computation.

1)   Interest point extraction: at this stage, the algorithm calculates the Hessian matrix determinant and extracting local maxima. The computation of Hessian Matrix is estimated with an amalgamation of filters of Haar basis in consecutively greater levels. As a result, this stage is $O$ $(mnlog2$ $(max$ $(m,$ $n)))$ for an $m \times n$ size image. At each scale, points that at the same time are local extrema of both the determinant and trace of the Hessian Matrix are considered as interest points, at location $x$, $y$, and scale $\sigma$ the Hessian Matrix can be described as.

$$H(x,y,\sigma)=\begin{bmatrix}L_{xx}(x,y,\sigma) & L_{xy}(x,y,\sigma)\\ L_{yx}(x,y,\sigma) & L_{yy}(x,y,\sigma)\end{bmatrix} \tag{1.2}$$

$Lxx(x,y, \sigma)$ is described as the Gaussian second order derivative convolution with the image $I$ at pixel $x$, $y$.  The result of this stage will be interest points and their scales.

2)   Repeatable Angle Computation: the repeatable angle can be extracted for each interest point before calculating the feature descriptor. This stage attempts at calculating the gradients angle neighbouring the interest points and the response of maximum angular is selected to be the feature direction.

In this work, the clustering of SURF features naively combined with HOG which is used to construct more invariant features to recognize multi class visual places. The idea is to have the amalgamation of several features of the same image which will give more robust feature and will be more reliable for these applications.

# 5    Histogram of Gradient

The image is divided into image windows (cells) or small spatial regions (cells), then a local HOG directions will be added for each region. The amalgamation of these histograms for each region will form the representation of the whole image. The normalization of these histograms may be used or accumulating local histogram measurements to contrast the local responses is considered in order to practise better features invariance to illumination, shadowing,,.., etc. This leads to larger spatial region that can be used for normalized descriptors like Histogram of Oriented Gradient (HOG). The technique has been used by [13,4,5], in addition to SIFT approach. The promising results and success of these descriptors led the authors to use it and combining it naively with SURF features after some process to make more distinct descriptor for visual place recognition.

# 6    Combining KPCA and HOG Naively

Learning visual recognition is considered as a process of clustering image features for some types of structural image contents. Essentially, this approach provides unsupervised semantic for increasing the reliability of image classification and the matching process. Therefore, a set of local patches *(I1….n)* is used, where each patch is expressed via a 64-bin of SURF features, the SURF grid approach is used in this paper to extract the local features *fs* for the images, where it extracts more features than the standard SURF approach (notice: the matlab code for  Lazebnik is used) [23].

   Research investigation demonstrated that SURF grid produced a more informative description of points for the scalable space.    Note that each image *Ij={f1,f2,…..fm}* and each feature *fi* contains 64 elements. A distance vector and its size is the length of the codebook *(sb)* that is used, which is the distance for each feature *fi* of any given image from the codebook *(B)*, this can be expressed as follows:-

$$D(x,\ B_{i=1:sb}) = \sqrt{\sum\nolimits_{l=1}^{sd}(B(i)^{\,1} - x^{1})^{2}} \qquad (1.3)$$

   Keeping in mind that *sd* is the features' length used, and as SURF has been used, thus, *sd*=64. However, for SIFT 128 is used. The distance for all features of any selected image from the codebook is characterised via a distance table containing *m* of distance vectors of size *(sb)* and it can be expressed as follows:-

$$D(x_{j=1:m}, B_{i=1:sb}) = \sqrt{\sum\nolimits_{l=1}^{sd}(B(i)^{1} - x^{1})^{2}} \qquad (1.4)$$

The minimum distance *(mD)* for the table *Dm, sb* presents a row of minimum values for each column in the table, consequently, for image *I*, the *mD* for the table *Dm, sb* can be expressed as follows :-

$$\mathbf{mD}_i = \min(\mathbf{D}_m, sb) = \{\min(\mathbf{D}_1 : sb, 1),$$
$$\{\min(\mathbf{D}_1 : sb, 2), \ldots \{\min(\mathbf{D}_1 : sb, sb)\} \quad (1.5)$$

As a matter of fact, KPCA is resulted via a fact which states that PCA is conducted based on the dot product matrix as an alternative for covariance matrix [16, 19, 24, 25]. Assuming that $\{d_1, \ldots d_N\}$, is a set of distance features data, let each $d_i$ belongs to one image in the data set and the dimension of this data set is the number of clusters used. The nonlinear representation of *Φ(d) is used, d* is represented into *F,* which is defined as a nonlinear feature space, and at that point via applying standard linear PCA on the mapped data, nonlinear principal components can be achieved. Thus, the covariance *matrix(C)* for the mapped data is computed in order to compute the Eigen values as:-

$$C = \frac{1}{N} \sum_1^N \Phi(d_i) \Phi\Phi(_i)^T \quad (1.6)$$

The *C* Eigen values and Eigenvectors are determined via explaining the problem of Eigen value. Notice that the kernel matrix *(K)* of $N \times N$ is represented as follows:-

$$K = \frac{1}{N} \Phi(d_j) \Phi\Phi(_i) = \frac{1}{N} d^T d \quad (1.7)$$

Assuming that $\lambda = \{\lambda_1, \lambda_2, \lambda_3, \ldots \lambda_p,\}$, is a set of nonzero Eigen values of *K*, λ, is sorted in descend order where $u = \{u_1, u_2, u_3, \ldots u_p,\}$ relates to Eigenvectors. Note that *C* has the same values of a one-to-one correspondence Eigen values and Eigenvectors.

Polynomial kernels and Gaussian are verified as a non-linear kernel in this work, furthermore to linear approach, several evaluation and choosing processes are conducted to select the optimum. Despite the fact that the effectiveness of image version of BOW is intended to signify images via a set of features using the number of their repetition, nonetheless, these globalizations of features are not adequate to characterize the spatial environment as they have no order. Consequently, it is required to decrease the error of unordered features matching especially in image scene for place recognition for enhancing BOW performance. Thus, undesired features are removed via using the PCA scheme. The features are combined together to give more variant features as expressed in equation below:-

$$NV(x, B_{i=1:sb}) = \{KPCMD(1 : sb), Hg(1 : sb)\} \quad (1.8)$$

The *KPCMD* kernel principle component of minimum distance for the image and *Hg* is the histogram of oriented gradient for the same image. The image's minimum distance features in both datasets Caltech101 and IDOL have great disparity, this is simply caused by the minimum distance in each image; in return, this reduces the PCA performance. The features have been separated by its spatial norm of the resulted features for each 10 bins from the length of the features in advance, before using the PCA or KPCA for processing them. Evidently, the implementation tests demonstrate that the normalized features would improve the classification results more than the non- normalized.

# 7 Classification and Optimization

The recent research studies about SVM show that it produces good performance for object classification [12]. This paper uses SVM and KNN for evaluating the classification performance and similarly for analysing the results for BOW and KPCA of minimum distance features. The process of PCA space optimization is carried out via deciding on the best number of Eigenvectors which provides the best filtering of the invariant features to help make the image scene more discriminate from each other. The trial and error methods are used to conduct the process of PCA space prediction. Then, the calculation of the average precision is achieved. The equation below describes the precision *(p)* of the first $N$ retrieved images for the query $Q$:-

$$p(Q,N) = \frac{|Ir \mid Rank(Q, Ir) \leq N \, such \quad that \quad Ir \in g(Q)\} \parallel}{N} \qquad (1.9)$$

*g(Q)* represents the group category for the query image and *Ir* is the retrieved image. The acquired results for all created feature values are sorted; then, the minimum values are taken to be the best matching visual place. This is called as *K* nearest neighbour (KNN).

# 8 Simulation and Implementation

Several experiments are conducted on two different types of data sets namely; IDOL and Caltech101.

## 8.1 IDOL Dataset

The first part of the experiments is conducted using the dataset of IDOL which is introduced by Pronobis et al., [26, 27]. The SURF features were extracted using a SURF grid algorithm. The size of each frame image is *230×340*. All the experiments were conducted via a laptop computer with these specifications: speed 2.2 GHz core 2 Duo and 3GB memory.

The feature vectors are used for machine learning using K-means algorithm where a set of different cluster numbers namely; 260, 275, 350, 400, 450, 500, 520,600,650 are employed. Next, the best *KB* is used to classify the features for the test images for different groups of the environmental navigation places, these places namely are; a one-person office, a corridor, a two-people office, a kitchen and a printer area.

The best projection space for the PCA is selected to filter the invariant features for the classification process (see Figure 1), the figure shows the effect of Eigen Vectors for optimum selected *KB* on the results of the classification. Details of the experiments are presented in the following subsections:-

### 8.1.1   Experimental Setup

The practical experiment is implemented using SURF grid algorithm. The feature vectors are quantized using *KB=260* clustering and the best Eigen Vectors used is *72*. The whole images are divided in IDOL data set into two groups to evaluate the proposed approaches. Different running tests are used in 5 times. Then, the performances for the two data subsets are reported using the average of the obtained classification results.

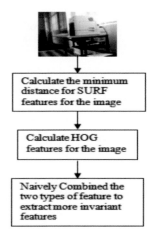

**Fig. 1** Feature vector for input image

### 8.1.2   Results on IDOL

Visual place recognition is implemented in IDOL for the proposed approaches which consistently show on-line performances for recognizing environments when using KPCA approach. To demonstrate the performance, the algorithm is implemented to recognize further 15 places including one-person office, corridor, two-people office, and kitchen and printer area respectively.

Compared with the other approaches, Table 1.1 shows the comparative experiment results of the proposed approach which is implemented on IDOL dataset. The groups in the table correspond to places; the total of successive frames used.

The results in the table are constructed according to the KNN and SVM for the PCA, KPCA and KPCHOG approaches. Three kernel functions (Linear, Polynomial, and RBF) are used for the KPCA. Polynomial and RBF as nonlinear approaches are outperformed the linear approaches.

**Table 1** Comparison results for different groups of Idol.

| Classification | KNN | SVM |
|---|---|---|
| BOW | 50.8 | 58.6 |
| min dist | 91.35 | 90.7 |
| PCA | 93.4498 | 90 |
| KPCA | 93.8048 | 90.6 |
| KPCHOG(NV) | 94.882 | 91.62 |

## 8.2 Caltech101 Dataset

The Caltech dataset images are more difficult to analyse than those of the IDOL dataset due to the difference of background texture which are used with the same object, particularly in the case of objects with different intensities and sizes. The experiment is done on 10, 20 classes from Caltech101. These classes are selected as a total of 300 jpeg images for 10 classes, 600 images for 20 classes. The images for groups consisted of airplane, cameras, cars, cell phones, cups, helicopters, motorbikes, scissors, umbrellas, and so on.

### 8.2.1 Experimental Setup

The experiments implemented the SURF grid algorithm with all scales of the features for evaluating the proposed approaches. The feature vectors are quantized using K-means clustering. The results of the KNN and SVM classification for 10 and 20 classes are verified. 15 training and 15 testing images, for each image class are used via 5-times of different running tests. The results are then described using mean and standard deviation to validate impacts of the obtained classification results.

### 8.2.2 Results of Caltech101

Several $K$'s have been used for 101 classes. In this work, the best one, 1024, was used. Table 2 shows the comparative results of the 5 different runs of the groups.

The above table shows that the performances of KPCHOG, KPCA and PCA approaches for visual object recognition are better than the other approaches, but the performance of the proposed approach using KNN is more than SVM. The implementations give good performances for object and place recognition purposes with more accurate results.

**Table 2** Comparison of Classes of Caltech10.

| Classification | KNN | SVM |
|---|---|---|
| BOW | 36.32 | 45.27 |
| min dist | 50.6 | 57.92 |
| PCA mind | 54.221 | 64.486 |
| KPCA mind | 56.92 | 65.686 |
| KPCHOG(NV) | 57.341 | 66.932 |

## 9    Discussion

For Idol dataset, KPCA with HOG and the polynomial KPCA for the minimum distance approach give better performance, which are better than BOW or other approaches. The aim of using a grid SURF, non-standard one is to take more informative features for the image scene to construct the cloud of the features. The utilized image sight was about *320×240* with a jpg arrangement, despite low quality, an accurate recognition rate arrived (96%), depended on problems of particular environment.

The features became more informative with less dimensionality when represented by the PCA and KPCA feature vectors with the spatial edge histogram. One important issue to achieve the best place recognition is the way in which a codebook for the training features is built or the way in which $K$ is optimized. To do so, in this work, a trial and error process is implemented on BOW algorithm.

The best performance of BOW indicated that $K$ is the best cluster for building a codebook. According to this criterion, $K$ is selected to be used in the proposed algorithm. The error for place recognition was increased and decreased due to the value of the selected $K$ and the best projection PCA space. The best $K$ and the best value of Eigen vectors gave the best performance of place recognition. The criterion used for decision making for the image scene recognition is determined by the majority of retrieved image for each group.

In visual object recognition, SURF grid features are used for describing the objects with texture that is more complicated to recognize than visual place, since for each object, there are various characteristics such as size, intensity and texture. Caltech101 is used for this purpose to verify the proposed approach. Several $K$'s were used for 101 classes. It is clear that the approach KPCA with HOG is suitable for both visual place recognition and object recognition.

## 10    Conclusion

Extracting KPCA or PCA features for the minimum distance vectors and then combined with HOG for the same images gives a decent method for visual place and object recognition in comparison with the other approaches. The experimental results show that spatial Histogram of oriented gradient with minimum distance using KPCA to recognize the visual object and place significantly outperforms linear KPCA, PCA and BOW approaches.

The approach can be used with other features to increase the speed of recognition. The results also show that the soft assignment features approach is better than HBOF, the two approaches are mainly depended on the optimized clustering features where the best $K$ gives the best result of recognition.

# References

1. Wenyu, C., Wenzhi, X., Ru, Z.: Method of item recognition based on SIFT and SURF, Mathematical Structures in Computer Science **24**(5) (2014)
2. Suaib, N.M., Marhaban, M.H., Saripan, M.I., Ahmad, S.A.: Performance evaluation of feature detection and feature matching for stereo visual odometry using SIFT and SURF. In: 2014 IEEE Region 10 Symposium, pp. 200–203 (2014)
3. Sivic, J., Zisserman, A.: Video google: a text retrieval approach to object matching in videos. In: ICCV 2003: Proceedings of the Ninth IEEE International Conference on Computer Vision, p. 1470 (2003)
4. Jiang, Y.-G., Ngo, C.-W., Yang, J.: Towards optimal bag-of-features for object categorization and semantic video retrieval. In: CIVR 2007: Proceedings of the 6th ACM International Conference on Image and Video Retrieval, pp. 494–501 (2007)
5. Philbin, J., Chum, O., Isard, M., Sivic, J., Zisserman, A.: Lost in quantization: improving particular object retrieval in large scale image databases. In: Proceedings of the IEEE Conference on Computer Visionand Pattern Recognition (2008)
6. Huang, J., Kumar, S.R., Mitra, M., Zhu, W.J., Zabih, R.: Image indexingusing color correlograms. In: IEEE Computer Society Conference on Computer Vision and Pattern Recognition, p. 762 (1997)
7. Gandhali, P.S., Debasis, M.: Correlogram Method for Comparing Bio-Sequences. Technical Report FIT-CS-2006-01, Master's Thesis, Florida Institute of Technology (2006)
8. Csurka, G., Dance, C., Fan, L., Bray, C.: Visual categorization with bag of keypoints. In: The 8th European Conference on Computer Vision, pp. 513–516 (2004)
9. Perronnin, F., Dance, C., Csurka, G., Bressan, M.: Adapted vocabulariesfor generic visual categorization. In: European Conference on Computer Vision (ECCV 2006), pp. 464–475 (2006)
10. Jain, A.K., Murty, M.N., Flynn, P.J.: Data clustering: a review. ACM Computing Surveys **31**(3), 264–323 (1999)
11. Forstner, W., Moonen, B.: A metric for covariance matrices. Technical report, Dept. of Geodesy and Geoinformatics, Stuttgart University (1999)
12. Tian, J., Qiuxia, H., Xiaoyi, M., Mingyu, H.: An Improved KPCA/GA-SVM Classification Model for Plant Leaf Disease Recognition. Journal of Computational Information Systems **8**(18), 7737–7745 (2012)
13. Schölkopf, B., Smola, A.J., Müller, K.-R.: Nonlinear component analysis as a kernel eigenvalue problem. Neural Computation **5**, 1299–1319 (1998)
14. Baudat, G., Anouar, F.: Generalized Discriminant Analysis Using a Kernel Approach. Neural Computation **12**(10), 2385–2404 (2000)

15. Artač, M., Jogan, M., Leonardis, A.: Mobile robot localization using an incremental eigenspace model. In: IEEE International Conferenceon Robotics and Automation, Washington, D.C., pp. 1025–1030 (2002)
16. Dzati, A.R., Salwani, I., Haryati, J.: Robust palm print verification system based on evolution kernel principal component analysis. In: IEEE International Conference on Control System, Computing and Engineering 2014 (ICCSCE 2014) (2014)
17. Jogan, M., Leonardis, A., Wildenauer, H., Bischof, H.: Mobile robot localization under varying illumination. In: The 16th International on Pattern Recognition, pp. 2385–2404 (2000)
18. Kröse, B., Bunschoten, R.: Probabilistic localization by appearance models and active vision. In: Proceedings of the IEEE International Conference on Robotics and Automation, pp. 2255–2260 (1999)
19. Hong, M.L., Dong, M.Z., Ren, C.N., Xiang, L., Hai, Y.D.: Face Recognition Using KPCA and KFDA. AMM, pp. 380–384:3850–3853 (2013)
20. Sim, R., Dudek, G.: Learning landmarks for robot localization. In: Proceedings of the National Conference on Artificial Intelligence SIGART/AAAI Doctoral Consortium, Austin, TX, SIGART/AAAI, pp. 1110–1111. AAAI Press (2000)
21. Phiwmal, N., Sanguansat, P.: An Improved Feature Extraction and Combination of Multiple Classifiers for Query-by-Humming. The International Arab Journal of Information and Technology 11(1) 103–110 (2014)
22. Bay, H., Tuytelaars, T., Van Gool, L.: Speeded up robust features. ETH Zurich, Katholieke Universiteit Leuven, vol. 3951, pp 404–417. Springer, Heidelberg (2006)
23. Lazebnik, S., Schmid, C., Ponce, J.: Beyond bags of features: spatial pyramid matching for recognizing natural scene categories. In: Proc. of CVPR 2006 (2006)
24. Suvi, T., Kai, N., Mikko, T., Antti, K., Tapio, S.: ECG-derived respiration methods: Adapted ICA and PCA. Medical Engineering & Physics (2015)
25. Vipsita, S., Shee, B.K., Rath, S.K.: Protein superfamily classification using kernel principal component analysis and probabilistic neural networks. In: 2011 Annual IEEE India Conference (INDICON) (2011)
26. Pronobis, A., Caputo, B., Jensfelt, P., Christensen, I.: A realistic benchmark for visual indoor place recognition. Robotics and Autonomous System 58(1), 81–96 (2009)
27. Lu, L., Jianhua, Y., Evrim, T., Ronald, M.S.: Multilevel Image Recognition using Discriminative Patches and Kernel Covariance, SPIE Medical Imaging (2014)

# Hardware Accelerator for Facial Expression Classification Using Linear SVM

Sumeet Saurav, Sanjay Singh, Ravi Saini and Anil K. Saini

**Abstract** In this paper, we present hardware accelerator for Facial Expression Classification using One-Versus-All (OVA) linear Support Vector Machine (SVM) classifier. The motivation behind this work is to perform real-time classification of facial expressions into three different classes: neutral, happy and pain, which could be used in an embedded system to facilitate automatic patient monitoring in ICUs of hospitals without any personal assistance. Pipelining and parallelism (inherent qualities of FPGAs) have been utilized in our architecture to achieve optimal performance. For achieving high accuracy, the architecture has been designed using IEEE-754 single precision floating-point data format. We performed the SVM training offline and used the trained parameters to implement its testing part on Field Programmable Gate Array (FPGA). Synthesis result shows that the designed architecture is operating at a maximum clock frequency of 200 MHz. Classification accuracy of 97.87% has been achieved on simulating the design with different test images. Thus, the designed architecture of the OVA linear SVM shows good performance in terms of both speed and accuracy facilitating real-time classification of the facial expressions.

**Keywords** Support Vector Machine · Linear SVM · Adaboost · Gabor filter

## 1 Introduction

Recent advances in the Computer Vision algorithms equipped with Machine Learning has open up wide areas of research in the design of Automatic Vision Systems deploying a blend of these algorithms. An important application, which comes under this category, is the development of a Facial Expression Recognition (FER) system. Automated recognition of human facial expression is an active research area with a wide variety of potential application in human-computer interaction, surveillance, customized consumer products and many more.

S. Saurav(✉) · S. Singh · R. Saini · A.K. Saini
CSIR-Central Electronics Engineering Research Institute, Pilani 333031, India
e-mail: {sumeetssaurav,sanjay.csirceeri}@gmail.com

© Springer International Publishing Switzerland 2016                    39
S.M. Thampi et al. (eds.), *Advances in Signal Processing and Intelligent Recognition Systems*,
Advances in Intelligent Systems and Computing 425,
DOI: 10.1007/978-3-319-28658-7_4

Although, software implementation of FER system is feasible, but they suffer from the real-time constraints. For any FER system classification step is the final and most important block which decides the overall performance of the system, and for this the most widely used classifiers are Support Vector Machines (SVM). Thus, in this work our aim is to facilitate real-time classification of the facial expressions using dedicated hardware block of the SVM classifier.

There are a number of works available in literature dealing with SVM, as a classifier for different applications related with pattern recognition and classification. However, most of them are based on its software implementation i.e. both training and testing phase of the algorithm is done on software [1]-[8]. On the other hand in terms of hardware implementation, from our knowledge the first significant work on SVM has been reported in [9]. Here, the authors have proposed a digital architecture for SVM training and classification using both linear and RBF kernels on Xilinx Virtex-II FPGA. The major drawback of this implementation is that it utilized a lot of FPGA resources and the clock frequency (35 MHz) obtained is also not optimal. With further advancement of technology and increasingly use of SVM in a number of pattern classification problems over the time, there have also been a number of works reported in literature dealing with efficient hardware implementation of SVM classifiers. These works can be found in [10]-[19].

The remainder of the paper is organized as follows: In section 2, we described the software implementation of the algorithm. Detailed description of the proposed hardware architecture is discussed in section 3. Simulation and synthesis results have been discussed in section 4 followed by conclusion in section 5.

## 2    Software Implementation of OVA Linear SVM

Since, testing phase of linear SVM classifier require parameters obtained from the training phase, therefore before going for the actual hardware implementation of the SVM architecture, we performed both its training and testing with facial features using libSVM library in Matlab environment. In order to generate the facial features we performed the following sequence of steps: face detection, cropping and resizing of detected faces to 128 by 128 dimension and finally feature extraction for both training and test image samples. For feature extraction we have used Gabor filters whose spatial representation is given in (1). From the mathematical point of view, a two-dimensional Gabor filter is basically a Gaussian kernel function modulated by a complex sinusoidal plane wave [20].

$$G(x,y) = \frac{f^2}{\pi\gamma\eta}\exp\left(-\frac{x'^2+\gamma^2 y'^2}{2\sigma^2}\right)\exp\left(j2\pi f x'+\phi\right) \tag{1}$$

where,
$$x' = x\cos(\theta) + y\sin(\theta)$$
$$y' = -x\sin(\theta) + y\cos(\theta)$$

In our analysis, we have employed forty Gabor filters corresponding to five scales and eight orientations. The number of features obtained after feature extraction and down-sampling by a factor of 4 is 40960, which is still very large. Therefore, we performed feature selection using AdaBoost algorithm [21]. This significantly reduced the feature size to 51 without any compromise in the accuracy of the classifier. The value of γ and η in (1) is taken as $\sqrt{2}$ . The scales (f) and orientations (θ) are selected according to (2).

$$f_g = \frac{f_{max}}{\left(\sqrt{2}\right)^g}$$

$$f_{max} = 0.25 \ and \ g \in \{0,1,2,3,4\} \tag{2}$$

$$\theta_h = \frac{h}{8}\pi \ and \ h \in \{0,1,2,3,4,5,6,7\}$$

For performing the experiments we used our own database consisting of 56 neutral, 77 happy and 49 pain faces of different individuals. The datasets has been divided into two parts, training and test sets whose distribution in shown in table 1. Sample images of the three expressions from our database is shown in the fig. 1.

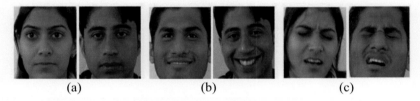

(a)                    (b)                    (c)

**Fig. 1** Samples of three Facial Classes (a) Neutral (b) Happy (c) Pain

**Table 1** Distribution of Training and Test Image Samples

| S.No | Expressions | Training Set Size | Test Set Size |
|------|-------------|-------------------|---------------|
| 1 | Neutral | 40 | 16 |
| 2 | Happy | 55 | 22 |
| 3 | Pain | 40 | 09 |

## 2.1   Classifier Training and Testing Using LibSVM Library

Linear classification using OVA linear classification strategy on AdaBoost reduced dataset has been done with LibSVM [22] library in Matlab. The classifier parameters corresponding to different values of the margin parameter C has been shown in table 2-table5. The parameters obtained after training are used in the testing part of the SVM using the designed architecture.

**Table 2** Classifier Parameter Corresponding to the Value of C=0.01

| S.No | Class1 | Class2 | Class3 |
|------|--------|--------|--------|
| No. of SVs | 47 | 55 | 48 |
| Bias | -1.3309 | 0.6229 | -0.1708 |
| Accuracy (%) | | 97.87 | |

**Table 3** Classifier Parameter Corresponding to the Value of C=0.02

| S.No | Class1 | Class1 | Class2 |
|------|--------|--------|--------|
| No. of SVs | 35 | 46 | 39 |
| Bias | -1.2950 | 0.6704 | -0.2994 |
| Accuracy (%) | | 97.87 | |

**Table 4** Classifier Parameter Corresponding to the Value of C=0.03

| S.No | Class1 | Class1 | Class2 |
|------|--------|--------|--------|
| No. of SVs | 33 | 38 | 36 |
| Bias | -1.3924 | 0.6862 | -0.4095 |
| Accuracy (%) | | 97.87 | |

**Table 5** Classifier Parameter Corresponding to the Value of C=0.039

| S.No | Class1 | Class1 | Class2 |
|------|--------|--------|--------|
| No. of SVs | 29 | 37 | 33 |
| Bias | -1.4608 | 0.6985 | -0.4154 |
| Accuracy (%) | | 97.87 | |

Since, the number of support vectors (SVs) directly affects the FPGA resources, as less SVs requires less storage and computation, therefore from the resource optimization point of view we selected the classifier trained with the margin parameter C=0.039. This is because, for C=0.039 the number of SVs obtained after training is lesser compared to the SVs obtained after training the classifier with other values of the margin parameter. The confusion matrix for test image samples obtained after testing the trained classifier (with C=0.039) is shown in table 6.

**Table 6** Confusion Matrix Corresponding to Classifier Trained with C=0.039

| | Neutral | Happy | Pain |
|---------|---------|-------|------|
| Neutral | 16 | 0 | 0 |
| Happy | 1 | 21 | 0 |
| Pain | 0 | 0 | 9 |

# 3    Proposed Hardware Architecture of OVA Linear SVM

The block diagram of the proposed VLSI architecture of OVA linear Support Vector Machine is shown in fig. 2. Here, we have designed architecture which results in the computation of the decision function (2). Details related with (2) can be found in [17].

$$D(x) = \sum_{i \in S} \alpha_i y_i x_i^T x + b \qquad (2)$$

Detailed description of blocks used in the design is discussed below.

## 3.1    *Input Feature Storage Block*

The size of input feature obtained after feature extraction and selection is 1x51 and is stored in the FPGA block memory. One block RAM is used to store the complete input data in 32-bit IEEE-754 single precision floating point format.

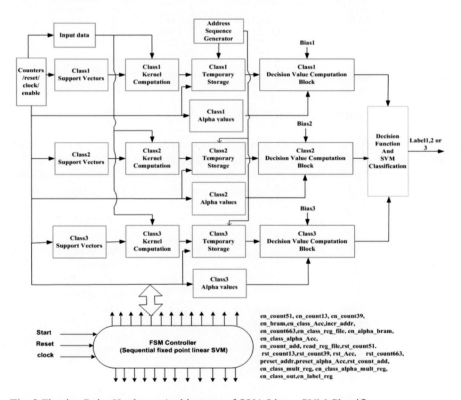

**Fig. 2** Floating Point Hardware Architecture of OVA Linear SVM Classifier

## 3.2   Support Vector Storage Block

This block is used to store Support Vectors (SVs) obtained after SVM training using block RAMs available in FPGA. The width of each block RAM used is 663 (corresponding to 13 rows and 51 columns of SVs matrix) and the corresponding depth is 32-bits taken in IEEE-754 single precision floating-point format. As shown in fig. 3, nine block RAMs have been used for the storage of SVs for all the three classes.

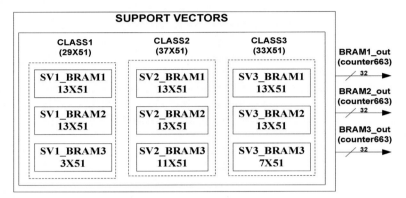

**Fig. 3** Support Vector Storage Block

## 3.3   Yalpha Values Storage Block

Yalpha values corresponding to different classes are also stored in FPGA block RAMs. In this case, the width of each block RAM used is 39 with a depth of 32-bits taken in IEEE-754 single precision floating-point format. Three block RAMs (one corresponding to each class) has been used to store the Yalpha (αy) values as shown in fig. 4.

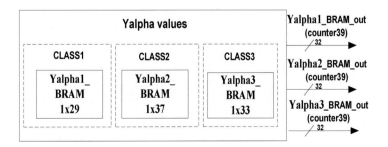

**Fig. 4** Yalpha Storage Block

## 3.4    Kernel Computation Block

This block is used to perform inner product operation between the support vectors corresponding to each classes and the input feature vector. As shown in fig. 5, we are performing kernel computation for three different rows in parallel. Three such blocks have been employed in the design (one corresponding to each class).

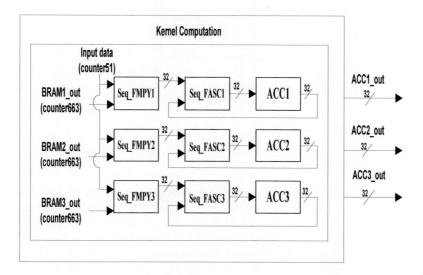

**Fig. 5** Kernel Computation Block

## 3.5    Temporary Storage Block

To store the intermediate results obtained after the kernel computation operation, register files as shown in fig. 6 have been used in the design as a temporary data storage block. Three register files corresponding to three classes have been used each storing results from their respective Kernel Computation Block. Temporary storage block consists of three register files to store the temporary results obtained after kernel computation for each class. The size of each register file is 39 with depth of 32-bits. Addresses for the register file are generated by the address sequence generator block. It generates 3 addresses namely address0, address13 and address26 for a register file at a fixed interval of 13 clock cycles. Initially address0 points to the $0^{th}$ location of register file, address13 points to the 13th location of register file and address 26 points to the 26th location. When counter13 is incremented, all the addresses are also incremented after storing the intermediate results of first row.

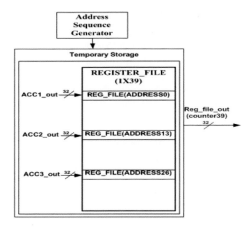

**Fig. 6** Temporary Storage Block

## 3.6    Counters Block

Four different counters have been used in the design to give address to the different storage blocks with the help of Controller. The Counter51 is used to count 51 support vectors stored in the SV_BRAMs (one row of support vector matrix). This counter is also used to give address to the input data storage block. Since 13 rows of the support vectors are stored in one block RAM so a counter called, Counter13 is used to keep track of the rows, which are to be processed. The count of this counter is incremented by one each time when the count value of counter51 becomes 51. Since the size of each block RAM is 663, therefore a global counter called Counter663 has been used to provide access to one support vector every clock cycle. To access the contents of the register files and Yalpha block RAMs, another counter (Counter39) has been used. When the count of this counter becomes 39, it indicates that the decision value of each class has been calculated which are then fed to the Decision Function and SVM classification block.

## 3.7    Decision Value Computation Block

This block performs vector dot product computation between the temporary stored results (Reg_file_out) and Yalpha values (Yalpha_BRAM_out) stored in the YAlpha Storage Block as shown in fig. 7. Similar to all other blocks discussed above, the designed architecture of SVM also uses three such Decision Value Computation Blocks (one for each class). From the figure we find that, a sequential floating-point multiplier Seq_FMPY accepts two inputs one from the register file and other from the Yaplha_BRAM. Counter39 is used to access the contents of these storage elements. The output of the multiplier is given as input to the sequential floating-point adder Seq_FASC who's another input comes from the accumulator

Aalpha1_ACC. Initially, the accumulator is preset with bias value of Bias1, which is equivalent to addition of 'b' as shown in (2). When the count of the Counter39 becomes equal to 39 we get Clas1_out as the output of the Decision Value Computation block.

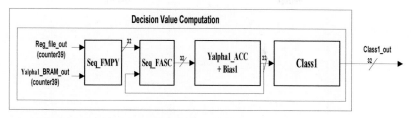

**Fig. 7** Decision Value Computation Block

## 3.8   Decision Function and SVM Classification Block

This is the final block in our design, which takes output from three Decision Value Computation block, decides the label of the input test data, and provides it the corresponding label (label1, label2, or label3). Decision is made as follows: When the value of class_1out is positive and that of class_2out and class_3out is negative, then label1 is assigned to the input test data, label2 is selected when class_2out is positive and other two (class_1out and class_3out) is negative, and finally label3 is selected when class_3out is positive and class_1out and class_2out is negative. Finally, one of the labels among label1, label2 and label3 will be the output of our proposed OVA Linear Support Vector machine classifier.

## 3.9   Controller Block

A top-level controller as shown in fig. 2 is designed to send the appropriate control signals to the blocks in the execution unit. Execution of operations are achieved by moving data into the input register(s) of a functional block, triggering of the functional block, waiting for the completion of operation of the functional block and then moving the generated result from the output register of the functional block to the specified destination address in the register files. Controls are required for the necessary data movement to/from the functional units and their triggering come from the top level controller. A state sequencer generates the sequences of the control states. The sequence of steps (states) that are performed for the implementation of the OVA linear SVM classifier is shown in fig. 8.

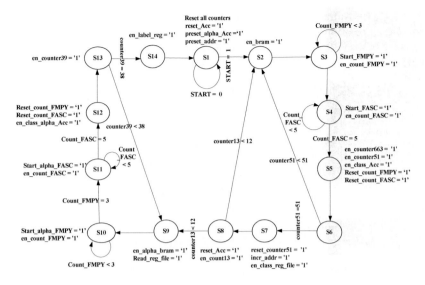

**Fig. 8** Controller State Diagram

# 4    Simulation and Synthesis Result

All the modules of the proposed architecture are coded in VHDL and simulated using ModelSim 10.1C. Synthesis is carried out using Xilinx ISE14.2 tool chain. We have used Xilinx ML510 (Virtex-5 FXT) FPGA platform for synthesizing the design. IEEE-Single precision floating-point data format has been used in the design. The hardware utilization report has been listed in table 7 and the simulation result has been shown in the fig. 9. Simulation result shows the final resulted output for 3 classes stored in class1_out, class2_out and class3_out corresponding to the input test image. Notice the positive value of class1_out (i.e. label1 is the correctly classified resulted output) and also, all the counters attained their maximum value as count13 is 13, count39 is 39 and count663 is 663.

**Table 7** Hardware Synthesis Result

| Logic Utilization | Available | Used | Utilization Rate (%) |
|---|---|---|---|
| Slice LUTS | 81920 | 11811 | 14 |
| Flip Flop Pairs | 81920 | 13065 | 15 |
| Occupied Slice | 20480 | 4143 | 20 |
| Slice Registers | 81920 | 8146 | 9 |
| RAMB36 | 298 | 13 | 4 |
| DSP48 | 320 | 24 | 7 |

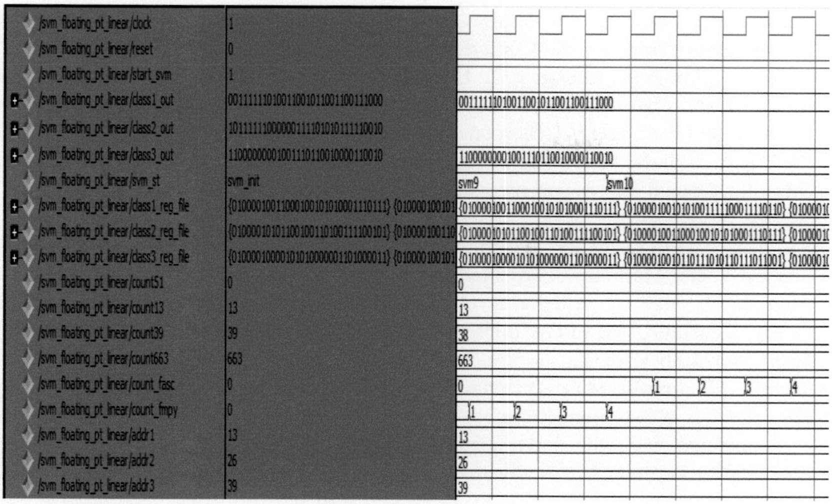

**Fig. 9** Simulation Result

# 5    Conclusion

In this paper, the hardware architecture of OVA linear Support Vector Machine classifier for facial expression classification has been presented. The design has been realized using VHDL and implemented on Xilinx ML510 FPGA platform. The architecture has been designed using IEEE-Single precision floating-point data format. The implemented architecture can perform real-time classification of the facial expression operating at a clock frequency of 200 MHz. Classification accuracy of 97.87% has been achieved. Thus, the designed architecture of the OVA linear SVM classifier shows good performance both in terms of speed and accuracy. The designed architecture can be integrated with the FER system as a hardware accelerator for performing real-time facial expression classification.

# References

1. Hirata, W., Tan, J.K., Kim, H., et al.: Recognizing facial expression for man-machine interaction. In: Proceedings of the IEEE ICCAS–SICE (2009)
2. Beszédeš, M., Culverhouse, P., Oravecm, M.: Facial emotion classification using active appearance model and support vector machine classifier. Machine Graphics and Vision International Journal **18**(1), 21–46 (2009)
3. Kotsia, I., Pitas, I.: Facial expression recognition in image sequences using geometric deformation features and support vector machines. IEEE Transactions on Image Processing **16**(1), 172–187 (2007)
4. Kotsia, I., Pitas, I.: Real time facial expression recognition from image sequences using support vector machines. In: Visual Communications and Image Processing, International Society for Optics and Photonics (2005)

5. Dumas, M.: Emotional expression recognition using support vector machines. In: Proceedings of International Conference on Multimodal Interfaces (2001)
6. Tsai, H.H., Lai, Y.S., Zhang, Y.C.: Using SVM to design facial expression recognition for shape and texture features. In: Proceedings of the IEEE International Conference on Machine Learning and Cybernetics (ICMLC), vol. 5 (2010)
7. Patil, R.A., Sahula, V., Mandal, A.S.: Features classification using support vector machine for a facial expression recognition system. Journal of Electronic Imaging **21**(4) (2012)
8. Visutsak, P.: Emotion Classification through Lower Facial Expressions using Adaptive Support Vector Machines. Journal of Man, Machine and Technology **2**(1), 12–20 (2013)
9. Anguita, D., Boni, A., Ridella, S.: A digital architecture for support vector machines: theory, algorithm, and FPGA implementation. IEEE Transaction on Neural Networks **14**(5), 993–1009 (2003)
10. Khan, F.M., Arnold, M.G., Pottenger, W.M.: Hardware-based support vector machine classification in logarithmic number systems. In: Proceedings of the IEEE International Symposium on circuits and systems, vol. 5, pp. 5154–5157 (2005)
11. Irick, K.M., DeBole, M., Narayanan, V., et al.: A hardware efficient support vector machine architecture for FPGA. In: Proceedings of the 16th IEEE International Symposium on Field-Programmable Custom Computing Machines, pp. 304–305 (2008)
12. Hsu, C.F., Ku, M.K., Liu, L.Y.: Support vector machine FPGA implementation for video shot boundary detection application. In: Proceedings of the IEEE International Conference on SOC, pp. 239–242 (2009)
13. Bauer, S., Kohler, S., Doll, K., et al.: FPGA-GPU architecture for kernel SVM pedestrian detection. In: Proceedings of the IEEE Computer Society Conference on Computer Vision and Pattern Recognition Workshop, pp. 61–68 (2010)
14. Pina-Ramirez, O., Valdes-Cristerna, R., Yanez-Suarez, O.: An FPGA implementation of linear kernel support vector machines. In: Proceedings of the IEEE International Conference on Reconfigurable Computing and FPGA's, pp. 1–6 (2006)
15. Nie, Z., Zhang, X., Yang, Z.: An FPGA implementation of multi-class support vector machine classifier based on posterior probability. In: Proceedings of 3rd International Conference on Computer and Electrical Engineering, pp. 296–302 (2012)
16. Anguita, D., Ghio, A., Pischiutta, S., et al.: A hardware-friendly support vector machine for embedded automotive applications. In: Proceedings of the IEEE International Joint Conference on Neural Networks (2007)
17. Mahmoodi, D., Soleimani, A., Khosravi, H., et al.: FPGA simulation of linear and nonlinear support vector machine. Journal of Software Engineering and Applications **4**, 320–328 (2011)
18. Ruiz-Llata, M., Guarnizo, G., Yebenes-Calvino, M.: FPGA implementation of a support vector machine for classification and regression. In: Proceedings of the IEEE International Joint Conference on Neural Networks, pp. 1–5 (2010)
19. Ruiz-Llata, M., Yébenes-Calvino, M.: FPGA implementation of support vector machines for 3D object identification. In: Proceedings of the International Conference on Artificial Neural Networks, pp. 467–474 (2009)
20. Shen, L., Bai, L., Fairhurst, M.: Gabor wavelets and general discriminant analysis for face identification and verification. Image and Vision Computing **25**(5), 553–563 (2007)
21. Shen, L., Bai, L.: AdaBoost Gabor feature selection for classification. In: Proceedings of Image and Vision Computing, New Zealand, pp. 77–83 (2004)
22. Chang, C.C., Lin, C.J.: LIBSVM: A library for support vector machines. ACM Transactions on Intelligent Systems and Technology (TIST) **2**(3) (2011)

# Enhancing Face Recognition Under Unconstrained Background Clutter Using Color Based Segmentation

Ankush Chatterjee, Deepak Mishra and Sai Subrahmanyam Gorthi

**Abstract** Face recognition algorithms have been extensively researched for the last 3 decades or so. Even after years of research, the algorithms developed achieve practical success only under controlled environments. Their performance usually takes a dip under unconstrained scene conditions like the presence of background clutter, non-uniform illumination etc. This paper explores the contrast in performance of standard recognition algorithms under controlled and uncontrolled environments. It proposes a way to combine 3 subspace learning algorithms, namely Eigenfaces (1DPCA), 2 dimensional Principal Component Analysis (2DPCA) and Row Column 2DPCA (RC2DPCA) with a color-based segmentation approach in order to boost the recognition rates under unconstrained scene conditions. A series of steps are performed that extract all possible facial regions from an image, following which the algorithm segregates the largest candidate for a probable face, and puts a bounding box on the blob in order to isolate only the face. It was found that the proposed algorithms, formed by the combination of such segmentation methods obtain a higher level of accuracy than the standard recognition techniques. Moreover, it serves as a general framework wherein much more robust recognition techniques could be combined to achieve boosted accuracies.

## 1 Introduction

The concept of face recognition has aroused the interests of computer scientists for a very long time. Even though alternate recognition schemes have provided better

A. Chatterjee(✉)
Indian Institute of Technology, Kharagpur, India
e-mail: ankushchatterjee.iitkgp@gmail.com

D. Mishra · S.S. Gorthi
Indian Institute of Space Science and Technology, Thiruvananthapuram, India
e-mail: deepak.mishra@iist.ac.in, saisubrahmanyam.gorthi@gmail.com

© Springer International Publishing Switzerland 2016
S.M. Thampi et al. (eds.), *Advances in Signal Processing and Intelligent Recognition Systems*,
Advances in Intelligent Systems and Computing 425,
DOI: 10.1007/978-3-319-28658-7_5

accuracy, facial recognition has always been a topic of interest among scientists, mostly because face is usually the fastest means through which humans recognize a person and secondly, the other alternate approaches come with the associated hurdles of data acquisition.

In Principal Component Analysis (PCA) (L. Sirovich et al. [1]), (Matthew Turk [2]), given a set of training images in high dimensional space, PCA finds a lower dimensional subspace whose basis vectors correspond to directions of maximum variance in the original sub-space. Independent Component Analysis (Bartlett et al.[3]) attempts to find the basis along which the projected data are statistically independent. Linear Discriminant Analysis (LDA) (Peter N. Belhumeur et al. [4]), finds the vectors in the underlying space that best discriminate among classes. Kernel methods (Bernhard Scholkopf et al. [5]), (M.H. Yang [6]), are based on the argument that the face manifold in subspace need not be linear. Direct non-linear manifold schemes are explored to learn this non-linear manifold.

But one major point to be noted is that the algorithms work pretty well only under controlled backgrounds. They suffer significantly once the conditions under which the images are captured are distorted. Such distortions may include variations in illumination or inclusion of complex objects in the background. Overcoming this particular challenge was the prime objective of the research. A deduction was made that a pre-processing step that would isolate the face from the rest of the image would essentially be nullifying the effects of the inclusions in the background. Experiments performed on a database proved the assumptions correct, the results of which have been discussed in Section 4. The details of the recognition algorithms on which the experiments were performed have been briefly explained in section 2. The modified version of the segmentation algorithm, has been explained in section 3.

## 2 Face Recognition Algorithms

A few of the face recognition techniques have been discussed below.

### 2.1 Eigenfaces

Developed by Sirovich and Kirby in 1987 [1] and used by Matthew Turk and Alex Pentland [2], the steps involved in this algorithm are explained below.

Given $M$ training images in the form of $n$-dimensional vectors, the mean-subtracted images are obtained and their covariance matrix $C$ constructed. The eigenvectors corresponding to the top $k$ eigenvalues are retained as they represent the directions of maximum variance. These eigenvectors, being of the same dimension as the input images can be visualized as images themselves, popularly known as Eigenfaces.

Once the eigenvectors are obtained, any face can be approximated as a linear combination of the $k$ eigenfaces. The weights given to to each eigenvector can be calculated and stored in the form of a weight vector $\Omega^T = [w_1, w_2, \ldots, w_k]$ which serves as the feature vector for the particular image. Any person in the training set is represented as the average of the weight vectors $\Omega$ of all sample images of that particular class.

A new image is classified as a member of class $k$ if $k$ minimizes the Euclidean distance $\varepsilon_k = ||(\Omega - \Omega_k)||$, and $\varepsilon_k$ is less than a certain threshold $\Theta_\varepsilon$. If it is more than the threshold, the face is classified as "unknown".

Any image can be reconstructed by the following formula

$$\tilde{A} = \Psi + w_1 V_1 + w_2 V_2 + \cdots + w_k V_k$$

where $\Psi$ is the mean image, $w_i$ is the weight assigned to the $i^{th}$ eigenvector $V_i$.

## 2.2   2-Dimensional PCA

Where the Eigenfaces approach treats the images as 1-dimensional vectors, its 2-dimensional counterpart [7] operates on the image without any conversion. Instead of a feature vector (weight vector) in Eigenfaces approach, a feature matrix is obtained in this case.

Consider the projection of an $m \times n$ image $A$ onto a vector $X$ to yield a feature vector $Y = AX$. A good projection vector $X$ would essentially maximize the total scatter of all projected samples. The total scatter of the projected samples can be characterized by the trace of the covariance matrix of the projected feature vectors of all samples. It is simplified by

$$J(x) = tr\{S_y\} = X^T E\{(A - EA)^T (A - EA)\}X = X^T G X$$

where $G$ is the image scatter matrix/image covariance matrix. The best directions of projection are along the the eigenvectors corresponding to the largest eigenvalues. So the steps of 2DPCA are briefly listed below.

Given $M$ training images in the form of $m \times n$ matrices their covariance $G$ is constructed. The top $d$ eigenvectors of $G$ are retained as $[X_1, X_2, \ldots, X_d]$. The principal components are calculated by projecting an image on each of these $d$ eigenvectors as $Y_k = AX_k$ for $k = 1 : d$. The feature matrix of an image $A$ is thus given by $B = [Y_1, Y_2, \ldots, Y_d]$. The feature matrices of all the input images are calculated.

Given a test image $A$, its feature matrix $B$ is calculated by projecting the sample $A$ on each of the $d$ eigenvectors $X_k$. The Euclidean distance of this feature matrix from the feature matrices of all training images are calculated.

$$\varepsilon_i^2 = \sum_{x=1}^{m} \sum_{y=1}^{d} |B_{xy}^{Test} - B_{xy}^{Train_i}|^2$$

where $\varepsilon_i$ is the distance between the feature matrices of the test sample $B^{Test}$ and the the $i^{th}$ training sample $B^{Train_i}$. The image is classified as a member of the same class as $i$, which minimizes $\varepsilon_i$. The 2DPCA method requires less computation time, even when the number of training samples is very large (in case of Eigenfaces approach computation time increases with the number of training samples used). This is primarily because the dimensions of the covariance matrix is independent of the number of samples used in the case of 2DPCA.

## 2.3   Row Column 2DPCA (RC2DPCA)

RC2DPCA [8] performs 2DPCA twice, once along the rows and a second time along the columns of an image. After the first 2DPCA transform in row direction, the feature matrix $B$ of sample $A$ has the same number of rows as $A$ but fewer columns, as the image is compressed along each row. Hence the transpose of this feature matrix is fed into the 2DPCA framework again so as to facilitate compression along each column. Thus, constructing the covariance matrix of $B^T$, we have

$$G_2 = \frac{1}{M} \sum_{j=1}^{M} (B_j - \bar{B})(B_j - \bar{B})^T$$

where $B_j = A_j X$, $\bar{B} = \bar{A} X = \Psi X$ and $G_2$ is the $m \times m$ covariance matrix, where $m$ is the number of rows in the $m \times n$ dimensional input image.
Suppose $[u_1, u_2, \ldots, u_q]$ are the eigenvectors of $G_2$ corresponding to the $q$ largest eigenvalues. Let $U = [u_1, u_2, \ldots, u_q]$ the second 2DPCA feature matrix $C$ is obtained by

$$C^T = B^T U$$

Thus

$$C = U^T B = U^T A X$$

The resulting feature matrix is thus $q \times d$ dimensional (where d is the number of eigenvectors retained by the first 2DPCA operation) and hence much smaller than the feature matrix of 2DPCA. The identification is done using this feature matrix in a similar fashion as the 2DPCA.

   The RC2DPCA method takes slightly more training time as it performs 2DPCA twice, but the time taken for classification reduces along with the memory usage as the feature matrix is smaller in dimensions than the 2DPCA algorithm.

## 2.4 Other Forms of PCA

There are many variations of the original form of PCA. They are for the purpose of betterment and for removing drawbacks in the traditional one. Kwang et al. [9] proposed a kernel principal component analysis (KPCA), which was a nonlinear extension of PCA. The basic idea was to map the input space into a feature space via nonlinear mapping and then computing the principal components in the feature space. Through adopting a polynomial kernel, the principal components can be computed within the space spanned by high-order correlations of input pixels making up a facial image, thereby producing a good performance. KPCA can be applied in supervised and unsupervised learning. Rajkiran Gottumukkal et al.[10] proposed a face recognition algorithm based on modular PCA approach. The PCA based face recognition considers the global information of each face image and represents them with a set of weights. In modular PCA, the face images are divided into smaller sub-images and the PCA approach is applied to each of these sub-image. The modular PCA method outperforms most PCA techniques when subjected to large variations in illumination and facial expression.

## 3 Color Segmentation

The methods discussed above try to find the directions in which the samples differ from the mean image, and then project the samples onto those directions. Such an approach works fine as long as the variance in an image represents mostly the variance caused due to the presence of a face. The databases on which such methods were tested usually contain pictures taken under controlled environments and a simple background. But what happens when the background contains complex objects. One can obviously infer that the directions in which the samples differ from the mean image will not just represent the facial part but the impact of the background in the picture has to be accounted for too. In such cases, it becomes difficult for the above discussed methods to recognize a face properly. In such cases where a complex background hinders the recognition process, it is only intuitive to isolate the facial region from the rest of the image. Various segmentation methods have been already researched. The paper proposes an algorithm to merge the recognition algorithms with a modified version of one such segmentation algorithm proposed in [11]. Our contributions to the existing segmentation algorithm are twofold.

- Firstly, the original paper proposed a density regularization step, which assumed that the background in an image generally contains a uniform distribution of pixel intensities and the variation is contained only in the facial part. But the objective at hand clearly takes into consideration the cases where the background is not free of variance, and under such unconstrained backgrounds, the idea is not of any practical use. Hence, this particular step of the algorithm was bypassed in the proposed algorithm for better results.

- Secondly we have implemented a series of methods to extract the largest possible candidate for a face in an image. The detailed implementation and advantage of this method is explained later.

The sequential steps of the modified segmentation algorithm are explained in detail below:-

1. Human faces have a special color distribution that differs significantly (although not entirely) from those of the background objects. It has been found that the chrominance values of the skin pixels in YCrCb space are tightly bound within a small range. The regions having Cr and Cb values within the provided thresholds are represented with the color white and the other regions by the color black. If $R_{Cr}$ and $R_{Cb}$ denote the regions bounded by the thresholds $R_{Cr} = \{\Theta_{min}^{Cr}, \Theta_{max}^{Cr}\}$ and $R_{Cb} = \{\Theta_{min}^{Cb}, \Theta_{max}^{Cb}\}$, then the output of stage 1 is

$$O_1(x, y) = \begin{cases} 255, & \text{if } [\, Cr(x, y) \in R_{Cr} \,] \\ & \cap [\, Cb(x, y) \in R_{Cb} \,] \\ 0, & \text{otherwise} \end{cases}$$

Figure 2 shows the bitmap obtained after isolating the pixels whose Cr and Cb intensities lie in the desired ranges.

**Fig. 1** Input image          **Fig. 2** Stage 1 output

2. The first step of the second stage of this algorithm is to calculate a density map. Given an $M \times N$ image an $M/8 \times N/8$ density map is calculated by the following algorithm.

$$D(x, y) = \frac{\sum_{i=0}^{7} \sum_{j=0}^{7} O_1(8x + i, 8y + j)}{64}$$

This gives us a gray scale image (Figure 3) in a reduced dimension representing the density of pixel intensities in $8 \times 8$ sub-matrices of the Stage 1 output. A series of noise reduction techniques are performed on this density map which are listed here as follows-

   i All points lying at the borderline of the density map is set to 0, i.e. $D(0, y) = D(M/8 - 1, y) = D(x, 0) = D(x, N/8 - 1) = 0$
   for $x = 0, 1 \ldots, M/8 - 1$ and $y = 0, 1, N/8 - 1$

ii Any full density point (255) is set to 0 if it is surrounded by less than 4 full density (255) points in its $3 \times 3$ neighborhood i.e., it erodes small white patches.

iii Any point with density $< 255$ is set to 255 if there are more than 2 full-density points in its local $3 \times 3$ neighborhood i.e., it dilates any non-black patches.

After the noise reduction steps, a bitmap is obtained using the following mathematical equation

$$O_2(x, y) = \begin{cases} 255, & \text{if } D(x, y) = 255 \\ 0, & \text{otherwise} \end{cases}$$

**Fig. 3** Density map of Stage 2          **Fig. 4** Output bitmap of Stage 2

3. The 3rd step of the segmentation algorithm employs minor noise reduction techniques, which remove any odd structures or projections. This is done by horizontal and vertical scanning of the image, the details of which are explained below.

   i We scan all the horizontal pixels, one row at a time. Once a white pixel is located, we calculate the number of pixels it is connected to on the same row. If the number of connected components is less than 4 pixels, we discard such a small run of pixels and blacken all the pixels.

   ii The same procedure is repeated in the vertical direction, scanning vertical pixels, one column at a time, giving us Figure 5.

4. In the 4th step, the output of stage 1 and 3 are used to get a contour image of $M \times N$ dimensions. We iterate through the pixels of the stage 3 bitmap. If a pixel is not on the boundary of black and white, it is mapped to a corresponding $8 \times 8$ sub-matrix with the same pixel value as in the stage 3 bitmap. But for the pixels on the boundary, they are mapped into the corresponding $8 \times 8$ sub-matrix with the value of each pixel being the same as the corresponding sub-matrix pixel intensities of stage 1.

5. In the 5th stage, we attempt to further reduce noise by running a Breadth First Search algorithm to detect non-black blobs in the output image of Stage 4. Among the blobs detected would be the desired face along with small blobs in the background that have close resemblance to human skin color. The number of connected pixels of each blob is calculated. The blob with the largest number of connected components is retained as it is most likely to contain the face. All smaller blobs are discarded (pixels assigned to 0).

**Fig. 5** Bitmap of Stage 3          **Fig. 6** Bitmap of Stage 4          **Fig. 7** Contour extracted

---

**Algorithm 1.** Step 5 Part 1

1: **procedure** LARGEST BLOB RETENTION
2:    *blobs.index* ← indices of blobs returnded by BFS
3:    *maxIndex* ← index of blob with maximum size
4:    **for** all pixels in image **do**
5:        **if** image pixel is not in the blob *blobs(maxIndex)* **then**
6:            make pixel black

---

**Algorithm 2.** Step 5 Part 2

1: **procedure** BOUNDING BOX
2:    *corners* ← 4 corner points of bounding box of non-black blob
3:    *m,n* ← dimensions of the image
4:    *image* ← *crop(image, corners)*
5:    resize *image* to dimensions *m, n*

---

Then, a bounding box is placed on the detected blob and the box is resized to the dimensions of the source image (Fig 8). The rationale behind this is that even the largest blob might only occupy a small fraction of the area of the image. Thus putting a bounding box on it and resizing it essentially magnifies the face. In such a situation the variance will correspond to the facial part only and the resizing part makes sure that only the face is available for recognition.

**Fig. 8** The final output          **Fig. 9** Sample image from with clutter

## 4   Experimental Results and Discussion

To validate our theories, we first compared the recognition rates of 1DPCA, 2DPCA and RC2DPCA on the ORL database. Convinced that all these method were good enough (with about 95 % accuracies) to recognize the faces we went on to test the recognition rates on the CVL database. The images have been provided by the Computer Vision Laboratory, University of Ljubljana, Slovenia [12, 13]. Using GIMP photo editor, we manually introduced brush strokes of different color to recreate the effect of background clutter. We introduced this clutter for 13 classes only as we saw that the accuracy of recognition is almost always increased by segmentation.

We evaluated all 3 algorithms under 3 different conditions on the subset of the CVL database as listed below :

i 13 classes with no synthetic clutter. Segmentation was not used either.
ii 13 classes with manually added clutter. Segmentation was not used to remove the clutter.
iii 13 classes with manually added clutter. Segmentation used to remove clutter and extract facial region, on which training and testing was done.

Our motivation behind adding such clutter was to simulate a real world scenario as the situations would be very different from laboratory controlled environments. Presence of cars, buildings or trees behind the face would affect the subspace learning algorithms and hence to capture the effect of such complex backgrounds, we have manually added synthetic clutter in all the images.

- It is to be noted that the CVL database contains 7 samples per person, out of which 4 images are turned by more than or equal to 45 degrees. This makes it almost impossible for either of the 3 recognition algorithms to classify these particular images, with or without segmentation. Hence the low recognition rates are on account of this important fact about the structure of data, and not due to any major fault of the segmentation algorithm. Our reason for choosing the CVL data-set was primarily because of the colored photographs, which were needed for testing our segmentation results.
- In Figure 10, we have plotted the accuracy of the 1DPCA (Eigenfaces) approach against the number of eigenvectors retained for projection. The green line (with crosses) represents the accuracy attained by the 1DPCA approach on the subset (13 classes) of the CVL data-set without any synthetic clutter. Segmentation was not used for this data-set. We observe roughly 50 % images are classified which amounts to roughly 3.5 images per class, which was according to our expectations.

When we introduce some synthetic clutter in the background, the performance dips drastically, which is shown by the red line (with circles). This is mainly because the variance in the background affects the covariance matrix. The directions in which the samples deviate from the mean image now are no longer limited to the facial region. In other words, the variance present in the background hinders the recognition process, and the effects can be seen in all of the three plots.

Finally we have used the segmentation approach to remove the background clutter and extract the facial region from the image. We expected the recognition rates to be at least as good as the data-set with no clutter at all. But the results achieved

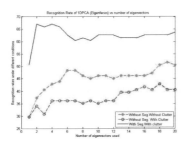

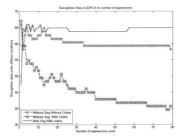

**Fig. 10** 1DPCA                          **Fig. 11** 2DPCA

were surprising at first as the curve, shown in blue solid line, was much higher than the green (crossed) line, which showed that the combination of segmentation with standard recognition algorithms can help us achieve superior recognition rates. We concluded that the improvement in the accuracies was probably because when we performed the segmentation, we not only reduced the variance in the background, we also magnified the variance important for recognition, as we roughly register the faces. When we cropped the largest blob and re-sized it to the size of the image, we extracted only the facial part. Hence the directions in which the samples deviate from the mean image now mostly represent the facial region, which boosts the accuracies.

- Figure 11 shows a similar plot, only trained with a 2DPCA model. Similar to what we saw in the case of the 1DPCA algorithm, the accuracy without any synthetic clutter (represented by the green crossed line), is around 60 %. This falls down when synthetic clutter is introduced, as shown by the red line (with circles). The segmentation improves the recognition rate to almost 70 %.

- Figure 12 shows a very similar plot as the previous one, except for the fact that it is trained on an RC2DPCA model. The x-axis shows the number of eigenvectors retained along the row and column directions i.e., if eigenvectors retained is 2, it means that the final feature matrix is of dimensions $2 \times 2$ One may recall that RC2DPCA performs compression along the rows as well as the columns. Hence with lower number of eigenvector retention, the feature matrix tends to be very low in dimensions. Such small matrices may not be enough for an accurate recognition, hence we observe low accuracies initially but as we progress towards the right hand side of the plot, the recognition gets boosted to as high as 68 %. As with the other plots the relative position of the blue, green (crossed) and red (circles) lines are the same, which goes on to show that segmentation boosts up recognition rates in an uncontrolled and complex background scenario.

- Figure 13 compares the recognition rates achieved by the hybrid models for different classes.

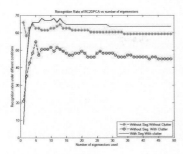

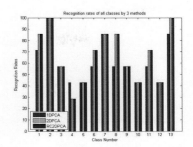

**Fig. 12** RCDPCA

**Fig. 13** Class Accuracies

## 5 Conclusion

In this paper, we proposed an effective way to boost the accuracies of standard face
recognition algorithms under unconstrained scenes. This paper is focused on the negative impact of the presence of background clutter on the performance of standard
recognition algorithms. Our experiments showed the contrast between the performances of such algorithms under a near-perfect situation and a real life situation. In
our efforts to find a way around the problem, we decided to modify a color-based
segmentation algorithm proposed in [11]. The algorithm works by deriving the regions in the image with a close resemblance to human skin color. Sequential noise
removal techniques are deployed and a final contour is extracted. A bounding box is
put around the face and the face is isolated. Our contributions to the existing segmentation algorithm were mainly twofold- Firstly, we bypassed a density regularization
step proposed in the original paper. The original algorithm assumed that the uniform
distribution of pixel intensities in the background could be exploited to remove the
background, which in our experiments would not work well due to variance in the
background pixel intensities. Secondly, the final step of the algorithm was a major
contribution, in which a Breadth First Search blob detection was deployed to locate
the candidate for a face with maximum area, and then a bounding box was put around
the box, after discarding all the smaller blobs. Thereafter the box was resized to mag-
nify the face so the directions of projection in the subspace correspond to variations
primarily due to the face. For a data set of 13 classes of 7 samples each, an average
of 15% increase in accuracies was observed. Moreover, it also provides a general
framework to improve other subspace learning methods by combining them with the
segmentation algorithm. As an extension of this work, we believe that slightly more
robust segmentation algorithms that take into account the geometrical properties of
a face may be experimented with, which would convey a more accurate description
of the facial region in an image.

# References

1. Sirovich, L., Kirby, M.: Low dimensional procedure for the characterization of human faces. J. Opt. Soc. Am. A (1987)
2. Turk, M., Pentland, A.: Eigenfaces for recognition. Journal of Cognitive Neuroscience (1991)
3. Movellan, J.R., Bartlett, M.S., Sejnowski, T.: Face recognition by independent component analysis. IEEE Transactions on Neural Networks (2002)
4. Belhumeur, J.P.N., Kriegman, D.: Eigenfaces vs. fisherfaces: Recognition using class specific linear projection. IEEE Transactions on PAMI (1997)
5. Bülthoff, A., Bernhard, S., Smola, A.J.: Non-linear component analysis as a kernel eigenvalue problem. Neural Computation
6. Yang, M.H.: Kernel eigenfaces vs. kernel fisherfaces: face recognition using kernel methods. In: IEEE International Conference on Face and Gesture Recognition (2002)
7. Firangi, A.F., Yang, J., Zhang, D., Yang, J.-y.: Two-dimensional pca: A new approach to appearance-based face representation and recognition. IEEE Transactions on PAMI (2004)
8. Yang, W., Ricanek, K.: Sequential rowcolumn 2dpca for face recognition. Neural Computing and Applications (2011)
9. Jung, K., Kim, K.I., Kim, H.J.: Face recognition using kernel principal component analysis. IEEE Signal Processing Letters (2002)
10. Gottumukkal, R., Asari, V.K.: An improved face recognition technique based on modular pca approach. Pattern Recognition Letters (2004)
11. Chai, D., Ngan, K.N.: Face segmentation using skin-color map in videophone applications. IEEE Transactions on Circuits and Systems for Video Technology (1999)
12. Peer, P.: Cvl face database: http://www.lrv.fri.uni-lj.si/facedb.html
13. Batagelj, B., Juvan, S., Kovac, J., Solina, F., Peer, P.: Color-based face detection in the'15 seconds of fame' art installation. In: Conference on Computer Vision /Computer Graphics Collaboration for Model-based Imaging, Rendering, ImageAnalysis and Graphical Special Effects, March 2003

# Emotion Recognition from 3D Images with Non-Frontal View Using Geometric Approach

D. KrishnaSri, P. Suja and Shikha Tripathi

**Abstract** Over the last decade emotion recognition has gained prominence for its applications in the field of Human Robot Interaction (HRI), intelligent vehicle, patient health monitoring, etc. The challenges in emotion recognition from non-frontal images, motivates researchers to explore further. In this paper, we have proposed a method based on geometric features, considering 4 yaw angles (0°, +15°, +30°, +45°) from BU-3DFE database. The novelty in our proposed work lies in identifying the most appropriate set of feature points and formation of feature vector using two different approaches. Neural network is used for classification. Among the 6 basic emotions four emotions i.e., anger, happy, sad and surprise are considered. The results are encouraging. The proposed method may be implemented for combination of pitch and yaw angles in future.

**Keywords** BU3DFE database · Emotion · Euclidean distance · 3D images · Classification · Neural network

## 1    Introduction

Human express emotions in their everyday life and recognizing it improves the communication in interactions. Emotion recognition is very important in human computer interactions and research on human emotion recognition can be quoted back to 19th century [6] where, Darwin stated all human beings express emotions, which convey distinct information. The different modalities of emotion recognition are speech, biological signals or facial expressions. In a communication which is human to human, the verbal communiqué divulges 7% of the meaning; vocal communiqué 38%; and physiognomy 55% [1]. In the past few decades most of the

D. KrishnaSri(✉) · P. Suja · S. Tripathi
Amrita Robotic Research Centre, Amrita School of Engineering,
Amrita Vishwa Vidyapeetham, Bengaluru 560 035, Karnataka, India
e-mail: krishnasri_d@yahoo.co.in

© Springer International Publishing Switzerland 2016                                   63
S.M. Thampi et al. (eds.), *Advances in Signal Processing and Intelligent Recognition Systems*,
Advances in Intelligent Systems and Computing 425,
DOI: 10.1007/978-3-319-28658-7_6

research is focused on emotion recognition from facial expressions in which facial pose is constrained to frontal or near frontal poses [3]. Very less work is reported in the literature on non-frontal poses.

The fundamental component in human communication [11] is change in the speaker's affective mental state which can be inferred from the expressions displayed by them in every aspect of their life. Emotion recognition accounts to the changes in the position of the facial muscles of eyebrows, eyes, nose, mouth and chin. For emotion 'anger', the muscles in the eyebrows, eyes and mouth differ from that of the neutral expression. For emotion 'happy' the lip muscles stretch indicating different position from that of other emotions.

The challenges in emotion recognition from 2D facial images which could not support pose and illumination variations, are solved by Lijun Yin in 2006 by proposing BU3DFE database [5] consisting of 3D facial images. This database leads many researchers to do research in emotion recognition from 3D images, with various combinations of pitch and yaw angles. Sample images from the database are shown in fig. 1. In this work, four emotions are considered as there is confusion between emotions disgust and fear in the non-frontal poses and we have proposed a novel approach for feature vector formation for the same.

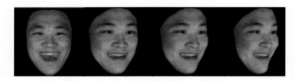

**Fig. 1** Non-frontal poses of BU3DFE database (0, +15˚, +30˚, +45˚).

The remaining part of this paper is organized as follows: Background work is discussed in section 2, proposed methods in section 3. Results and analysis are reported in section 4. The last section gives the conclusion and future work.

## 2    Background Work

Analysis of facial expressions and research involved in recognizing emotions using 2D/3D images have become an active area of research over a decade due to its importance in applications like human-computer interaction, affective computing, patient health monitoring etc., . Emotion recognition involves feature extraction and classification.   The geometric based approach and appearance based approach are the common methods used for feature extraction [12]. Spatial domain and transform domain techniques are used in appearance based approach for feature vector formation. Transform domain techniques gives better results than spatial domain techniques [8]. A number of pre-processing steps are required for analyzing the emotional expression of a human face [4]. Yuxiao Hu [11] investigated the performance for 5 yaw angles with various classifiers and concluded that 45˚ is the best view angle. Yuxiao Hu [10] by using local patch descriptors for feature extraction and classifier fusion for classification achieved an average recognition rate of 73.46% for 5 yaw angles. Zheng et al. [9] proposed a method using upper bound of Bayes error and Rank-One Update (ROU) technique. Moore [7]   investigated effects of pose in emotion recognition in BU3DFE and CMU-MultiPIE using LBPs at different resolutions and found that optimal view is feature dependent achieving a recognition rate of 78.65%.

Limited work has been reported related to emotion recognition from illumination and pose variations. In this paper, we have considered 3D images from BU3DFE database and proposed two methods for feature extraction. In the first method, Euclidean distance between the feature points of neutral and emotion image is calculated and also the angle measures between feature points in an emotion image are calculated. The Euclidean distance and the angle vector together form the feature vector and given to neural network to classify emotions. In the second method, the Euclidean distance is calculated between the points in the neutral and basic emotion. This distance is normalized using mean and standard deviation and formed as a feature vector. The proposed methods yielded good results compared to those in literature.

# 3    Proposed Methods for Emotion Recognition

We have used BU3DFE database, which contains 100 samples with occlusions such as hair, wrinkles and dark spots on their faces. It also contains the co-ordinate locations of 83 feature points marked on the face as shown in figure 2. These feature points for frontal view image are given in the BU3DFE database.  In this work we have considered 80 subjects, 4 poses (0°, +15°, +30°, +45°) and 4 emotions. We have used a reduced set of 24 feature points out of 83, which is circled in fig. 2. Our approach uses the geometric features that convey the maximum and most precise information, which works well for all the non-frontal poses that are under consideration. We have considered two approaches for feature vector formation: one using a concatenated distance and angle vectors, other using normalized Euclidean distances. Both the approaches give good results compared to literature. Out of the two proposed approaches, the method utilizing normalized feature vector outperformed the distance and angle vector approach. Our proposed method includes feature points on chin region of the face, which contributed to increase in recognition rate compared to literature.

## 3.1   *Distance and Angle Vector Method*

The algorithm for forming feature vector using distance and angle vector method is shown in figure 3 and explained as follows:

D. KrishnaSri et al.

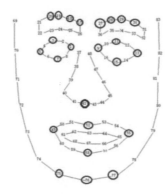

**Fig. 2** Fiducial points used in the proposed work.

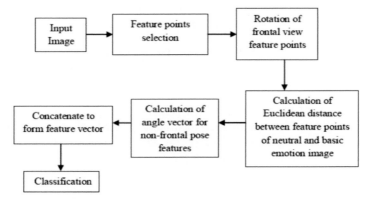

**Fig. 3** Block diagram of our proposed work using distance and angle vector.

Step 1: An image that expresses one of the basic emotions in frontal pose is given as input. Out of the 83 feature points we have selected 24 points on the image as given in figure 2. Those points are rotated with respect to a pivot point, which is considered as the tip of the nose, i.e., the $42^{nd}$ feature point. The feature points for the non-frontal poses, which are +15°, +30°, +45° (yaw angles) are obtained by rotating through respective angles from the frontal pose. For frontal pose the feature points are taken directly as there they are provided in the database.

Step 2: Feature vectors are formed by calculating the Euclidean distance. For a basic emotion like anger, the feature points that are obtained by rotation, and the feature points in a neutral emotion are considered. Corresponding 24 feature points in both these images are paired and Euclidean distance between them is determined using equation (1). For distance vector formation we make use of 24 feature points out of 83 feature points as shown in fig. 4.

$$D= \sqrt{(x_i - x_j)^2 + (y_i - y_j) + (z_i - z_j)^2} \text{ , where } i, j = 1, 2, 3 \dots 83. \qquad (1)$$

The raising of eyebrow, opening of eye and mouth and the distance between lower lips and chin of a person signifies the expression of an emotion. For example the distance calculated from $d_{17}$, $d_{18}$, $d_{19}$, $d_{20}$ indicates the raising of left eyebrow, which could imply expressing emotion surprise.

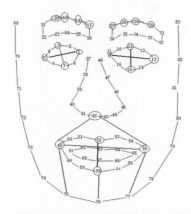

**Fig. 4** Diagrammatic representation of location of feature points in the facial image for calculating distance vectors.

In a similar way other distances are determined. The combination of distances for other feature points is given as follows:

Stretching of left eyebrow: $\delta 1 = d_{17}+d_{18}+ d_{19}+d_{20}$
Stretching of right eyebrow: $\delta 2 = d_{27}+d_{28}+ d_{29}+d_{30}$
Opening of left eye: $\delta 3 = d_{3-7}$
Opening of right eye: $\delta 4 = d_{11-15}$
Average eye width: $\delta 5 = 0.5(d_{1-5}+d_{9-13})$
Opening of the mouth: $\delta 6 = d_{52-58}$
Vertical movement of mouth corners: $\delta 7 = 0.2(d_{42-49}+d_{42-45}+d_{49-75}+d_{58-76}+d_{55-77})$
Mouth width: $\delta 8 = d_{49-55}$

After calculating the entire relevant distances $\delta i$ (i=1, 2, 3 ..... 8), a distance vector $D_i$ is formed as: $D_i = [\delta 1, \delta 2, \delta 3, \delta 4, \delta 5, \delta 6, \delta 7, \delta 8]$. The distance vector thus formed for all the samples in the database for the 4 poses and 4 emotions.

Step 3: We make use of 24 feature points out of 83 feature points for calculating angle vector. The angle between the 24 feature points is calculated and an angle vector is formed. The diagrammatic representation of locations of angle vectors determined in a facial image is given in fig. 5. The cosine angle is calculated as:

$$\theta = arcCos\left[\frac{s_j s_k}{\| s_j \| \cdot \| s_k \|}\right]$$, where $s_j$, $s_k$ are the slopes of the vectors on which

the ends of the angles are straddled and $\|s_j\|$ and $\|s_k\|$ are their respective lengths. These angles are now formed into an angle vector. Using 18 feature points out of 24, 18 angle vectors are determined for one image. $\Theta_i = [\Theta_1, \Theta_2, \Theta_3 ,.......\Theta_{18}]$. Similarly, it is repeated for all the samples in the database for 4 poses and 4 emotions.

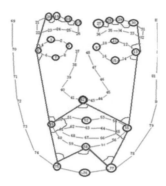

**Fig. 5** Representation of location of angle vectors in the facial image.

Step 4: For a particular image, the calculated distance vectors and angle vectors are concatenated to form a single feature vector as [δ1, δ2, δ3, δ4, δ5, δ6, δ7, δ8, $\Theta_1$, $\Theta_2$....$\Theta_{18}$]. The feature vector includes both the distance and angle vectors which capture the essential information of a 3D image. Thus a complete feature vector is formed for all the samples in the database, which is now used for classification purpose.

Step 5: The obtained feature vector for 4 poses, 80 samples and 4 emotions comprising a total of 1280 feature vectors and their corresponding target vectors are given as input to the Neural Pattern Recognition Tool (NPR Tool) which uses a Scaled Gradient Conjugate algorithm. The network yields good accuracy which is discussed in section 4.

## 3.2    *Normalized Euclidean Distance Method*

The algorithm for feature vector formation using normalized Euclidean distance method is shown in fig. 6.

Step1: An image that expresses one of the basic emotions in frontal pose is given as input. Out of the 83 feature points we have selected 24 points on the image as given in fig. 2. Those points are rotated with respect to a pivot point, which is considered as the tip of the nose, i.e., the 42nd feature point. The feature points for the non-frontal poses, which are +15°, +30°, +45° (yaw angles) are obtained by rotating through respective angles from the frontal pose. For frontal pose the feature points are taken directly as there they are provided in the database.

Step 2: The Euclidean distances are calculated for all 4 poses and 4 emotions for each subject. For a basic emotion like anger, the feature points that are obtained by rotation, and the feature points in a neutral emotion are considered. For distance vector formation we make use of 24 feature points out of 83 feature points as shown in fig. 4. Corresponding 24 feature points in both these images are paired and Euclidean distance between them is determined using equation (1).

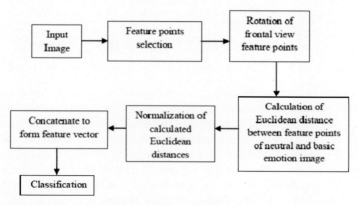

**Fig. 6** Block diagram for proposed work using approach 2.

Step 3: Now the Euclidean distances are normalized to zero mean and unit variance. Now for all the poses, 4 emotions, the Euclidean distances are normalized.

Step 4: These normalized distances form the feature vector. The obtained feature vector for 4 poses, 80 samples and 4 emotions comprising a total of 1280 feature vectors and their corresponding target vectors are given as input to the Neural Pattern Recognition Tool (NPR Tool) which uses a Scaled Gradient Conjugate algorithm. The network yields good accuracy compared to the previous proposed method. The comparison of the results is shown in the next section.

## 4    Results and Analysis

In this work, we have considered 3D images of 4 yaw angles. In the literature, most of the work has been reported on 3D images using all the 83 points, very less work has been reported in optimizing the number of feature points. Our proposed work shows good results for both the approaches using optimized set of 24 feature points.

The result obtained using the concatenated distance and angle vector method, is shown in table 1.

**Table 1**  Accuracy obtained by distance and angle vector method.

| Angle / Emotion | 0° | +15° | +30° | +45° | Overall accuracy % |
|---|---|---|---|---|---|
| Anger | 86 | 100 | 88.4 | 85.7 | **90.02** |
| Happy | 88.2 | 87.4 | 100 | 86.4 | **90.5** |
| Sad | 89.2 | 83.3 | 87.6 | 90.5 | **87.65** |
| Surprise | 85.8 | 86.8 | 85.4 | 92.6 | **87.65** |
| Overall accuracy % | **87.3** | **89.37** | **90.35** | **88.8** | **88.95** |

The overall accuracy achieved when distance and angle vectors used is 88.95%, as the angle vectors contributes to the orientation of eyebrow, eye and mouth when an image is rotated at yaw angles. The angle subtended by a feature point with reference to other feature points in a rotated image, when an emotion is expressed at maximum intensity, is well captured by the angle vector. Emotions 'anger' gives 100% accuracy for +15°, 'happy' gives 100% accuracy for 30° pose. Compared to the frontal poses, the non-frontal poses show good recognition accuracy. Emotions 'sad' and 'surprise' show highest recognition rate at +45° view angle. Emotion 'happy' recorded highest results compared to all other emotions. Out of all the 4 non-frontal angles, +30° view exhibited good results. The positive view angles are only considered as the features are symmetric on the face.

The result for the normalized Euclidean distance feature vector method is tabulated in table 2.

**Table 2** Accuracy of normalized Euclidean distance feature vector method.

| Angle / Emotion | 0° | +15° | +30° | +45° | Overall accuracy % |
|---|---|---|---|---|---|
| Anger | 88.4 | 100 | 100 | 92.9 | **95.32** |
| Happy | 86 | 87.4 | 89.9 | 89.24 | **88.13** |
| Sad | 84 | 83.3 | 85.9 | 84.5 | **84.42** |
| Surprise | 87.8 | 86.8 | 90.2 | 89.2 | **88.5** |
| Overall accuracy % | **86.55** | **89.37** | **91.5** | **88.96** | **89.09** |

The overall accuracy achieved using the normalized Euclidean distance feature vector method is 89.09%. Normalization helps in reducing redundancy of data thereby improving the performance of the network. Emotion 'anger' shows 100% recognition rate at both +15° and +30° poses, resulting highest recognition rate compared to other emotions. At +30° pose all the emotions show a good recognition accuracy compared to other non-frontal angles. Both the approaches yield encouraging accuracy for all non-frontal poses and thus making the algorithm pose independent. The feature points include points on the chin region also which contributes to a great variation in recognition accuracy as compared to the literature. Comparison of our proposed methods with literature is shown in table 3.

**Table 3** Comparison of results of the proposed methods with literature.

| Reference | Method | Database | Feature points | Number of angles | Accuracy % |
|---|---|---|---|---|---|
| Yuxaio Hu, 2008 [4] | Normalization of feature points | BU-3DFE | 25 | 5 yaw angles (0°, 30°, 45°, 60°, 90°) | 71.41 |
| Yuxaio Hu, 2008 [7] | Local patch descriptors and classifier fusion | BU-3DFE | 83 | 5 yaw angles (0°, 30°, 45°, 60°, 90°) | 73.46 |
| Wenming Zheng, 2009 [8] | Multi-class Bayes error | BU-3DFE | 83 | 5 yaw angles (0°, 30°, 45°, 60°, 90°) | 78.65 |
| S. Moore, 2011 [9] | Local Binary Pattern | BU-3DFE | Histogram of LBP features | 5 yaw angles (0°, 30°, 45°, 60°, 90°) | 77.67 |
| Proposed Methods | Distance and angle vectors method and normalize Euclidean distance method | BU-3DFE | 24 | 4 yaw angles(0°, 15°, 30°, 45°) | 88.95 and 89.09 |

By considering the feature points on chin, we can measure the variation in distance between chin and mouth corners for each emotion which contributes to the effective emotion recognition for non-frontal poses.

# 5    Conclusion and Future Work

Emotion recognition is a current area of research, and it is desired that more research should be done using varying poses. In this paper two methods are proposed for the 3D images of BU3DFE. The first method calculates distance and angle vectors and the second uses the normalized Euclidean distance. The non-frontal poses provide face depth information. In the second method as normalization separates the data into distinct unique sets, the proposed method achieves good results compared to the proposed distance and angle vectors method. The difference in the location of the feature points between lip corners and chin differs from emotion to emotion which provide the peak information when a person expresses an emotion. The contribution of this distance improves recognition rate of each emotion for all poses. The normalized feature vector achieved a recognition rate of 89.09% compared to the distance and angle vectors method which makes the system pose independent.

As future work, the proposed method can be carried out for images which include variations in illumination, age groups, combination of various yaw and pitch angles and subjects wearing glasses, cap, etc.

# References

1. Mehrabian, A.: Communication without Words. Psychology Today **2**(4), 53–56 (1986)
2. Rabiu, H.: 3D facial Expression Recognition using Maximum Relevance Minimum Redundancy Geometric Features. EURASIP, Advances in Signal Processing, 1–8 (2008)
3. Tang, H., Hasegawa-Johnson, M., Huang, T.: Non-frontal view facial expression recognition based on ergodic hidden markov model supervectors. In: IEEE International Conference on Multimedia and Expo (ICME), Suntec City, pp. 1202–1207 (2010)
4. Wang, J, Yin, L., Wei, X., Sun, Y.: 3D facial expression recognition based on primitive surface feature distribution. In: IEEE Computer Society Conference on Computer Vision and Pattern Recognition (CVPR), New York, vol. 2, pp. 1399–1406 (2006)
5. Yin, L., Wei, X., Sun, Y., Wang, J.: A 3D facial expression database for facial behavior research. In: 7th International Conference on Automatic Face and Gesture Recognition, Southampton, pp. 211–216 (2006)
6. Ioannou, S.V., Raouzaiou, A.T., Tzouvaras, V.A., Mailis, T.P., Karpouzis, K.C., Kollias, S.D.: Emotion recognition through facial expression analysis based on a neuro-fuzzy network. Neural Networks **18**(4), 423–435 (2005)
7. Moore, S., Bowden, R.: Local Binary Patterns for Multi View Facial Expressions. Computer Vision and Image Understanding **115**(4), 541–558 (2014)
8. Suja, P., Tripathi, S.: Analysis of Emotion Recognition from Facial Expressions using Spatial and Transform Domain Methods. International Journal of Advanced Intelligence Paradigms (IJAIP) **7**(1), 57–73 (2015)
9. Zheng, W., Tang, H., Lin, Z., Huang, T.S.: A novel approach to expression recognition from non-frontal face images. In: IEEE 12th International Conference on Computer Vision, Kyoto, pp. 1901–1908 (2009)

10. Hu, Y., Zeng, Z., Yin, L., Wei, X., Zhou, X., Huang, T.S.: Multi-view facial expression recognition. In: Proc. International Conference on Automatic Face and Gesture Recognition, Amsterdam, pp. 1–6 (2008a)
11. Hu, Y., Zeng, Z., Yin, L., Wei, X., Tu, J., Huang, T.: A study of non-frontal-view facial expressions recognition. In: Proc. in International Conference on Pattern Recognition, Tampa FL , pp. 1–4 (2008b)
12. Tian, Y., Kanade, T., Cohn, J.F.: Facial Expression Analysis, Handbook of face Recognition. Springer, October 2003

# Real-Time Automatic Camera Sabotage Detection for Surveillance Systems

K. Sitara and B.M. Mehtre

**Abstract** Video surveillance is very common for security monitoring of premises and sensitive installations. Criminals tamper the surveillance camera settings so that their (criminal) activities in the scene are not recorded properly, thereby making the captured video frames useless. Various camera tampering/sabotage include - changing the normal view of the camera by turning the camera away from the scene, obstructing the camera lens by placing some objects in front of the camera or spraying paint on it and defocusing the camera lens by changing the camera focus settings, spraying water or some viscous fluid on it. Manual monitoring of the surveillance systems have many limitations - human fatigue, lack of continuous monitoring, etc. Hence, real-time automated analysis and detection of suspicious events have gained importance. In this paper, we propose an efficient algorithm for camera tamper detection based on background modeling, edge details, foreground object size and its movement. In our testing or experimental setup, the results are encouraging with high precision and low false alarm rate. As the proposed method can process $320 \times 240$ resolution videos at $60-70$ frames/sec, it can be implemented for real-time applications.

**Keywords** Video surveillance · Camera tampering · Camera sabotage · Camera occlusion · Real-time video processing · Camera defocusing · Camera displacement

K. Sitara(✉) · B.M. Mehtre
Institute for Development and Research in Banking Technology (IDRBT),
Hyderabad 500057, India
e-mail: {ksitara,bmmehtre}@idrbt.ac.in

K. Sitara
School of Computer Science and Information Sciences (SCIS),
University of Hyderabad, Hyderabad 500046, India

© Springer International Publishing Switzerland 2016                     75
S.M. Thampi et al. (eds.), *Advances in Signal Processing and Intelligent Recognition Systems*,
Advances in Intelligent Systems and Computing 425,
DOI: 10.1007/978-3-319-28658-7_7

# 1   Introduction

Video surveillance systems are implemented in public and private places like railway stations, airports, bank ATMs, office buildings, restaurants, home surveillance etc. In all these situations, it is implemented for ensuring the safety or monitoring the areas of interest which forms the recorded frames or images from the surveillance camera. These recordings are often produced as evidence in case of crimes. Surveillance system operators have to manually go through these recordings from several cameras at a time. This manual monitoring process has limitations (like, the operators may loose their concentration in surveillance videos by themselves in short time intervals or someone can intentionally distract the operators from monitoring) and operators may miss important events in the scene. Surveillance camera sabotage/tampering is a deliberate action on a surveillance camera to change the proper working of the camera thereby altering the images recorded and processed by the surveillance systems. These actions include: turning the camera to a different direction from the field of view (i.e., camera moved/displaced), partial or full covering of the camera lens by placing an object in front of the camera and very close to it or spraying paint (i.e., camera occlusion), defocusing the camera lens by changing the camera focus settings or using water droplets thereby blurring the frames(i.e., camera out-of-focus/defocus) and so on. Camera displacement or occlusion will produce newer frames which are partially or totally distinct to the intended scene. Some of these may occur accidentally, but detecting them is equally important as the video quality and scene/area under surveillance have to be properly recorded.

Automatic analysis and detection of camera tampering in real-time is very important as it helps to prevent crime and alert the operator so that he/she may not miss any suspicious events. Apart from detection accuracy, the time complexity and false alarm rate of the camera tamper detection algorithms are other factors to be taken care of. In most cases, this method has to be implemented as part of a surveillance system which has multiple cameras. Since, real-time detection of suspicious activities is required in such systems, the time complexity of the proposed method should be low. Coming to false alarm rate, if it is high the operator may loose interest in the automated system and he/she may ignore warnings from the system. Reducing false alarm rate is a challenge to the automated system as false alarms can be easily triggered due to illumination changes, crowd, large objects passing through the scene etc.

The proposed method can detect camera tampering due to camera occlusion, defocus and displacement in static surveillance cameras. Background modeling is the first operation performed, adopting a state of the art technique from [1]. The method is slightly modified so as to make it robust for camera tamper detection. Foreground objects are extracted from newer frames utilizing information from the background model. The size of the foreground object and the edge details in the newer frames are parameters for deciding the suspicious activity. The alarm is triggered for sustained events which is present in more than 12 frames to reduce false alarm rate.

The rest of the paper is organized as follows. Section 2 deals with related works in the area. Background modelling is described in Section 3. Section 4 discusses the proposed method for camera tamper detection. Demonstration on experimental results regarding performance of the proposed method is provided in Section 5. Conclusion and recommendations on future research are summarized in Section 6.

## 2    Related Works

Most research on automatic camera tamper detection is based on the changes of the currently captured frames to a statistically-modeled background template of the scene of surveillance. These methods are capable of detecting the three types of camera tampering discussed in Sect. 1.

Entropy, edge information and zero-mean normalized cross correlation computed from the background model are used in [2], but it generates false alarms for crowd and large objects passing through the scene. A learned background model and wavelet transform [3] were used for detecting obscured camera view and camera defocus based on the assumption that an occlusion will always result in large homogenous area in the current frame. It may fail in situations where a textured object is used for occlusion. In [4], recent and older frames are kept in separate pools and three dissimilarity functions are calculated for every frame in these pools. Due to high computation, this method is not suitable for real-time operations. Saglam and Temizel [5] used discrete Fourier transform for comparing the high frequency components of the current frame and its background to detect camera defocus, 32 bin histogram for camera occlusion and comparison of current background with delayed background to detect displaced camera. In [6], edges is used for background estimation, histogram comparison and minimum squared difference is used for tamper detection. In [7], the authors utilized edge energy and standard deviation of whole image and image patches with Kalman filtering. The work in [8] is similar to [5] with FPGA implementation. Camera tamper detection algorithm implemented on TI DAVINCI DM6437 Digital Signal Processor based on edge difference and histogram comparison between current frame and background model is discussed in [9]. In [10], an adaptive background codebook model is used.

The difference of keypoints extracted using Scale Invariant Feature Transform (SIFT) between the current frame and background model is discussed in [11]. Tsesmelis et al. [12] used keypoints extracted from Speeded Up Robust Features (SURF) to address the three types of camera tampering in static as well as panning surveillance cameras. They developed a panoramic view of the scene as background for panning cameras. Deng-Yuan Huang et al. [13] proposed methods for camera tamper detection due to occlusion, motion, defocus, screen shaking, color cast, fogging and screen flickering based on edge details, brightness, high-frequency information and histogram analysis.

# 3   Background Modelling

The Local Illumination based Background Subtraction (LIBS) scheme [1] for object detection from videos consists of two stages. In the first stage, the stationary pixels in the frames taken for constructing the background model is found. In the second stage, new frames are compared with the background model using local thresholding to find the objects in the scene. Since the method used for background estimation and object extraction is built on this, we provide a brief explanation of it and its slight improvement in this section. The RGB video frame sequences are converted into gray-scale frames. These frames are also scaled to half its size for reduced time complexity.

## 3.1   Development of Background Model

In order to make the estimated background robust to illumination variation, dynamic objects and small changes in the scene $n$ frames are taken to find the stationary pixels in the scene. A good background model will give better result in foreground object extraction which makes the tamper detection system robust to suspicious events. So instead of taking the first $n$ frames in the video, it is better to choose frames with illumination variation from previously recorded video sequences if possible. This will give better range of intensity values corresponding to each background pixel location. $n$ can take values between 20 to 30.

This frame sequence is partitioned into overlapping subsequence having odd number of frames say W. The mean of the values at the same pixel location $(i, j)$ of all the frames in a subsequence starting at time $t$ can be computed by,

$$m(i, j) = \frac{1}{W} \sum_{k=t}^{t+W} frame_k(i, j) \tag{1}$$

where $frame_k$ indicates each frame in the subsequence. Let us keep the values at $frame_k(i, j)$ where $k \in [t, t + W)$ to a 1-D vector $V$.

Though, [1] used standard deviation of intensity values in the subsequence for background modelling, we adopted mean of intensity values as it was found to give better results in our camera tamper detection phase.

Compute the absolute difference between the first and the middle elements in $V$. Perform the same for all the values in $V$, i.e., absolute difference between the second and the middle and so on, except for the middle one and store in another 1-D vector $D$. This is done to find the variation among pixel values at $(i, j)$. Take the sum of the lowest $\lfloor \frac{W}{2} \rfloor$ values in $D$. If this sum is less than or equal to $\lfloor \frac{W}{2} \rfloor * m(i, j)$, then mark $frame_{k+\lfloor \frac{W}{2} \rfloor}(i, j)$ as stationary else mark it as non-stationary. It is done for every pixel positions in the subsequence, where $i \in [0, m_1 - 1]$ and $j \in [0, m_2 - 1]$.

Once it is completed, the same process is repeated for the next subsequence consisting of $W$ frames from $t + 1$. The entire process is repeated until no more subsequence of length $W$ is possible in the $n$ frame sequence. Construct two matrices $M$ and $N$ which forms the background model of size $m_1 \times m_2$ to keep the maximum and minimum pixel intensity values respectively, corresponding to each stationary pixel in the sequence from $\lceil \frac{W}{2} \rceil$ to $n - \lfloor \frac{W}{2} \rfloor$.

## 3.2 Object Extraction

To extract the foreground object in the current frame $f$, compute the local lower and upper thresholds $(T_L)$ and $(T_U)$ respectively using a constant $C$. For a pixel at position $(i, j)$ in $f$ denoted by $f(i, j)$, the threshold values $T$, $T_L$ and $T_U$ are computed by,

$$T = \frac{1}{C}(M(i, j) + N(i, j)) \tag{2}$$

$$T_L = M(i, j) - T \tag{3}$$

$$T_U = N(i, j) + T \tag{4}$$

If $T_L <= f(i, j) <= T_U$ is satisfied, then it is a background pixel (indicate with 0), otherwise mark it as foreground pixel (indicate with 1). This is performed for each pixel in the current frame.

After object extraction using [1], a binary image $obj\_det$ having the same dimensions of the current frame is obtained. We found that certain pixels or small groups of pixels were wrongly identified as objects in $obj\_det$. As this might affect the efficacy of our camera tamper detection algorithm, we eliminated them using the following decision strategy proposed by us. Compute the areas of each and every object in $obj\_det$ individually and objects having area less than 8 are eliminated, thereby updating $obj\_det$. This reduces noise residues and false negatives.

## 4 Proposed Method

Each of the new frames have to undergo the object detection phase explained in Sect. 3 which produce a binary image, $obj\_det$. Camera defocus leads to reduced visibility, which in turn leads to less edge pixels in the acquired frames. Edge pixels are computed from the background model $M$ and the current frame $f$ for its detection. If the number of edge pixels, $f_{edge\_count}$, in $f$ is less than a threshold times that in the background $M_{edge\_count}$, it indicates camera defocus, concept borrowed from [3]. Hence, check the following condition,

$$\frac{f_{edge\_count}}{M_{edge\_count}} \leq Th_1 \tag{5}$$

where $Th_1$ is the threshold value.

Camera occlusion will produce newer frames which are partially or totally distinct from the background scene. If any of the foreground object appears to be bigger than $(\frac{1}{3})^{rd}$ of the frame, then there is a chance that this object is suspicious. Foreground objects placed in front of the camera very close to it, to obstruct the camera view may appear bigger and it may occupy at least $(\frac{1}{3})^{rd}$ of the frame. Thus, the size of each foreground object in the current frame is taken as a parameter for camera occlusion detection. The area occupied by each of the foreground objects are computed (considering 8-neighborhood) and stored to a 1-D vector $Area$. Find the maximum value in $Area$ and check the condition

$$\frac{max(Area)}{(m_1 \times m_2)} \geq Th_2 \tag{6}$$

where $Th_2$ is the threshold value. If (6) is satisfied, it indicates camera occlusion.

Camera displacement will also produce new frames which are partially or totally distinct from the background scene. When the camera view is changed, object detection algorithm will return it as a large object in the scene due to the difference in the current view with the estimated background. Hence, (6) can also be used for camera displacement detection.

So, if (5) or (6) is satisfied, it may indicate a camera tamper event. We have to check for the persistence of these events to avoid false alarms that may occur due to crowd and large objects passing through the scene. If it is crowd, at least one person will make a movement in between as it is human nature. Hence, the object position or object boundary will differ in the previous, current and the next frames. The same concept is also applicable for moving large objects. For this, we compute the absolute difference between the $obj\_det$ of previous and current frame denoted as $diff_p$. The same is computed between the $obj\_det$ of current and next frame denoted as $diff_n$. If the object is moving, then the movement will be recorded in the difference images $diff_p$ and $diff_n$. Since $obj\_det$ is a binary image, $diff_p$ and $diff_n$ are also binary images. If the object is stationary, then the number of non-zero elements in $diff_p$ and $diff_n$ is 0 or almost 0 (considering noisy pixels). The number of non-zero elements in $diff_p$ and $diff_n$ are computed, denoted by $diff_{p_{count}}$ and $diff_{n_{count}}$. If these values are less than a threshold value $Th_3$, it indicates camera tamper.

$$\frac{diff_{p_{count}}}{(m_1 \times m_2)} \leq Th_3 \quad \&\& \quad \frac{diff_{n_{count}}}{(m_1 \times m_2)} \leq Th_3 \tag{7}$$

Check whether it is a sustained event or not to reduce the false alarms. A variable $tamper_{count}$ initialized to 0 is used to keep track of this. If (7) is satisfied for the first time, then $tamper_{count}$ is incremented to 1. Whenever (7) is satisfied we check whether the same condition is satisfied in the previous iteration also. If so, increment

$tamper_{count}$ by 1 or else reset it back to 0. When $tamper_{count} > Th_4$ is satisfied, where $Th_4$ is a threshold value, alarm is triggered.

To reduce computation, we skip one frame at a time. i.e, after processing frame $t$, the next frame taken for camera tamper detection is $t + 2$. Since we skip one frame on each iteration, $Th_4$ is set to 5 which ensures that the suspicious event is present in at least 12 continuous frames.

## 5 Experimental Results and Discussion

The proposed method is implemented in MATLAB R2013a(8.1.0.604), performed on a 2.50 GHz Intel Core i5 PC. The parameters $W$, $C$, $Th_1$, $Th_2$, and $Th_3$ are set to 9, 9, 0.7, 0.3, and 0.02 respectively.

Experiments are conducted on recorded videos simulating various challenging situations and also on the camera tamper detection dataset provided by the authors in [5]. We created 10 test videos corresponding to each camera anomaly totaling 30 videos. These are HD videos with resolution of $1280 \times 720$ pixels at a frame rate of 30 frames/s. The frames of some of the experimental videos are shown in Fig.1. Sample frames under normal surveillance is shown in Fig. 1(a), (b), (c), (g), (h) and (i). Camera view after turning the camera away from the scene is shown in Fig. 1(d) and Fig. 1(j). Camera defocus by spraying aerosol is given in Fig. 1(e). Camera defocus by placing water drop in the camera lens is given in Fig. 1(k). Camera occlusion by placing opaque objects very close to the surveillance camera lens is shown in Fig. 1(f) and Fig. 1(l). To get an idea on false alarm rate of our algorithm, we took 6 typical surveillance videos without tamper events from the i-LIDS dataset [14] and also captured 10 non-tamper videos, totaling 16 videos. Among the 30 videos, the proposed method missed 1 event in camera defocus and 1 in camera occlusion. All camera displacement events are detected successfully. No alarm is triggered for non-tamper events.

The performance of the proposed method is discussed here in terms of precision rate $(R_p)$, recall rate $(R_r)$, false alarm rate $(R_{fa})$ and accuracy $(Acc)$ according to the following equations.

$$R_p = \frac{TP}{TP + FP} \times 100\% \qquad (8)$$

$$R_r = \frac{TP}{TP + FN} \times 100\% \qquad (9)$$

$$R_{fa} = \frac{FP}{TN + FP} \times 100\% \qquad (10)$$

$$Acc = \frac{TP + TN}{TP + TN + FP + FN} \times 100\% \qquad (11)$$

**Fig. 1** Example frames showing camera tamper. (a), (b), (c), (g), (h) and (i) are the original intended camera scene images, (d) view after camera displacement (e) view after camera defocus (f) obstructed view by placing object infront of camera (j) view after camera displacement (k) view after camera defocus and (l) obstructed view by placing object infront of camera

where $TP$ - the number of true positives, i.e., camera tamper events detected correctly and alarm triggered, $TN$ - the number of true negatives, i.e., camera non-tamper events not detected as a suspicious event, $FP$ - the number of false positives, camera non-tamper events detected as anomaly, and $FN$ - the number of false negatives, camera tamper events not detected. $R_p$, $R_r$ and $R_{fa}$ of the proposed method are 100%, 93.33%, and 0% respectively. The accuracy of our system is 95.65%.

A comparison of performance of the proposed method with works in [5], [12], and [8] is shown in Table 1. Dataset provided by the authors in [5] is used for comparison. We received only 39 videos out of the original 54 sequences used in [5]. In [5], the authors compared their work with [2], [3], and [4] on the same dataset. From this, we can say that the proposed system has high precision and low false alarm rate. Updating the background model with background information from recent frames in specific time interval will keep the system robust. Our method may fail in the following situations 1) object used for camera occlusion is moving/shaking rapidly

**Table 1** Performance Comparison of the Proposed Method with [2], [3], [4], [5], [8] and [12]

| Tamper type | Approach | Tamper events | TRUE Positives | FALSE Negatives | FALSE Positives |
|---|---|---|---|---|---|
| | | **Results** | | | |
| Defocus | Gil-Jimnez et al. [2] | 35 | 22 | 13 | 1 |
| | Aksay et al. [3] | 35 | 28 | 7 | 5 |
| | Ribnick et al. [4] | 35 | 21 | 14 | 0 |
| | Saglam and Temizel [5] | 35 | 29 | 6 | 0 |
| | Tsesmelis et al. [12] | 13 | 11 | 2 | 5 |
| | Kryjak et al. [8] | 14 | 14 | 0 | 0 |
| | **Proposed method** | **13** | **12** | **1** | **0** |
| Occlusion | Gil-Jimnez et al. [2] | 40 | 35 | 5 | 14 |
| | Aksay et al. [3] | 40 | 33 | 7 | 2 |
| | Ribnick et al. [4] | 40 | 37 | 3 | 1 |
| | Saglam and Temizel [5] | 40 | 38 | 2 | 0 |
| | Tsesmelis et al. [12] | 17 | 17 | 0 | 2 |
| | Kryjak et al. [8] | 11 | 11 | 0 | 4 |
| | **Proposed method** | **17** | **15** | **2** | **0** |
| Displacement | Gil-Jimnez et al. [2] | 12 | 8 | 4 | 3 |
| | Ribnick et al. [4] | 12 | 7 | 5 | 0 |
| | Saglam and Temizel [5] | 12 | 11 | 1 | 0 |
| | Tsesmelis et al. [12] | 12 | 12 | 0 | 3 |
| | Kryjak et al. [8] | 18 | 18 | 0 | 2 |
| | **Proposed method** | **12** | **11** | **1** | **0** |

and (2) scene background containing highly smooth regions with very less edge components.

Coming to time complexity, [5], [7], [8], [9] and [13] can process video sequences with resolutions of VGA@12.8 fps, VGA@20.8 fps, VGA@60 fps, $320 \times 240$@26 fps and $320 \times 240$@30 fps respectively. The proposed method can process video sequences with resolutions $320 \times 240$, VGA and HD on an average of $60 - 70$ fps, $25 - 30$ fps and $10 - 13$ fps . This shows that our method is computationally efficient compared to other methods and can be implemented for real-time processing. The proposed method could also detect fogging, as fog causes reduced visibility.

## 6 Conclusion and Future Scope

A simple and efficient method for surveillance video camera tamper detection due to camera defocus, occlusion and displacement is presented. The algorithm is based on background modeling, edge components, foreground object size, its relative position in the current, previous and the future frames. Experimental results on video sequences containing tamper events and non-tamper events demonstrate the effectiveness of the proposed system. Video sequences with resolution of $640 \times 480$ pixels are processed at $25 - 30$ fps. Hence, the proposed method can be implemented in real-time with high precision and low false alarm rate. Future work could concentrate on improving the precision of the system on smooth background surveillance scene and camera occlusion by rapidly moving/shaking objects. Testing the robustness of the method on more real-world surveillance videos and extending this work to panning surveillance cameras are also under consideration.

# References

1. Hati, K.K., Sa, P.K., Majhi, B.: Intensity range based background subtraction for effective object detection. Signal Processing Letters **20**(8), 759–762 (2013)
2. Gil-Jimnez, P., et al.: Automatic control of video surveillance camera sabotage. In: NatureInspired Problem-Solving Methods in Knowledge Engineering, pp. 222–231. Springer, Heidelberg (2007)
3. Aksay, A., Temizel, A., Cetin, A.E.: Camera tamper detection using wavelet analysis for video surveillance. In: Proc. IEEE Conf. Adv. Video Signal Based Surveillance, pp. 558–562 (2007)
4. Ribnick, E., Atev, S., Masoud, O., Papanikolopoulos, N., Voyles, R.: Real-time detection of camera tampering. In Proc. IEEE Conf. Adv. Video Signal Based Surveillance, p. 10 (2006)
5. Saglam, A., Temizel, A.: Real-time adaptive camera tamper detection for video surveillance. In: IEEE Conference on Advanced Video and Signal Based Surveillance, Genova, pp. 430–435 (2009)
6. Ellwart, D., Szczuko, P., Czyżewski, A.: Camera sabotage detection forsurveillance systems. In: Security and Intelligent Information Systems, pp. 45–53. Springer, Heidelberg (2012)
7. Wang, Y.K., Fan, C.T., Cheng, K.Y., Deng, P.S.: Real-time camera anomaly detection for real-world video surveillance. In: International Conference on Machine Learning and Cybernetics, Guilin, China, pp. 1520–1525 (2011)
8. Kryjak, T., Komorkiewicz, M., Gorgon, M.: FPGA implementation of camera tamper detection in real-time. In: 2012 Conference on Design and Architectures for Signal and Image Processing (DASIP). IEEE (2012)
9. Lin, D.-T., Wu, C.-H.: Real-time active tampering detection of surveillance camera and implementation on digital signal processor. In: Eighth International Conference on Intelligent Information Hiding and Multimedia Signal Processing (IIH-MSP). IEEE (2012)
10. Tung, C.-L., Tung, P.-L., Kuo, C.-W.: Camera tamper detection using codebook model for video surveillance. In: 2012 International Conference on Machine Learning and Cybernetics (ICMLC), vol. 5. IEEE (2012)
11. Yi, H., Jiao, X., Luo, X., Yi, C.: Sift-based camera tamper detection for video surveillance. In 25th Chinese Control and Decision Conference, pp. 665–668 (2013)
12. Tsesmelis, T., Christensen, L., Fihl, P., Moeslund, T.B.: Tamper detection for active surveillance systems. In: 2013 10th IEEE International Conference on, Advanced Video and Signal Based Surveillance (AVSS), pp. 57–62 (2013)
13. Huang, D.-Y., Chen, C.-H., Chen, T.-Y., Hu, W.-C., Chen, B.-C.: Rapid detection of camera tampering and abnormal disturbance for video surveillance system. Journal of Visual Communication and Image Representation **25**, 1865–1877 (2014)
14. i-LIDS dataset for AVSS (2007). http://www.elec.qmul.ac.uk/staffinfo/andrea/avss2007_d. html (cited July 8, 2015)

# An Improved Approach to Crowd Event Detection by Reducing Data Dimensions

Aravinda S. Rao, Jayavardhana Gubbi and Marimuthu Palaniswami

**Abstract** Crowd monitoring is a critical application in video surveillance. Crowd events such as running, walking, merging, splitting, dispersion, and evacuation inform crowd management about the behavior of groups of people. For an effective crowd management, detection of crowd events provides an early sign of the behavior of the people. However, crowd event detection using videos is a highly challenging task because of several challenges such as non-rigid human body motions, occlusions, unavailability of distinguishing features due to occlusions, unpredictability in people movements, and other. In addition, the video itself is a high-dimensional data and analyzing to detect events becomes further complicated. One way of tackling the huge volume of video data is to represent a video using low-dimensional equivalent. However, reducing the video data size needs to consider the complex data structure and events embedded in a video. To this extent, we focus on detection of crowd events using the Isometric Mapping (ISOMAP) and Support Vector Machine (SVM). The ISOMAP is used to construct the low-dimensional representation of the feature vectors, and then an SVM is used for training and classification. The proposed approach uses Haar wavelets to extract Gray Level Coefficient Matrix (GLCM). Later, the approach extracts four statistical features (contrast, correlating, energy, and homogeneity) at different levels of Haar wavelet decomposition. Experiment results suggest that the proposed approach is shown to perform better when compared with existing approaches.

A.S. Rao(✉)
Department of Electrical and Electronic Engineering, The University of Melbourne,
Parkville Campus, Melbourne, VIC 3010, Australia
e-mail: aravinda@student.unimelb.edu.au

J. Gubbi
TCS Innovation Labs, Tata Consultancy Services, Bengaluru, Karnataka 560066, India
e-mail: j.gubbi@tcs.com

M. Palaniswami
Department of Electrical and Electronic Engineering, The University of Melbourne,
Parkville Campus, Melbourne, VIC 3010, Australia
e-mail: palani@unimelb.edu.au

© Springer International Publishing Switzerland 2016
S.M. Thampi et al. (eds.), *Advances in Signal Processing and Intelligent Recognition Systems*,
Advances in Intelligent Systems and Computing 425,
DOI: 10.1007/978-3-319-28658-7_8

85

# 1  Introduction

Crowd behavior analysis of video surveillance applications has gained significant interests in computer vision research. Analyzing crowd behavior requires several levels of video processing: low- and high-level [8, 11]. The low-level processing involves extracting features at pixel levels and high-level semantics are often derived by combining other features relevant to crowd behavior. Analyzing individual movements is important for applications where tracking of an individual in required. However, analyzing crowd requires a holistic view of the scene. In the case of crowds, multi-person interaction must be considered and this becomes a challenging task because of undefined motion patterns involved in crowd movements. The human visual system's advancements mask the difficulty and challenges for humans in finding the motion patterns. In contrast, crowd video analytics is still under development to reaching the stage of being fully automated.

The need for automated crowd behavior analysis stems from the fact that the crowd monitoring applications are highly important. Applications of crowd monitoring include counting, density estimation, detecting crowd events (such as walking, running, etc.), and understanding crowd behavior. These applications often find their importance in crowd monitoring at public places, such as public transport hubs, airports, subway stations, and other. The fundamental steps for any crowd analytics algorithms include preprocessing the video frames, object detection and tracking. However, the challenges of object detection and tracking include [33]: (1) unavailability of depth information, (2) unaccounted video noise generated during the formation of video frames, (3) articulated motions of the moving objects, (4) occluded scenarios, and (5) illumination variations. Compared with computer vision algorithms developed for detecting rigid objects, crowd analytics suffers severely because of articulated human motions and occlusions.

The crowd behavior analysis provides a holistic view of crowd under surveillance. Several features are aggregated at different levels of processing and behavior is inferred. One of the indicators used in behavior are the crowd events. Crowd events consist of interaction of people and their activities [16] and one of the main end-user application for officials concerned with crowd management [23]. Analyzing crowd motion pattern is one of highly challenging tasks in computer vision. To the best our knowledge, the literature consists of six crowd events defined using the Performance Evaluation of Tracking and Surveillance (PETS) 2009 dataset [10]: walking, running, merging, splitting, dispersion, and evacuation. Fig. 1 shows the six crowd events from PETS 2009 dataset. The objective of event detection at a broader level is to analyze the video and detect events such that frames with similar events are clustered together. However, object detection itself has been a research challenge for many years. Although several detection methods have been proposed in the literature, crowd event detection has received less attention. This is primarily due to the complexities involved in detection humans and the crowd.

Video frames are a source of high-dimensional data. Let $x$ and $y$ denote position of a pixel in a video frame $I$. Then a pixel location can be denoted as $I(x, y)$. Let

**Fig. 1** Examples of crowd events from the PETS 2009 dataset [10]: (a) walking, (b) running, (c) merging, (d) splitting, (e) dispersion, and (f) evacuation.

$m \in \mathbb{R}$ and $n \in \mathbb{R}$ denote the number of rows and columns of the video frame $I(x, y)$. Let $(r, g, b) \in \mathbb{R}^3$ denote the three-tuple notation for red, green and blue channels of the camera. Then, the dimensions of a video frame is $\mathbb{R}^{m \times n \times 3}$. When $m$ and $n$ increase considerably, the processing of such high-dimensional frames and later performing the analytics (counting, detecting events) coupled with complex human motions becomes extremely difficult. Dimensionality reduction is a much sought-after approach to reduce video dimensions without losing much of the pertinent data. The core of dimensionality reduction is to represent the high dimensional data in a low-dimensional form. The high-dimensional features are represented in a low-dimensional format such that they can be used to distinguish objects and track them. These features are then used for applications such as counting, density estimation, tracking, detecting crowd events and behavior analysis.

In this work, we focus on detection of crowd events using the Isometric Mapping (ISOMAP) [28] and Support Vector Machine (SVM). In particular, we use the PETS 2009 dataset [10] and extract the Gray Level Coefficient Matrix (GLCM) using the Haar wavelet [14] with up to eight levels of decomposition. With this, the feature vectors span a 100-dimensional features space. The approach uses ISOMAP to find the low-dimensional representation of the feature vectors by constructing a graph distance matrix with feature vectors as nodes of the graph. The approach learns the mapping from the low-dimensional space in the event classes using the SVM. The experiments were conducted for both linear and Radial Basis Function (RBF) kernels. These two different dimensional data were trained and classified using both linear and RBF kernels. The output from the ISOMAP were mapped to one-dimensional and ten-dimensional embedded feature space.

## 2   Related Work

### 2.1   Dimensionality Reduction Methods

There are two main types of dimensionality reduction methods: (1) linear and (2) non-linear. Linear methods assume the feature subspace to be linear, i.e., the feature vectors satisfy the linearity property of the subspace. Principal Component Analysis (PCA) [15] is a good example of linear dimensionality reduction. PCA assumes the data vectors to be lying in a linear feature space. PCA endeavors to maximize the global variance while discarding the order of the vectors in the put feature space. In addition, the relationship among feature vectors are ignored

On the other hand, the nonlinear methods approximate the global space by locally augmenting the linear subspaces. These algorithms are also called as manifold learning algorithms. They endeavor to find the embedded (lower) dimensions of the high-dimensional features. Manifold is intuitively thought to be a point in some topological space that can be reached without ambiguity [18, 24]. The three common steps involved in finding the low-dimensional feature vectors by nonlinear methods are [20]: (1) construct a weighted graph with nodes denoted by feature input vectors and connections using the neighborhood information, (2) convert the weighted graph to a form suitable to find the low-dimensional embedding, i.e., find the graph distances, and (3) solve the eigenfunction ("spectral embedding") to obtain a set of low-dimensional embedding vectors.

The ISOMAP [28] is an example of nonlinear dimensionality reduction algorithm. Classical Multidimensional Scaling (MDS) [31] aims at maintaining inter-point distances from high-dimensional input to low-dimensional space. The inter-point distances represent objects and MDS calculates the proximity matrix based on the Euclidean distances between points. ISOMAP [28] uses classical MDS that attempts to identify a low-dimensional subspace while preserving the isometry of the input data points. ISOMAP assumes that the points are invariant under transformation and geodesic distances are used. The ISOMAP employs a $k$-Nearest Neighbor ($k-$NN) approach, followed by MDS to construct the graph distances and then apply eigen-decomposition. This method is a global approach to finding the low-dimensional embedding and the other manifold learning methods approach the dimensionality reduction from a local observer perspective. Work presented in [28] showed the effectiveness of the ISOMAP to applications, such as handwriting recognition and head-pose estimation. Locally Linear Embedding (LLE) [25] assumes the local geometry of data points to be linear coefficients such that the patches are reconstructed by its neighbors, where it employs a $k$-Nearest Neighbor ($k-$NN) approach.

Souvenir and Pless [22, 26, 27] proposed image distances based on manifold learning that are similar to ISOMAP. The authors contributed to highlight the natural parameterization space of image manifolds. Guo et al. [13] proposed a method to learn the "age manifold" from the set of training images. The learned manifold would serve as model to estimate and predict the age of the people from images. Chang et al. [5] proposed a new method to model, track and recognize facial expressions using a low-dimensional manifold. Instead of learning a manifold from images, the authors use the facial contours.

## 2.2 Crowd Event Detection

Chen et al. [6] included Haar features related to head and then tracked the objects in the scene. The objects were then treated as agents and information such as direction and speed was derived. Objects were tracked using template matching and Kalman filtering. Feature vectors were constructed based on agents' movements such as walking, running, jumping and stopping. An SVM was trained using known samples to recognize the actions from a local dataset. Ke *et al.* [17] proposed an approach to recognize events in crowd videos by identifying actions. Event detection was achieved by matching the shapes in spatio-temporal volumes. First, the shape contours are extracted from spatial-temporal patches and next, these shapes were matched in spatial-temporal patches using optical flow features. As a final step in recognizing the events, part-matching of shape templates was performed: instead of matching the whole templates, Ke *et al.* [17] allowed parts to be matched independently.

Gárate *et al.* [12] proposed to use 2D Histogram of Gradients (HOG) descriptors as features. In particular, they utilized the features around a block of nine cells ($3 \times 3$). They computed gradient vector magnitude and orientation for all the pixels within the block. Then, the orientations were binned and 2D feature vectors were constructed. These feature vectors were later tracked in the next frame using the object speed and a time window. Finally, the crowd events were recognized based on the tracked features over frames.

Li *et al.* [19] focused their work on multi-object activities to characterize the group motion. In this aspect, they proposed a data-driven Discriminative Temporal Interaction Manifold (DTIM) framework. The framework established probability densities on the DTIM based on the interactions between the objects. They used discriminative temporal interaction matrix to arrive at the probability densities. The method was tested on a soccer game video. Benabbas *et al.* [2] approached the problem of crowd event detection by building a direction and a magnitude model using the optical flow motion vectors. Circular clustering was performed to learn the prominent directions. The motion vectors were refined using the online Gaussian Mixture Model (GMM). The magnitude and direction patterns were used to track the objects in the neighborhood. They were then clustered and tracked to detect the crowd events.

Thida *et al.* [29] used Histogram of Optical Flow (HOOF) as features to detect crowd events. They divided each video frame into blocks and extracted the HOOF features. These features were later supplied to LE [1] to find a low-dimensional representation of the features. Using a similarity measure, the crowd events were represented in a low-dimensional representation.

# 3   Methodology

## 3.1   Preprocessing and Feature Extraction

Raw frames from the cameras contain high-frequency noise and would contribute to generating errors when information is passed from low-level processing to semantic-level decisions. In addition, low-frequency image signals contain most information and this can be verified by performing a wavelet decomposition of an image. Therefore, in the first step, a Gaussian low-pass filter of size $7 \times 7$ is used to smoothen the images spatially. Next, the bilateral filter [30] is applied to preserve the edges. The literature provides features such as edge, texture, color, motion (optical flow), shapes, and other. In this work, feature extraction included the Haar features [14] through Haar Wavelet Transform (HWT), where the video frames were analyzed at multiple resolutions using the Haar functions. Previous research on people detection (e.g. [7, 9]) provide extensive evidence to use the Haar wavelet as crowd features in this work.

## 3.2 Haar Wavelet Transform

The Haar piecewise constant function (wavelet) is given by [14, 21]

$$\Psi(t) = \begin{cases} 1 & \text{if } 0 \leq t < 1/2 \\ -1 & \text{if } 1/2 \leq t < 1 \\ 0 & \text{otherwise} , \end{cases} \tag{1}$$

which generates an orthonormal basis on dilations (scalings) and translations (shiftings) given by([14, 21])

$$\Psi_{j,n}(t) = \frac{1}{\sqrt{2^j}} \Psi\left(\frac{t - 2^j n}{2^j}\right), \tag{2}$$

where $(j, n) \in \mathbb{Z}^2$. The orthonormal basis also forms the basis of a space of finite energy signals given by

$$\|f\|^2 = \int_{-\infty}^{\infty} |f(t)|^2 \, dt \quad < \infty, \tag{3}$$

which can be represented in the inner product form as

$$\langle f, \Psi_{j,n} \rangle = \int_{-\infty}^{\infty} f(t) \Psi_{j,n} \, dt, \tag{4}$$

where $\langle f, \Psi_{j,n} \rangle$ are called the Haar coefficients. In the case of images, the Haar coefficients correspond to different levels of decomposition. Using the Haar Wavelet, each frame was decomposed into eight levels. For each level, horizontal, vertical and diagonal components were computed. Furthermore, for each of the three components, a GLCM was calculated. The four statistical measures—contrast, homogeneity, energy and correlation—were extracted from the GLCM. In addition, the four statistical measures of the approximation coefficient were extracted. Therefore, for each frame, the feature vector length would be $N_D \times N_S \times N + N_A \times N_S = 3 \times 4 \times 8 + 1 \times 4 = 100$, where $N$ is the number of levels of decomposition, $N_D$ is the number of detailed coefficients for each level, $N_A$ is the number of approximation coefficient of the $N^{\text{th}}$ level and $N_S$ is the statistical measures for each of $N_D$ or $N_A$. A high dimensional feature matrix ($\mathbb{R}^{m \times n}$) was constructed using the texture features, where $m$ represents the number of features (texture) for each frame and $n$ denotes the frames.

## 3.3 Dimensionality Reduction and Classification

The general dimensionality reduction algorithm can be described as follows: given a set (data) $X = [\mathbf{x}_1, \mathbf{x}_2, \ldots, \mathbf{x}_n]$ with $\mathbf{x}_i \in \mathbb{R}^m$, find a set $Y = [\mathbf{y}_1, \mathbf{y}_2, \ldots, \mathbf{y}_n]$ with $\mathbf{y}_i \in \mathbb{R}^d$ that represents $X$ such that $d \ll m$ [1]. Most of the dimensionality reduction methods are proposed based on meeting certain objectives. Hence, most the methods employ a certain form of optimization. ISOMAP [28] uses three steps to complete feature mappings: (1) determine neighbors based on the distance $d(i, j)$ between points $(i, j)$ in the input space ($X \in \mathbb{R}^m$) are represented as a weighted graph. (2) geodesic distance between all the points on the manifold are computed using the shortest path over weighted graph, and (3) apply classical MDS to the graph distance matrix to construct a $d$-dimensional embedding from the $m$-dimensional input space $X$, where $d \ll m$. The neighborhood can be in the first step can be either fixed $\epsilon$-neighborhood or $k$-NN.

Support Vector Machine (SVM) is a supervised learning algorithm used for binary classification. The idea behind the SVM is to find the hyperplane such that the distance between two classes separated from the closest points to the hyperplane is maximized and is given by

$$\min_{\omega, b, \xi} \frac{1}{2} \|\omega\|^2 + \frac{C}{n} \sum_{i=1}^{n} l(\xi^k), \tag{5}$$

subject to $y_i(\omega \cdot x_i + b) \geq 1 - \xi_i, \forall\ i \in 1, \ldots, n$, where $C$ is a positive regularization constant and $\xi$ is the slack term. The points that lie on the boundaries are called as *support vectors*. In cases where the data are not linearly separable using the hyperplane, the kernel techniques are used to project the data points to high-dimensional space so that the data can be separated. In our approach, the output of the ISOMAP algorithms were used to train the SVM. Both linear and nonlinear approaches were used. For nonlinear approach, the RBF kernel was used to project the data to higher dimensions and is given by

$$K(x, x') = e^{\left(-\frac{\|x - x'\|^2}{2\sigma^2}\right)}, \tag{6}$$

where $K(\cdot, \cdot)$ is the RBF kernel, $x$ and $x'$ are feature samples in the input space, and a free parameter $\sigma$. The positive regularization constant $C$ and the kernel parameter $\sigma$ were learned using the grid search optimization algorithm [4].

## 4   Results and Discussion

The proposed method was implemented in MATLAB on a Windows 7 machine (64-bit) equipped with 4 GB RAM and Intel® i7 − 2600 CPU running at 3.4 GHz.

### 4.1   Dataset

The PETS 2009 [10] dataset was used to evaluate the proposed method. The crowd events are categorized under the PETS 2009 [10] dataset (S3) with four different timings ($14-16$, $14-27$, $14-31$ and $14-33$). The timings refer to the hour-minute format. For each timing, there are four different views ($001, 002, 003,$ and $004$). To the best of our knowledge, only the PETS 2009 [10] dataset has the events where all six crowd events can be clearly evaluated based on human motion analysis. The proposed approach was evaluated on a total of 16 different sequences. The annotations were done manually for all the video sequences. The video frames were prepossessed based on the approach provided in [23].

### 4.2   Experiment

Table 1 provides the results of the proposed approach. The data were split into 70% training and 30% testing sets. Tenfold cross validation was performed to train the model. The columns with † indicate the approach using one-dimensional vector inputs to SVM. The ISOMAP neighborhood parameter $k$ was set 7. The texture 100-dimensional feature vectors are represented as one-dimensional vectors using the ISOMAP. These one-dimensional vectors represent the video frames, which are used for learning the SVM mapping function. Similarly, the columns with ‡ indicate the approach using 10-dimensional vector input to SVM. Furthermore, the SVM-Linear indicates the use of SVM with linear kernel and SVM-RBF indicates the use of SVM with RBF kernel.

   From Table 1, the ISOMAP+SVM-Linear (†) and ISOMAP+SVM-RBF (†), both performed equally well in classifying the merging events (precision: 0.78, recall: 0.99, $F$-score: 0.87). However, ISOMAP+SVM-RBF (†) showed better performance in detecting and classifying the splitting events (precision: 1.00, recall: 0.78, $F$-score: 0.87). Overall, both ISOMAP+SVM-Linear (†) and ISOMAP+SVM-RBF (†) performances were similar. The SVM-Linear (‡) and SVM-RBF (‡) both performed equally well in detecting and classifying the splitting events with SVM-RBF (‡) performing slightly better than SVM-Linear (‡). However, the ISOMAP+SVM-Linear (‡) performed better in case of dispersion and evacuation events. The ISOMAP+SVM-RBF (‡) showed excellent performance in the case of merging event as compared with ISOMAP+SVM-Linear (‡).

Compared with existing methods from Table 1, running event was best detected by statistical filtering approach [32] with precision, recall, and $F$-score of 0.99 each. The statistical filtering approach [32] also achieved a recall score of 1 in splitting events. One of the critical point to note here about the approach presented in [32] is that the experiment design included only two classes—running and splitting. This established a high score for running and splitting as both are clearly distinguishable among crowd events. The Random Forest and Motion Pattern approach described in [2] performed well to detect walking events in crowd videos. The method presented in [2] models the crowd motion using the optical flow features and refines the motion patterns. This enables the method to provide better walking detection scores.

In [24], the authors presented a crowd event detection approach using Riemannian manifolds. In [24] the approach models the crowd movement based on the optical flow features. The scheme, which is an unsupervised approach, finds the location of the crowd groups at any time based on the temporal evolution of the crowd locations, which is determined by the localization of crowd groups on Riemannian manifolds. This localization provided better results to detect dispersion events. In contrast, the supervised approach presented in this work, performed well in detecting merging, splitting, dispersion, and evacuation.

One of the main problems with the crowd event detection is the processing of large amounts of video data. Linear dimensionality reductions such as PCA violate the data nonlinearity. Nonlinear approach such as ISOMAP preserve the nonlinear structure of the data by first constructing the graph using frames as vertices and then finding the low-dimensional embedding based on global optimization such that the isometry of the data is preserved in both higher and lower dimensions. Previous research on people detection (e.g. [7, 9]) provide extensive evidence to use the Haar wavelet as crowd features. The work presented here reinstates that using statistical features from Haar wavelet is suitable to detect crowd events. This work also shows that using manifold learning algorithms, the high-dimensional feature space of videos can be reduced to a low-dimensional representation and still detect the events with a trained classifier. Future work includes experiments to have an unsupervised learning system that can automatically detect the manifold parameters, such as $k$ (neighborhood), and also to detect the events from low-dimensional feature vectors without using supervised classifiers. More research is also required to determine as to what kind of features are better suited to detect crowd events and in general crowd monitoring. In this work, only the PETS 2009 dataset was used to detect crowd events as only the PETS 2009 dataset had the crowd events defined in the literature. The research community should also focus on sharing annotated crowd movement data to develop robust algorithms.

**Table 1** The last four columns of the table provide the results of the proposed approach. The bold fonts indicate the best results obtained for each event. The (†) indicates the approaches using the one-dimensional vector input to SVM and (‡) indicates the approaches using a ten-dimensional vector to SVM.

| Crowd Event | Measure | Statistical Filters [32] | Holistic Approach [3] | Random Forest [2] | Motion Pattern [2] | Riemannian Manifolds [24] | ISOMAP + SVM-Linear (†) | ISOMAP + SVM-RBF (†) | ISOMAP + SVM-Linear (‡) | ISOMAP + SVM-RBF (‡) |
|---|---|---|---|---|---|---|---|---|---|---|
| Walking | Precision | - | 0.87 | 0.96 | **0.97** | 0.61 | 0.61 | 0.64 | 0.65 | 0.86 |
| | Recall | - | - | **0.99** | 0.96 | 0.75 | 0.81 | 0.76 | 0.81 | 0.87 |
| | F-score | - | - | **0.97** | 0.96 | 0.67 | 0.69 | 0.68 | 0.8 | 0.86 |
| Running | Precision | **0.99** | 0.75 | 0.86 | 0.75 | 0.78 | 0.81 | 0.76 | 0.87 | 0.87 |
| | Recall | **0.99** | - | 0.68 | 0.81 | 0.63 | 0.61 | 0.64 | 0.86 | 0.86 |
| | F-score | **0.99** | - | 0.75 | 0.77 | 0.69 | 0.69 | 0.68 | 0.86 | 0.86 |
| Merging | Precision | - | 0.68 | 0.65 | 0.59 | 0.85 | 0.78 | 0.78 | 0.88 | **0.96** |
| | Recall | - | - | 0.46 | 0.45 | 0.88 | **0.99** | **0.99** | 0.97 | 0.98 |
| | F-score | - | - | 0.53 | 0.51 | 0.86 | 0.87 | 0.87 | 0.93 | **0.96** |
| Splitting | Precision | 0.65 | 0.74 | 0.73 | 0.47 | 0.66 | 0.99 | **1.00** | 0.97 | 0.98 |
| | Recall | **1.00** | - | 0.92 | 0.47 | 0.6 | 0.78 | 0.78 | 0.88 | 0.96 |
| | F-score | 0.78 | - | 0.81 | 0.47 | 0.62 | 0.87 | 0.87 | **0.96** | **0.96** |
| Dispersion | Precision | - | 0.8 | 0.58 | 0.67 | 0.9 | 0.59 | 0.79 | **0.98** | 0.96 |
| | Recall | - | - | 0.48 | 0.45 | **0.94** | 0.84 | 0.64 | 0.92 | 0.90 |
| | F-score | - | - | 0.52 | 0.53 | 0.91 | 0.69 | 0.70 | **0.95** | 0.93 |
| Evacuation | Precision | - | **0.94** | 0.83 | 0.69 | 0.75 | 0.84 | 0.64 | 0.92 | 0.90 |
| | Recall | - | - | **1.0** | 0.82 | 0.65 | 0.59 | 0.79 | 0.96 | 0.96 |
| | F-score | - | - | 0.90 | 0.74 | 0.69 | 0.69 | 0.70 | **0.93** | 0.92 |

# 5   Conclusion

In this work, an approach to detect crowd events such as running, walking, merging, splitting, dispersion and evacuation, was presented. However, crowd event detection using videos is a highly challenging task, and the video itself is a high-dimensional data and analyzing to detect events becomes further complicated. In this work, the huge volume of video data was represented using low-dimensional equivalent. The ISOMAP was used to reduce the high-dimensional nature of the video to a low-dimensional representation. Later, an SVM was trained to detect the six crowd events based on the low-dimensional feature vectors. The experiment used the PETS 2009 dataset, which consists of the six crowd events identified in the literature. The proposed approach used Haar wavelets to generate GLCM from which the four statistical features (contrast, correlating, energy, and homogeneity) at different levels of Haar wavelet decomposition were extracted. The proposed approach performed better in detecting merging, splitting, dispersion, and evacuation, compared with existing approaches. The work presented reinstates that using statistical features from Haar wavelet is suitable to detect crowd events. This work also shows that using manifold learning algorithms, the high-dimensional feature space of videos can be reduced to a low-dimensional representation and still detect the events with a trained classifier.

# References

1. Belkin, M., Niyogi, P.: Laplacian eigenmaps for dimensionality reduction and data representation. Neural Computation **15**(6), 1373–1396 (2003). doi:10.1162/089976603321780317
2. Benabbas, Y., Ihaddadene, N., Djeraba, C.: Motion pattern extraction and event detection for automatic visual surveillance. J. Image Video Process. **2011**, 1–15 (2011). doi:10.1155/2011/163682
3. Chan, A.B., Morrow, M., Vasconcelos, N.: Analysis of crowded scenes using holistic properties. In: Performance Evaluation of Tracking and Surveillance workshop at CVPR, pp. 101–108. IEEE (2009)
4. Chang, C.C., Lin, C.J.:
5. Chang, Y., Hu, C., Feris, R., Turk, M.: Manifold based analysis of facial expression. Image and Vision Computing **24**(6), 605–614 (2006)
6. Chen, Y., Zhong, Z., Ka Keung, L., Yangsheng, X.: Multi-agent based surveillance. In: 2006 IEEE/RSJ International Conference on Intelligent Robots and Systems, pp. 2810–2815. IEEE (2006). DOI: 10.1109/iros.2006.282064
7. Dalal, N., Triggs, B., Schmid, C.: Human detection using oriented histograms of flow and appearance. In: Computer Vision–ECCV 2006, pp. 428–441. Springer (2006)
8. Ekin, A., Mehrotra, R., et al.: Automatic soccer video analysis and summarization. IEEE Transactions on Image Processing **12**(7), 796–807 (2003)
9. Enzweiler, M., Gavrila, D.M.: Monocular pedestrian detection: Survey and experiments. IEEE Transactions on Pattern Analysis and Machine Intelligence **31**(12), 2179–2195 (2009)
10. Ferryman, J.: PETS 2009 benchmark data (2009). http://www.cvg.rdg.ac.uk/PETS2009/a.html (accessed May 19, 2014)
11. Foresti, G.L., Micheloni, C., Snidaro, L., Remagnino, P., Ellis, T.: Active video-based surveillance system: the low-level image and video processing techniques needed for implementation. IEEE Signal Processing Magazine **22**(2), 25–37 (2005)

12. Gárate, C., Bilinsky, P., Bremond, F.: Crowd event recognition using HOG tracker. In: 2009 Twelfth IEEE International Workshop on Performance Evaluation of Tracking and Surveillance (PETS-Winter), pp. 1–6. IEEE (2009). doi:10.1109/pets-winter.2009.5399727
13. Guo, G., Fu, Y., Dyer, C.R., Huang, T.S.: Image-based human age estimation by manifold learning and locally adjusted robust regression. IEEE Transactions on Image Processing **17**(7), 1178–1188 (2008)
14. Haar, A.: Zur theorie der orthogonalen funktionensysteme. Mathematische Annalen **69**(3), 331–371 (1910)
15. Hotelling, H.: Analysis of a complex of statistical variables into principal components. Journal of Educational Psychology **24**(6), 417–441 (1933). doi:10.1037/h0071325
16. Hughes, R.L.: A continuum theory for the flow of pedestrians. Transportation Research Part B: Methodological **36**(6), 507–535 (2002)
17. Ke, Y., Sukthankar, R., Hebert, M.: Volumetric features for video event detection. Interantional Journal of Computer Vision **88**(3), 339–362 (2010). doi:10.1007/s11263-009-0308-z
18. Lee, J.M.: Riemannian Manifolds: An Introduction to Curvature, vol. 176. Springer (1997)
19. Li, R., Chellappa, R., Zhou, S.K.: Learning multi-modal densities on discriminative temporal interaction manifold for group activity recognition. In: IEEE Conference on Computer Vision and Pattern Recognition, pp. 2450–2457. IEEE (2009). DOI: 10.1109/cvpr.2009.5206676
20. Ma, Y., Fu, Y.: Manifold Learning Theory and Applications. CRC Press, Inc. (2011)
21. Mallat, S.: A wavelet tour of signal processing: the sparse way. Academic press (2008)
22. Pless, R., Souvenir, R.: A survey of manifold learning for images. IPSJ Transactions on Computer Vision and Applications **1**, 83–94 (2009)
23. Rao, A.S., Gubbi, J., Marusic, S., Palaniswami, M.: Estimation of crowd density by clustering motion cues. The Visual Computer, 1–20 (2014). DOI: 10.1007/s00371-014-1032-4
24. Rao, A.S., Gubbi, J., Marusic, S., Palaniswami, M.: Probabilistic detection of crowd events on riemannian manifolds. In: 2014 International Conference on Digital Image Computing: Techniques and Applications (DICTA), pp. 1–8. IEEE (2014). DOI: 10.1109/DICTA.2014.7008124
25. Roweis, S.T., Saul, L.K.: Nonlinear dimensionality reduction by locally linear embedding. Science **290**(5500), 2323–2326 (2000). doi:10.1126/science.290.5500.2323
26. Souvenir, R., Pless, R.: Manifold clustering. In: Tenth IEEE International Conference on Computer Vision (ICCV), vol. 1, pp. 648–653. IEEE (2005)
27. Souvenir, R., Pless, R.: Image distance functions for manifold learning. Image and Vision Computing **25**(3), 365–373 (2007)
28. Tenenbaum, J.B., Silva, V.D., Langford, J.C.: A global geometric framework for nonlinear dimensionality reduction. Science **290**(5500), 2319–2323 (2000). doi:10.1126/science.290.5500,2319
29. Thida, M., How-Lung, E., Remagnino, P.: Laplacian eigenmap with temporal constraints for local abnormality detection in crowded scenes. IEEE Transactions on Cybernetics **43**(6), 2147–2156 (2013). doi:10.1109/TCYB.2013.2242059
30. Tomasi, C., Manduchi, R.: Bilateral filtering for gray and color images. In: Sixth International Conference on Computer Vision, pp. 839–846. IEEE (1998)
31. Torgerson, W.: Multidimensional scaling: I. theory and method. Psychometrika **17**(4), 401–419 (1952). doi:10.1007/BF02288916
32. Utasi, Á., Kiss, Á., Szirányi, T.: Statistical filters for crowd image analysis. In: Proceedings of the 11th IEEE International Workshop on Performance Evaluation of Tracking and Surveillance (in conjunction with CVPR 2009). IEEE (2009)
33. Yilmaz, A., Javed, O., Shah, M.: Object tracking: A survey. ACM Computing Surveys 38(4) (2006). DOI: 10.1145/1177352.1177355

# Art of Misdirection Using AES, Bi-layer Steganography and Novel King-Knight's Tour Algorithm

Sudharshan Chakravarthy, Vishnu Sharon, Karthikeyan Balasubramanian and V. Vaithiyanathan

**Abstract** Nowadays, the need for of data-integrity and privacy is higher than ever. From small bank transactions to large-scale classified military information, these factors have been prerequisites. This classified information can be faked deliberately by the sender with the help of a pseudo secret in order to increase the level of security. Cryptography makes this information unreadable and steganography hides the same. This paper brings together the paths of steganography and Rijndael encryption, which is commonly known as the AES. The secret image or text-file is encrypted using AES algorithm and is then camouflaged into the pseudo-secret image using a modified proposed Knight's Tour-inspired chess-based algorithm that enhances data diffusion. Then the pseudo-secret image/text is once again encrypted using the AES algorithm and is masked into the cover image/text using a checkerboard-based location map. Thus lossless, highly secure, nested-layer steganography is achieved with high complexity and high PSNR for various extension types of images as illustrated in the paper.

**Keywords** Pseudo secret · 2-bit LSB · Rijndael · King-Knight's tour · Checkerboard location map

## 1 Introduction

With massive improvements in communication, the need for securing the data has also increased. Data security confirms the protection of digitized data from exploitation. A primary branch of data security is cryptography [1, 2, 3, 4, 5, 6]. Ensuring data integrity through insecure channels pertains to cryptography. Retrieving

S. Chakravarthy(✉) · V. Sharon · B. Balasubramanian · V. Vaithiyanathan
School of Computing, SASTRA University, Thanjavur 613401, India
e-mail: {chakravarthy.sudharshan,viskes95,mbalakarthi}@gmail.com, vvn@it.sastra.edu

© Springer International Publishing Switzerland 2016
S.M. Thampi et al. (eds.), *Advances in Signal Processing and Intelligent Recognition Systems*,
Advances in Intelligent Systems and Computing 425,
DOI: 10.1007/978-3-319-28658-7_9

the original data can only be done when the key is available. On the other hand, steganography is the branch of data security that enables hiding the data within a cover in order to avoid exposure, interruption and modification. Thus, by extracting the advantageous attributes of these prospects, data security is imminent [7, 8, 9, 10, 11, 12,13].

This paper has brought together steganography (2-bit LSB substitution) and cryptography (AES). The AES algorithm is used to encrypt the secret image, as done previously by Islam et al[10], Saini et al[14] and Thanikaiselvan et al [15]. The AES is chosen because of its colossal key permutation size of 3.4 x $10^{38}$, which amounts to $2^{128}$ different permutations [16], and has a higher computational complexity than other techniques like the DES, 3DES, etc. [17]. A well-known and prominent method in steganography is the Least Significant Bit (LSB) substitution [18, 19, 20, 21] wherein the least significant n-bits of a pixel can be chosen for embedding. This paper proposes to embed in the least significant 2-bits of a pixel to achieve an embedding rate of 25% and a high value of PSNR. The traditional Knight's tour algorithm [22, 23] involves moving the knight on N x N board from one square till all the other squares are toured. This paper has embraced a novel knight's tour inspired chess-based algorithm called the King-Knight's Tour algorithm as a scanning path for embedding the secret image into the pseudo secret image. The AES algorithm is once again applied to encrypt the pseudo secret image. The checkerboard location map is used for embedding the pseudo secret image into the cover image. The decryption process is done in the reverse routine.

## 2    Proposed Methodology

### 2.1    Encryption and Embedding Phase

**STEP 1:**   Choose a secret text or image to embed (fig. 1(a)).
**STEP 2:**   Apply AES algorithm to the secret (fig. 1(a) becomes fig. 1(b)).

The Rijndael algorithm, also famously known as the AES [24], is a prevalent hexadecimal oriented encryption algorithm and has been used to encrypt all forms of data. As illustrated by Christof et al [16], It is a non-Feistel algorithm which banks on data diffusion using its shift-row and mix-columns sub-procedures. In AES, the entire input data block used several permutations and is processed in parallel. In this procedure a 128 bit key-size and 16 byte input data is selected. The flow diagram is given below and the explanations are as given by Daemen et al [24]:

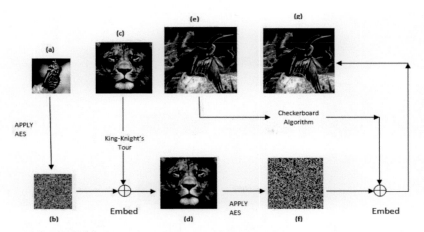

**Fig. 1** Flow diagram for encryption, embedding and transmission: (a) Secret image to be transmitted    (b) The AES encrypted secret image    (c)   Pseudo-secret image where the secret is embedded using King-Knight's tour    (d)   Embedded Pseudo-secret  (e) Cover image   (f)   Encrypted pseudo-secret  (g)   The final Stego-Image.

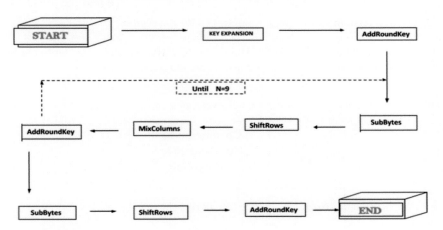

**Fig. 2** Flowchart for AES.

### 2.1.1    Encryption Illustration

One single implementation of AES provides encryption for 16 pixel values of the input image. The results shown below show that the AES is better than other encryption algorithms: ('bird', 'lion', 'butterfly' and their encrypted counterparts).

#### 2.1.1.1    Histogram and Contour Analysis of AES-Image Encryption

The altered histogram and their contour diagrams are as shown:- ('bird', 'lion' and 'butterfly' histograms, contours and their encrypted counterpart's histogram and contour):

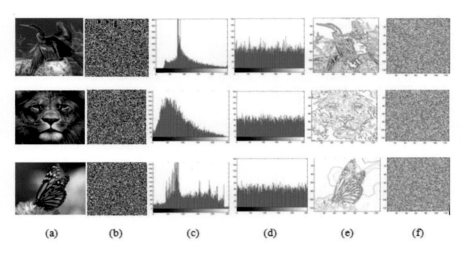

**Fig. 3** Input images and their encryption analysis: (a) - set of input images, (b) - the AES encrypted images  (c) - original images' histograms , (d) - encrypted images' histograms , (e) - original images' contour , (f) - encrypted images' contours

**2.1.1.2    Correlation and Entropy Analysis of AES Encryption**

In ideal circumstances, correlation should be low after encryption.  Utilizing the concept of image entropy, following results are obtained:

**Table 1** Correlation between the images before and after encryption.

| Image input | Correlation be-fore and after encryption |
|---|---|
| Bird | -0..000641 |
| Lion | -0.006652 |
| Butterfly | 0.0058000 |
| Average | -0.0004976 |

**Table 2** Comparison between the AES and Narendra's [25] proposed methodology.

| Image Entropy | | | | | |
|---|---|---|---|---|---|
| AES Entropy | | | Narendra's [25] | | |
| Image | Original entropy | Entropy after AES | Image | Original entropy | Encrypted Entropy |
| Butterfly | 7.6702 | 7.9882 | Tiger | 7.2367 | 7.7315 |
| Lion | 7.4735 | 7.9895 | Lena | 7.4551 | 7.7333 |
| Bird | 6.8288 | 7.9841 | Water | 7.7754 | 7.7365 |

For a similar set of images, the AES has produced entropy around 7.99 which is very close to the ideal value 8. Narendra's [25] encryption method yields only entropy of around 7.73 which is far from the results obtained by the AES.

***STEP 3:*** Choose a pseudo secret text or image (fig. 1(c)).
***STEP 4:*** Apply King-Knight's Tour algorithm to fig. 1(c) and embed fig. 1(b) into fig. 1(c) (becomes fig. 1(d)).

This proposed algorithm is inspired from the Knight's Tour algorithm [22,23]. But the issue with this conventional method is that it needed just the starting pixel position with just N x N possibilities as to which pixel to start from. With the help of this starting pixel position it is easy to extract the entire secret from the image. This paper proposes the King-Knight's Tour algorithm based on a set of priorities that are customizable by the sender and the receiver. These priorities involve the various positions that a king or knight can visit and that gives us a maximum of 16 priorities for any pixel.

The procedure for the King-Knight's Tour Algorithm is as follows -:

*Step i:* The starting pixel location is arbitrarily chosen by the sender.
*Step ii*: The sender defines a set of 16 priorities according to which the remainder of the secret will be embedded. Here, S is the current pixel location and the numbers 1-16 are the next possible locations for embedding, as shown in Fig. 4.

a) Green - King's Tour     b) Yellow - Knight's Tour

**Fig. 4** Candidate locations for embedding.

*Step iii:* There will arise a stalemate or a situation where in a pixel faces the possibility of all its 16 priority locations already embedded. In this case, the remainder of the secret is embedded in raster order.

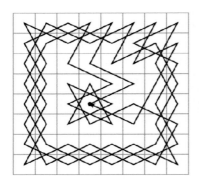
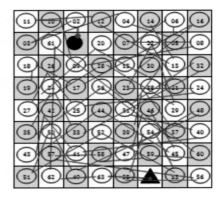

**Fig. 5** Original knight's tour             **Fig. 6** Proposed algorithm

Thus this algorithm undoubtedly increases the perceivable data diffusion in the image after embedding and also works for rectangular images.

**STEP 5:**  Apply AES algorithm to fig. 1(d) (fig. 1(d) becomes fig. 1(e)).

**STEP 6:** Choose a cover Image (fig. 1(f)).

**STEP 7:**  Apply checkerboard location map algorithm to fig. 1(f) and embed fig. 1(e) into fig. 1(f) (becomes fig. 1(g)).

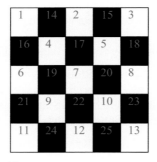

**Fig. 7** Checkerboard-like algorithm

The following steps show the procedure for the checkerboard algorithm (Fig. 7)-:

*Step i:* Start embedding the pseudo secret from the first pixel.

*Step ii:* Then, embed the remaining part of the pseudo secret in the alternating pixels that follow, i.e. like in the same colors in a checkerboard as the first pixel, in raster order.

*Step iii:* Repeat the same process for the other colored pixels starting from the second pixel.

This algorithm is simple to implement and has a 100% embedding rate. The reasons for choosing this algorithm to embed the pseudo secret into the cover image will be discussed later in the Extraction and Decryption Phase.

**STEP 8:** fig. 1(g) is our stego image.

## 2.2 Extraction and Decryption Phase

**STEP 1:** From the stego image (fig. 8(a)) extract the pseudo-secret using the checkerboard algorithm (fig. 8(a) becomes fig. 8(b)). Repeat the same procedure as explained in the Encryption and Embedding phase. The extraction part is deliberately made easier for the hacker to extract the pseudo secret. The main idea is to provide some false, yet meaningful data that can convince the intruder into believing that the pseudo secret is indeed the original secret. Furthermore, once the meaningful pseudo secret is extracted, it wouldn't make sense for the intermediary to check for any other hidden secrets thereby leaving the original secret undiscovered.

**STEP 2:** Apply inverse-AES to fig. 8(b) and get fig. 8(c). Reverse the same procedure as explained in the Encryption and Embedding phase.

**STEP 3:** From fig. 8(c), extract the secret using the King-Knight's tour algorithm (fig. 8(c) becomes fig. 8(d)). Repeat the same procedure as explained in the Encryption and Embedding phase.

**STEP 4:** Apply inverse-AES to fig. 8(d) to obtain the decrypted secret image or text (fig. 8(e)). Reverse the same procedure as explained in the Encryption and Embedding phase.

**STEP 5:** fig. 8(e) is our obtained secret.

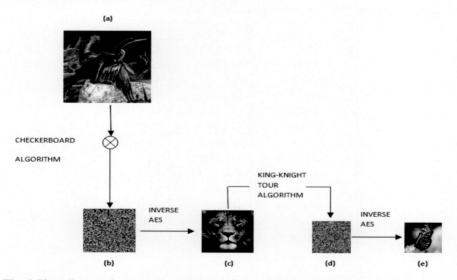

**Fig. 8** Flow diagram for reception and decryption: (a) Stego image obtained at the receiver's end (b) The extracted, encrypted Pseudo-secret image obtained from using checkerboard algorithm. (c) Original Pseudo-Secret obtained by Inverse-AES (d) The extracted, encrypted secret image obtained from using King-Knight tour algorithm. (e) The secret image retrieved.

## 3    Results and Discussions

To depict the performance of the proposed methodology, five result sets are displayed in a table. The original secret will depend on any generic non-negative integer n, which is chosen by the dispatcher.

The original secret to be sent can be an image (for instance, 128 x 128) or text (a maximum of 16,384 (which is $n^2$) bytes, when n = 128; which is 131,072 bits). This corresponds to a lofty embedding data of 2731 words, averaging 5 letters a word.

Moreover, the embedding rate used in this paper is the maximum possible threshold in the 2-bit LSB methodology, i.e. 25% of the stego (or the pseudo-secret) image.

Furthermore, the pseudo-secret is conscripted as a 256 x 256 image (since n is originally chosen as 128). Finally, the stego-image is taken as a 512 x 512 gray scale image. Though the payload is 25% of the size in each level, the integrity of the original secret is unprecedentedly overwhelming.

The result-sets are segmented and presented below and demonstrates the desired high PSNR and low MSE values -:

**Table 3** PSNR and MSE results for various extensions

| EXTENSION | PSNR | MSE |
|-----------|------|-----|
| COVER 1 | 44.54294 | 2.2845 |
| PSEUDO 1 | 44.21291 | 2.2845 |
| INNER 1 | N/A | N/A |
| COVER 2 | 44.14183 | 44.14183 |
| PSEUDO 2 | 44.15103 | 44.15103 |
| INNER 2 | N/A | N/A |
| COVER 3 | 44.02921 | 2.5713 |
| PSEUDO 3 | 44.12042 | 2.5713 |
| INNER 3 | N/A | N/A |
| COVER 4 | 44.0891 | 2.531 |
| PSEUDO 4 | 44.1229 | 2.531 |
| INNER 4 | N/A | N/A |
| COVER 5 | 44.1531 | 2.499 |
| PSEUDO 5 | 44.1056 | 2.499 |
| INNER 5 | N/A | N/A |

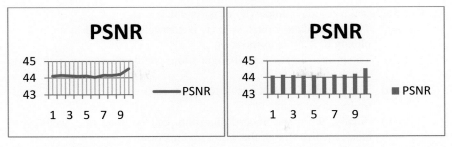

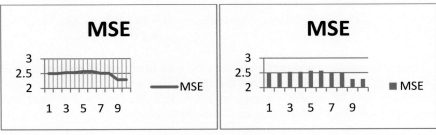

**Fig. 9** (a, b) The line and bar graphs of PSNR values    (c, d) The line and bar graphs of MSE values.

Embedding the covert data directly into the cover image [26] makes it highly vulnerable to steganalysis; thus the embedded information is potentially encrypted twice, both times by the AES. This shields the data with four layers whose complexity is as discussed below.

Furthermore, the introduction of location map does play an important role in optimal PSNR and MSE, as proposed by Karthikeyan et al [27].

   A.  Ali et al [28] methodology's complexity of permutations:

      (i)   The number of ways of choosing two bits at the output of the circuit is 4

      (ii)  The two bits of message for embedding are chosen in 4 ways.

      (iii) The final complexity of embedding at least two bits is 4+4

Final complexity in [28] = 4

   B.  Complexity of permutations   in Janakiraman et al[29] 's methodology :

      (i)   Since the data is encrypted with DES then it would introduce $2^{64}$ choices.

      (ii)  The three bits at the output of the circuit is one of the possible 8 ways.

      (iii) Selecting the combination of three bits of message is 8

      (iv) The final complexity of embedding is $2^{64} * (8+8)$.

$$\text{Final complexity in [29]} = \mathbf{16\ (2^{64})}$$
$$= \mathbf{2.951 \times 10^{20}}$$

C.  Proposed method's complexity of permutations is:
  (i)  The secret data (image or text) is encrypted with AES; so, a complexity of $2^{128}$ is introduced.
  (ii)  The priority of the King-Knight's tour can be chosen in $C_1^{16}$ ways, which amounts to 16.
  (iii)  The pseudo-secret is now encrypted, so a complexity of $2^{128}$ is introduced yet again.
  (iv)  Finally, the checkerboard algorithm can be done in 2 ways, with black squares first or the white squares first.

$$\text{Final complexity (proposed method)} = 16\ (\ 2^{128}\ ) + 2\ (2^{128})$$
$$= 2^{132} + 2^{129}$$
$$= 6.125 \times 10^{39}$$

The implementation was performed on a PC with Intel Core i5 processor, 8GB RAM and 1.7 GHz Clock Speed and the elapsed time for the entire process was found to be 186.39 seconds (average of 100 trials). The proposed method was found to be robust against cropping and Gaussian filter attack.

## 4    Conclusions

A new combination of encryption and embedding is proposed with the help of a pseudo secret. The proposed methodology uses both AES and 2-bit LSB substitution for 2 levels of steganography. The first layer makes use of the innovative alternative to the already existing Knight's tour algorithm as a touring path and the layer uses a checkerboard-like algorithm. Based on the analysis's results, it was found that the proposed methodology has enhanced data diffusion and PSNR values and high payload of 25%. Thus it is safe to conclude that the proposed methodology is ideal for lossless, highly secure multi-level steganography and also compatible for all texts and image extensions.

## References

1.  Katz, J., Lindell, Y.: Introduction to modern cryptography. CRC Press, London (2014)
2.  Zhang, Y.P., Zhai, Z.J., Liu, W., et al.: Digital image encryption algorithm based on chaos and improved DES. In: IEEE International Conference on Systems, Man and Cybernetics (2009)
3.  Schneier, B.: Applied cryptography: protocols algorithms and source code in C. Wiley, New York (1996)
4.  Menezes, A.J., Oorschot, P.C.V., Vanstone, S.A.: Handbook of applied cryptography. CRC Press, Boca Raton (1997)
5.  Rajput, A.S., Mishra, N., Sharma, S.: Towards the growth of image encryption and authentication schemes. In: IEEE International Conference on Advances in Computing, Communications and Informatics (ICACCI) (2013)

6. Qiu, J., Wang, P.: An image encryption and authentication scheme. In: IEEE Seventh International Conference on Computational Intelligence and Security (CIS) (2011)
7. Krishnagopal, S., Pratap, S., Prakash, B.: Image encryption and steganography using chaotic maps with a double key protection. In: Proceedings of Fourth International Conference on Soft Computing for Problem Solving. Springer, India (2015)
8. Nassar, S.S., Ayad, N.M., Kollash, H.M., et al.: Multi-level Security Technique Using Steganography with Chaotic Encryption. Digital Image Processing **6**(3) (2014)
9. Patel, F.R., Cheeran, A.N.: Performance Evaluation of Steganography and AES encryption based on different formats of the Image. Performance Evaluation **4**(5) (2015)
10. Islam, M.R., Siddiqa, A., Uddin, M.P., et al.: An efficient filtering based approach improving LSB image steganography using status bit along with AES cryptography. In: IEEE International Conference on Informatics, Electronics and Vision (ICIEV) (2014)
11. Mare, S.F., Vladutiu, M., Prodan, L.: Secret data communication system using Steganography, AES and RSA. In: 17th IEEE International Symposium for Design and Technology in Electronic Packaging (SIITME) (2011)
12. Kadam, P., Nawale, M., Kandhare, A., et al.: Separable reversible encrypted data hiding in encrypted image using AES Algorithm and Lossy technique. In: International Conference on Pattern Recognition, Informatics and Mobile Engineering (PRIME) (2013)
13. Peinado, A., García, A.O.: Reducing the key space of an image encryption scheme based on two-round diffusion process. In: International Joint Conference. Springer International (2015)
14. Saini, J.K., Verma, H.K.: A hybrid approach for image security by combining encryption and steganography. In: IEEE Second International Conference on Image Information Processing (ICIIP) (2013)
15. Thanikaiselvan, V., Arulmozhivarman, P., Amirtharajan, R., Rayappan, J.B.B.: Horse riding and hiding in image for data guarding. In: Procedia Engineering (2012)
16. Paar, C., Pelzl, J.: Understanding cryptography: a textbook for students and practitioners. Springer Science & Business Media (2009)
17. Sanchez-Avila, C., Sanchez-Reillo, R.: The Rijndael block cipher (AES proposal): a comparison with DES. In: IEEE 35th International Carnahan Conference on Security Technology (2001)
18. Chan, C.-K., Cheng, L.-M.: Hiding data in images by simple LSB substitution. Pattern recognition **37**(3) (2004)
19. Bender, W., Gruhl, D., Morimoto, N., et al.: Techniques for data hiding. IBM Systems Journal **35** (1996)
20. Amirtharajan, R., Anushiadevi, R., Meena, V., et al.: Image Hides Image: A Secret Stego Tri-layer Approach. Research Journal of Information Technology (2013)
21. Jung, K.H., Ha, K.J., Yoo, K.Y.: Image data hiding method based on multi-pixel differencing and LSB substitution methods. In: International Conference on Convergence and Hybrid Information Technology (ICHIT) (2008)
22. Ian, P.: An efficient algorithm for the Knight's tour problem. Discrete Applied Mathematics (1997)
23. Gotelli, N.J., Entsminger, G.L.: Swap and fill algorithms in null model analysis: rethinking the knight's tour. Oecologia (2001)
24. Daemen, J., Rijmen, V.: The design of Rijndael: AES-the advanced encryption standard. Springer Science & Business Media (2013)
25. Pareek Narendra, K.: Design and analysis of a novel digital image encryption scheme. arXiv preprint arXiv:1204.1603 (2012)

26. Manisha, M., Malvika, S.S., Karthikeyan, B., et al.: Devanagari text embedding in a gray image: an offbeat approach. In: 2nd International Conference on Electronics and Communication Systems (ICECS) (2015)
27. Karthikeyan, B., Chakravarthy, J., Ramasubramanian, S.: Amalgamation of scanning paths and modified hill cipher for secure Steganography. Australian Journal of Basic and Applied Sciences (2012)
28. Daneshkhah, A., Aghaeinia, H., Seyedi, S.H.: A more secure steganography method in spatial domain. In: International Conference on Intelligent Systems, Modelling and Simulation (ISMS) (2011)
29. Janakiraman, S., Anitha Mary, A., Chakravarthy, J., et al.: Smart bit manipulation for K bit encoded hiding in K-1 pixel bits. In: 3rd International Conference on Trendz in Information Sciences and Computing (TISC) (2011)

# Digital Watermarking Using Fractal Coding

Rama Seshagiri Rao Channapragada and Munaga V.N.K. Prasad

**Abstract** This paper presents a watermarking method using fractals. In the method discussed here, the host image is encoded by the proposed fractal coding method. To embed the watermark evenly over the whole host image, specific Range blocks are selected. Then, the scrambled watermark is inserted into the selected Range blocks. Finally, the watermarked image is obtained by the fractal decoding method. Simulation results have proven the imperceptibility of the proposed scheme and shown the robustness against various attacks.

**Keywords** Watermarking · Fractals · Similarities · Scrambling

## 1 Introduction

The introduction of World Wide Web has led to the digital information sharing across the world extremely. Due to easy availability of Internet and digital image processing devices, the digital image sharing and accessing has become very simple and convenient. Besides many advantages, such extensive content transmission threatens to destroy ownership and copyright protection [3]. Thus protection of the digital images has gained main focus of the active researchers. The protection of the digital images can be enforced through encryption and decryption, steganography or digital Watermarking [7, 9]. Watermarking has appeared as a method to embed specific data as a watermark into a host media like text, image, video, and audio to identify malicious actions and authenticates ownerships. Mostly the previous methods on digital watermarking insert the watermark into color and gray level images [3]. According to the needs, developers use different

R.S.R. Channapragada(✉)
CMR Institute of Technology, R.R.Dist., Telangana, India
e-mail: crsgrao@yahoo.com

M.V.N.K. Prasad
IDRBT, Hyderabad, Telangana, India
e-mail: mvnkprasad@idrbt.ac.in

© Springer International Publishing Switzerland 2016
S.M. Thampi et al. (eds.), *Advances in Signal Processing and Intelligent Recognition Systems*,
Advances in Intelligent Systems and Computing 425,
DOI: 10.1007/978-3-319-28658-7_10

109

watermarking schemes, such as Blind, Semi-blind and Non-blind watermarking schemes [4, 10]. To achieve high protection of intellectual properties it has to satisfy several requirements like effectiveness, fidelity, robustness etc [8, 12]. Fractals are irregular and fragmented shapes surrounding an image. The term fractal was formed by mathematician Benoit Mandelbrot in 1975. Mandelbrot derived it from the Latin word fractus, which means irregular, broken or fractured. The two main properties of fractals are self-similarity and self-reference [6].

Rao and Prasad have presented four techniques based on magic square and curvelets transformation techniques [11]. They have resized the given watermark using magic square technique and embedded into curvelt coefficients of original image. SoheilaKiani et al. have proposed a multi-purpose digital image watermarking method using fractal block coding to fulfill both authentication and verification purposes simultaneously [2]. This technique uses a new method for fractal coding with a local search region having contrast scaling and mean of range block as its parameters. Marek Candik and Zita Klenovicova proposed a Fractal Image coding with Digital Watermarks [5]. The range blocks are encoded using contrast scaling and brightness, and watermark is inserted during encoding in each range block using nearest neighbor method. A novel watermarking method based on fractal theory was proposed by Fatemeh Daraee, Saeed Mozaffari named Watermarking in binary document images using fractal codes [3]. In this method watermark is inserted into fractal code which contains range block. Digital watermarking using fractal compression technique [6] was proposed by Hsien-chu Wu and Chin-Chen Chang, that efficiently protects the intellectual property rights of digital images. Host image is compressed in order to embed all bits of watermark. A novel watermarking method [13] was proposed by Ming Hong Pi et al. to hide a binary watermark into image files compressed by fractal block coding. This technique uses orthogonalization fractal coding scheme where the fractal affine transform is determined by the range block mean and contrast scaling. Watermarking using similarities based on fractal codification [9] was proposed by Pedro Aaron Hernandez-Avalos et al. to decrease the distortion and robustness of watermarked image. Here, K embedding regions are selected using characteristic points of image, and these are divided into overlapping and non-overlapping blocks called range and domain blocks. Range blocks are modified and watermark bit is inserted based on its value. A robust fractal color image watermarking algorithm [1] was proposed by Jian Lu, Yuru Zou, Chaoying Yang, and Lijing Wangis to achieve a better tradeoff between robustness and high visual quality of a host image. Rao and Prasad have presented a method based on ridgelets and discrete wavelet transform techniques [14]. They have applied fractal coding technique on watermark for embedding into ridgelet coefficients of original image.

The rest of the paper is organized as follows: the section 2 explains the proposed watermarking technique in detail. The section 3 shows the experimental results of proposed method and comparison with previous methods and the section 4 discusses conclusions.

## 2    Proposed Method

In this section, the proposed watermarking technique is explained. First, a color image (I) as shown in Fig. 1(a) of size M × N is taken and divided into RGB components. From this Blue component of the image as shown in Fig. 1(b) is considered for embedding watermark as it is less prone towards noise attacks[15]. Next, this blue component undergoes bit plane slicing in which 8 bit planes are formed since each pixel contains only 8 bits. In this method 4th and 5th bit planes as shown in Fig. 1(c), Fig. 1(d) respectively are used for fractal coding and watermark insertion.

a                                                      b

C                                                      d

**Fig. 1** (a) Lena Color image (512 × 512), (b) Blue component of Lena color image, (c) 4$^{th}$ bit plane of blue component, (d) 5$^{th}$ bit plane of blue component.

### 2.1    Procedure for Embedding Watermark

The procedure consists of four major steps: fractal encoding of host image, scrambling the watermark, embedding into fractal code and fractal decoding to obtain the watermarked image.

Step 1. Fractal block coding involves selecting suitable range and domain block pairs and manipulating various parameters like luminous offset (g) contrast scaling (s) etc.. The process of converting fractal code from the given image is called fractal encoding or fractal image compression. In this method, range blocks are taken from the 4th bit plane and domain block pool is from the 5th bit plane.

Both the range and domain blocks are of equal sizes, i.e. b × b. The numbers of range and domain blocks obtained are M/b × N/b. Next for each Range block (Rb), a block is searched in the domain block pool (Db) which is most similar to the range block. Here, similarity of the range and domain blocks is calculated using the correlation. If the correlation is more, similarity is more. From this, the best range-domain block pair is found and the range block is modified as per Eq.1.

$$R = s \times \arg\max\big(corr2(Rb, Db)\big) + g \times U) \tag{1}$$

Where s stands for contrast scaling, g stands for luminous offset, U stands for unit matrix. For each range block of the image, the fractal code is calculated using Eq.1 and stored as {Rb, R}, where Rb is original Range block and R is the modified Range block.

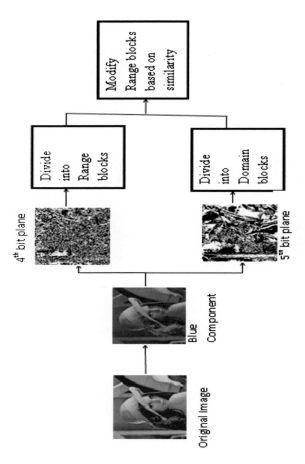

**Fig. 2** Proposed Fractal encoding procedure for color host image

Step 2. Watermark (Wm) of size wm × wn is shuffled first as shown in Fig. 3(b) before being embedded into 4th bit plane of blue component of color image. The shuffling process is employed by the scrambling with a seed K, where K plays the role of secret key. Therefore, K secretly belongs to the legal owner. Scramble (K, Wm) generates a random number of sequences of permutation positions for all bits of Wm with each bit w(p, q)(where p, q represent position). Wm is rearranged into another position, so that the rearranged Wm is Wm' = {w' (p, q) | 0 ≤ p, q ≤ 64, w' (p, q) = 0 or 1}. Because of this, the watermarked image becomes more robust when it undergoes a cropping attack.

 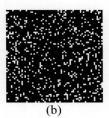

     (a)            (b)

**Fig. 3** (a) Original watermark image, (b) Scrambled watermark image

Step 3. For watermark embedding consider 4th bit plane which was undergone fractal compression and all range blocks of this bit plane were modified and represented as R where size of each range block is b × b and number of range blocks are M/b × M/b. Divide R into m × n sets called S such that wm=m, wn=n and each set contains 2 × 2 Range blocks. Wm' is embedded into S, which is of same size. For every set of 4 range blocks of R one bit of watermark is embedded. Find the blocks $B_{max}$ and $B_{min}$ for every set, where $B_{max}$ and $B_{min}$ represents the blocks having maximum and minimum number of 1's present respectively. Block (WB) is selected from a set of 4 blocks and watermark bit is inserted into the first pixel using XOR operator. This WB block is determined based on watermark to be inserted into that block i.e. if,

$$w(p, q) = \begin{cases} 0, & WB = B_{min} \text{ of } S(p, q) \\ 1, & WB = B_{max} \text{ of } S(p, q) \end{cases} \qquad (2)$$

Select WB block for each watermark bit of Wm'. The indexes of WB are stored as a secret key which is used during watermark extraction process. Then the watermarked block is formed using Eq. 3 by changing the original block. Remaining blocks of set are not modified, and fractal decoding is used for watermarked image construction. This will result the watermarked image as shown in Fig. 4.

$$WB(1, 1) = w(p, q) \, XOR \, WB(1, 1) \qquad (3)$$

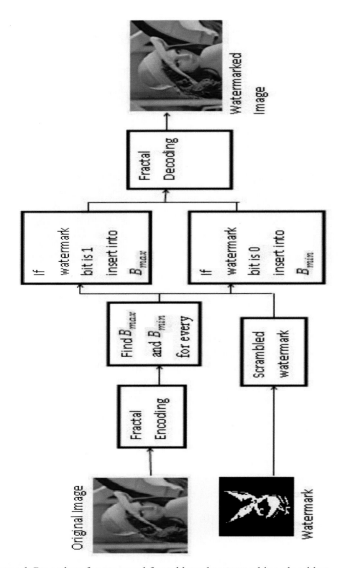

**Fig. 4** Watermark Procedure for proposed fractal based watermarking algorithm

Step 4. The process of image reconstruction from fractal codes is called fractal decoding. Decoding of original image from fractal codes is an iterative process of
reconstruction of range blocks from the set of fractal code parameters. Since it
is a binary image decoding, it takes one iteration only. After getting the watermarked
range blocks, image is constructed using these blocks. 4th bit plane of watermarked
image is constructed using fractal decoding technique. Then combine this bit
plane with the remaining 7 bit planes of original blue component of color image.

This watermarked blue component is again combined with Red and Green components of original color image. Thus watermarked color image (WI) is formed.

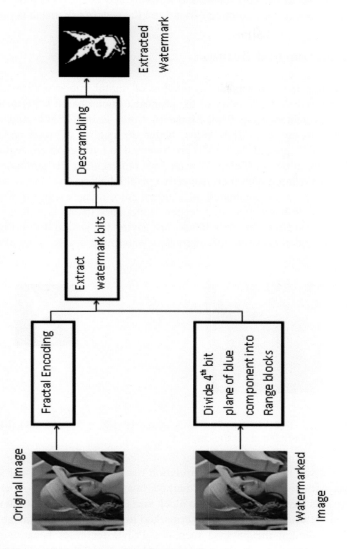

**Fig. 5** Proposed procedure for extraction of watermark

## 2.2 Procedure for Watermark Extraction

To extract the watermark image, both watermarked image and the host image are needed. First, Range blocks of both host and watermarked images are calculated. Fractal codes are calculated for host image using the proposed method shown in

Fig. 5. Using the indexes that stored in secret key range blocks having watermark bit are found and watermark bits are extracted by applying XOR operator between the range blocks of the host image and watermarked image. Using the seed K that stored in secret key descrambling is done and the watermark image is produced.

## 3    Experimental Results

In this section, the performance evaluation of digital image watermarking algorithms is discussed. The quality of the proposed fractal based watermarking technique was evaluated using Peak Signal to Noise Ratio (PSNR) and correlation coefficient. Greater the PSNR value, better the quality of watermarked image. Generally, PSNR greater than 40dB are assumed to have good concealing properties. The correlation coefficient is a number representing the similarity between two images in relation with their respective pixel intensity.

All experiments are conducted on standard color images with the size of 512 × 512 like Lena, Baboon, Barbara, Pepper. Watermark image is a binary image of size 64 × 64. Range and domain blocks are having the size of 4 × 4. Fig. 6 shows the watermarked image and the extracted watermark image after applying the proposed procedure.

a                                                              b

**Fig. 6** (a) 512 × 512 Watermarked Lena image (PSNR = 51.6582 dB) (b) Extracted watermark with Correlation coefficient = 1.0.

Table 1 gives the PSNR and Correlation results when the original image was compared with the obtained watermarked image. The table also demonstrates various results based on varied range block size. Whereas the results show that the correlation between the obtained watermarked image and original image is highest value. Even though the payload increases PSNR increases because the Range block size is decreasing i.e. the most similar block which had found during fractal encoding will be almost equal. Table 2 gives the comparative study of the proposed method with other methods discussed by Jian Lu et al. [1] and SoheilaKiani et al. [2].

**Table 1** PSNR of the watermarked image against respective original image.

| Host Image | Range block Size | Size of watermark (bits) | PSNR of watermarked image |
|---|---|---|---|
| | $2 \times 2$ | $128 \times 128$ | 56.4008 |
| Lena | $4 \times 4$ | $64 \times 64$ | 51.6582 |
| | $8 \times 8$ | $32 \times 32$ | 48.2552 |
| | $2 \times 2$ | $128 \times 128$ | 56.9801 |
| Barbara | $4 \times 4$ | $64 \times 64$ | 51.9967 |
| | $8 \times 8$ | $32 \times 32$ | 48.6438 |
| | $2 \times 2$ | $128 \times 128$ | 56.6184 |
| Baboon | $4 \times 4$ | $64 \times 64$ | 52.0403 |
| | $8 \times 8$ | $32 \times 32$ | 49.1262 |
| | $2 \times 2$ | $128 \times 128$ | 56.6184 |
| Pepper | $4 \times 4$ | $64 \times 64$ | 51.6822 |
| | $8 \times 8$ | $32 \times 32$ | 48.2985 |

**Table 2** Comparison of obtained PSNR between original image and watermarked image of different methods.

| Methods | Payload(bits) | Lena | Baboon | Barbara | Pepper |
|---|---|---|---|---|---|
| Jian Lu et al. [1] | $128 \times 128$ | 34.71 | | 30.37 | 32.01 |
| SoheilaKiani et al. [2] | 512 | 49.86 | | | 47.65 |
| Proposed Method | $64 \times 64$ | 51.6582 | 51.9967 | 52.0403 | 51.6822 |
| | $128 \times 128$ | 56.4008 | 56.6184 | 56.9801 | 56.3412 |
| | 512 | 51.9927 | 52.3482 | 52.6715 | 51.8213 |

    The watermarked image's robustness was evaluated by attacking the image through various image manipulation procedures like cropping, adding noise and compression etc. Then the watermark was extracted from these manipulated images and compared with the original watermark image. Table 3 gives the correlation results when original watermark was compared with extracted watermark from various attacked watermarked images.

**Table 3** Correlation coefficient of original watermark and extracted watermark after various attacks on watermarked image.

| Attacks | Lena | Baboon | Barbara | Pepper |
|---|---|---|---|---|
| Cropping | | | | |
| Type-1 | 0.7210 | 0.7292 | 0.7488 | 0.7555 |
| Type-2 | 0.6848 | 0.7043 | 0.6980 | 0.7260 |
| Type-3 | 0.4906 | 0.4860 | 0.4880 | 0.5188 |
| Salt and Pepper | | | | |
| Type-1 | 0.9784 | 0.9840 | 0.9720 | 0.9804 |
| Type-2 | 0.9188 | 0.9098 | 0.9054 | 0.9075 |
| Type-3 | 0.8507 | 0.8330 | 0.8720 | 0.8362 |
| Jpeg Compression | 0.9880 | 0.9853 | 0.9807 | 0.9895 |
| TIF Compression | 1 | 1 | 1 | 1 |

# 4    Conclusion

In this work, a fractal color image watermarking method is developed and assessed its performance. The main objectives of watermarking are imperceptibility
and to achieve robustness against attacks. In this method fractal coding is applied
to the bit planes of blue component of the color image. After fractal coding
scrambled watermark is embedded. Proposed method is robust against various
attacks. Both the cover image and watermarked image are used for extracting
binary watermark. The robustness of the algorithm is shown in PSNR calculations.
The results show that the method is comparable to the methods discussed by Jian
Lu et al. [1] and SoheilaKiani et al. [2].

# References

1. Lu, J., Zou, Y., Yang, C., Wang, L.: A Robust Fractal Color Image Watermarking
   Algorithm. Mathematical Problems in Engineering 3, 1–12 (2014)
2. Kiani, S., Moghaddam, M.E.: A multi-purpose digital image watermarking using
   fractal block coding. The Journal of Systems and Software 84, 1550–1562 (2011)
3. Daraee, F., Mozaffari, S.: Watermarking in binary document images using fractal
   codes. Pattern Recognition Letters 35, 120–129 (2014)
4. Arun, Kabi, K.K., Saha, B.J., Pradhan, C.: Enhanced digital watermarking scheme
   using fractal images in wavelets. In: International Conference on Computing, Communication and Networking Technologies (ICCCNT), pp. 1–6 (2014)
5. Candik, M., Klenovicova, Z.: Fractal Image Coding with Digital Watermarks. Radio
   engineering 9(4), 22–26 (2000)
6. Wu, H.-C., Chang, C.-C.: Hiding Digital Watermarks Using Fractal Compression
   Technique. Fundamental Informatica 58, 189–202 (2003)
7. Gupta, P.: Cryptography based digital image watermarking algorithm to increase security of watermark data. International Journal of Scientific and Engineering Research
   3(9), 1–4 (2012)
8. Ramana Reddy, P., Prasad, M.V.N.K., Sreenivasa Rao, D.: Robust Digital Watermarking of Color Images under Noise attacks. International Journal of Recent Trends in
   Engineering 1(1), 334–338 (2009)
9. Hernandez-Avalos, P.A., Feregrino-Uribe, C., Cumplido, R.: Watermarking using similarities based on fractal codification. Digital Signal Processing (22), 324–336 (2012)
10. Tang, C.-W., Hang, H.-M.: A Feature-Based Robust Digital Image Watermarking Scheme.
    IEEE Transactions On Signal Processing 51(4), 950–959 (2003). Fellow, IEEE
11. Channapragada, R.S.R., Prasad, M.V.N.K.: Watermarking techniques in curvelet
    domain. In: International Conference on Computational Intelligence in Data Mining
    (ICCIDM–2014), vol. 1, pp. 199–212 (2014)
12. Li, Y., Du, S.: Research on Wavelet Domain Fractal Coding in Digital Watermarking.
    Multimedia and Expo, 61–64 (2005)
13. Pi, M.H., Li, C.H., Li, H.: A Novel Fractal Image Watermarking. IEEE Transactions
    on Multimedia 8(3), 488–499 (2006)
14. Rao, C.R.S.G., Prasad, M.V.N.K.: Digital image watermarking based on fractal image
    coding. In: Pal, R. (ed.) Innovative Research in Attention Modeling and Computer
    Vision Applications, pp. 388–399 (2015)
15. Kutter, M., Petitcolas, F.A.P.: A fair Benchmark for image Watermarking Systems.
    Proceedings of SPIE in Security and Watermarking of Multimedia Contents 3657, 1–14
    (1999)

# A Cheque Watermarking System Using Singular Value Decomposition for Copyright Protection of Cheque Images

Sudhanshu Suhas Gonge and Ashok Ghatol

**Abstract** Recently, all bank system has online money transfer facility. It may be either by online transactions or by NEFT cheques. Bank system pays money through cash, demand draft, and online money transfer or by cheques. Few days ago, there was a flow of the physical cheques issued by bank customer from drawer to the drawee branch. This process consumes more time and increase physical work of the employee. To overcome this problem cheque truncation system (CTS) comes into picture. This system provides faster clearing and transmission of bank cheques issued by bank customer from drawer branch account to the drawee branch account.CTS transfers the cheque in image format. Due to this reason, there is a need of copyright protection to the cheque image. It can be provided by digital watermarking techniques. The digital watermarking techniques can be applied using various methods. Such as, frequency domain, spatial domain, computational intelligence technique, etc. In this paper, it is going to discuss, "A Cheque Watermarking System Using Singular Value Decomposition for Copyright Protection of Cheque Images". The aim of this paper is to obtain secure cheque image & watermark after extraction process of watermark at drawee branch.

**Keywords** Digital watermarking technique · Cheque truncation system · Frequency domain · SVD

S.S. Gonge(✉)
Research Scholar, Faculty of Engineering and Technology,
Santa Gadge Baba Amravati University, Amravati, India
e-mail: sudhanshu1984gonge@rediffmail.com

A. Ghatol
Former Vice-Chancellor, Dr. Babasaheb Ambedkar Technological University,
Lonere, Maharashtra, India
e-mail: vc_2005@rediffmail.com

© Springer International Publishing Switzerland 2016
S.M. Thampi et al. (eds.), *Advances in Signal Processing and Intelligent Recognition Systems*,
Advances in Intelligent Systems and Computing 425,
DOI: 10.1007/978-3-319-28658-7_11

# 1    Introduction

Noise Signal can be useful when it is used as information for embedding in data signal, then it called as digital watermark. Digital watermark is used for embedding in digital data like audio, video, and image. On the basis of digital data, watermarking process is classified as:-

1.    Digital audio watermarking.
2.    Digital video watermarking.
3.    Digital image watermarking.

The life cycle of digital watermarking process consist of three phases:-

1.    Embedding phase.
2.    Attack phase.
3.    Extraction phase.

Embedding phase in which watermark is embedded in digital data. After successful embedding process of digital watermark into digital data, watermarked digital data is transferred through proper channel. Transmission of watermarked digital data takes place from source to destination. Attack phase is insecure part of channel for digital data while transmission process. Extraction phase also called as detection phase of watermark. Digital watermark is detected and removed from watermarked data. In this case watermark is extracted and compare with original watermark to get similarity of watermark. CTS of bank transfer cheques in image format from drawer branch to drawee branch [1-3]. Hence there is need of copyright protection to bank cheque image. This can be provided by embedding watermark into cheque image. The aim of this paper is to obtain secures watermarked cheque image as well as watermark after extraction process at drawee branch.

# 2    Various Technique Used For Digital Image Watermarking

There are various technique used for digital image watermarking as shown in following figure 1.

## 2.1    Frequency Domain

There are many frequency domain techniques like discrete wavelet transform, Fast Fourier transform, discrete cosine transform or combination of one or more frequency domain transforms. Digital image in frequency domain means "the spectral density estimation of that particular image." However, in this paper working of singular value decomposition transform is explained which is used for copyright protection of digital bank cheque images [1-2].

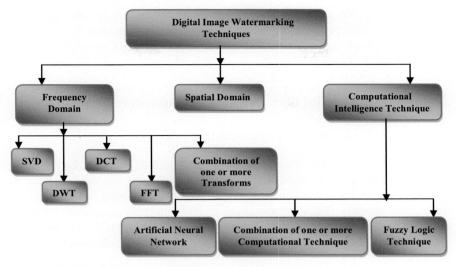

**Fig. 1** Various Techniques used for digital watermarking process.

## 2.2   *Computational Intelligence Techniques*

Computational intelligence technique is computational part of artificial intelligent. It is associate with various elements of learning, adaptation, evaluation and fuzzy linguistic to create programs in some sense intelligent. It also deals with various heuristic algorithms such as in fuzzy systems, neural networks and evolutionary computation [1-7].

It also includes swarm intelligence, fractal and chaos theory, artificial immune systems, wavelets, etc. Computational intelligence technique does not reject statistical method but it always gives a complementary view in the case of fuzzy system. Artificial neural network is a branch of computational intelligence which is related to machine learning, soft computing, scruffy artificial intelligence, connectionist system and cybernetics [1-8].

Computational intelligence technique is an adaptive mechanism to enable or facilitates intelligent behaviors in complex and changing environment [1-5].

## 2.3   *Spatial Domain*

In Spatial domain, pixel positions of image are changed for object representation. It is also used for  manipulating pixels position of the image using techniques like smoothing, un-sharp masking, laplacian filtering, etc.Generally, it is used for filtering.There are various methods that are directly applied on image pixels[1-5]. Mathematically, spatial domain image processing function can be expressed as:

$$F(x, y) = M [h(x, y)].\qquad(1)$$

Where h(x,y) is an input image, F(x,y) is output image obtained after processing and M represent an image processing function operation on 'h' defined over some neighbourhood of (x,y) [7,8,9,10,11,13].Following figure shows various techniques used for digital image watermarking.

# 3      Working of Singular Valued Decomposition Transform

## 3.1    Singular Valued Decomposition Transform

Singular valued decomposition transform is a technique used for handling matrices. These matrices are consists of set of equations that do not have an inverse. This includes the square matrices whose determinant is zero and all rectangular matrices. The common usage of SVD includes computing the least-squares solutions, rank, range i.e. (column space), null space and Pseudo inverse of matrix. The main advantage of SVD method is used for noise reduction [12, 13, 14 and 15].

Thus, SVD Transform is selected for digital watermarking of cheques image.

Mathematically, it is expressed as:-

➢   Any matrix A of mix size can be decomposed uniquely as:

$$A=U.D.V^T \qquad\qquad (2)$$

Where, U is m x n matrix and its columns are orthogonal i.e. the value of its columns are Eigen vector of $A.A^T$

$$A.A^T=U.D.V^T.V.D.U^T=U.D^2.U^T \qquad\qquad (3)$$

➢   V is n x n matrix and matrix is orthogonal in nature i.e. the value of its columns are Eigen vector of $A^T.A$

$$A^T.A= V.D.U^T.U.D.V^T=V.D^2.V^T \qquad\qquad (4)$$

➢   D is n x n diagonal matrix i.e. non-negative real values called singular values of matrix D [3, 7].
➢   If, D= diag (W1, W2,..............., Wn) are ordered so that: $W1{\geq}W2{\geq}W3{\geq}.......{\geq}Wn$. If, W is a singular value of matrix A, then its square value is an Eigen value of $A^T.A$
➢   If U=(U1,U2,U3,....,Un) and V=(V1,V2,V3,.....,Vn), then

$$A= \sum\nolimits_{(i=0 \text{ to } n)} U_i * W_i * V^T_i \qquad\qquad (5)$$

➢   Columns Ui of U and Vi of V are called as left and right singular vectors.

A working of cheque watermarking system is explained in following embedding and extraction algorithm with the help of SVD transform [12, 13, 14, and 15].

## 3.2 Watermark Embedding Algorithm Using Singular Value Decomposition Transform

The working of algorithm is as shown in following fig. 2.

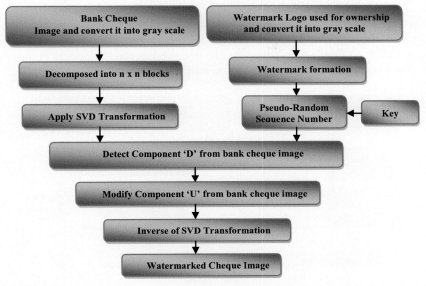

**Fig. 2** Watermark embedding procedure for a cheque watermarking systems using SVD Transform.

> ➢ Step 1: Select the cheque image that has to be transfer from drawer branch to drawee branch.
> ➢ Step 2: Convert this cheque image into gray scale image.
> ➢ Step 3: Gray scale cheque image is decomposed into n x n block of pixels.
> ➢ Step 4: Apply singular value decomposition transform is applied on every n x n blocks of pixels.
> ➢ Step 5: Determined the complexity of the block, by calculating number of non-zero co-efficient in D component of each block.
> ➢ Step 6: Generate two pseudo-random number sequence i.e.W_0 and W_1which is not similar to each other for embedding the watermark logo bit by selecting greater complexity blocks by using the attributes of component D. Apply the following equation, if the watermark bit is 1, then:

$$D_w = D + \alpha * W\_1 = U_{w\_1} * D_{w\_1} * V^T_{w\_1} \tag{6}$$

Otherwise,

$$D_w = D + \alpha * W\_0 = U_{w\_0} * D_{w\_0} * V^T_{w\_0} \tag{7}$$

Where, $U_w$ and $V_w$ are private orthogonal matrix and $D_w$ is a diagonal matrix formed from the cheque image. '$\alpha$' is constant [6, 7, 14, and 15].

➢ Step 7: Modify the component of U and V with the help of step 6.
➢ Step 8: Take the inverse of SVD transform on cheque image.
➢ Step 9: Finally, watermarked image is obtained.

This watermarked cheque image is send from drawer branch to drawee branch for payment through the clearing house. However, the extraction process of watermark logo from watermarked cheque image takes place at drawee branch for checking ownership or copyright protection of bank cheques.

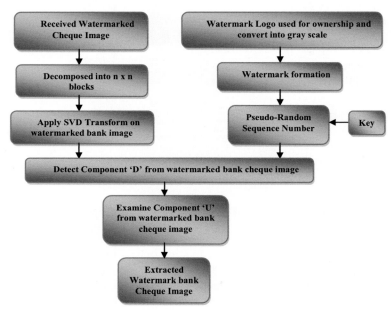

**Fig. 3** Watermark embedding procedure for a cheque watermarking systems using SVD Transform.

## 3.3 *Watermark Extraction Algorithm Using Singular Value Decomposition Transform*

➢ Step 1: Select received watermarked cheque image.
➢ Step 2: Decomposed that image into n x n blocks of pixels.
➢ Step 3: Apply SVD transform on each n x n blocks of pixels.
➢ Step 4: Determine the complexity of the block by calculating the non-zero co-efficient in 'D' component for each block.
➢ Step 5: The same seeds of image pixels are used to create again two pseudorandom sequences W_0 and W_1 which were used for embedding process.

> ➤ Step 6: Extract the watermark bit '1' if, the correlation with W_0 is less than W_1,if not extracted bit of watermark is consider as '0'.
> ➤ Step 7: Similarity between original watermark and extracted watermark from cheque image is calculated by reconstruction of watermark with the help of extracted watermark bits.

# 4 Calculating Imperceptibility of Watermarked Cheque Image

Imperceptibility is term related to peak-signal to noise ratio value of image. To Maintain the quality of watermarked bank cheque image; imperceptibility is calculated. Cover image of bank cheque quality should not get disturbed due to presence of watermark[4,8,12,13].However, imperceptibility is calculated in terms of peak signal to noise ratio i.e.in decibels(db).The equations used for calculating imperceptibility of watermarked cheque image are given below:

$$\text{PSNR}_{db} = 10 \text{Log}_{10} \{\text{Max}^2_1 / \text{MSE}\} \tag{8}$$

$$= 20 \text{Log}_{10} \{\text{Max}_1 / \text{MSE} \wedge (1/2)\}$$

Where $\text{Max}_1$ is maximum pixel value of host or input image and MSE is mean square error between original image and watermarked image.

# 5 Experiment and Results

Observation Table 1:-For corresponding PSNR values and elapsed time before attack on watermarked cheque image.

| Alpha value or Gain factor α | Embedded watermark by SVD PSNR(dB) | Extracted watermark by SVD Before Attack PSNR(dB) | Elapsed Time for complete process. |
|---|---|---|---|
| 1 | 20.640 | 284.714 | 5.16 sec. |
| 2 | 11.153 | 284.69 | 5.47 sec. |
| 3 | 6.527 | 281.356 | 5.07 sec. |
| 4 | 3.602 | 287.713 | 5.23 sec. |
| 5 | 1.880 | 289.460 | 4.71 sec. |
| 6 | 0.687 | 289.288 | 5.76 sec. |
| 7 | 0.516 | 291.353 | 4.88 sec. |
| 8 | -1.708 | 289.682 | 5.09 sec. |

Above observation table shows the calculated values of peak Signal to the noise ratio for watermarked cheque image on embedding watermark and after extracting watermark from cheque image for different gain factor i.e. alpha also called as constant. This constant is used for controlling the strength of watermark signal used for embedding process.

In this experiment, image used for watermarking purpose is of 24bit depth and it is in bitmap format. This image is convert into gray scale image and result into 8bit depth. The value of constant is considered from 1 to 8. Since the image bit depth is 8bit. Also calculate the elapsed time required for complete process.

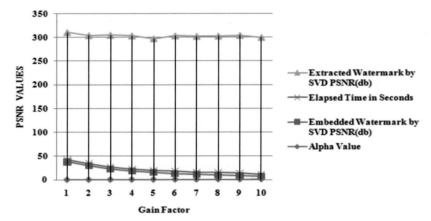

**Fig. 4** Graphical Representation for a cheque watermarking systems between corresponding PSNR and Co-relation Co-efficent values before attack.

Graphical representation shows a corresponding PSNR and elapsed time before attack on cheque watermarking systems. Fig. 4 explains the different reading taken for eight different gain factor .It also explains, as increases in the gain factor, watermark embedding strength increases. Graph explains different PSNR for embedding and extraction process before attack with respect to time and gain factor. The singular valued transform has properties to reduce noise.

It helps in maintaining the quality of watermarked cheque. Input image of bank cheque and watermark used as logo is as shown in fig. 5. & fig. 6 and fig. 7 & fig. 8 are gray scale conversion of input image:

**Fig. 5** Original bank cheque image.

**Fig. 6** Original watermark image used as logo.

**Fig. 7** Grayscale bank cheque image.

**Fig. 8** Grayscale watermark image used as logo.

## 5.1 For Gain Factor or value of alpha as constant =1

Above figures explain the output image obtain from the cheque watermarking system for scaling factor 1. Fig. 9 is a watermarked image of bank cheque. Fig. 10 is extracted gray scale watermark before attack. For experiment purpose, in this paper two different attacks are being considered i.e. Poisson attack and salt & pepper noise attack for 0.05 value. Fig. 11 is watermarked cheque image after applying Poisson noise attack. Fig. 12 is an extracted gray scale watermark from the watermarked cheque image on which Poisson noise attack is applied. It is observed that the quality of extracted watermark is degraded as compare to original gray scale watermark. Fig. 13 explains the watermarked image of bank after applying salt and pepper noise attack. Due to this attack quality of watermarked cheque image as well as extracted watermark gets degraded as shown in Fig. 14.

**Fig. 9** Watermarked image of bank cheque before attack.

**Fig. 10** Extracted Gray Scale watermark before attack.

**Fig. 11** Watermarked image of bank cheque after applying poisson noise attack.

**Fig. 12** Extracted Gray Scale watermark after applying piosson noise attack.

**Fig. 13** Watermarked image of bank cheque after applying salt and pepper noise attack. (0.05)

**Fig. 14** Extracted Gray Scale watermark after applying salt and pepper noise attack. (0.05)

*Observation Table 2:-For corresponding PSNR values and elapsed time after attack on watermarked cheque image.*

| Alpha value or Gain factor $\alpha$ | Embedded watermark by SVD PSNR (dB) | Watermark embedding Process in Elapsed Time in (Second) | Extracted watermark by SVD After attack PSNR(dB) | Elapsed Time for watermark extraction After attack in (Second) | Correlation Co-efficient with After Attack | |
|---|---|---|---|---|---|---|
| | | | | | Salt &Pepper Noise For(0.05) | Poisson Noise |
| 1 | 20.640 | 5.48 | 1.305 | 1.46 sec. | 0.9539 | 0.9746 |
| 2 | 11.153 | 4.25 | 6.580 | 1.26 sec. | 0.9269 | 0.9476 |
| 3 | 6.527 | 6.095 | 9.103 | 1.13 sec. | 0.9213 | 0.9430 |
| 4 | 3.602 | 4.806 | 10.511 | 1.21 sec. | 0.8753 | 0.8796 |
| 5 | 1.880 | 5.210 | 11.363 | 1.26 sec. | 0.8948 | 0.8983 |
| 6 | 0.687 | 5.234 | 11.910 | 1.19sec. | 0.8937 | 0.8979 |
| 7 | 0.516 | 5.248 | 12.278 | 1.22 sec. | 0.8935 | 0.8977 |
| 8 | -1.708 | 6.055 | 12.535 | 1.14 sec. | 0.8930 | 0.8975 |

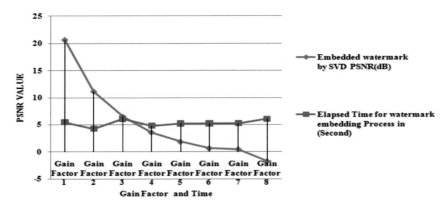

**Fig. 15** A cheque watermarking systems graph between embedded watermark psnr values and elapsed time after attack.

Above observation table-II explain different PSNR values, elapsed time for embedding & extraction process for different gain factor. It also explains co-relation obtain for different gain factor after applying different attacks. Fig. 15 explains the graph plot between PSNR values for embedded watermark versus time required for embedding process for eight different gain factors. It is observed from the graph, as increase in the gain factor Peak signal to noise ratio reduces due to SVD transform. Gain factor used for controlling the strength of watermark signal for embedding purpose. Thus, Singular valued transform satisfies the property of noise reduction. Following Fig. 16 graph explains the PSNR value after watermark extraction versus time required for extraction process using SVD transform.

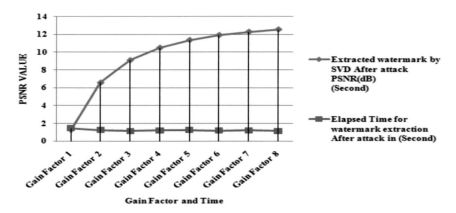

**Fig. 16** A cheque watermarking systems graph between extracted watermark psnr values and elapsed time after attack.

It is observed that due to increase in gain factor psnr values increases and quality of watermark get degraded. PSNR value for extraction process is less as compare to embedding process. Time required for extraction process is also reduced as compare to time required for embedding process. Following Fig. 17 shows watermark obtain after applying poisson noise attack and salt & pepper noise attack during extraction of watermark from watermarked bank cheque image for different gain factor. The correlation co-efficient is calculate for finding the similarity between original watermark and extracted watermark. Following graph explains the different values of correlation co-efficient for different gain factor. The value of correlation co-efficient varies between 0&1.If value of correlation co-efficient is 1, then extracted watermark is secured and robust. If value of correlation coefficient is between 0 and 1 then quality of watermark is degraded. It is observed that correlation co-efficient value of extracted watermark after attacks is above 0.8750.The graph shows that 87% of watermark quality has been maintain with the help of SVD transform. The following graph also shows that the poisson attack is more effective than salt & pepper noise.

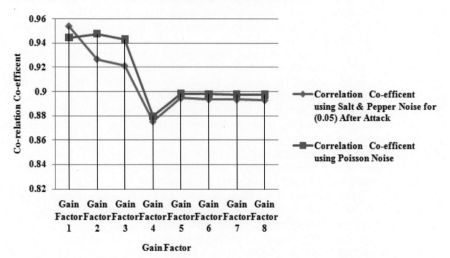

**Fig. 17** A cheque watermarking systems graph between corresponding correlation co-efficient values and gain factor after attack.

## 6    Conclusion

This paper explains working of digital cheque image watermarking system by using SVD transform for copyright protection of bank cheque images.SVD explains many good mathematical characteristics for image processing. SVD having singular value of a cheque image has very good stability even if any watermark is added to cheque image. This is because there is no rapid variation in singular value of cheque image.SVD transforms explains the internal algebraic properties of cheque image which is not visible. 87% of watermark quality is maintain using SVD transform However, as increase in the gain factor embedding power of watermark logo in cheque image increases.Furthur increase in value of gain factor i.e. more than '8',it may disturbed the quality of watermark logo and cheque image but PSNR value get reduced. In this way, one can provide copyright protection to bank cheque image. It also helps in providing secure transmission of cheque from drawer to the drawee branch using cheque truncation system for faster clearance of cheques. It also helps to reduce the physical work of bank employee.

# References

1. Shnayderman, A., Gusev, A., Eskicioglu, A.M.: An SVD-Based Gray scale Image Quality Measure for Local and Global Assessment. IEEE Transaction on Image Processing **15**(2), 422–429 (2006)
2. Deepa Mathew, K.: SVD based Image Watermarking Scheme. IJCA (2010). Special Issue on Evolutionary Computation for Optimization Techniques ECOT
3. Hsu, C.T., Wu, J.L.: Hidden Digital Watermarks In Images. IEEE Trans. Image Process. **8**(1), 58–68 (1999)
4. Liu, R.Z., Tan, T.N.: An SVD-Based Watermarking Scheme for Protecting Rightful Ownership. IEEE Trans. On Multimedia **4**(1), 121–128 (2002)
5. Wang, B., Ding, J., Wen, Q., Liao, X., Liu, C.: An image watermarking algorithm based on DWT DCT and SVD. In: Proceedings of IC-NIDC 2009 (2009)
6. Bedi, S.S., Kumar, A., Kapoor, P.: Robust Secure SVD Based DCT – DWT Oriented Watermarking Technique for Image Authentication. Special Issue of the International Journal of the Computer, the Internet and Management **17**(SP1), March 2009
7. Hernandez, J.R., Perez Gonzalez, F., Amado, M.: DCT-Domain Watermarking Techniques for Still Images: Detector Performance Analysis and a New Structure. IEEE Transaction on Image Processing **9**(1), January 2000
8. Panada, J., Bisht, J., Kapoor, R., Bhattacharyya, A.: Digital image watermarking in integer wavelet domain using hybrid techniques. In: International Conference on Advances in Computer Engineering (ACE), pp, 163–167 (2010)
9. Chan, C., Cheng, L.: Hiding Data in Images by Simple LSB Substitution. Pattern Recognition **37**(3), 469–474 (2004)
10. Na, W., Chiya, Z., Xia, L., Yunjin, W.: Enhancing iris-feature security with steganography. In: The Fifth IEEE Conference on Industrial Electronics and Applications (ICIEA), pp. 2233–2237 (2010)
11. Wang, S., Lin, Y.: Wavelet Tree Quantization for Copyright Protection Watermarking. IEEE Trans. Image Processing **13**(2), 154–164 (2004)
12. Nikolaidis, A., Pitas, I.: Asymptotically optimal detection for additive watermarking in the DCT and DWT domains. IEEE Trans. Image Processing **2**(10), 563–571 (2003)
13. Lin, S., Chin, C.: A Robust DCT-based Watermarking for Copyright Protection. IEEE Trans. Consumer Electronics **46**(3), 415–421 (2000)
14. Wu, Y.: On the Security of an SVD-Based Ownership Watermarking. IEEE Transactions on Multimedia **7**(4), 624–627 (2005)
15. Liu, R., Tan, T.: An SVD-based watermarking scheme for protecting rightful ownership. IEEE Trans. Multimedia **4**(1), 121–128 (2002)

# Biometric Watermarking Technique Based on CS Theory and Fast Discrete Curvelet Transform for Face and Fingerprint Protection

Rohit Thanki and Komal Borisagar

**Abstract** Nowadays, multibiometric system is employed to overcome limitation of the unimodal biometric system. The problem associated with multibiometric system is to design technique which offers security for biometric data. A biometric watermarking technique using compressive sensing (CS) theory and Fast Discrete Curvelet Transform is proposed for face and fingerprint protection. Compressive sensing has provided computational security to watermark fingerprint image and used for generation of sparse measurements of the watermark fingerprint image. This proposed watermarking technique embeds sparse measurements of the watermark fingerprint image into high frequency curvelet coefficients of host face image. The quantitative measure such as a structural similarity index measure is used for cross verification between original watermark fingerprint image and reconstructed fingerprint image. The watermarked face image and reconstructed watermark fingerprint image are formed face-fingerprint based multibiometric system which is used for two levels of verification of individuals. The experimental results demonstrate that proposed watermarking technique does not affect verification and authentication performance of multibiometric system.

**Keywords:** CS theory · Fast discrete curvelet transform · Fragile · Multibiometric system · Watermarking

R. Thanki(✉)
Ph.D. Research Scholar, Faculty of Technology and Engineering,
C. U. Shah University, Wadhwan City, Gujarat, India
e-mail: rohitthanki9@gmail.com

K. Borisagar
E.C. Department, Atmiya Institute of Technology and Science, Rajkot, Gujarat, India
e-mail: krborisagar@aits.edu.in

© Springer International Publishing Switzerland 2016
S.M. Thampi et al. (eds.), *Advances in Signal Processing and Intelligent Recognition Systems*,
Advances in Intelligent Systems and Computing 425,
DOI: 10.1007/978-3-319-28658-7_12

# 1    Introduction

Nowadays, the automatic biometric authentication system has been used at several establishments, federal agencies, railway stations and airports for security pur-
posed. This automatic biometric authentication system recognizes an individual
based on its biometric characteristics [1]. In most of the automatic biometric system, used fingerprint, face and iris biometric features for individual recognition
because of these three biometric features are easy to available and accepted for
worldwide [1]. The automatic biometric authentication system has advantages
compared to traditional authentication system, but some restrictions are connected
with this automatic biometric system such as noisy sensor, intra class variation,
distintivness, Nonuniversality and spoof attacks [1 – 3]. In 2001, Ratha was
pointed out several limitations and vulnerable areas in an automatic biometric
authentication system. Some limitations of the automatic biometric systems such
as modification of the template at system database and attack on the template on
the communication channel can be overcome by using multiple biometric characteristics of individuals. This system is called as multimodal or multibiometric
system which used multiple biometric features for individual recognition [4, 5].

There are two issues such as designing of fusion technique and designing of
template protection technique are related to any multibiometric system when it
design [6]. For template protection in any multibiometric system, one of the solution is a watermarking technique is proposed by Jain and its research team [7].
They are proposed fingerprint minutiae based watermarking technique for protection of facial features of individual [7]. This is a first watermarking technique
which is used for template protection in biometric system. The interest of researchers in designing biometric template protection using watermarking technique began to mushroom after 2003. There are various watermarking techniques
are proposed by researchers using biometric features such as fingerprint [7, 8, 9,
10, 11], iris [12, 13, 14, 15, 16, 17], voice [16, 18, 19, 20], face [7, 8, 12, 20, 21,
22] and signature [18, 23] as watermark data inserted into other biometric features
for template protection are available in the literature.

The first watermarking technique based on curvelet transform is proposed by
Zhang in 2008 for protection of images against copyright issues [24]. In this technique, image hash is computed using curvelet transform and inserted into different
level of curvelet transform of the image. Xu [25] proposed watermarking technique based on Fast Discrete Curvelet Transform for standard image protection
against watermark attacks where binary watermark is inserted into the curvelet
coefficients of the image. Bazargani [26] makes a comparison of robust
watermarking techniques in wavelet, contourlet and curvelet for standard image
protection. These all proposed curvelet based techniques are robust against watermarking attacks.

A biometric watermarking technique based on compressive sensing (CS) theory
and Fast Discrete Curvelet Transform is proposed in this paper. In this technique,
sparse measurements of the watermark fingerprint image are generated using CS

theory. These sparse measurements of watermark fingerprint are inserted into curvelet coefficients of host face image such way that watermarked face image is used for individual identification. Then, extracted sparse measurements of watermark fingerprint image after applying attacks on watermarked face image. Then reconstructed watermark fingerprint image from extracted sparse measurements using the CS theory recovery procedure. The reconstructed watermark fingerprint image is used for cross verification of the individual and make decisions about modification of biometric templates. In this technique, same size of watermark biometric data can be inserted into the host image which is not possible with an existing watermarking technique in transform domain.

The rest of the paper is organized such that section 2 gives Fast Discrete Curvelet Transform, section 3 gives proposed watermarking technique, section 4 gives experimental results, section 5 gives effect of proposed watermarking technique on the performance of multibiometric system. Lastly, section 6 gives conclusions of the paper.

## 2 Fast Discrete Curvelet Transform (FDCT)

In 2005, Donoho and Candes described new image transform, namely curvelet transform based on sparsity theory [27]. Curvelet transform is calculated the inner relationship between the image and its curvelet function to realize sparse representation of the image. There are two types of curvelet transform such as continuous curvelet transform and discrete curvelet transform available in the literature. Discrete curvelet transform is used most of image processing applications such as compression, watermarking and edge detection [27].

There are two types of fast discrete curvelet transform, namely unequispaced fast Fourier transform (USFFT) and frequency wrapping. The frequency wrapping based discrete curvelet transform technique is easy to implement, less computation time and easy to understand compared to the USFFT technique [27, 28]. Therefore, frequency wrapping based curvelet transform technique is used in many image processing applications. When the frequency wrapping based curvelet transform [27, 28] applied to an image, then the image is converted into low frequency coefficients and high frequency coefficients.

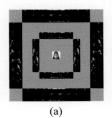
(a)

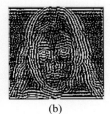
(b)

**Fig. 1** (a) Curvelet Coefficients (b) High Frequency Curvelet Coefficients $C\{1,4\}$

The curvelet decomposition of image by frequency wrapping based curvelet transform with 4 scales and 16 orientation parameter are given as $C$ {1, 1}; $C$ {1, 2}; $C$ {1, 3}; $C$ {1, 4}. Given the decomposition scale 4, the curvelet coefficients $C$ {1, 1} is the low frequency coefficients and the other coefficients $C$ {1, 2}; $C$ {1, 3}; $C$ {1, 4} is the high frequency coefficients. The curvelet decomposition of the image is shown in Figure 1.

## 3    Proposed Watermarking Technique

This section describes the proposed watermarking technique based on frequency wrapping based fast discrete curvelet transform and CS theory [29, 30]. This proposed watermarking technique is divided into two procedures such as watermark preparation & embedding and watermark extraction & reconstruction.

### 3.1    Watermark Preparation and Embedding

1. Take a biometric image with size of N×N as a watermark data and compute the size of the image.
2. Generate measurement matrix A with size of N×N using normal distribution with mean =0 and variance = 1. This measurement matrix A is same for embedder and decoder side.
3. Generate sparse measurements of a watermark biometric image by multiplication of the measurement matrix with image pixels.
4. Then multiply sampling factor with sparse measurements of a watermark biometric image which is denoted as $W_{sparse}$. This sampling factor range should be 0.01 to 0.001. This is same for embedder and detector side.
5. Take another biometric image with size of N×N as a host image and compute the size of the image.
6. Then the frequency wrapping based fast discrete curvelet transform with 4 scales and 16 orientation parameter is applied to host image convert into various curvelet coefficients such as $C$ {1,1}, $C$ {1,2}, $C$ {1,3}, $C$ {1,4}. The reason behind choosing in high frequency coefficients is that the size of high frequency curvelet coefficients $C$ {1, 4} is same as size of $W_{sparse}$.
7. Then high frequency curvelet coefficients $C$ {1, 4} of the host image are modified according to values of $W_{sparse}$ and gain factor using Cox equation [31].

$$C_W\{1,4\} = C\{1,4\} * (1 + K \times W_{sparse})   \tag{1}$$

Where $C_W$ {1, 4} = Modified high frequency curvelet coefficients; $C$ {1, 4} = original high frequency curvelet coefficients; $K$ = gain factor which is varying from 0.2 to 0.5.
8. Applied inverse frequency wrapping based fast discrete curvelet transform on modified curvelet coefficients with another unmodified curvelet coefficient to get watermarked biometric image.

## 3.2 Watermark Extraction and Reconstruction

1. Take watermarked biometric image which many corrupted or degrade by the imposter. Then the frequency wrapping based fast discrete curvelet transform (FDCT) with 4 scales and 16 orientation parameter is applied on watermarked biometric image convert into various curvelet coefficients such as $C_W\{1, 1\}$, $C_W\{1, 2\}$, $C_W\{1, 3\}$, $C_W\{1, 4\}$.
2. Take an original host biometric image with size of N×N. Then the frequency wrapping based fast discrete curvelet transform (FDCT) with 4 scales and 16 orientation parameter is applied to host biometric image convert into various curvelet coefficients such as $C\{1, 1\}$, $C\{1, 2\}$, $C\{1, 3\}$, $C\{1, 4\}$.
3. Extracted sparse measurements of a watermark biometric image using the reverse procedure of embedding.

$$W_{Extracted} = \frac{(\frac{C_W\{1,4\}}{C\{1,4\}} - 1)}{K} \tag{2}$$

Where $C_W \{1, 4\}$ = Watermarked high frequency curvelet coefficients; $C \{1, 4\}$ = original high frequency curvelet coefficients; $K$ = gain factor which is varying from 0.2 to 0.5.
4. Then extracted sparse measurements of the watermark biometric image are divided sampling factor which is used at embedder side to get actual sparse measurements of the watermark biometric image. After getting actual sparse measurements of the watermark biometric image, then reconstructed actual biometric image from its sparse measurements using CS theory recovery process.
5. For reconstruction of watermark biometric image, first generate DCT basis with size of N×N and then this DCT basis, multiply by measurement matrix A to generate a Theta matrix.
6. Then apply orthogonal matching pursuit (OMP) [32] algorithm on actual sparse measurements of the watermark biometric image along with Theta matrix. The output of OMP algorithm is sparse coefficients of a watermark biometric image.
7. Finally, the DCT basis matrix is multiplied with sparse coefficients of a watermark biometric image to get reconstructed watermark biometric image at detector side.

## 4    Experimental Results and Discussion

For testing of proposed watermarking technique, 8 bit grayscale face images with size of 128×128 pixels of 50 images from Indian face database [33] and 110 face images from FEI face Database [39] as authentic face images as the host images and 8 bit gray scale fingerprint image with size of 128×128 pixels of 80 images from FVC2002 DB3_ set B [34] and 80 images from FVC2004 DB4_B [34] as watermark is taken. For a generation and embedding of sparse measurements of watermark fingerprint below procedure is adopted.

First generate a DCT basis matrix with size of 128×128 using basic cosine transform in MATLAB. The measurement matrix A with size of 128×128 is generated using normal distribution with mean = 0 and variance = 1. Theta matrix with size of 128×128 is generated by multiplying DCT basis matrix with measurement matrix. The sparse measurements of fingerprint image with size of 128×128 are generated using $W_{128×128}=A_{128×128}×f_{128×128}$. These sparse measurements of fingerprint image are multiplying by sampling factor 0.001 to get $W_{sparse}$. Then these sparse measurements $W_{sparse}$ of fingerprint image are inserted into high frequency curvelet coefficients C $\{1, 4\}_{128×128}$ of host face image with gain factor of 0.2.

For reconstruction of watermark fingerprint image from its actual sparse measurements at the detector, a CS recovery algorithm such as orthogonal matching pursuit (OMP) is applying for actual sparse measurements of watermark fingerprint. The input of OMP algorithm is sparse measurements of watermark fingerprint with size of 128×128; Theta matrix with size of 128×128; level of sparsity is 128. The output of OMP algorithm is sparse coefficients of watermark fingerprint with size of 128×128. Then the DCT basis matrix with size of 128×128 is multiply with sparse coefficients of a watermark fingerprint image to get reconstructed watermark fingerprint image at detector side. Figure 2 shows original & watermarked face image; original & reconstructed watermark fingerprint image; sparse & extracted sparse measurements of watermark fingerprint.

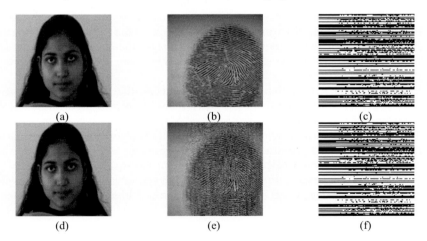

(a)                              (b)                              (c)

(d)                              (e)                              (f)

**Fig. 2** (a) Original Face Image, (b) Watermark Fingerprint Image, (c) Sparse Measurements at Embedder Side, (d) Watermark Face Image (e) Reconstructed Fingerprint Image (f) Extracted Sparse Measurements at Detector Side

This watermarking technique is tested by applying various watermarking attacks such as JPEG compression, addition of noise, applied different image filter, rotation and cropping on it. For quality measure such as peak signal to noise ratio (PSNR), normal cross correlation (NCC) is used for quality check between original face and watermarked face image while the structural similarity index measure (SSIM) is used for quality check between original fingerprint and reconstructed fingerprint image in proposed watermarking technique. The results are summarized in Table 1.

**Table 1** Quality Measures for Proposed Watermarking Technique

| Attacks | NCC | PSNR (dB) | SSIM | Decision |
|---|---|---|---|---|
| No Attack | 1.00 | 66.77 | 0.987 | Authentic |
| JPEG Compression (Q = 90) | 1.00 | 42.59 | 0.005 | Unauthentic |
| Gaussian Noise ($\mu$ =0, $\sigma$=0.001) | 0.99 | 29.88 | 0.004 | Unauthentic |
| Salt & Pepper Noise (Noise Density = 0.005) | 0.99 | 27.93 | 0.012 | Unauthentic |
| Speckle Noise (Variance = 0.004) | 0.99 | 29.81 | 0.012 | Unauthentic |
| Median Filter (size = 3 × 3) | 1.00 | 40.05 | 0.005 | Unauthentic |
| Mean Filter (size = 3 × 3) | 0.99 | 27.18 | 0.003 | Unauthentic |
| Gaussian Low Pass Filter (size = 3 × 3) | 1.00 | 36.40 | 0.004 | Unauthentic |
| Histogram Equalization | 0.98 | 21.18 | 0.021 | Unauthentic |
| Cropping | 0.88 | 17.86 | 0.001 | Unauthentic |

For user authentication, SSIM value between the watermark and extracted watermark fingerprint images must be greater than 0.95. SSIM values in Table 1 are indicated that when attacks is applied on watermarked face image, then the watermark fingerprint image can't reconstruct successfully at detector side. This situation indicated that the proposed watermarking technique is fragile against watermarking attacks.

# 5 Effect of Proposed Watermarking Technique for Performance of Multibiometric System

When designing any template protection based on watermarking technique for multibiometric system, then verification and authentication performance of the system should not be degraded by template protection technique. The watermarked face image and reconstructed fingerprint image are made face-fingerprint based multibiometric system. In this section, analysis of the effect of proposed watermarking techniques for the performance of this multibiometric system is given. For analysis of watermark effect on face image, face matching algorithm developed by Moon and Philips is used [35]. For analysis of CS recovery procedure on fingerprint image, fingerprint matching algorithm developed by Prabhakar and Jain is used [36, 37]. These two algorithms are selected because of there are given the Euclidean distance between the query image and its closest match in the database.

For calculation of verification and authentication performance of face-fingerprint based multibiometric system, 160 watermarked face images and 160 reconstructed fingerprint images are stored in the system database. We have taken 50 images from Indian face database [33] and 110 face images from FEI face

Database [39] as authentic face images, 50 images from FEI face Database and 110 face images from CVL face database [40, 41] as fake face images. We have taken 80 images from from FVC2002 DB3_ set B [34] and 80 images from FVC2004 DB4_B [34] as authentic fingerprint images, 80 images from FVC2002 DB4_B [34] and 80 images from FVC2004 DB3_B as fake fingerprint images. The verification accuracy of face-fingerprint based multibiometric system can be computed using equation 3, 4 and 5. The verification accuracy of multibiometric systems should be changed from 0 to 1.

$$V(Face) = \frac{(No.ofMatching\_Score) < \tau_{selected\_Threshold}}{TotalNo.ofMatching\_Score} \tag{3}$$

$$V(Fingerpr\,int) = \frac{(No.ofMatching\_Score) < \tau_{Selected\_Threshold}}{TotalNo.ofMatching\_Score} \tag{4}$$

$$V(Multibiom\acute{a}ric) = \frac{V(Face) + V(Fingerpr\,int)}{2} \tag{5}$$

Where **Matching_Score** = Euclidean distance between query biometric image, and authenticate biometric image, **V(Face)** = Verification Accuracy of Face System, **V (Fingerprint)** = Verification Accuracy of Fingerprint System, **V (Multibiometric)** = Verification Accuracy of Multibiometric System.

The results of the verification accuracy of proposed watermarking technique for face-fingerprint based multibiometric system is summarized in Table 2. The ROC Curve for verification accuracy of proposed watermarking technique for face-fingerprint based multibiometric system is shown in Figure 3 (a).

**Table 2** Verification Accuracy of Proposed Watermarking Technique for Face-Fingerprint based Multibiometric System

| Threshold | Watermarked Face System | Reconstructed Fingerprint System | Multibiometric System |
|---|---|---|---|
| 0 | 0.000 | 0.000 | 0.000 |
| 0.1 | 0.000 | 0.050 | 0.025 |
| 0.2 | 0.056 | 0.150 | 0.103 |
| 0.3 | 0.569 | 0.344 | 0.456 |
| 0.4 | 0.831 | 0.506 | 0.669 |
| 0.5 | 0.888 | 0.700 | 0.794 |
| 0.6 | 0.938 | 0.831 | 0.884 |
| 0.7 | 0.975 | 0.925 | 0.950 |
| 0.8 | 0.975 | 0.975 | 0.975 |
| 0.9 | 0.988 | 0.994 | 0.991 |
| 1.0 | 1.000 | 1.000 | 1.000 |

For the authentication performance of face-fingerprint based multibiometric system, results obtained by matching algorithm [35 – 37] based on various threshold distance two probabilities with names such as False Rejection Ratio (FRR) and False Acceptance Ratio (FAR) is calculated by equation described by Giot and its research team [38]. The results of authentication performance are summarized in Table 3 and based on this value, plot ROC curve for authentication performance of Face-Fingerprint based multibiometric system is shown in Figure 3(b).

**Table 3** Authentication Performance of Proposed Watermarking Technique for Face-Fingerprint based Multibiometric System

| Threshold | FRR-Face System | FAR-Face System | FRR-Fingerprint System | FAR-Fingerprint System | FRR - Multibiometric System | FAR-Multibiometric System |
|---|---|---|---|---|---|---|
| 0 | 1.000 | 0.000 | 1.000 | 0.000 | 1.000 | 0.000 |
| 0.1 | 0.994 | 0.019 | 0.950 | 0.000 | 0.972 | 0.009 |
| 0.2 | 0.919 | 0.075 | 0.850 | 0.000 | 0.884 | 0.038 |
| 0.3 | 0.425 | 0.131 | 0.656 | 0.000 | 0.541 | 0.066 |
| 0.4 | 0.156 | 0.206 | 0.494 | 0.056 | 0.325 | 0.131 |
| 0.5 | 0.113 | 0.250 | 0.300 | 0.200 | 0.206 | 0.225 |
| 0.6 | 0.063 | 0.338 | 0.169 | 0.481 | 0.116 | 0.409 |
| 0.7 | 0.025 | 0.556 | 0.075 | 0.706 | 0.050 | 0.631 |
| 0.8 | 0.025 | 0.769 | 0.025 | 0.894 | 0.025 | 0.831 |
| 0.9 | 0.013 | 0.963 | 0.001 | 0.981 | 0.007 | 0.972 |
| 1.0 | 0.000 | 1.000 | 0.000 | 1.000 | 0.000 | 1.000 |

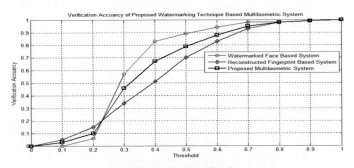

**(a)   ROC Curve for Verification Accuracy**

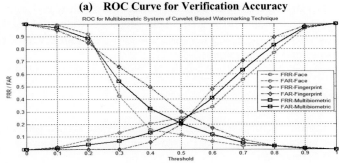

**(b) ROC Curve for Authentication Performance**

**Fig. 3** ROC Curve for Performance of Face-Fingerprint based Multibiometric System

Based on the graphs shown in Figure 3, indicated that ROC Curve of verification and authentication performance of face-fingerprint based multibiometric system with proposed watermarking technique indicated that the proposed watermarking technique provide security to template protection technique without affecting performance of multibiometric system.

The qualitative comparison of the proposed watermarking technique with existing watermarking techniques in literature is as shown in Table 4.

**Table 4** Comparison of Proposed Watermarking Technique with Existing Watermarking Techniques in Literature

| Features | Zhang (2008) et al. [24] | Xu (2010) et al. [25] | Bazargani (2012) et al. [26] | Proposed Watermarking Technique |
|---|---|---|---|---|
| Type | Robust | Robust | Robust | Fragile |
| Used Host Medium | Standard Image | Standard Image | Standard Image | Face Image |
| Used Watermark | Standard Image | Screambled Binary Image | Logo | Sparse Measurements of Fingerprint Image |
| Used Curvelet Coefficients | Low Frequency | Low Frequency | Low Frequency | High Frequency |
| Computational Security Achieved | No such scope | Arnold Transform | No such scope | Compressive Sensing (CS) Theory |
| PSNR (dB) | 43.18 | 39.85 | 60.80 | 66.77 |

# 6 Conclusion

A biometric watermarking technique using the CS theory and high frequency curvelet coefficients for hiding a fingerprint image into face image has been proposed in this paper. This technique has been proposed for face and fingerprint protection in the multibiometric system against various watermarking attacks. The proposed watermarking algorithm does not change verification and authentication performance of multibiometric system. This proposed watermarking technique provides additional security using CS theory with higher PSNR values compared to existing watermarking techniques.

# References

1. Jain, A., Kumar, A.: Biometric recognition: An overview, second generation biometrics: The ethical, legal and social context. In: Mordini, E., Tzovaras, D. (eds.), pp. 49–79. Springer 2012

2. Jain, A., Ross, A., Pankanti, S.: Biometrics: A Tool for Information Security. IEEE Transactions on Information Forensics and Security **1**(2), 125–143 (2006)
3. Ratha, N., Connell, J., Bolle, R.: Enhancing Security and Privacy in Biometric Based Authentication Systems. IBM Systems Journal **40**(3), 614–634 (2001)
4. Hong, L., Jain, A., Pankanti, S.: Can multibiometrics improve performance?. In: Proceedings of AutoID, pp. 59–64 (1999)
5. Jain, A., Ross, A., Prabhakar, S.: An Introduction to Biometric Recognition. IEEE Transactions on Circuits and Systems for Video Technology, Special Issue on Image and Video Based Biometrics **14**(1), 4–20 (2004)
6. Paunwala, M., Patnaik, S.: Biometric Template Protection with DCT based Watermarking. Machine Vision and Applications **25**(1), 263–275 (2013)
7. Jain, A., Uludag, U.: Hiding Biometric Data. IEEE Transactions on Pattern Analysis and Machine Intelligence **25**(11), 1494–1498 (2003)
8. Bedi, P., Bansal, R., Sehgal, P.: Multimodal Biometric Authentication using PSO based Watermarking. Procedia Technology **4**, 612–618 (2012)
9. Zebbiche, K., Khelifi, F., Bouridane, A.: An Efficient Watermarking Technique for the Protection of Fingerprint Images. EURASIP Journal of Information Security, 1–20 (2008)
10. Noore, A., Singh, R., Vatsa, M., Houck, M.: Enhancing security of fingerprint through contextual biometric watermarking. In: Proceedings of Forensic Sci. Int., vol. 169, pp. 188–194 (2008)
11. Vatsa, M., Singh, R., Noore, A.: Robust Biometric Image Watermarking for Fingerprint and Face Template Protection. IEICE Electronics Express **3**(2), 23–28 (2006)
12. Edward, S., Sumanthi, S., Ranihemamalini, R.: Person authentication using multimodal biometrics with watermarking. In Proceedings of 2011 ICSCCN, pp. 100–104 (2011)
13. Feng, G., Lin, Q.: Iris feature based watermarking algorithm for personal identification. In: Proceedings of MIPPR Remote Sensing and GIS Data Proc. Appl., 6790(5), 2007
14. Park, K.R., Jeong, D.S., Kang, B.J., Lee, E.C.: A study on iris feature watermarking on face data. In: Beliczynski, B., Dzielinski, A., Iwanowski, M., Ribeiro, B. (eds.) ICANNGA 2007. LNCS, vol. 4432, pp. 415–423. Springer, Heidelberg (2007)
15. Ko, T.: Multimodal biometric identification for large user population using fingerprint, face and iris recognition. In: Proceedings of the IEEE 34th AIRP 2005 (2005)
16. Giannoula, A., Hatzinakos, D.: Data hiding for multimodal biometric recognition. In: Proceedings of the 2004 IEEE ISCS, pp. 11–16 (2004)
17. Vatsa, M., Singh, R., Mitra, P., Noore, A.: Digital watermarking based secure multimodal biometric system. In: Proceedings of the 2004 IEEE ICSMC, pp. 2983–2987 (2004)
18. Inamdar, V., Rege, P.: Dual Watermarking Technique with Multiple Biometric Watermarks. Sadhana© Indian Academy of Science **29**(1), 3–26 (2014)
19. Motwani, R.: A Voice Based Biometric Watermarking Scheme for Digital Rights Management of 3D Mesh Models, Ph.D. Thesis, University of Nevada, Reno (2010)
20. Vatsa, M., Singh, R., Noore, A.: Feature based RDWT Watermarking for Multimodal Biometric System. Science Direct, Image and Vision Computing **27**, 293–304 (2009)
21. Naik, A., Holambe, R.: Blind DCT Domain Digital Watermarking for Biometric Authentication. International Journal of Computer Applications **16**(1), 11–15 (2010)

22. Cao, Y., Gong, W., Cao, M., Bai, S.: Robust biometric watermarking based on contourlet transform for fingerprint and face protection. In: Proceedings of 2010 IEEE ISPACS, pp. 1–4 (2010)
23. Inamdar, V., Rege, P., Arya, M.: Offline Handwritten Signature based Blind Biometric Watermarking and Authentication Technique Using Biorthogonal Wavelet Transform. International Journal of Computer Applications 11(1), 19–27 (2010)
24. Zhang, C., Cheng, L.L., Zhengding, Q., Cheng, L.M.: Multipurpose Watermarking based on Multiscale Curvelet Transform. IEEE Transactions on Information Forensics and Security 3(4), 611–619 (2008)
25. Xu, J., Pang, H., Zhao, J.: Digital Image Watermarking Algorithm Based on Fast Curvelet Transform. Journal Software Engineering and Applications 3, 939–943 (2010)
26. Bazargani, M., Ebrahimi, H., Dianat, R.: Digital Image Watermarking in Wavelet, Contourlet and Curvelet Domains. Journal of Basic and Applied Scientific Research 2(11), 11296–11308 (2012)
27. Candès, E., Demanet, L., Donoho, D.: Fast Discrete Curvelet Transforms. Applied and Computational Mathematics, pp. 1–44 (2005)
28. Ying, L.: Curve Lab 2.1.2, California Institute of Technology, USA (2005)
29. Candès, E.: Compressive sampling. In: Proceedings of the International Congress of Mathematicians, Madrid, Spain, pp. 1–20, (2006)
30. Donoho, D.: Compressed Sensing. IEEE Transactions on Information Theory 52(4), 1289–1306 (2006)
31. Cox, I., Kilian, J., Shamoon, T., Leighton, F.: Secure Spread Spectrum Watermarking for Multimedia. IEEE Transactions on Image Processing 6(12), 1673–1687 (1997)
32. Tropp, J., Gilbert, A.: Signal Recovery from Random Measurements via Orthogonal Matching Pursuit. IEEE Transactions on Information Theory 53(12), 4655–4666 (2007)
33. Indian Face Database (2002). http://vis-www.cs.umass.edu/~vidit/IndianFaceDatabase
34. Fingerprint Database. http://csr.unibo.it/fvc2004/, http://csr.unibo.it/fvc2002/
35. Moon, H., Phillips, P.: Computational and Performance aspects of PCA based Face Recognition Algorithm. Perception 30, 303–321 (2001)
36. Jain, A., Prabhakar, S., Pankanti, S.: A Filterbank based representation for classification and matching of fingerprints. In: International Joint Conference on Neural Networks, pp. 3284–3285 (1999)
37. Prabhakar, S.: Fingerprint Classification and Matching Using a Filterbank, Ph.D. Thesis, Michigan State University, USA (2001)
38. Giot, R., El-Abed, M., Rosenberger, C.: Fast Computation of the Performance Evaluation of Biometric Systems: Application to Multibiometrics. Future Generation Computer Systems 1, 1–30 (2012)
39. FEI Face Database. http://fei.edu.br/~cet/facedatabase.html
40. Peter Peer, CVL Face Database. http://www.lrv.fri.uni-lj.si/facedb.html
41. Solina, F., Peer, P., Batagelj, B., Juvan, S., Kova, J.: Color-based face detection in the 2015 seconds of fame' art installation. In: Mirage 2003, Conference on Computer Vision / Computer Graphics Collaboration for Model-based Imaging, Rendering, Image Analysis and Graphical special Effects, March 10-11, INRIA Rocquencourt, France, Wilfried Philips, Rocquencourt, INRIA, 2003, pp. 38–47 (2003)

# Cancelable Fingerprint Cryptosystem Based on Convolution Coding

Mulagala Sandhya and Munaga V.N.K. Prasad

**Abstract** In this paper we propose a fingerprint bio-cryptosystem with cancelability using fuzzy commitment scheme. The minutiae of a fingerprint are transformed using Delaunay triangulation net constructed from fingerprint minutiae. Further, these transformed features are encrypted using convolution coding. During decoding phase, the Viterbi algorithm is used to get the codeword. Experimental evaluation done on FVC 2002 databases show the credibility of the proposed method. The EER obtained for the proposed method is 1.66%, 1.89% and 6.87% for FVC 2002 DB1, DB2, and DB3 respectively.

**Keywords** Template protection · Fuzzy commitment scheme · Convolution coding · Viterbi algorithm · Delaunay triangulation

## 1 Introduction

The raising popularity of biometrics and it's widespread use in recent years made template protection in biometrics an important task. Unlike token or knowledge based authentications systems, biometric based authentication systems could not reissue a new one [18]. Due to large intra-class variations, fingerprint template protection is still a challenging task. Template protection schemes are broadly classified into three categories namely features transformation approach, biometric cryptosystem, and hybrid system [4]. In features transformation approach, an irreversible transformation

M. Sandhya(✉) · M.V.N.K. Prasad
Institute for Development and Research in Banking Technology, Road No.1, Castle Hills,
Masab Tank, Hyderabad 500057, India
e-mail: msandhya.phd@gmail.com, mvnkprasad@idrbt.ac.in

M. Sandhya
School of Computer and Information Sciences, University of Hyderabad, Prof. C. R. Rao Road,
Gachibowli, Hyderabad 500046, India

© Springer International Publishing Switzerland 2016
S.M. Thampi et al. (eds.), *Advances in Signal Processing and Intelligent Recognition Systems*,
Advances in Intelligent Systems and Computing 425,
DOI: 10.1007/978-3-319-28658-7_13

145

is applied to biometric data at enrollment stage. At verification stage, the same transformation is applied to the query image. However, the raw image could not be obtained by cross-matching various applications. Biometric cryptosystem binds a cryptographic key to biometric features or generates a key using biometric features. Hybrid systems incorporate both these approaches. The cancelable template is encrypted to generate a secure template. A valid template protection scheme should satisfy four requirements namely diversity, irreversibility, changeability, and accuracy [17]. A fuzzy commitment scheme takes binary biometric feature vector and XORed with a codeword chosen randomly from the error correcting codes. This is stored as a secure template along with the hash value of codeword. During decoding the query binary feature vector is XORed with the stored template. Now a decoding attempt is made to retrieve the codeword [7]. A fuzzy vault represents biometric data as a unordered set of points. Vault is constructed by taking a key which is encoded into a polynomial. The minutiae are evaluated on the polynomial and stored as tuples. Now these tuples are stored among randomly generated chaff points. In this way, the vault is protected [6].

## *Our Contribution*

We computed transformed features using Delaunay triangulation net constructed from fingerprint minutiae. Then a bio-cryptosystem is developed using convolution coding to generate a codeword that is used in fuzzy commitment scheme. The use of Viterbi algorithm to decode the codeword maintained a good balance between security and accuracy of the system. We used Secure Hash Algorithm(SHA-1) to compare the hash value of codeword used for encoding and decoded codeword. Also, proposed bio-cryptosystem is cancelable. The requirements of biometric template protection schemes namely diversity, changeability, irreversibility, accuracy are clearly analyzed in experimental evaluation.

Rest of the paper is organized as follows: Section 2 reviews the literature of existing template protection methods for fingerprints. Section 3 presents the proposed method using convolution coding and decoding in fuzzy commitment scheme. Section 4 provides experiment results, discussion and security analysis for the proposed method. Section 5 describes the comparison of proposed method with the existing methods of fingerprint template protection. Concluding remarks are discussed in Section 6.

## 2   Related Work

Nagar et al. [15]designed method that incorporate minutiae descriptors, which capture ridge orientation and frequency information in a minutia's neighborhood in the vault construction using the fuzzy commitment approach. Yang et al. [21] developed a fingerprint bio-cryptosystem using modified Voronoi neighbor structures(VNS). The modified VNSs are represented as bit strings. An encrypted matching is performed using secure sketch called Pinsketch. Benhammadi et al. [1] built a bio-cryptographic system that combines transformed minutiae pairwise feature and user-generated password fuzzy vault. Isa et al. [3] developed a framework that uses random projections to biometric data by secure keys derived from passwords to generate revocable biometric templates for verification. Mihailescu et al. [12] designed a new enrollment scheme for biometric template based on hash chaos-based cryptography.

Prasad et al. [13] presented a method for fingerprint template protection by constructing minimum spanning tree of minutiae points. Liu et al. [11] developed a fingerprint-based key generation system under the framework of fuzzy extractor by fusing two kinds of features: minutia-based features and image-based features. Minutiae-based sketch, modified Biocode based sketch, and combined feature based sketch are constructed to deal with the feature differences. P.Li et al. [9] built a binary fixed feature generation method of fingerprint then biometric cryptosystems are constructed by using BCH code, a concatenated code of BCH code and Reed–Solomon code, and LDPC code. Prasad et al. [14] built a cancelable fingerprint generation method using multi-neighboring relation. Sandhya et al. [19] designed a method by using k-Nearest Neighborhood Structure(k-NNS)for minutiae points in the fingerprint image. Further, a bit string representation of k-NNS is presented. With the review of existing work, we designed a hybrid system that encypts the cancelable template using convolution coding discussed in section 3.

## 3    Proposed Method

The schematic diagram for proposed method is shown in figure 1. It contains the following steps:

1. Fingerprint image preprocessing and feature extraction
2. Computation of transformed features using Delaunay triangulation net constructed from minutiae points.
3. Representing the transformed features as fixed length bit string
4. Encode the bit string using convolution coding in fuzzy commitment scheme.
5. Decode the codeword from query binary feature vector and Viterbi algorithm.
6. Matching.

### 3.1    Fingerprint Image Preprocessing and Feature Extraction

The preprosessing of fingerprint image before feature extraction is important because it can infuence the speed and accuracy of the entire system. It includes seperation of foreground and background regions, image enhancement, binarization and thinning.

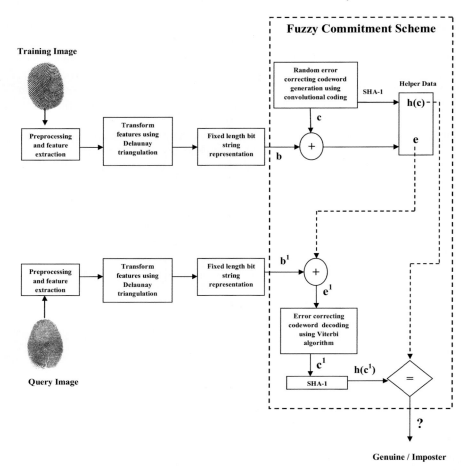

**Fig. 1** Schematic diagram for the proposed method

After preprosessing, We extract minutiae points from the image that are represented as $MP_i = (x_i, y_i, \theta_i)_1^t$, where t is the number of minutiae in the fingerprint image, $(x_i, y_i)$ represent x and y coordinates of minutiae point, $\theta_i$ represent the orientation of minutiae point.

## 3.2 Computation of Transformed Features Using Delaunay Triangulation Net Constructed from Minutiae Points

Figure 2 shows the delaunay triangulation net formed by minutiae points of a fingerprint image. For each triangle in the Delaunay tringulation net, say $\triangle ABC$,

**Fig. 2** Delaunay triangulation formed from fingerprint minutiae('*' represents minutiae point, '+' represents the incircle center of each Delaunay triangle)

we compute 9 features, i.e., the length from minutiae point(vertex) to incircle center($L_A$, $L_B$, $L_C$) of the triangle, original orientation of minutiae points($\theta_A$, $\theta_B$, $\theta_C$) and the internal angles of the triangle($\alpha_A$, $\alpha_B$, $\alpha_C$).

The transformed feature vector, say TFV, computed for a triangle shown in figure 3 is represented as follows:

$$TFV_{ABC} = (L_A, L_B, L_C, \theta_A, \theta_B, \theta_C, \alpha_A, \alpha_B, \alpha_C) \tag{1}$$

Let k be number of triangles in the Delaunay triangulation net. Then the total number of transformed features computed are $k \times 9$ represented as follows:

$$TFV = (TFV_1, TFV_2, TFV_3, .....TFV_k) \tag{2}$$

## 3.3 Representing the Transformed Features as Fixed Length Bit Strings

Take a predefined 3D array, say $A_1$ and divide it into cells of size $c_x$, $c_y$, $c_z$ [8]. The transformed feature vector TFV computed in equation 2 is projected onto $A_1$ by

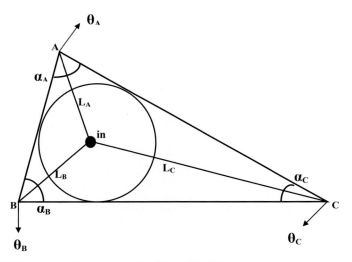

**Fig. 3** $\triangle ABC$ considered for computation of transformed features

taking lengths on x-axis(ranging from 0 to $max(L_i)$), orientation on y-axis(ranging from 0 to $2\pi$ ) and internal angles on z-axis(ranging from 0 to $\pi$). Now find cells in the array that includes the point $(L_i, \theta_i, \alpha_i)$ as follows:

$$\begin{Bmatrix} x_i \\ y_i \\ z_i \end{Bmatrix} = \begin{Bmatrix} \lfloor L_i/c_x \rfloor \\ \lfloor \theta_i/c_y \rfloor \\ \lfloor \alpha_i/c_z \rfloor \end{Bmatrix} \tag{3}$$

where $x_i$, $y_i$, $z_i$ represents indices of $A_1$. If a point or more than one point falls in a cell its value is set to 1 otherwise 0. Now, visit the cells of $A_1$ in a sequential manner. Hence, a fixed length bit string $b$ is obtained. As many-to-one mapping is used for cells in the array, the irreversibility of transformed template is ensured.

### 3.4   Encode the Bit String Using Convolution Coding in Fuzzy Commitment Scheme

### 3.5   Convolution Coding and Decoding

A convolution code is an error correcting code represented by (n,k,m), where n is a number of output bits, k is number of input bits and m is number of memory registers. Convolution code generates parity symbols by the sliding application of a boolean polynomial function to a bit stream [10]. The decoding of convolutional codes can be done by sequential decoding and maximum likelihood decoding. Viterbi algorithm is the best-known implementation of maximum likelihood decoding. Figure 4 shows a (3,1,3) convolution encoder.

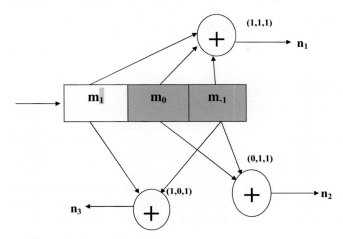

**Fig. 4** A (3,1,3)convolution encoder

## 3.6 Fuzzy Commitment Scheme

A fuzzy commitment scheme [7] of bio-cryptosystem takes a fixed length binary feature vector b as input. In the encoding phase, a random codeword c from error-correcting codes is selected. A random key is taken and is encoded using convolution coding to generate the codeword c. Now an encrypted template 'e' is obtained by performing XOR operation between 'b' and 'c', $e = b \oplus c$. The hash value of c, say h(c) is computed and stored along with e. h(c) and e are termed as helper data. In the decoding phase, a binay feature vector $b^1$ of query image is presented. $b^1$ is XORed with 'e', $e^1 = b^1 \oplus e$. A viterbi decoding algorithm is applied to $e^1$ to get $c^1$. The hash value of $c^1$ is computed. If $h(c) = h(c^1)$ then $c = c^1$. We used Secure Hash Algorithm(SHA-1 ) to compute the hash value of the codeword. This is shown in figure 1.

## 3.7 Matching

Matching is the process of comparing the secure enrolled template(e) and secure query template($e^1$). The hamming measure is the percentage of bits that differ between e and $e^1$. To compute the match score(ms) we used thresholding($th_1$)of hamming measure, say HM as follows:

**if** $HM(e, e^1) < th_1$ **then**
  accept
  ms= $HM(e, e^1)$
**else**
  reject
  ms= $HM(e, e^1)$
**end if**


## 4  Experiment, Analysis and Discussion

The proposed method is tested on FVC 2002 databases [2]. Neurotechnology Verifinger SDK [16] is used to extract minutiae points from the fingerprint images in the databases. These databases contain 800 images of 100 users with 8 samples for each user. We used the first sample of each user for training the system and second sample for testing the system. Hence, 100 genuine matching tests and 9900 imposter matching tests were conducted. For performance evaluation we computed Equal Error Rate(EER) and separability value($d^1$).


### 4.1  Accuracy

The proposed method is tested by tuning the cell sizes $c_x$, $c_y$ and $c_z$ in the quantization step as shown in table 1. For cell sizes $c_x = 15$, $c_y = 15$ and $c_z = 15$ the proposed method show optimal result, i.e. an EER of 2.66%, 1.89% and 6.87% for FVC 2002 DB1, DB2 and DB3 respectively. Also the $d^1$ values shown in the table shows better seperation between genuine and imposter scores.

**Table 1**  EER and $d^1$ values obtained for proposed method

| Cell Sizes of 3D array | | | FVC 2002 | | | | | |
|---|---|---|---|---|---|---|---|---|
| | | | DB1 | | DB2 | | DB3 | |
| $c_x$ | $c_y$ | $c_z$ | EER | $d^1$ | EER | $d^1$ | EER | $d^1$ |
| 10 | 9 | 9 | 3.44 | 2.83 | 2.88 | 2.92 | 10.85 | 2.34 |
| **15** | **15** | **15** | **2.66** | **3.53** | **1.89** | **3.12** | **6.87** | **2.65** |
| 20 | 20 | 20 | 3.43 | 2.83 | 3.34 | 3.06 | 7.55 | 2.71 |
| 25 | 25 | 25 | 3.26 | 3.44 | 1.98 | 3.08 | 7.78 | 2.63 |

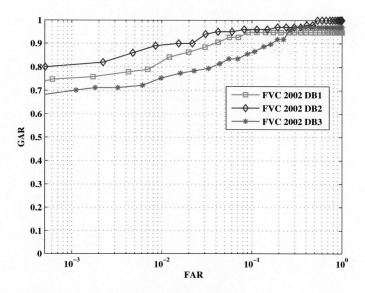

**Fig. 5** ROC curves of proposed method for FVC 2002 databases

Figure 5 and 6 shows the ROC curves and EER obtained for proposed method for databases in FVC 2002 respectively.

## 4.2   Changeability

To test changeability of proposed method, pseudo-imposter distribution is plotted in 7. To obtained pseudo-imposter distribution, we generate 100 transformed templates using the same fingerprint. Now, match the transformed templates with the secure template that is enrolled in training the system. From figure 7, it is noticed that pseudo-imposter distribution is close to imposter distribution and far from genuine distribution. This shows that transformed templates are dissimilar to compromised ones.

## 4.3   Security Analysis

We considered the following cases for performing security analysis of our proposed method

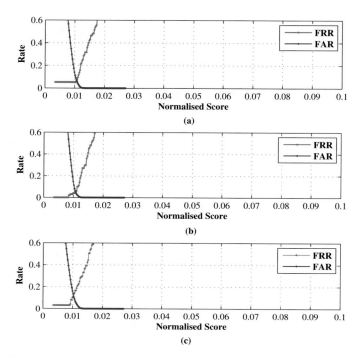

**Fig. 6** EER obtained of proposed method for databases FVC 2002 (a)DB1 (b)DB2 (c)DB3

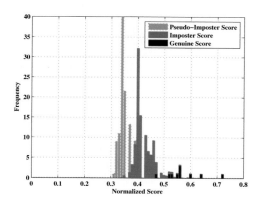

**Fig. 7** Genuine, Imposter, and Pseudo-imposter score distribution of proposed method for database FVC 2002 DB3

- **Case 1** : Assume attacker obtained information stored in the template of a valid user, i.e., the threshold value that is user-specific, the XORed binary vector and the hash value of codeword. It is hard for the intruder to guess the original fingerprint minutiae information from the stored XORed template 'e' because of many-to-one mapping done to a cell in 3D array after quantization of cells. Hence, the proposed feature generation process is irreversible.
- **Case 2** :Hill-climbing procedure described in [20] generate a fixed number of synthetic templates and attack the biometric system, then accumulates corresponding match scores and picks the best guess, modifies the initial set and finds the best score accepted by the matcher. For the proposed method, it is hard to estimate the minutiae locations even through hill-climbing attacks since the transformed feature set does not store the location of minutiae points.

## 5  Performance Comparison with Existing Fingerprint Template Protection Schemes

We compared our result with other existing fingerprint template protection methods. Table 2 shows the comparison for FVC 2002 databases. The proposed method shows better performance in terms of EER compared to [19, 21, 22] for all databases in FVC 2002 and shown better EER for FVC2002 DB1 compared to [5].

**Table 2**  Performance comparision in terms of EER for proposed method

| Method | FVC 2002 | | |
|---|---|---|---|
| | DB1 | DB2 | DB3 |
| Yang et al.(2013) [22] | 5.93 | 4 | - |
| Jin et al. (2014) [5] | 4.36 | 1.77 | - |
| Yang et al.(2014) [21] | 3.38 | 0.59 | 9.8 |
| Sandhya et al.(2015) [19] | 4.71 | 3.44 | 8.79 |
| **Proposed Method** | 1.66 | 1.89 | 6.87 |

## 6  Conclusion

We developed a bio-cryptosystem with cancelability based on Delaunay triangulation constructed from fingerprint minutiae points. The use of convolution coding in fuzzy commitment scheme and Viterbi algorithm for decoding the codeword maintained a good balance between accuracy and security. Also, the bio-cryptosystem developed is cancelable and is analyzed in the experimental results.

# References

1. Benhammadi, F., Bey, K.B.: Password hardened fuzzy vault for fingerprint authentication system. Image and Vision Computing **32**(8), 487–496 (2014)
2. Fingerprint Verification Competetion: http://bias.csr.unibo.it/fvc2002 (accessed January 30, 2014)
3. Isa, M.A.M., Ahmad, M.M., Sani, N.F.M., Hashim, H., Mahmod, R.: Cryptographic key exchange protocol with message authentication codes (mac) using finite state machine. Procedia Computer Science **42**, 263–270 (2014)
4. Jain, A.K., Nandakumar, K., Nagar, A.: Biometric template security. EURASIP J. Adv. Signal Process., 113:1–113:17 (2008)
5. Jin, Z., Lim, M.H., Teoh, A.B.J., Goi, B.M.: A non-invertible randomized graph-based hamming embedding for generating cancelable fingerprint template. Pattern Recognition Letters **42**, 137–147 (2014)
6. Juels, A., Sudan, M.: A fuzzy vault scheme. In: Proceedings of the 2002 IEEE International Symposium on Information Theory, 2002, p. 408 (2002)
7. Juels, A., Wattenberg, M.: A fuzzy commitment scheme. In: Proceedings of the 6th ACM Conference on Computer and Communications Security, CCS 1999, pp. 28–36. ACM (1999)
8. Lee, C., Kim, J.: Cancelable fingerprint templates using minutiae-based bit-strings. Journal of Network and Computer Applications **33**(3), 236–246 (2010)
9. Li, P., Yang, X., Qiao, H., Cao, K., Liu, E., Tian, J.: An effective biometric cryptosystem combining fingerprints with error correction codes. Expert Systems with Applications **39**(7), 6562–6574 (2012)
10. Lin, S., Costello, D.J.: Error control coding. Prentice Hall (1982)
11. Liu, E., Liang, J., Pang, L., Xie, M., Tian, J.: Minutiae and modified biocode fusion for fingerprint-based key generation. Journal of Network and Computer Applications **33**(3), 221–235 (2010). Recent Advances and Future Directions in Biometrics Personal Identification
12. Mihailescu, M.I.: New enrollment scheme for biometric template using hash chaos-based cryptography. Procedia Engineering **69**, 1459–1468 (2014)
13. Prasad, M.V.N.K., Swathi, P., Rao, C.R., Deekshatulu, B.L.: Minimum spanning tree (MST) based techniques for generation of cancelable fingerprint templates. International Journal of Pattern Recognition and Artificial Intelligence **28**(06), 1456013 (2014)
14. Prasad, M.V.N.K., Kumar, C.S.: Fingerprint template protection using multiline neighboring relation. Expert Systems with Applications **41**(14), 6114–6122 (2014)
15. Nagar, A., Nandakumar, K., Jain, A.K.: A hybrid biometric cryptosystem for securing fingerprint minutiae templates. Pattern Recognition Letters **31**(8), 733–741 (2010)
16. Neurotechnology VeriFinger SDK: http://www.neurotechnology.com (accessed December 11, 2013)
17. Ratha, N., Chikkerur, S., Connell, J., Bolle, R.: Generating cancelable fingerprint templates. IEEE Transactions on Pattern Analysis and Machine Intelligence **29**(4), 561–572 (2007)
18. Ratha, N., Connell, J., Bolle, R.: An analysis of minutiae matching strength. In: Audio- and Video-Based Biometric Person Authentication. Lecture Notes in Computer Science, vol. 2091, pp. 223–228. Springer, Heidelberg (2001)
19. Sandhya, M., Prasad, M.V.N.K.: k-nearest neighborhood structure (k-NNS) based alignment-free method for fingerprint template protection. In: International Conference on Biometrics (ICB), 2015, pp. 386–393 (2015)

20. Uludag, U., Jain, A.K.: Attacks on biometric systems: a case study in fingerprints. In: Delp III, E.J., Wong, P.W. (eds.) Security, Steganography, and Watermarking of Multimedia Contents VI. Society of Photo-Optical Instrumentation Engineers (SPIE) Conference Series, vol. 5306, pp. 622–633 (2004)
21. Yang, W., Hu, J., Wang, S., Stojmenovic, M.: An alignment-free fingerprint bio-cryptosystem based on modified voronoi neighbor structures. Pattern Recognition **47**(3), 1309–1320 (2014)
22. Yang, W., Hu, J., Wang, S., Yang, J.: Cancelable fingerprint templates with delaunay triangle-based local structures. In: Cyberspace Safety and Security. Lecture Notes in Computer Science, vol. 8300, pp. 81–91. Springer International Publishing (2013)

# A Bio-cryptosystem for Fingerprints Using Delaunay Neighbor Structures(DNS) and Fuzzy Commitment Scheme

Mulagala Sandhya and Munaga V.N.K. Prasad

**Abstract** The emergence of biometrics and it's widespread use in authentication applications made template protection in biometrics an important task to be considered in recent years. In this paper we propose a fingerprint bio-cryptosystem using fuzzy commitment scheme. We constructed Delaunay Neighbor Structures(DNS) from fingerprint minutiae points. Then a bit string representation of DNSs was developed. Bose, Chaudhuri, and Hocquenghem(BCH) error correction code is used along with bit string in the fuzzy commitment scheme to generate a secure template. During decoding phase, BCH decoding is performed to get the codeword. The EER obtained for proposed method on FVC 2002 DB1, DB2 and DB3 are 1.43%, 1.79%, and 5.89% respectively. This shows the credibility of the proposed method.

**Keywords** BCH code · Delaunay triangulation · Template protection · Fuzzy commitment scheme

## 1 Introduction

In traditional cryptographic systems, a user is authenticated by a token or knowledge such as pin or password. As tokens may be lost or passwords may be forgotten/guessed, a convenient technology called biometric recognition replaced the traditional systems. But the major drawback that lied in biometric-based authentication systems is a biometric can't be replaced like a token or password [16].

M. Sandhya(✉) · M.V.N.K. Prasad
Institute for Development and Research in Banking Technology, Road No.1, Castle Hills,
Masab Tank, Hyderabad 500057, India
e-mail: msandhya.phd@gmail.com, mvnkprasad@idrbt.ac.in

M. Sandhya
School of Computer and Information Sciences, University of Hyderabad, Prof. C. R. Rao Road,
Gachibowli, Hyderabad 500046, India

© Springer International Publishing Switzerland 2016
S.M. Thampi et al. (eds.), *Advances in Signal Processing and Intelligent Recognition Systems*,
Advances in Intelligent Systems and Computing 425,
DOI: 10.1007/978-3-319-28658-7_14

159

A biometric template without protection introduces several security and privacy threats. Hence, template protection schemes such as cancelable biometrics, fuzzy extractors, fuzzy vault and fuzzy commitment were developed [3]. Fingerprint template protection is still a challenging task because of noise introduced while capturing image, finger placement and large intra-class variations. Template protection schemes of biometrics fall under three categories: Features transformation, biometric cryptosystem, and hybrid system [3]. A template protection method should satisfy four requirements: diversity, revocability, irreversibility, and accuracy. Bio-cryptography which combines research in cryptography and biometrics provides security by binding cryptographic key to biometric features or generating the key using biometric features. The physical identity of a person is stored in encrypted domain [19]. A binary biometric feature vector is XORed with a codeword chosen randomly from the error correcting codes in fuzzy commitment scheme. This is stored as a secure template along with the hash value of codeword. During decoding the query binary feature vector is XORed with the stored template. Now a decoding attempt is made to retrieve the codeword [6]. In the fuzzy vault, biometric data is represented as a unordered set of points. Vault is constructed by taking a key that is encoded into a polynomial. The minutiae are evaluated on the polynomial and stored as tuples. Now these tuples are stored among randomly generated chaff points [5].

## *Our Contribution*

We developed a bio-cryptosystem by computing Delaunay Neighbor Structures and then combining them with error correction code in fuzzy commitment scheme. We used Secure Hash Algorithm(SHA-1) to compare the hash value of codeword used for encoding and decoded codeword. The requirements of biometric template protection schemes namely diversity, changeability, irreversibility, accuracy are clearly analyzed in experimental evaluation.

Rest of the paper is organized as follows: Section 2 reviews the literature of existing template protection methods. Section 3 presents the proposed method using BCH coding in fuzzy commitment scheme. Section 4 shows the experiment evaluation and analysis for the proposed method. Section 5 describes the comparison of proposed method with the existing methods of fingerprint template protection. Concluding remarks are discussed in Section 6.

## 2  Related Work

Arathi et al. [1] used bit strings generated applied to an existing secure sketch. Global features and local features are extracted and used in this scheme. Li et al. [9] used minutiae descriptors and minutiae local structure then fused them in the fuzzy vault construction. Yang et al. [19] developed bio-cryptosystem by constructing modified

Voronoi neighbor structures(VNS). From modified VNS, a fixed length bit string is obtained. Further encrypted matching is performed by using fuzzy vault construction using Pinsketch that uses syndrome encoding. Prasad et al. [14] generated cancelable templates for fingerprints by drawing M rectangles around each reference minutiae and generating a multi-neighboring relation. Leng et al. [8] designed a texture coding approach for palmprint, dubbed row-alone and row-co-occurrence fuzzy vaults. Isa et al. [2] developed a framework that applies random projections to biometric data using secure keys derived from passwords to generate revocable biometric templates. As described in Prasad et al. [13] cancelable fingerprints can be generated using Minimum Spanning Tree constructed from fingerprint minutiae. Liu et al. [12] developed a fingerprint-based key generation system under the framework of fuzzy extractor by fusing two kinds of features: minutia-based features and image-based features. Three types of sketch, including minutiae based sketch, modified Biocode based sketch, and combined feature based sketch, are constructed to deal with the feature differences. P.Li et al. [10] built a binary fixed feature generation method of the fingerprint. The binary feature vector is used in the fuzzy commitment scheme and a biometric cryptosystem is built by combining various error correction codes. Sandhya et al. [17] computed k-Nearest Neighborhood Structure(k-NNS)for minutiae points in the fingerprint image and represented it as bit strings to generate cancelable templates. With the review of existing work, we developed a fingerprint cryptosystem by computing a novel structure called Delaunay Neighbor Structure(DNS) and combined it with BCH error correcting code discussed in section 3.

## 3    Proposed Method

The flow diagram for the proposed method is shown in figure 1. It contains the following steps:

1. Computation of Delaunay Neighbor Structures(DNS) from fingerprint minutiae
2. Computation of bit string from DNS
3. Apply Dimensianality reduction matrix to the bit string
4. By using BCH code in fuzzy commitment scheme generate a secure template
5. Decode the codeword from query binary feature vector in BCH decoding
6. Matching

### 3.1    Computation of Delaunay Neighbor Structures(DNS) from Fingerprint Minutiae

The fingerprint minutiae points are extracted from fingeprint image and represented as $MP_i = (x_i, y_i, \theta_i)_1^t$, where t is the number of minutiae in the fingerprint image, $(x_i, y_i)$ represent x and y coordinates of minutiae point, $\theta_i$ represent the orientation

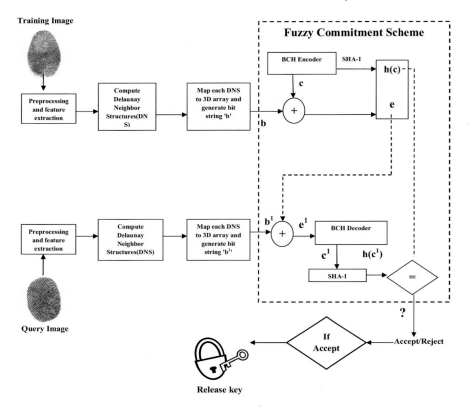

**Fig. 1** Flow diagram for the proposed method

of minutiae point. From the minutiae points construct a Delaunay tringulation net as shown in figure 2. The Delaunay Neighbor Structure(DNS) centered around minutiae 'r' contains neighboring minutiae as shown in figure 3(a). This is represented as follows:

$$DNS_r = (r, a, b, c, d, e, f) \qquad (1)$$

For all minutiae points in the fingerprint image compute the DNS. Hence the DNS for entire image is represented as follows:

$$DNS = (DNS_1, DNS_2, DNS_3.....DNS_t) \qquad (2)$$

where 't' is number of minutiae points in the image.

**Fig. 2** Delaunay triangulation formed from fingerprint minutiae and DNS centered around minutia point 'r'

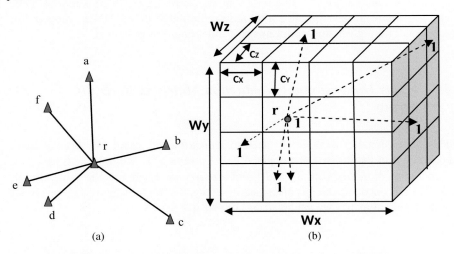

**Fig. 3** (a)DNS centered around minutiae point 'r' (b)3D array to which DNS is mapped

## 3.2    Computation of Bit String for Each DNS

We take a 3D array of size $W_x$, $W_y$, $W_z$ as shown in figure 3(b), where $W_x$ ranges from 0 to $max$(x-coordinate value), $W_y$ ranges from 0 to $max$(y-coordinate value), $W_z$ ranges from 0 to $2\pi$. Divide the array into cells of size $c_x$, $c_y$, $c_z$ [7]. The central minutiae 'r' of each DNS is taken as reference. It is located at center of first layer of 3D array, i.e., $(W_x/2, W_y/2, 0)$. The remaining minutiae in the DNS are rotated and transformed to $(x_i^1, y_i^1, \theta_i^1)$ as follows:

$$\begin{bmatrix} x_i^1 \\ y_i^1 \end{bmatrix} = \begin{bmatrix} cos\theta_r & -sin\theta_r \\ sin\theta_r & cos\theta_r \end{bmatrix} \begin{bmatrix} (x_i - x_r) \\ -(y_i - y_r) \end{bmatrix} + \begin{bmatrix} W_x/2 \\ W_y/2 \end{bmatrix} \tag{3}$$

$$\theta_i^1 = \begin{Bmatrix} \theta_i - \theta_r & if \theta_i \geq \theta_r \\ 2\pi + \theta_i - \theta_r & if \theta_i < \theta_r \end{Bmatrix} \tag{4}$$

Now a quantization process is applied by the cell sizes $c_x, c_y, c_z$ to $(x_i^1, y_i^1, \theta_i^1)$ as follows:

$$\begin{bmatrix} x_i^{11} \\ y_i^{11} \\ \theta_i^{11} \end{bmatrix} = \begin{bmatrix} x_i^1/c_x \\ y_i^1/c_y \\ \theta_i^1/c_z \end{bmatrix} \tag{5}$$

Now we project $(x_i^{11}, y_i^{11}, \theta_i^{11})$ onto 3D array. If a point or more than one point falls in a cell its value is set to 1 otherwise 0 as shown in figure 3(b). Visiting the cells in a sequential manner produce a bit string. Similarly, the 't' DNSs computed in equation 2 for a given fingerprint image 't' bit strings are generated.

## 3.3    Apply Dimensionality Reduction Matrix to Each Bit String

This dimensionality reduction is used to reduce the bit string length to fit BCH encoding. Let the bit strings computed are of size $m \times 1$. Multiply each bit string with a random matrix(with 0's and 1's)of size $n \times m$, where n is the value used in [n,k,t] BCH encoder which is discussed in section 3.4. Now by applying a threshold to the resultant matrix we get bit strings of size $n \times 1$.

## 3.4    Encode Each Bit String Using BCH Code in Fuzzy Commitment Scheme

The Bose, Chaudhuri, and Hocquenghem (BCH) is a random error-correcting cyclic code represented as BCH(n,k,t), where n is the codeword length, t is the error correction capability [11]. For prime power 'q', positive integers 'm' and 'd' such that $d <= q^m - 1$, BCH code defined over finite field GF(q) with code length $n = q^m - 1$ and distance atleast 'd' is given as follows:

1. Let a be a primitive element of $GF(q^m)$.
2. Let $m_i(x)$ be the minimal polynomial of $a_i$ over GF(q) for positive integer 'i'.
3. The generator polynomial of the BCH code is given by the least common multiple $g(x) = LCM(m_1(x), ..., m_{d-1}(x))$.
4. Now, g(x) is a polynomial with coefficients in GF(q) and divides $x^n - 1$. Hence, the polynomial code defined by g(x) is cyclic. This is called narrow sense BCH code.

Decoding can be done by calculating syndrome values for recived vector. A fuzzy commitment scheme [6] takes a binary feature vector b as input. In the encoding phase, a random key is taken and encoded using BCH coding to produce a codeword c. Now an encrypted template e is obtained by performing XOR operation between b and c, $e = b \oplus c$. The hash value of c, say h(c) is computed and stored along with e. For our method we used cryptographic hash function SHA-1 to compute the hash value. In the decoding phase, a binay feature vector $b^1$ of query image is presented. $b^1$ is XORed with e, $e^1 = b^1 \oplus e$. A decoding algorithm is applied to $e^1$ to get $c^1$. The hash value of $c^1$ is computed. If $h(c) = h(c^1)$ then $c = c^1$ as shown in figure 1.

## 3.5   Matching

Let 't' be the number of minutiae in the training image. Hence, 't' DNSs are formed in training image to produce 't' bit strings. The 't' bit strings are encoded using BCH code to produce 't' secure templates represented as $T = (T_1, T_2...T_t)$. Let 'q' be the number of minutiae in the testing image. Hence 'q' DNSs formed gets 'q' bit strings to produce 'q' secure templates represented as $\vartheta = (\vartheta_1, \vartheta_2, ...\vartheta_q)$. Matching is the process of comparing the $T$ and $\vartheta$. The hamming measure, say HM is the percentage of bits that differ between $T_i$ and $\vartheta_j$. We employ a local matching procedure as shown in figure 4. The local score between $T_i$ and $\vartheta_j$ can be given by

$$LS(T_i, \vartheta_j) = HM(T_i, \vartheta_j) \tag{6}$$

We matched every $T_i$ in $T = (T_1, T_2...T_t)$ with every $\vartheta_j$ in $\vartheta = (\vartheta_1, \vartheta_2, ...\vartheta_q)$ that results in a similarity matrix as in equation 7.

$$S(T_i, \vartheta_j) = \begin{cases} max(LS(T_i, \vartheta_j)) \ \forall i \epsilon [1, t], \forall j \epsilon [1, q] \\ 0 \qquad\qquad\qquad Otherwise \end{cases} \tag{7}$$

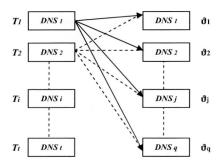

**Fig. 4** Local matching between encrypted templates of trained and testing images

From this similarity matrix, the final matching score is generated by equation 8.

$$MS = \frac{\sum_{i=1}^{t} \sum_{j=1}^{q} S(T_i, \vartheta_j)}{\eta} \tag{8}$$

where $\eta$ represents the number of non-zero elements in $S(T_i, \vartheta_j)$.

## 4  Experiment, Analysis and Discussion

### 4.1  Experiment Setup and Performance Measures

Neurotechnology Verifinger SDK [15] is used to extract minutiae points from the fingerprint images in the databases. The proposed method is tested on databases in FVC 2002 with different values of (n,k,t) of BCH encoder. These databases contain 800 images of 100 users with 8 samples for each user. We used the first sample of each user for training the system and second sample for testing the system. Hence, 100 genuine matching tests and 9900 imposter matching tests were conducted. For performance evaluation, we considered Equal Error Rate(EER).

### 4.2  Accuracy

We tested our proposed method by tuning the cell sizes $c_x$, $c_y$ and $c_z$ in the quantization step and different values of (n,k,t) in BCH encoding as shown in table 1. For cell sizes $c_x = 20$, $c_y = 20$ and $c_z = 20$ the proposed method show optimal result, i.e. an EER of 1.43%, 1.79% and 5.89% for FVC 2002 DB1, DB2 and DB3 respectively for (1023,11,255) BCH encoder.

**Table 1** EER obtained for proposed method for different values of (n,k,t) of BCH encoder

| (n,k,t) | (255,63,33) | | | (511,259,30) | | | (1023,11,255) | | |
|---|---|---|---|---|---|---|---|---|---|
| $c_x$  $c_y$  $c_z$ | DB1 | DB2 | DB3 | DB1 | DB2 | DB3 | DB1 | DB2 | DB3 |
| 15 15 15 | 4.64 | 5.13 | 14.96 | 6.57 | 5.89 | 12.41 | 4.12 | 4.15 | 11.87 |
| 15 20 15 | 5.76 | 7.67 | 15.87 | 5.13 | 6.12 | 17.97 | 3.84 | 3.44 | 11.86 |
| **20 20 20** | **3.26** | **3.98** | **9.85** | **3.14** | **3.48** | **8.89** | **1.43** | **1.79** | **5.89** |
| 20 25 20 | 5.14 | 7.63 | 10.43 | 4.63 | 4.93 | 14.96 | 1.89 | 2.12 | 6.89 |
| 25 25 25 | 6.57 | 5.89 | 12.43 | 5.74 | 7.63 | 15.82 | 2.98 | 3.02 | 7.16 |

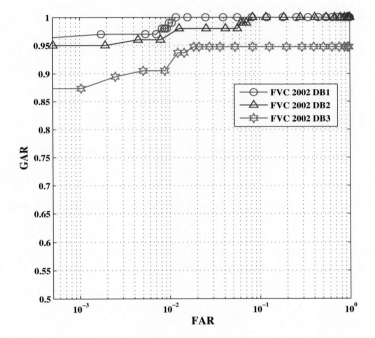

**Fig. 5** ROC curves of proposed method for FVC 2002 databases

Figure 5, 6 and 7 depicts the ROC curves, EER obtained and score distribution for proposed method of FVC 2002 databases respectively.

## 4.3 Revocability

The Dimensionality reduction matrix used in section 3.3 reduces the size of the bit string. This is revocable. The use of a key to generate the codeword through BCH

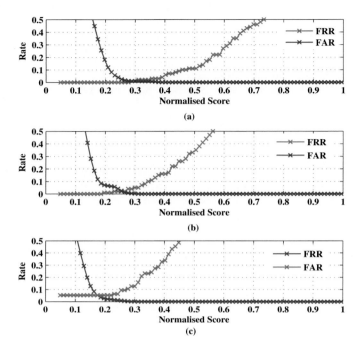

**Fig. 6** EER obtained of proposed method for databases FVC 2002 (a)DB1 (b)DB2 (c)DB3

encoding is also revocable. Hence for same fingerprint image we can get many secure templates that are different from one another.

## 4.4 Security Analysis

Assume that the intruder obtained information stored in the template of the valid user, i.e., the threshold value that is user-specific, the XORed binary vector and the hash value of codeword. It is hard for him to guess the original fingerprint minutiae information because a many-to-one mapping is done to a cell in 3D array after quantization of cells. Also, the dimensionality reduction matrix is user specific. Hence, the proposed feature generation process is irreversible. The hill-climbing method described in [18] estimates the minutiae locations by generating synthetic templates. For the proposed method, it is hard to estimate the minutiae locations since the bit string generated is reduced in size through Dimensionality reduction matrix and hence does not store the location of minutiae points.

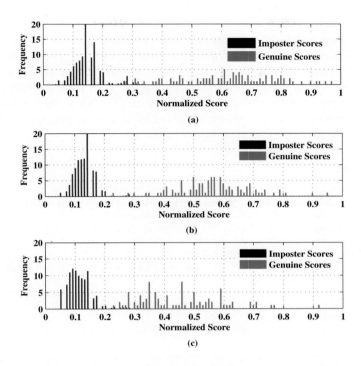

**Fig. 7** Genuine and imposter score distribution of proposed method for databases FVC 2002 (a)DB1 (b)DB2 (c)DB3

**Table 2** Performance comparison in terms of EER for proposed method

| Method | FVC 2002 | | |
|---|---|---|---|
| | DB1 | DB2 | DB3 |
| Yang et al.(2013) [20] | 5.93 | 4 | - |
| Jin et al. (2014) [4] | 4.36 | 1.77 | - |
| Yang et al.(2014) [19] | 3.38 | 0.59 | 9.8 |
| Prasad et al.(2014) [13] | 1.8 | 1.2 | - |
| Sandhya et al.(2015) [17] | 4.71 | 3.44 | 8.79 |
| **Proposed Method** | 1.43 | 1.79 | 5.89 |

## 5  Performance Comparison with Existing Fingerprint Template Protection Schemes

Table 2 shows the comparison for FVC 2002 databases. A better EER is obtained by the proposed method compared to [17, 19, 20] for all databases in FVC 2002 and shown better EER for FVC2002 DB1 compared to [4, 13].

# 6  Conclusion

A fingerprint bio-cryptosystem is developed by computing a novel structure called Delaunay Neighbor Structures(DNS)from the minutiae points of a fingerprint image. Dimensionality reduction matrix was used to reduce the size of the bit string to be XORed with error correcting BCH code. The use BCH error correcting code improved the performance of system thereby reducing the Equal Error Rate(EER). Also, the proposed system satisfies the requirements of template protection schemes and is clearly analyzed in the experimental evaluation.

# References

1. Arakala, A., Jeffers, J., Horadam, K.: Fuzzy extractors for minutiae-based fingerprint authentication. In: Advances in Biometrics, vol. 4642, pp. 760–769 (2007)
2. Isa, M.A.M., Ahmad, M.M., Sani, N.F.M., Hashim, H., Mahmod, R.: Cryptographic key exchange protocol with message authentication codes (mac) using finite state machine. Procedia Computer Science **42**, 263–270 (2014)
3. Jain, A.K., Nandakumar, K., Nagar, A.: Biometric template security. EURASIP J. Adv. Signal Process., 113:1–113:17 (2008)
4. Jin, Z., Lim, M.H., Teoh, A.B.J., Goi, B.M.: A non-invertible randomized graph-based hamming embedding for generating cancelable fingerprint template. Pattern Recognition Letters **42**, 137–147 (2014)
5. Juels, A., Sudan, M.: A fuzzy vault scheme. Designs, Codes and Cryptography **38**(2), 237–257 (2006)
6. Juels, A., Wattenberg, M.: A fuzzy commitment scheme. In: Proceedings of the 6th ACM Conference on Computer and Communications Security, CCS 1999, pp. 28–36. ACM (1999)
7. Lee, C., Kim, J.: Cancelable fingerprint templates using minutiae-based bit-strings. Journal of Network and Computer Applications **33**(3), 236–246 (2010)
8. Leng, L., Teoh, A.B.J.: Alignment-free row-co-occurrence cancelable palmprint fuzzy vault. Pattern Recognition **48**(7), 2290–2303 (2015)
9. Li, P., Yang, X., Cao, K., Tao, X., Wang, R., Tian, J.: An alignment-free fingerprint cryptosystem based on fuzzy vault scheme. Journal of Network and Computer Applications **33**(3), 207–220 (2010)
10. Li, P., Yang, X., Qiao, H., Cao, K., Liu, E., Tian, J.: An effective biometric cryptosystem combining fingerprints with error correction codes. Expert Systems with Applications **39**(7), 6562–6574 (2012)
11. Lin, S., Costello, D.J.: Error control coding. Prentice Hall (1982)
12. Liu, E., Liang, J., Pang, L., Xie, M., Tian, J.: Minutiae and modified biocode fusion for fingerprint-based key generation. Journal of Network and Computer Applications **33**(3), 221–235 (2010)
13. Prasad, M.V.N.K., Swathi, P., Rao, C.R., Deekshatulu, B.L.: Minimum spanning tree (MST) based techniques for generation of cancelable fingerprint templates. International Journal of Pattern Recognition and Artificial Intelligence **28**(06), 1456013 (2014)
14. Prasad, M.V.N.K., Kumar, C.S.: Fingerprint template protection using multiline neighboring relation. Expert Systems with Applications **41**(14), 6114–6122 (2014)
15. Neurotechnology VeriFinger SDK: http://www.neurotechnology.com (accessed December 11, 2013)
16. Ratha, N., Chikkerur, S., Connell, J., Bolle, R.: Generating cancelable fingerprint templates. IEEE Transactions on Pattern Analysis and Machine Intelligence **29**(4), 561–572 (2007)

17. Sandhya, M., Prasad, M.V.N.K.: k-nearest neighborhood structure (k-NNS) based alignment-free method for fingerprint template protection. In: International Conference on Biometrics (ICB), 2015, pp. 386–393 (2015)
18. Uludag, U., Jain, A.K.: Attacks on biometric systems: a case study in fingerprints. In: Security, Steganography, and Watermarking of Multimedia Contents VI, vol. 5306, pp. 622–633 (2004)
19. Yang, W., Hu, J., Wang, S., Stojmenovic, M.: An alignment-free fingerprint bio-cryptosystem based on modified voronoi neighbor structures. Pattern Recognition **47**(3), 1309–1320 (2014)
20. Yang, W., Hu, J., Wang, S., Yang, J.: Cancelable fingerprint templates with delaunay triangle-based local structures. In: Cyberspace Safety and Security, vol. 8300, pp. 81–91 (2013)

# Part II
# Image/ Video Processing

# Improving the Feature Stability and Classification Performance of Bimodal Brain and Heart Biometrics

Ramaswamy Palaniappan, Samraj Andrews, Ian P. Sillitoe, Tarsem Shira and Raveendran Paramesran

**Abstract** Electrical activities from brain (electroencephalogram, EEG) and heart (electrocardiogram, ECG) have been proposed as biometric modalities but the combined use of these signals appear not to have been studied thoroughly. Also, the feature stability of these signals has been a limiting factor for biometric usage. This paper presents results from a pilot study that reveal the combined use of brain and heart modalities provide improved classification performance and further-more, an improvement in the stability of the features over time through the use of binaural brain entrainment. The classification rate was increased, for the case of the neural network classifier from 92.4% to 95.1% and for the case of LDA, from 98.6% to 99.8%. The average standard deviation with binaural brain entrainment using all the inter-session features (from all the subjects) was 1.09, as compared to

R. Palaniappan(✉)
School of Computing, University of Kent, Chatham, UK
e-mail: r.palani@kent.ac.uk

S. Andrews
Department of Information Technology, Mahendra Engineering College, Salem, India
e-mail: andrewsmalacca@gmail.com

I.P. Sillitoe
School of Engineering, University of Wolverhampton, Telford, UK
e-mail: I.Sillitoe@wlv.ac.uk

T. Shira
Department of Engineering and Mathematics, Sheffield Hallam University, Sheffield, UK
e-mail: T.H.Sihra@shu.ac.uk

R. Paramesran
Department of Electrical Engineering, University of Malaya, Kuala Lumpur, Malaysia
e-mail: ravee@um.edu.my

© Springer International Publishing Switzerland 2016
S.M. Thampi et al. (eds.), *Advances in Signal Processing and Intelligent Recognition Systems*,
Advances in Intelligent Systems and Computing 425,
DOI: 10.1007/978-3-319-28658-7_15

1.26 without entrainment. This result suggests the improved stability of both the EEG and ECG features over time and hence resulting in higher classification performance. Overall, the results indicate that combining ECG and EEG gives improved classification performance and that through the use of binaural brain entrainment, both the ECG and EEG features are more stable over time.

## 1    Introduction

Identification and authentication of individuals using various biometric modalities is an active research area due to its importance in everyday activities such as banking transactions and computer logins. In addition, the worldwide interest in security has further raised the importance of this topic.

Some of the common modalities used in biometrics are fingerprint, palmprint, face and iris [1], but other less common biometric modalities such as keystroke dynamics [2], gait [3] and ear shape [4] have also been proposed. More recently, biometrics based on the electrical activity of the heart (electrocardiogram, ECG) [5-7] and brain (electroencephalogram, EEG) [7-9] have emerged. Traditionally, ECG and EEG are used for medical diagnosis but these have also found use in biometrics, since they have been shown to be less prone to counterfeit (in supervised conditions such as in research lab environments).

An overview of the use of non-fiducial features such as Hjorth parameters, fiducial features and hybrid combinations of both for ECG biometrics have been explored using various classifiers, such as nearest neighbour and neural networks in [6, 7]. When compared to EEG, ECG is easier to record since in practice most biometric studies use only a single lead signal (from two active electrodes) [5], whereas EEG can require up to 64 channels [8]. ECG signals vary with the physical conditions under which the readings are taken, and thus recording is normally restricted to resting conditions only. On the other hand, EEG recording must be conducted under specific mental conditions [10] to reduce their variability. Typical features derived from EEG signals for biometric applications include spectral [8] and autoregressive (AR) [10] features.

Most of the previous studies have investigated the use of ECG and EEG features separately rather than employing them in combination. However, given that neural responses (i.e. EEG) also influence the cardiac rhythms (i.e. ECG) it would seem the use of their combined use might lead to improved classification rates. However, whilst it is generally known that ECG and EEG biometric offer fraud resistance due to their uniqueness for each individual, the features obtained from such measures are not stable over time [6, 11], which is perhaps why most of the studies only report performance within sessions rather than inter-sessions. For example, in [12], the authors have studied using ECG and EEG for biometrics that gave perfect reliability but it is not clear whether the system's classification rate remained as high after a lapse of time. Another study [13] successfully utilised wavelet features from EEG recorded over a short period of two weeks but as the patterns were randomised for training and testing, stability of features would not possibly been established.

Therefore, this study has dual objectives. Firstly, to investigate the use of combined ECG and EEG features to improve the classification performance. Secondly, to investigate the use of a novel application of binaural brain entrainment to minimise the inter-session variability of the ECG and EEG features. It is first shown that combining AR and Hjorth features from ECG with EEG energies in the alpha and gamma bands, provides improved classification performance for both multilayer perceptron neural network (NN) and linear discriminant analysis (LDA) classifiers. Next, the variability of these features over time is shown to be less when the ECG and EEG signals are recorded under binaural brain entrainment, and this reduction in variability leads to even higher classification rates.

## 2    Methodology

Data from five subjects (one female and four males within the age range of 24-39) were recorded using a Biosemi Active Two device [14], over six sessions separated by monthly intervals. The study received ethical approval and the subjects signed consent forms after being briefed on the objective of the study. Subjects were paid a small honorarium for their time and agreed to attend six monthly sessions[1]. ECG data were recorded from left and right wrists whilst the EEG data were obtained from 19 locations (based on the standard 10-20 location [15]), as shown in Figure 1. In addition, two mastoid electrode locations were used as reference channels and used as described in another study [11]. Scalp and wrist preparation were not required due to the use of active electrodes, but water based gel was used to increase the contact between the scalp/skin and electrodes. The single lead ECG signal was obtained by subtracting the data from the two wrist electrodes, whilst the 19 channel EEG signals were derived by using the average of the mastoid channels as reference.

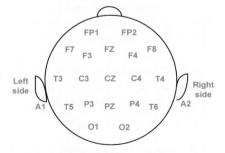

**Fig. 1** EEG electrode locations used in this study

---

[1] Many subjects dropped out in the initial stage as they could not commit for the six months period. Hence, the five subjects were the only available test population.

The sampling frequency was 256 Hz, which is sufficient to avoid aliasing in both the ECG and EEG signals. Subjects were asked to sit comfortably with their eyes closed and data obtained for two minutes each in the following conditions:

- Listening to relaxing music (i.e. waves hitting the beach); this is the control condition
- Listening to the same music but with masked binaural tones; this is the entrained condition

Two minutes recording was deemed sufficient to obtain responses based on a previous study [16]. Subjects were blind to the conditions (i.e. the subjects did not know which was the entrained and which was the control condition, as the music was the same in both cases) and the order of the conditions were alternated during the six monthly sessions. The music played was the same during each of the sessions, so as to reduce any effect the actual choice of music might have upon the classification performance. The binaural tones were generated by using two sinusoidal waves, one with frequency of 400 Hz (presented to the left ear) and another with frequency of 408 Hz (presented to the right ear). The tones were presented using Etymotic flat frequency response stereo ear phones (to avoid any spectral attenuation) with disposable ear plugs [17] and they were masked by the 'waves' music, hence they were not heard in the ordinary sense. It is known that such binaural tones can evoke a third pseudo-tone in the brain that differs in frequency of the evoking tones [18]. Hence, the brain will perceive also hearing a tone at 8 Hz, which is the difference of 400 Hz and 408 Hz. The 8 Hz was chosen as it falls within the alpha frequency region, commonly found in the EEG rhythm during eyes closed and relaxed situation [19].

## 2.1 ECG Signal Processing

The ECG signal was band-pass filtered in the range of 1-35 Hz using an Elliptic IIR filter (with forward-reverse filtering to avoid phase distortion) with a minimum stopband attenuation of 30 dB and maximum passband ripple of 0.1 dB. To reliably detect R peaks in the ECG signal, the signal was then high pass filtered with a cut-off at 10 Hz and then, in order to avoid spurious peaks [5], a R peak was detected to exist within the signal wherever the peak values exceeded the maximum amplitude multiplied by a threshold Th (for all subjects):

$$Th > 0.3 * \max\_amplitude\_ECG \qquad (1)$$

With the R peak locations identified, the average R-R interval length was computed and the 1-35 Hz band-pass filtered ECG signal was then further segmented into segments consisting of four R peaks plus half of an R-R interval on either side, as shown in Figure 2. Forty such segments were obtained for each session and each condition, for each subject.

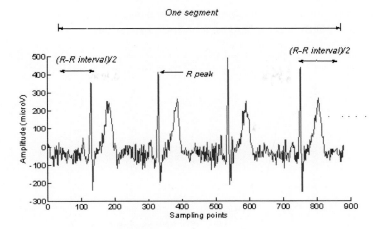

**Fig. 2** Example of a segment from ECG signal

An autoregressive (AR) model was used [20] to extract features from each segment, according to equation:

$$x(n) = -\sum_{k=1}^{p} a_k x(n-k) + e(n),$$ (2)

where $p$ is the model order, $x(n)$ is the segmented ECG signal at the sampled point $n$, $a_k$ are the real valued AR coefficients and $e(n)$ represents the white noise error term which is assumed to be independent of past samples.

Determining the appropriate order of the AR model is an important step in such an approach since if the model is too small it will not represent the signal in sufficient detail and if order is too large the representation will include the original signal noise. The model order was selected using the Akaike Information Citerion (AIC) which selects the model order to minimise the following function [21]:

$$AIC(p) = N \ln \sigma_e^2(p) + 2p,$$ (3)

where $p$ is the model order, $N$ is the length of the signal, $\sigma_e^2(p)$ is the estimated error variance for the model. The term $2p$ represents the penalty for selecting higher order models.

AIC was computed from order one to ten for all the ECG data segments and the average values of these were plotted for analysis, as shown in Figure 3. It can be seen that the gradient does not change significantly after order six. Hence, a sixth order model was selected. The six AR features for each ECG segment were obtained using Burg's method, which is more accurate than the Levinson-Durbin method because it uses the data points directly, unlike the Levinson-Durbin method, which relies upon the estimation of the autocorrelation function of the data, which is generally erroneous for small data segments [20]. Burg's method also uses more data points simultaneously by minimising not only a forward error (as in the Levinson-Durbin case) but also a backward error [20].

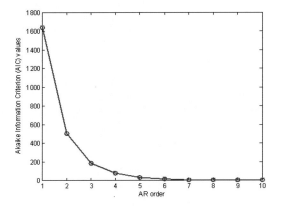

**Fig. 3** Averaged AIC values vs AR order

In addition to the AR features, two Hjorth features namely mobility and complexity [22] were also used in the feature set:

$$MOB = \sqrt{\text{var}(x')/\text{var}(x)} \tag{4}$$

$$Complexity = \sqrt{(\text{var}(x'')/\text{var}(x'))/(\text{var}(x')/\text{var}(x))} \tag{5}$$

where *var* denotes variance, $x$ is the ECG segment, $x'$ is the first derivative of the ECG segment, while $x''$ is the second derivative of the ECG segment. These particular Hjorth features were chosen based upon their reported performance in another ECG biometric study [5].

## 2.2  EEG Signal Processing

The EEG signals were segmented into intervals corresponding to those derived in the ECG processing. Thus forty segments for each of the 19 channels were obtained, for each condition and for each subject in each session. Next, these EEG signals were band-pass filtered in the alpha and gamma frequency ranges of 8-12 Hz and 30-50 Hz, respectively using a forward-reverse Elliptic IIR filter with a minimum stopband attenuation of 30 dB and a maximum passband ripple of 0.1 dB. The energy in each channel was obtained by computing the variance of the signal giving 19 energy features for both the alpha band and also for the gamma band. Both the gamma band [8] and the alpha band [23] have been previously employed successfully for biometrics applications and hence used here.

## 2.3    Classification

Two classifiers were used to classify the combined ECG and EEG feature vectors: NN with a single hidden layer (with architecture as shown in Figure 4) and LDA. These classifiers were used to classify the features into five categories representing each subject. The size for the NN hidden layer was varied from 5 to 150 in steps of 5 while the number of units in the input layer changed according to the number of utilized features. The number of output units was set to five as there were five classes. The activation functions were sigmoid and the learning method was resilient backpropagation [24] due to its quick speed and the training was conducted until the error limit fell below 0.0001 or reached epoch limit of 1000.

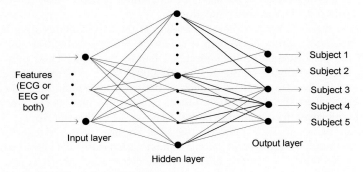

**Fig. 4** MLP NN architecture as used here

Despite its simplicity, LDA has been shown to give comparable or even better results than NN for EEG classification [25]. It is also advantageous due to its low computational resource consumption and simplicity of design and hence it was used to compare with the NN results. As there were five classes, the LDA was used with one-against-the-rest approach, where different discriminant functions classified the features (one function per class).

To analyse the results of the classifiers, a six fold cross validation test for inter-session classification was used, where features from five sessions were used for training and the remaining session for testing (i.e. 200 segments for training and 40 segments for testing from each subject, giving total of 1000 segments for training and 200 segments for testing). To further reduce the bias caused by the stochastic nature of NN, the training and testing was repeated ten times for each case, giving a total of 60 classifications for each hidden unit (HU) and the mean classification performance computed. The classifications were repeated for different combinations of ECG (six AR and two Hjorth) and EEG features (19 alpha and 19 gamma energies) and the results are reported in the next section.

## 3    Results and Discussion

Table 1 show the results of the ten runs of six-fold cross validation NN for different hidden unit (HU) sizes, for various feature combinations (ECG - AR, Hjorth
and combined, EEG - alpha, gamma and combined, all - ECG combined with
EEG) for the control and binaural conditions, respectively. To save space, only the
mean ± standard deviation of these classification performance values are shown.

In general, it can be seen that using a combination of both AR and Hjorth features improved the classification performance for majority of the HU sizes. An
ANOVA test (using MATLAB's anova1 function) gave a statistically significance
difference in this case with $F(2,87)=535.8$, $p<0.00001$. The use of AR features
from ECG performed better than the Hjorth features (paired t-test, $t(29)=19.6$,
$p<0.00001$, using MATLAB's *ttest* function).

Similarly to the ECG results, the combination of the EEG features significantly
improved the classification performance $(F(2,87)=190.9$, $p<0.00001)$ and when
comparing the performance of the individual bands, the alpha band performed
better than the gamma band $(t(29)=19.3$, $p<0.00001)$. This is likely to be due to
the fact that alpha band is more prominent during eyes closed and relaxed conditions but nevertheless, the classification results indicate that there is complementary biometric information contained in both the chosen EEG spectral bands.

Comparing ECG and EEG features, the ECG features (AR and Hjorth) performed better than EEG (alpha and gamma bands) with a statistical significance of
$t(29)=37.8$, $p<0.00001$. Combining all the available features from ECG and EEG
gave significantly improved performance $(F(6,203)=391.8$, $p<0.00001)$, which
illustrates that the combined features have complementary individual-specific
information which is useful for biometric purposes.

Considering the binaural conditions, once again the results indicate that the combined use of both the AR and Hjorth features improved classification performance
$(F(2,87)=534.4$, $p<0.00001)$, and that AR gave better performance than Hjorth features $(t(29)=21.9$, $p<0.00001)$ when they were used in isolation. More importantly,
comparing the NN classification performance of ECG features under control and
binaural conditions illustrates that a statistically significant higher classification rate
was obtained under binaural conditions $(t(29)=27.6$, $p<0.00001)$. It is speculated that
under entrained conditions, the sympathetic and parasympathetic systems are more
controlled, therefore leading to less variability in the heart rhythms.

Using both the alpha and gamma features from EEG improved classification
performance $((F(2,87)=288.6$, $p<0.00001)$ and similarly to the previous control
condition case, the alpha band EEG used alone provided a better classification rate
than using gamma band features alone $(t(29)=16.4$, $p<0.00001)$. Comparing EEG
features under both sets of experimental conditions indicates, that once again, a
higher classification performance was achieved when the measurements were
taken under binaural conditions $(t(29)=34.5$, $p<0.00001)$.

With combination of ECG and EEG features, the classification performance
was improved for the binaural condition $((F(6,203)=532.1$, $p<0.00001)$ and similar to the previous control condition results, the ECG features, rather than the EEG
features, provided the higher classification performance $(t(29)=30.6$, $p<0.00001)$.

In the control condition, using both AR and Hjorth ECG features gave an average classification performance of 91.5%, whilst using both alpha and gamma bands of EEG give 87.5%, however, when used in combination an improved classification of 92.4% was achieved. Under binaural brain entrainment conditions and when all features were used, a further improvement in the classification rate was achieved (95.1%). Using the final column results from Table 1 reveal that NN performances (when using all the available features) are superior under binaural conditions as compared to control ($t(29)=31.7, p<0.00001$).

The classification performance of the LDA classifier is given in Table 2. Similar to the NN classifier results, it can be seen that the use of both AR and Hjorth ECG features provides improved classification when compared to the use of either of the ECG features separately. Table 2 also illustrates that combining the EEG features from both alpha and gamma bands gave better performance, than using the spectral bands separately. This is the case for both the control and binaural entrained conditions. More importantly, a combination of all the available features led to a further improvement in classification performance. Finally, as hypothesised, the classification performances were higher for the binaural condition as compared to the control case for any feature combination. Statistical testing (ANOVA and paired t-test) that was done for NN results was not possible for LDA results as only six-fold cross validation results were available. The case was different with NN, where the use of several HU sizes led to availability of a higher number of classification performance values sufficient for statistical testing.

Comparing the classifiers, it is evident that LDA provides the better classification performance for most of the feature combinations. When all the features were used, LDA performed better than NN.

The results of analysing the standard deviation of the features across all the sessions from both sets of conditions are shown in Table 3, where it can be seen that for all the subjects, the average standard deviation of all the features across sessions is lower by 13.2% for the binaural condition as compared to the control condition. The lower standard deviation denotes that the feature values were more stable for the binaural condition even after a significant period of time (as the six sessions were conducted across monthly intervals).

**Table 1** NN classification performance (%, mean±std) for the control and binaural conditions

| | | | Features | | | | |
|---|---|---|---|---|---|---|---|
| HU size | AR | Hjorth | AR+Hjorth | Alpha | Gamma | Alpha+ Gamma | All (ECG +EEG) |
| Control | 88.6±2.5 | 78.7±0.5 | 91.5±1.1 | 85.9±0.7 | 80.6±2.2 | 87.5±1.0 | 92.4±0.8 |
| Binaural | 93.2±2.1 | 84.0±0.4 | 94.5±0.9 | 86.9±0.6 | 83.6±1.6 | 90.0±0.7 | 95.1±0.7 |

**Table 2** Mean and standard deviation of LDA classification (%) results for the different feature combinations

| | Features | | | | | | |
|---|---|---|---|---|---|---|---|
| | AR | Hjorth | ECG (AR, Hjorth) | Alpha | Gamma | EEG (Alpha, Gamma) | All (ECG and EEG) |
| **Control** | 94.1±0.02 | 80.3±0.08 | 96.6±0.01 | 87.5±0.02 | 76.9±0.06 | 92.9±0.02 | 98.6±0.02 |
| **Binaural** | 97.4±0.01 | 81.1±0.06 | 98.1±0.008 | 89.8±0.08 | 80.8±0.02 | 93.5±0.03 | 99.8±0.002 |

**Table 3** Standard deviation of features for each subject for both conditions

| | Control | | | | | Binaural | | | | |
|---|---|---|---|---|---|---|---|---|---|---|
| Feature | AR | Hjorth | Alpha energy | Gamma energy | All | AR | Hjorth | Alpha energy | Gamma energy | All |
| **Subject** | | | | | | | | | | |
| 1 | 0.4014 | 0.0831 | 0.1720 | 0.1217 | 0.4918 | 0.2895 | 0.0400 | 0.1520 | 0.1269 | 0.3959 |
| 2 | 1.5813 | 0.0714 | 0.1709 | 0.1575 | 1.6154 | 1.3620 | 0.0400 | 0.1463 | 0.1463 | 1.3936 |
| 3 | 1.3251 | 0.0608 | 0.1661 | 0.1497 | 1.3618 | 1.1257 | 0.0332 | 0.1517 | 0.1421 | 1.1628 |
| 4 | 1.1546 | 0.0529 | 0.2117 | 0.1490 | 1.2030 | 0.9836 | 0.0346 | 0.2133 | 0.1435 | 1.0374 |
| 5 | 1.2841 | 0.0510 | 0.2105 | 0.1597 | 1.3312 | 1.1518 | 0.0361 | 0.2117 | 0.1549 | 1.2020 |
| Overall average | 1.2165 | 0.0648 | 0.1873 | 0.1480 | 1.2590 | 1.0488 | 0.0374 | 0.1778 | 0.1432 | 1.0928 |

## 4    Conclusion

This study has attempted to combine features from ECG and EEG to obtain improved individual identification performance. In addition, a novel approach based on binaural brain entrainment was applied during ECG and EEG recording to analyse the stability of the extracted features. The results show that combining the features from both ECG and EEG modalities gave significantly improved classification performance as compared to using either of the modality alone. It can also be concluded from the results that the use of brain entrainment significantly improved the classification performance as compared to without entrainment. It is speculated that brain entrainment allows the subjects to be more relaxed and hence minimise the feature variability over the monthly recording sessions.

When using all the available features, the LDA classifier provided the better classification performance; therefore, the use of the LDA is suggested since it has the added advantage of being simpler to design and quicker to use.

It should be noted here that a comparative analysis of all the various features utilised in ECG and EEG is beyond the scope of this paper, and therefore it is possible that the use of other features and classifiers could result in further improved performance. There are practicalities such as requiring the subject to be still and also on the cumbersomeness for data enrollment that need to be overcome before such approaches can be compared to conventional biometrics. But nevertheless, this study has shown that combining features from ECG and EEG is useful for biometrics, and the binaural brain entrainment reduces the variability of these features over time, thereby increasing the potential for use in biometric applications.

**Acknowledgments** Part of R. Paramesran's work was funded by grant: UM.C/625/1/ HIR/MOHE/ENG/42.

# References

1. Jain, K., Ross, A.A., Nandakumar, K.: Introduction to Biometrics. Springer Science and Business Media, London (2011)
2. Karnan, M., Akila, M., Krishnaraj, N.: Biometric personal authentication using keystroke dynamics: A review. Applied Soft Computing **11**(2), 1565–1573 (2011)
3. Zhou, X., Bhanu, B.: Integrating face and gait for human recognition at a distance in video. IEEE Transactions on Systems, Man, Cybernetics B **37**(5), 1119–1137 (2007)
4. Pflug, A., Busch, C.: Ear biometrics: a survey of detection, feature extraction and recognition methods. IET Biometrics **1**(2), 114–129 (2012)
5. Palaniappan, R., Krishnan, S.M.: Identifying individuals using ECG signals. In: Proceedings of International Conference on Signal Processing and Communications, Bangalore, India, pp. 569–572 (2004)
6. Odinaka, I., Lai, P.-H., Kaplan, A.D., O'Sullivan, J.A., Sirevaag, E.J., Rohrbaugh, J.W.: ECG biometric recognition: A comparative analysis. IEEE Transactions on Information Forensics and Security **7**(6), 1812–1824 (2012)
7. Boulgouris, N.V., Plataniotis, K.N., Micheli-Tzanakou, E. (eds.): Biometrics: Theory, Methods, and Applications. IEEE Press/Wiley, USA (2010)
8. Palaniappan, R., Mandic, D.P.: Biometric from the brain electrical activity: A machine learning approach. IEEE Transactions on Pattern Analysis and Machine Intelligence (Special Issue on Biometrics) **29**(4), 738–742 (2007)
9. Gupta, C.N., Palaniappan, R., Swaminathan, S.: On the analysis of various techniques for a novel brain biometric system. International Journal of Medical Engineering and Informatics **1**(2), 266–273 (2008)
10. Palaniappan, R.: Two-stage biometric authentication method using thought activity brain waves. International Journal of Neural Systems **18**(1), 59–66 (2008)
11. Palaniappan, R., Revett, K.: PIN generation using EEG: A stability study. International Journal of Biometrics **6**(2), 95–105 (2014)
12. Riera, A., Dunne, S., Cester, I., Ruffini, G.: Starfast: a wireless wearable EEG/ECG biometric system based on the enobio sensor. In: Proceedings of the International Workshop on Wearable Micro and Nanosystems for Personalised Health, Valencia, Spain, pp. 21–23, May 2008
13. Abdullah, M.K., Subari, K.S., Loong, J.L.C., Ahmad, N.N.: Analysis of the EEG signal for a practical biometric system. World Academy of Science, Engineering and Technology **4**(8), 931–935 (2010)
14. http://www.biosemi.com/ (accessed June 20, 2012)
15. Jasper, H.: The ten twenty electrode system of the international federation. Electroencephalographic and Clinical Neurophysiology **10**, 371–375 (1958)

16. O'Kelly, J., Magee, W., James, L., Palaniappan, R., Taborin, J., Fachner, J.: Neurophysiological and behavioural responses to music therapy in vegetative and minimally conscious states. Frontiers in Neuroscience - Special Edition on Music, Brain, and Rehabilitation: Emerging Therapeutic Applications and Potential Neural Mechanisms **7**(00884) (2013)

17. http://www.etymotic.com/ (accessed June 20, 2012)

18. Schwarz, D.W.F., Taylor, P.: Human auditory steady state responses to binaural and monaural beats. Clinical Neurophysiology **116**(3), 658–668 (2005)

19. Kaul, P., Passafiume, J., Sargent, R.C., O'Hara, B.F.: Meditation acutely improves psychomotor vigilance, and may decrease sleep need. Behavioral and Brain Functions **6**(47) (2010). doi:10.1186/1744-9081-6-47

20. Shiavi, R.: Introduction to Applied Statistical Signal Analysis, 2nd edn. Academic Press, San Diego (1999)

21. Burnham, K.P., Anderson, D.R.: Model Selection and Multimodel Inference: A Practical Information-Theoretic Approach, 2nd edn. Springer-Verlag, New York (2002)

22. Hjorth, B.: EEG analysis based on time-domain properties. Electroencephalography and Clinical Neurophysiology **34**, 306–310 (1970)

23. Nakanishi, I., Baba, S., Miyamoto, C.: EEG based biometric authentication using new spectral features. In: Proceedings of International Symposium on Intelligent Signal Processing and Communication Systems, Kanazawa, Japan, pp. 651–654 (2009)

24. Riedmiller, M., Braun, H.: A direct adaptive method for faster backpropagation learning: the RPROP algorithm. In: Ruspini, H. (ed.) Proceedings of the IEEE International Conference on Neural Networks (ICNN), San Francisco, pp. 586–591 (1993)

25. Huan, N.-J., Palaniappan, R.: Neural network classification of autoregressive features from electroencephalogram signals for brain-computer interface design. Journal of Neural Engineering **1**, 142–150 (2004)

# Denoising Multi-coil Magnetic Resonance Imaging Using Nonlocal Means on Extended LMMSE

V. Soumya, Abraham Varghese, T. Manesh and K.N. Neetha

**Abstract** Denoising plays key role in the field of medical images. Reliable estimation and noise removal is very important for accurate diagnosis of the disease. This should be done in such a way that original resolution is retained while maintaining the valuable features. Multi-coil Magnetic Resonance Image(MRI) trails nonstationary noise following Rician and Noncentral Chi(nc-$\chi$) distribution. On using the modern techniques which make use of multi-coil MRI like in GRAPPA would yield nc-$\chi$ distributed data. There has been lots of research done on the Rician nature but only few for nc-$\chi$ distribution. The proposed method uses Nonlocal Mean(NLM) on extended Linear Minimum Mean Square Error(ELMMSE) for denoising multi-coil MRI having nc-$\chi$ distributed data. The performance of the nonlocal scheme on multi-coil MRI is evaluated based on PSNR, SSIM and MSE and the result indicates proposed scheme is better than the existing scheme including Non local Maximum Likelihood(NLML), adaptive NLML and ELMMSE.

## 1 Introduction

Noise is an important factor which decreases the quality of the image. It is an unwanted information that comes along the required data during the image acquisition and image transmission. Medical image plays an important part for diagnosing the disease. Medical images shows visual representation of the interior body which will be helpful for the medical analysis. Major medical imaging modalities include Radiography, MRI, Nuclear Medicine, Computed Tomography(CT), Ultrasound. For studying the soft tissues CT and MRI can be used.

V. Soumya(✉) · A. Varghese · K.N. Neetha
Department of Computer Science and Engineering,
Adi Shankara Institute of Engineering and Technology, Ernakulam, India
e-mail: {soumyav17,abrahamvarghese77}@gmail.com

T. Manesh
Prince Sattam Bin Abdul Aziz University, Al-kharj, Saudi Arabia

© Springer International Publishing Switzerland 2016
S.M. Thampi et al. (eds.), *Advances in Signal Processing and Intelligent Recognition Systems*,
Advances in Intelligent Systems and Computing 425,
DOI: 10.1007/978-3-319-28658-7_16

187

MRI provides higher details of soft tissues and is also preferred greatly for tumor detection. Even though the scanning duration is lesser for CT, it uses x-rays to construct the image and radiation from these are harmful. Presence of noise in medical images like MRI will result in poor visual quality and physician won't be able to diagnose the disease correctly. These noise can also affect the MRI post processing algorithms.

In single coil MRI the noise distribution will be stationary and magnitude data is coped with Rician, Gaussian and Rayleigh distribution. It is on an assumption that only single $\sigma$ value featuring the complete data. Modern techniques make use of multi-coil MRI in which noise varies with position in the image i.e. nonstationary data. Multi-coil magnitude data follows Rician and nc-$\chi$ distributed. Systems like Generalized Auto calibrating Partially Parallel Acquisitions(GRAPPA) trails nc-$\chi$ distributed magnitude data. There has been a lot of literature work done on the Rician nature of the noise but very few for the nc-$\chi$ distribution.

The objective of the MRI denoising technique is to remove the noisy information from the original image. Raw data acquired from MRI scanning are complex values and it can be represented in Fourier transform as:

$$M = x + iy \tag{1}$$

The noise added to the MRI can be expressed as:

$$M = (x + n_{re}) + i(y + n_{im}) \tag{2}$$

The magnitude of the noisy raw data can be given as:

$$|M| = \sqrt{(x + n_{re})^2 + i(y + n_{im})^2} \tag{3}$$

In order to reduce noise, methods like acquiring MR data repeatedly and then averaging these data can be used. But this technique increases the capturing time. Another way to reduce noise is, by utilizing post processing techniques. The denoising technique needs to be selected in such a way that it should not modify the original features of the MRI.

In Sections 2, the related work of multi-coil MRI denoising methods for nc-$\chi$ distributed data are discussed, section 3 explains the theory behind proposed method. In Section 4, experiments and results are discussed. Finally, conclusions are drawn in Section 5.

## 2   Related Work

Several Rician noise removal techniques are available in the literature for denoising MRI. For example linear estimators [1], Partial Differential Equations (PDE) [2, 3],

Nonlocal Mean(NLM) [4, 5], anisotropic diffusion [6, 7], wavelet based methods [8], wiener filter [9], statistical approach like maximum likelihood [10, 11, 12], LMMSE [13] and many other methods.

If we are considering denoising methods for nc-$\chi$ distributed noise elimination, only few techniques are available. For nc-$\chi$ distributed data, k-space should be totally sampled and there should not be any correlation among coil data [14, 15]. Number of coils L and $\sigma$ values are important parameters that needs to be considered. Here signals from different receiving coils are merged as Sum of Squares(SoS) [15, 16, 17, 18]. The coil configuration and MRI processing will indicate the probability density function(pdf) of noise [19].

An analytically exact correction method is introduced by Koay and Basser [18] to separate the true signal intensity and the noise. Fixed point formula of SNR and correction factor is used for the extraction of noise.

Koay et.al [20] transformed noisy nc-$\chi$ signals to noisy Gaussian signals using three stages. A framework has been set for creating Gaussian distributed magnitude signals. Least square method is used for analysing MRI information and least square is very potential in mathematical traceability for hypothesis testing, parameter assessment and confidence interval evaluation. Both [18, 20] repeated evaluation in large quantity is required so that implies long computation time.

Brion et.al [21] proposed a method to estimate nc-$\chi$ distributed noise by using an extended LMMSE. From the local estimates of the mean and variance true signal is calculated. Before starting the denoising process there is no need of complete data set capturing in all diffusion direction. Advantage of this technique would be easy to calculate and fast execution.

Maximum Likelihood(ML) estimation schemes were proved to be capable of denoising MR images. Based on ML, NLML estimation scheme was formulated [12] for single coil MRI following Rician distribution. This NLML method is extended for $nc - \chi$ distributed noise in the multi-coil MRI [22]. This extended NLML method is employed for estimating true signal from nc-$\chi$ distributed data. Non stationary noise nature and nc-$\chi$ distribution has been considered in this work. The main drawback here is the its long computation time.

The performance of the extended NLML scheme can be considerably enhanced by adaptively choosing the number of pixels k for the ML estimation [27] and this is done for single coil MRI. This is extended for multi-coil MRI and is proposed by Soumya et.al [23] as Adaptive Nonlocal Maximum likelihood(Adaptive NLML) for nc-$\chi$ distribution.

Traditional LMMSE which practice isotropic has been extended by using anisotropic method by Brion et.al [24]. Authors have utilized anisotropic local evaluation. Advantage is structural characters are well-maintained. Correlation between channels are taken into account here considering effective variance of noise and number of channels. This method shows high computational performance.

# 3   Proposed Approach

In the traditional LMMSE method, isotropic method of calculation is done. The improvement is done on the traditional method using nonlocal scheme. In this work ELMMSE is modified using non local means. Firstly noise variance $\sigma^2$ with the necessity of a background [21] is estimated. Then the application of nonlocal means on ELMMSE for nc-$\chi$ distributed noise removal in multi-coil MRI.

Non local means(NLM) and ELMMSE are shown in appendix 1 and 2 respectively. For NLM, Find out the mean values and gradients of the pixels in the search neighborhood. Compute the distance using euclidean distance(refer Eq. 10) and weights by using Eq. 8 then apply NLM filter. Filter the pixel using the Eq. 6 and is given as $NLM_{out}$. Finally perform filtering for non local ELMMSE by using Eq. 4.

$$S^2 = \langle M^2 \rangle - 2n\sigma^2 + K(NLM_{Out}^2 - M^2) \tag{4}$$

where S is the noise free magnitude, M is the measured magnitude which is shown in Eq. 12, n is the no of channel, $\sigma$ is the standard deviation. K and $NLM_{Out}$ can be expressed as shown in Eq. 5 and 6

$$K = \left( 1 - \frac{4\sigma^2[\langle M^2 \rangle - n\sigma^2]}{\langle M^4 \rangle - \langle M^2 \rangle^2} \right) \tag{5}$$

$$NLM_{Out} = \sqrt{(NLM_p - 2\sigma^2)} \tag{6}$$

where $NLM_p$ is the normalized value of the $NLM_{Out}$. Algorithm 1 depicts the proposed scheme.

# 4   Experiments and Results

In the experiment section, the nonlocal ELMMSE, ELMMSE, NLML and adaptive NLML methods are evaluated and compared. Data set used for the experiments purpose include the standard noise free MRI of the brain obtained from the Brain web database [28] and is depicted in Fig.1(a). Then to this noise free MRI, nc-$\chi$ distributed noise is added. This is done by generating noise artificially by means of MRI noisy phantom simulator available in [29]. The number of channels(n) used is 4,8 and 16. In this paper, denoising results are shown for number of channel,n=4. The standard deviation taken is on the range 5-40 and the window size used is 25.

Firstly nc-$\chi$ distributed noise is created by using phantom simulator [29]. Then noise estimation, that is noise variance with the necessity of a background is computed [21]. finally denoising stage, nonlocal means is calculated using Euclidean distance followed by calculation of the ELMMSE for denoising multi-coil MRI having $nc - \chi$ distributed noise.

---

**Algorithm 1.** Non-local means on ELMMSE

---

**Input:** nc-$\chi$ distributed noisy MRI

**Output:** denoised MRI

1. Find out the noise variance $\sigma^2$ from input image U.

2. for each pixel u(i) of U do

3.    Work out the similarity measure for N pixels in an extended neighbourhood of u(i) for finding the distance.

4.    Calculate weight, $w_{(i,j)}$

5.    for j = 1 $\rightarrow$ N do

6.      NLM$u_{(i)} = \sum_{j\in I} w_{(i,j)} u_{(i)}$ {Evaluate the nonlocal means}

7.    end for

8. end for

9. Apply ELMMSE on the NLM filtered value for denoising $nc - \chi$ distributed noise

---

The methods used for comparison includes NLML, Adaptive NLML and ELMMSE. NLML method is extended for $nc - \chi$ distributed noise in the multi-coil MRI [22]. For the purpose of experiment, number of non local neighbouring pixels used is 12. Adaptive NLML method is adopted by Soumya et.al [23] in order to improve the computational complexity of NLML. Another method employed to denoise nc-$\chi$ distributed noise in the multi-coil MRI is ELMMSE method [26] which is an extension of the LMMSE method.

Visual comparison has been made among the traditional and proposed approach. The proposed approach clearly shows a visual quality image denoising multi-coil MRI with $nc - \chi$ distributed noise. Fig.1 depicts denoising by different methods having (a)noise free MRI, (b)nc-$\chi$ noisy MRI for standard deviation 40 and n=4 (c)denoised by NLML(d)denoised by adaptive NLM (e)denoised by ELMMSE (f) denoised by Non-local ELMMSE method. It is detected that MRI denoised with proposed scheme is closer to the original image compared to the existing denoising methods.

Once phantom image is degraded with $nc - \chi$ distributed noise, the denoising efficiency of NLML, ELMMSE and proposed scheme were evaluated in accordance with the Peak Signal to Noise Ratio (PSNR), Mean Squared Error (MSE) and Structural SIMilarity (SSIM) Index [30]. Tests are repeated with several values of noise variance $\sigma$ ranging from 5 to 40 for quantitative examination. SSIM, MSE and PSNR and is depicted in Fig.2. Table 1, 2 and 3 shows the values for SSIM, MSE and PSNR evaluation.

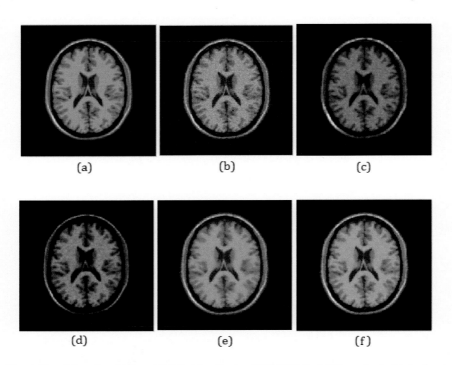

**Fig. 1** Denoising of multiple-coil MRI: (a) noise free MRI (b)Noisy MRI acquired with L=4 and $\sigma$=40 (c) denoised by NLML (d)denoised by adaptive NLML (e)denoised by ELMMSE (f) denoised by Non-local ELMMSE method.

SSIM is measured by verifying the difference in terms of structural information between noise free MRI and denoised MRI. SSIM Evaluation for NLML, adaptive NLML, ELMMSE and non-local ELMMSE is depicted in Fig.2(a).

MSE is the difference between original noise free image and the calculated denoised image. It can be estimated by taking average of the squares of errors and is shown in appendix 3. MSE for NLML, adaptive NLML, ELMMSE and non-local ELMMSE is depicted in 3(b) and 3(c).

PSNR is measured by taking proportion among the maximum probable power of a signal and the power of noise corrupted. Signal means original information which is noise free MRI and the noise introduced, which is to be nc-$\chi$ distributed. PSNR is expressed in appendix 4 and this can be represented in decibel(dB). PSNR for NLML and adaptive NLML, ELMMSE and non-local ELMMSE is depicted in 3(d) and 3(e).

**Table 1**  Structural SIMilarity Index evaluation

| Standard Deviation | NLML | Adaptive NLML | ELMMSE | NLM ELMMSE |
|---|---|---|---|---|
| 5 | 0.600501 | 0.758237 | 0.57507 | 0.766151 |
| 10 | 0.394449 | 0.734709 | 0.52927 | 0.74852 |
| 15 | 0.282625 | 0.633969 | 0.502364 | 0.717708 |
| 20 | 0.234509 | 0.518941 | 0.484923 | 0.687028 |
| 25 | 0.210544 | 0.427645 | 0.47241 | 0.666984 |
| 30 | 0.197135 | 0.364193 | 0.462652 | 0.651429 |
| 35 | 0.189679 | 0.323733 | 0.45393 | 0.6335 |
| 40 | 0.185601 | 0.296544 | 0.44691 | 0.623098 |

**Table 2**  Mean Squared Error evaluation

| Standard Deviation | NLML | Adaptive NLML | ELMMSE | NLM ELMMSE |
|---|---|---|---|---|
| 5 | 6735.160477 | 5299.724401 | 18.945772 | 12.662365 |
| 10 | 6737.495523 | 5314.426973 | 32.82623 | 14.233437 |
| 15 | 6759.439938 | 5329.975542 | 47.461703 | 16.040367 |
| 20 | 6767.551514 | 5340.882459 | 61.720324 | 18.364832 |
| 25 | 6789.609248 | 5344.110751 | 75.479523 | 19.940749 |
| 30 | 6844.796677 | 5373.186557 | 88.892737 | 21.824465 |
| 35 | 6868.649931 | 5373.477285 | 101.048587 | 23.966841 |
| 40 | 6902.419191 | 5397.921068 | 113.003862 | 25.682699 |

**Table 3**  Peak Signal-to-Noise ratio evaluation

| Standard Deviation | NLML | Adaptive NLML | ELMMSE | NLM ELMMSE |
|---|---|---|---|---|
| 5 | 19.76264 | 21.844533 | 70.779353 | 74.279302 |
| 10 | 19.759629 | 21.82047 | 66.005178 | 73.263403 |
| 15 | 19.731384 | 21.795094 | 62.802732 | 72.225313 |
| 20 | 19.720967 | 21.777338 | 60.521035 | 71.049859 |
| 25 | 19.692703 | 21.77209 | 58.773016 | 70.334769 |
| 30 | 19.622388 | 21.72496 | 57.352273 | 69.550726 |
| 35 | 19.592171 | 21.72449 | 56.238994 | 68.737383 |
| 40 | 19.549572 | 21.685068 | 55.267733 | 68.136785 |

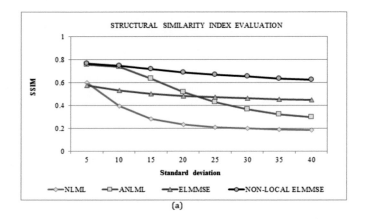

(a)

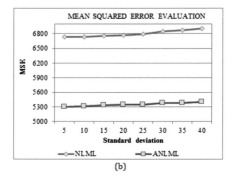

(b)

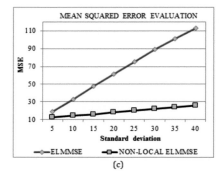

(c)

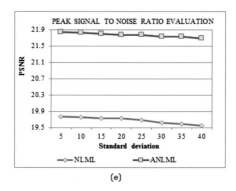

(e)

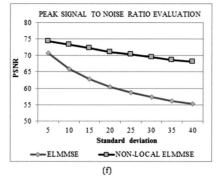

(f)

**Fig. 2** Quantitative analysis of methods with standard deviation($\sigma$) varying from 5 to 40. (a)SSIM Evaluation for NLML, Adaptive NLML, ELMMSE and Non-local ELMMSE (b)MSE for NLML and Adaptive NLML (c)MSE for ELMMSE and Non-local ELMMSE (d)PSNR for NLML and Adaptive NLML (f)PSNR for ELMMSE and Non local ELMMSE

# 5 Conclusion

An enhanced scheme to denoise multi-coil MRI having $nc - \chi$ distributed noise is explained in this paper. The improvement is reached by applying non local means on traditional extended LMMSE method. i.e. proposed scheme is an additional version of the extended LMMSE method which was proposed for denoising multi-coil MRI distorted by $nc - \chi$ distributed noise. Different stages in this work include noise variance estimation, non local means calculation using euclidean distance and finally calculation of extended LMMSE. Non local means on ELMMSE is compared using NLML, adaptive NLML and ELMMSE. These denoising methods are evaluated using MSE, PSNR and SSIM. For instance for $\sigma = 20$, SSIM evaluation has the value 0.234509 for NLML, 0.518941 for adaptive NLML, 0.484923 for ELMMSE and 0.687028 for non local ELMMSE. On considering MSE for $\sigma = 20$, MSE evaluation include 6767.551514 for NLML, 5340.882459 for adaptive NLML, 61.720324 for ELMMSE and 18.364832 for non local ELMMSE. In the case of PSNR for $\sigma = 20$, 19.720967 for NLML, 21.777338 for adaptive NLML, 60.521035 for ELMMSE and 71.049859 for non local ELMMSE. It is clear that the non local ELMMSE scheme is more effective than NLML, adaptive NLML and ELMMSE for denoising $nc - \chi$ distributed noise from multi-coil MRI.

**Acknowledgements** Authors would like to thank **Dr Santiago Aja-Fernndez**, Associate Professor, university of Valladolid, Spain for his valuable discussions, timely suggestions on multi-coil MRI noise estimation and denoising and also for supplying information about the multi-coil MRI data.

# Appendix 1. Nonlocal Means method

NLM can be measured using weighted average of the pixels in the MRI. Weights are considered by taking similarity of two pixels in the image. i.e. Mean of the similar neighbourhood pixel in the MRI is considered. Similarity is measured by Euclidean distance and the likeness between two pixels depends on the similarity of intensities in grey level [4, 25]. It is on the assumption that similar neighborhood pixels can happen any place in the image and denoising by nonlocal means produces better results. Let U be the image, at point i the filtered value can be calculated as weighted average of all pixels in the image which can be expressed as [4, 25]:

$$NLM[U](i) = \sum_{j \in I} w_{(i,j)} U(j) \tag{7}$$

here $w_{(i,j)}$ depends on the similarity between pixels i and j. The similarity between pixels neighbourhoods $N_i$ and $N_j$ of the pixels i and j. Condition to be satisfied is $0 \leq w_{(i,j)} \leq 1$ and $\sum_j w_{(i,j)} = 1$

$$w_{(i,j)} = \frac{1}{x_{(i)}} e^{\frac{-d(i,j)}{h^2}} \tag{8}$$

$$x_{(i)} = \sum_{j \forall I} e^{\frac{-d(i,j)}{h^2}} \tag{9}$$

where i represents the point being filtered, j depicts each one of the pixel in the image, $x_{(i)}$ is the normalizing constant. h is a exponential decay control parameter also termed as degree of filtering and d is Euclidean distance of all the pixels of each neighbourhood which can be expressed as:

$$d(i, j) = \|U(N_i) - U(N_j)\|_2^2 \tag{10}$$

## Appendix 2. Extended LMMSE Method

Brion et.al [21] proposed an extended version on LMMSE to estimate nc-$\chi$ distributed noise. Extended LMMSE [21] for nc-$\chi$ noise can be given as:

$$S^2 = \langle M^2 \rangle - 2n\sigma^2 + \left( 1 - \frac{4\sigma^2[\langle M^2 \rangle - n\sigma^2]}{\langle M^4 \rangle - \langle M^2 \rangle^2} \right) \times (M^2 - \langle M^2 \rangle) \tag{11}$$

where S is the noise free magnitude, expectation is denoted by $\langle . \rangle$, M is the measured magnitude, $\sigma$ is the standard deviation, n is the number of channel in the multi-coil MRI. M can be represented as:

$$M = \sqrt{\Sigma_{t=1}^{n}(S_{rt} + N_{rt})^2 + (S_{it} + N_{it})^2} \tag{12}$$

$S_{rt}$ and $S_{it}$ are the real and imaginary parts of the noise-free signal, $S_t$ received by the coil t. $N_{rt}$ and $N_{it}$ are the real and imaginary parts of the noise corrupting the signal $S_t$.

## Appendix 3. Mean Squared Error

MSE is estimated by using the below expression.

$$MSE = \Sigma\Sigma(dif^2)/n \tag{13}$$

here n is the number of elements and dif is the difference between original noise free MRI and denoised MRI.

# Appendix 4. Peak Signal to Noise Ratio

PSNR is estimated by using the below expression.

$$PSNR = 20.log_{10}(MAX_I/\sqrt{MSE}) \tag{14}$$

where $MAX_I$ is maximum possible value of the MRI used and MSE can be found from the appendix 3.

# References

1. McVeigh, E.R., Henkelman, R.M., Bronskill, M.J.: Noise and filtration in magneticresonance imaging. Med. Phys. **12**, 586–591 (1985)
2. Lysaker, M., Lundervold, A., Tai, X.-C.: Noise removal using fourth-order partial differential equation with applications to medical magnetic resonance images in space and time. IEEE Trans. Image Processing **12**, 1579–1590 (2003)
3. Assemlal, H., Tschumperl, D., Brun, L.: Estimation variationnelle robuste de modèles complexes de diffusion en IRM à haute résolution angulaire et tractographie. In: GRETSI, pp. 5–8 (2007)
4. Buades, A., Coll, B., Morel, J.: A non-local algorithm for image denoising. In: IEEE Computer Society Conference on CVPR 2005, vol. 2, pp. 60–65 (2005)
5. Coupé, P., Yger, P., Prima, S., Hellier, P., Kervrann, C., Barillot, C.: An optimized blockwise nonlocal means denoising filter for 3-D magnetic resonance images. IEEE Trans. Med. Imaging **27**, 425–441 (2008)
6. Krissian, K., Aja-Fernández, S.: Noise-driven anisotropic diffusion filtering of MRI. IEEE Trans. Image Processing **18**, 2265–2274 (2009)
7. Zhang, F., Ma, L.: MRI denoising using the anisotropic coupled diffusion equations. In: 3rd International Conference on BMEI 2010, vol. 1, pp. 397–401. IEEE (2010)
8. Nowak, R.D.: Wavelet-based Rician noise removal for magnetic resonance imaging. IEEE Trans. Image Processing **8**, 1408–1419 (1999)
9. Martín-Fernández, M., Muñoz-Moreno, E., Cammoun, L., Thiran, J.-P., Westin, C.-F., Alberola-López, C.: Sequential anisotropic multichannel Wiener filtering with Rician bias correction applied to 3D regularization of DWI data. Med. Image Analysis **13**, 19–35 (2009)
10. Sijbers, J., den Dekker, A.J., Scheunders, P., Van Dyck, D.: Maximum-likelihood estimation of Rician distribution parameters. IEEE Trans. Med. Imaging **17**, 357–361 (1998)
11. Clarke, R.A., Scifo, P., Rizzo, G., Dell'Acqua, F., Scotti, G., Fazio, F.: Noise correction on Rician distributed data for fibre orientation estimators. IEEE Trans. Med. Imaging **27**, 1242–1251 (2008)
12. He, L., Greenshields, I.R.: A nonlocal maximum likelihood estimation method for Rician noise reduction in MR images. IEEE Trans. Med. Imaging **28**, 165–172 (2009)
13. Aja-Fernández, S., Alberola-López, C., Westin, C.-F.: Noise and signal estimation in magnitude MRI and Rician distributed images: a LMMSE approach. IEEE Trans. Image Processing **17**, 1383–1398 (2008)
14. Dietrich, O., Raya, J.G., Reeder, S.B., Ingrisch, M., Reiser, M.F., Schoenberg, S.O.: Influence of multichannel combination, parallel imaging and other reconstruction techniques on MRI noise characteristics. MRI **26**, 754–762 (2008)
15. Aja-Fernández, S., Tristán-Vega, A., Alberola-López, C.: Noise estimation in single-and multiple-coil magnetic resonance data based on statistical models. MRI **27**, 1397–1409 (2009)

16. Constantinides, C.D., Atalar, E., McVeigh, R.: Signal-to-noise measurements in magnitude images from NMR phased arrays. Magn. Reson. Med. **38**, 852–857 (1997)
17. Aja-Fernández, S., Vegas-Sánchez-Ferrero, G., Tristán-Vega, A.: About the background distribution in MR data: a local variance study. MRI **28**, 739–752 (2010)
18. Koay, C.G., Basser, P.J.: Analytically exact correction scheme for signal extraction from noisy magnitude MR signals. J. Magn. Reson. **179**, 317–322 (2006)
19. Aja-Fernández, S., Pie, T., Vegas-Sánchez-Ferrero, G., et al.: Spatially variant noise estimation in MRI: A homomorphic approach. Med. Image Analysis **20**, 184–197 (2015)
20. Koay, C.G., Özarslan, E., Basser, P.J.: A signal transformational framework for breaking the noise floor and its applications in MRI. J. Magn. Reson. **197**, 108–119 (2009)
21. Brion, V., Poupon, C., Riff, O., Aja-Fernández, S., Tristán-Vega, A., Mangin, J.-F., Le Bihan, D., Poupon, F.: Parallel MRI noise correction: an extension of the LMMSE to non central $\chi$ distributions. In: MICCAI 2011, pp. 226–233. Springer (2011)
22. Rajan, J., Veraart, J., Van Audekerke, J., Verhoye, M., Sijbers, J.: Nonlocal maximum likelihood estimation method for denoising multiple-coil magnetic resonance images. MRI **30**, 1512–1518 (2012)
23. Soumya, V., Abraham V.: An adaptive maximum likelihood estimation method for denoising multi-coil magnetic resonance images. In: Recent Advances in Computing and Communication Systems, pp. 16–19. Tata McGraw Hill (2015)
24. Brion, V., Poupon, C., Riff, O., Aja-Fernández, S., Tristán-Vega, A., Mangin, J.-F., Le Bihan, D., Poupon, F.: Noise correction for HARDI and HYDI data obtained with multi-channel coils and Sum of Squares reconstruction: An anisotropic extension of the LMMSE. MRI **31**, 1360–1371 (2013)
25. Manjón, J.V., Carbonell-Caballero, J., Lull, J.J., García-Martí, G., Martí-Bonmatí, L., Robles, M.: MRI denoising using non-local means. Med. Image Analysis **12**, 514–523 (2008)
26. Aja-Fernández, S., Niethammer, M., Kubicki, M., Shenton, M.E., Westin, C.-F.: Restoration of DWI data using a Rician LMMSE estimator. IEEE Trans. Med. Imaging **27**, 1389–1403 (2008)
27. Rajan, J., Van Audekerke, J., Van der Linden, A., Verhoye, M., Sijbers, J.: An adaptive non local maximum likelihood estimation method for denoising magnetic resonance images. In: 9th IEEE International Symposium on ISBI (2012), pp. 1136–1139 (2012)
28. Collins, D.L., Zijdenbos, A.P., Kollokian, V., Sled, J.G., Kabani, N.J., Holmes, C.J., Evans, A.C.: Design and construction of a realistic digital brain phantom. IEEE Trans. Med. Imaging **17**, 463–468 (1998)
29. Aja-Fernández, S.: Parallel MRI noisy phantom simulator (2012). http://in.mathworks.com/matlabcentral/fileexchange/36893-parallel-mri-noisy-phantom-simulator
30. Wang, Z., Bovik, A.C., Sheikh, H.R., Simoncelli, E.P.: Image quality assessment: from error visibility to structural similarity. IEEE Trans. Image Processing **13**, 600–612 (2004)

# An Intelligent Blind Semi-fragile Watermarking Scheme for Effective Authentication and Tamper Detection of Digital Images Using Curvelet Transforms

**K.R. Chetan and S. Nirmala**

**Abstract** A novel intelligent semi-fragile watermarking scheme for authentication and tamper detection of digital images is proposed in this paper. This watermarking scheme involves embedding and extraction of the quantized first level Discrete Curvelet Transform (DCLT) coarse coefficients. The amount of quantization of the first level coarse DCLT coefficients of the input image is decided intelligently based on the energy contribution of the coefficients. At the receiver side, the extracted and generated first level coarse DCLT coefficients of the watermarked image is divided into blocks of uniform size. A feature similarity index value between each block of extracted and generated coefficients is compared and if the difference exceeds threshold, the block is marked as tampered. The watermarking scheme is blind and does not require any additional information to identify authenticity of the watermarked image. Experiments are conducted rigorously and the results reveal that the proposed method is robust than the existing method [1]. Better accuracy in localizing tampered regions is achieved compared to method [1].

**Keywords** Discrete curvelet transforms · Intelligent semi-fragile watermarking · Normalized correlation coefficient · Tamper detection · Incidental attacks · Intentional attacks · Feature similarity index

## 1 Introduction

Digital watermarking is mainly used for protection of the integrity and authenticity of the digital images. The watermarking schemes are broadly categorized as robust, fragile and semi-fragile. Robust watermarks thwart attempts to remove or destroy

K.R. Chetan(✉) · S. Nirmala
Department of CSE, JNN College of Engineering, Chennai, India
e-mail: {krc_555,nir_shiv_2002}@yahoo.co.in

© Springer International Publishing Switzerland 2016 199
S.M. Thampi et al. (eds.), *Advances in Signal Processing and Intelligent Recognition Systems*,
Advances in Intelligent Systems and Computing 425,
DOI: 10.1007/978-3-319-28658-7_17

the watermark. Unlike robust watermarks, fragile watermarks can be easily destroyed for the minor change on the watermarked image. They are also capable of detecting and localizing tampered regions in the watermarked image. A semi-fragile watermark combines the properties of fragile and robust watermarks [2].

Many works on semi-fragile watermarking schemes were reported in literature. Chang et.al. [3] developed a semi-fragile watermarking scheme to improve the accuracy of tamper detection by analyzing the impact of various image manipulations like noise, JPEG compression and median filtering on the wavelet transformed coefficients. Different semi-fragile watermarking methods based on the Computer Generated Hologram coding techniques were described in [4-6]. These methods were able to detect malicious tampering while tolerating some incidental distortions. Maeno et.al., [7] presented a semi-fragile watermarking scheme for authentication and tamper detection of document images. A non-blind semi-fragile watermarking scheme based on scrambling and Singular Value Decomposition was discussed in [8]. However, the localization of the tampered regions was not addressed in this scheme [8]. Some of the works in semi-fragile watermarking scheme based on parameterized Integer Wavelet transform has been discussed in [9-12].

A new watermarking approach based on Discrete Curvelet Transforms (DCLT) [13] was developed in [1]. In this method, the first level coarse Curvelet Transform coefficient of an image is obtained. A logo watermark is embedded into first level coarse DCLT coefficients using an additive embedding scheme controlled by visibility parameter. The existing method [1] has many limitations. The scheme is not blind as it requires the first level coarse DCLT coefficients of the input image and visibility factor for authentication and tamper detection. The Normalized Correlation Coefficient (NCC) values significantly reduces for a variance of less than 0.05 for the various types of noise namely Salt and Pepper, Speckle and Gaussian. The existing method [1] has a tradeoff between higher robustness and better imperceptibility. Thus higher robustness can be achieved with lesser imperceptibility of the watermarked image.

In this paper, an intelligent and blind semi-fragile watermarking scheme is proposed. All the above mentioned limitations are addressed in the proposed scheme. The watermarking scheme proposed in this paper, is completely blind and does not require any information of the input image. The watermark is generated from the first level coarse DCLT coefficients of the input image. Hence, no additional watermark in the form of logo image is required. An intelligent embedding of the watermark is designed based on the contribution of the watermark to the information content and structure of the input image. The authentication scheme is designed based on the analysis of the impact of the incidental and intentional attacks on the watermarked image. Incidental attacks do not change information content and structure in a block. However, the intentional attack do change the information content and/or structure substantially. Hence, in this work, tamper assessment is performed block- wise between generated and extracted watermarks using a feature similarity index [14]. The rest of the paper is organized as follows: Section 2 introduces Discrete Curvelet Transform.

The proposed methodology is dealt in Section 3. Section 4 presents the experimental results and comparative analysis. Discussions are carried out in Section 5. Conclusions are summarized in Section 6.

## 2    Discrete Curvelet Transform

Discrete Curvelet transform has been developed to overcome the limitations of wavelet and Gabor filters [15]. Wavelet transforms are not suitable for representing line singularities and thus do not represent objects containing randomly oriented edges and curves [15]. Gabor filters are found to perform better than wavelet transform in representing textures and in the content based retrieval of images [16]. However, an effective representation of images in spectral domain is not possible with gabor filters due to the loss of spectral information [16]. To achieve a complete coverage of the spectral domain and to capture more orientation details in few coefficients, curvelet transform has been developed [17].

The Digital Curvelet Transform (DCLT) is usually implemented in the frequency domain for higher efficiency reasons [18]. Curvelet transform is a multiscale transform with a pyramid structure consisting of many levels at each scale

## 3    Proposed Semi-fragile Watermarking Scheme

In this paper, an intelligent semi-fragile watermarking using DCLT for authentic-cation and effective localization of tampering of watermarked images is presented. This approach consists of two modules namely, watermark embedding and watermark extraction with tamper detection.

### 3.1    Watermark Embedding

The process of generating and embedding watermark intelligently is depicted in Fig. 1. The input image is a gray scale image and is transformed using DCLT. The watermark is generated from the first level coarse DCLT coefficients. These coefficients are quantized and amount of quantization depends on their energy contribution. They are embedded into the first level coarse DCLT coefficients at the locations determined using pseudo random method [19]. Inverse DCLT is applied to get watermarked image. The number of scales to be used in DCLT has an impact on the imperceptibility of the watermarked image. We have conducted experiments to find the appropriate number of scales to be used in DCLT. Suppose the size of the input image is $NXN$, then the number of scales used in DCLT is varied from $2$ $to$ $log_2(N)$ [18].

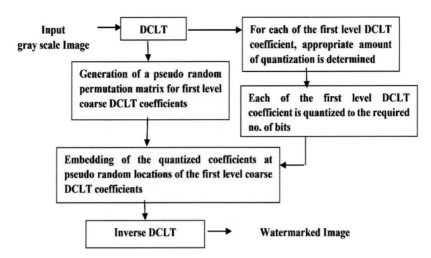

**Fig. 1** Semi-fragile watermark Embedding process

The perceptual quality of the watermarked image is measured using Peak Signal to Noise Ratio (PSNR) [20] and is computed using equations below:

$$MSE = \frac{1}{NXN} \sum_{i=0}^{N} \sum_{j=0}^{N} (I_s(i,j) - W_s(i,j))^2 \tag{1}$$

$$PSNR = 20 * log_{10}(MAX_{I_s}) - 10 * log_{10}(MSE) \tag{2}$$

where,

$I_s(i,j)$ – represents the gray scale intensity value of the pixel of the sample input image at location $(i,j)$, $W_s(i,j)$ – represents the gray scale intensity value of the pixel of the watermarked image at location $(i,j)$, $MAX_{I_s}$ - maximum intensity value of the pixel of a sample input image $I_s$.

The average PSNR values of all the images in the corpus for different dimensions of the input image at different scales used in DCLT are shown in Table 1. It is observed from the values shown in Table 1 that, if the value of scale is less than or equal to $(log_2(N) - 3)$, there is a significant increase in PSNR values. However, if value of scale exceeds $(log_2(N) - 3)$, there is a slight increase in the PSNR values. Hence, to preserve the imperceptibility of the watermarked image, we have arrived to the decision that it is enough to have number of scales equal to $(log_2(N) - 3)$. So in this work, the input image is transformed using DCLT at the number of scales equal to $(log_2(N) - 3)$.

**Table 1** PSNR values of the watermarked image at different scales of DCLT coefficients

| Size of the Input Image $N \times N$ | PSNR value (in dB) Range of the scale (2 to $log_2(N)$) | | | | | | | | |
|---|---|---|---|---|---|---|---|---|---|
| | 2 | 3 | 4 | 5 | 6 | 7 | 8 | 9 | 10 |
| 128 X 128 | 36.46 | 41.51 | 47.62 | 53.76 | 59.11 | 59.91 | | | |
| 256 X 256 | 37.19 | 42.61 | 48.50 | 54.15 | 58.12 | 61.14 | 62.07 | | |
| 512 X 512 | 37.72 | 43.61 | 49.02 | 55.13 | 59.05 | 62.12 | 63.11 | 63.46 | |
| 1024 X 1024 | 38.01 | 44.14 | 49.89 | 55.94 | 59.66 | 62.84 | 63.55 | 64.21 | 64.62 |

The coarse coefficients of first level DCLT are extracted. The coefficients are quantized and amount of quantization is not uniform. The rationale behind this decision is that the contribution of each first level coarse DCLT coefficient to the information content of the image is not uniform. Hence, an intelligent embedding criteria is designed in this work, which takes into account the energy contribution of each DCLT coefficient. The energy contribution of each DCLT coefficient is defined by the following equation:

$$E_c(i,j) = \frac{|D_c(i,j)|}{|D_{max}|} \tag{3}$$

where,

$E_c(i,j)$- Energy contribution of the first level DCLT coarse coefficient $c$ at the location $(i,j)$, $D_c(i,j)$ – Magnitude of the first level DCLT coarse coefficient $c$ at the location $(i,j)$, $D_{max}$ - First level DCLT coarse coefficient with maximum magnitude

We have conducted experiments for all the images in the corpus to decide on the appropriate amount of quantization required. The corpus consists of images with varying information content and structure. We have computed the energy contribution of the first level DCLT coarse coefficients of all the images in the corpus using Equation (3). It is found that the contribution of each coefficient is not same. Hence, the coefficients are classified into different energy levels based on their contribution. The energy contribution level and amount of quantization values are recorded in Table 2. It is inferred from the values shown in Table 2 that, the energy contribution at each energy level varies and accordingly the amount of quantization could be decided intelligently for each energy level. Thus, it is arrived to the decision to set appropriate number of bits for quantization of the coefficients at each energy level and values are shown in Table 2.

We have also conducted experiments to measure the perceptual quality of the watermarked image and the accuracy of tamper detection of the all the images in the corpus. The perceptual quality of the watermarked image is measured in terms of PSNR using equations (1) and (2). The accuracy of tamper detection is evaluated using the following equation:

**Table 2** Energy contribution levels of first level DCLT coarse coefficients and the amount of quantization

| Energy Contribution Levels (in %) | Percentage of coefficients in the energy contribution level | Amount of quantization (in bits) |
|---|---|---|
| 90-100 | 27 | 4 |
| 75-89 | 48 | 3 |
| 50-74 | 19 | 2 |
| 25-49 | 3 | 1 |
| <25 | 3 | 0 |

$$Accuracy\ of\ Tamper\ Detection \qquad\qquad (4)$$
$$= \frac{Average\ Number\ of\ bits\ identified\ as\ tampered}{Average\ Number\ of\ bits\ actually\ tampered}$$

The results of the accuracy of tamper detection and PSNR values for both fixed number of Least Significant Bits (LSBs) and variable number of LSBs used for quantization of first level DCLT coarse coefficients are computed and values are shown in Table 3. It is observed from the accuracy of tamper detection values in Table 3 that using variable number of LSBs for the quantization of first level DCLT coefficients yields high accuracy in tamper detection than fixed number of bits. Further, from the PSNR values shown in Table 3, it is evident that imperceptibility of the watermarked image is better for variable number of LSBs than fixed number of LSBs. Since, only required number of LSBs are used for quantization of each first level DCLT coarse coefficient, the embedding capacity is optimized and consequently there is no deterioration of the watermarked image. Further, size of the matrix used for detection of tampered regions is also reduced. Thus, more accurate localization of tampered regions is possible. To achieve high PSNR values (good imperceptibility) with better tamper detection and optimized embedding capacity, it is arrived to the decision that number of LSB's to be used for quantization should be variable.

**Table 3** Accuracy of the tamper detection and PSNR of the watermarked image

| No. of Least Significant Bits used | PSNR values | Accuracy of tamper detection (in %) |
|---|---|---|
| 2 | 66.62 | 84.95 |
| 3 | 59.61 | 90.12 |
| 4 | 53.15 | 94.08 |
| 5 | 36.24 | 96.51 |
| 6 | 26.58 | 98.38 |
| Variable | 62.97 | 97.19 |

Each of the first level coarse DCLT coefficient is quantized to appropriate number of bits according to its energy contribution and embedded into a coarse coefficient, whose location is decided by a pseudo random permutation algorithm [19]. The embedding of the watermark is performed according to the equation given below:

$$D_{c_k}(m, n) = D_{c_k}^q(i, j) \qquad (5)$$

where, $D_{c_k}(m,n) -$ k$^{th}$ Least Significant Bit of the coarse first level DCLT coefficient at the location $(m,n)$, $D_{c_k}^q(i,j) -$ k$^{th}$ Least Significant Bit of the quantized coarse first level DCLT coefficient at the location $(i,j)$,

The range of $k$ is decided based on the value of $E_c(i,j)$ using the Equation below:

$$k = \begin{cases} 1..4, & 0.90 \le E_c(i,j) \le 1.0 \\ 1..3, & 0.75 \le E_c(i,j) \le 0.89 \\ 1..2, & 0.50 \le E_c(i,j) \le 0.74 \\ 1 & 0.25 \le E_c(i,j) \le 0.49 \\ 0 & otherwise \end{cases} \qquad (6)$$

The matrices $D_{c_k}$ and $D_{c_k}^q$ represents source and quantized first level coarse DCLT coefficients respectively. $D_{c_k}$ and $D_{c_k}^q$ are of same size. $D_{c_k}(m,n)$ is mapped with $D_{c_k}^q(i,j)$ using pseudo random algorithm [19].

## 3.2    Watermark Extraction

The watermarked image can undergo the incidental or intentional attacks before reaching the receiver. Checking the integrity and authenticity of the watermarked image are performed during watermark extraction process. The process of watermark extraction is shown in Fig. 2. The watermarked image is transformed using first level DCLT. The watermark is intelligently extracted from each of the first level DCLT coarse coefficient based on their energy contribution. The watermark is inversely permuted and dequantized.

At the receiver side, tamper assessment is carried out by dividing the dequantized and generated first level coarse DCLT coefficients into blocks of uniform size. The size of the blocks depends on (i) accuracy of tamper detection and (ii) processing time for analysis. We have conducted experiments to determine the appropriate size of block for tamper assessment. The accuracy of tamper detection is computed using Equation (4). The average accuracy of tamper detection and processing time required for tamper detection for all the images in the corpus is shown in Table 4. It is observed from the values in Table 4 that, the accuracy of tamper detection is better when each of the block is of size 2X2. Hence, it is arrived to the decision that the size of the block is set to 2X2.

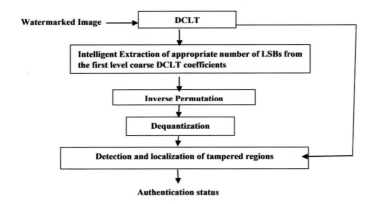

**Fig. 2** Semi-fragile Watermark extraction process

**Table 4** Accuracy of tamper detection and total processing time for varying size of the blocks

| Block Size | Tamper detection accuracy (in %) | Processing Time (in secs) |
|---|---|---|
| 2X2 | 97.19 | 4.91 |
| 3X3 | 96.34 | 4.46 |
| 4X4 | 92.87 | 4.07 |
| 5X5 | 89.44 | 3.48 |
| 6X6 | 86.29 | 3.05 |

The tamper detection and localization of tampered regions is performed using a Feature Similarity Index (FSI) based on both high and low level features of the image [14]. The highly informative features can be extracted at the points of high Phase Congruency (PC). Hence, PC is used as the primary feature in computing FSI. However, PC is invariant to contrast globally and sensitive to the local contrast of the image [21]. Thus, image Gradient Magnitude (GM) is computed as the secondary feature to encode contrast information. PC and GM are complementary and they reflect different aspects of the Human Visual System in assessing the local features of the input image [22]. There are different ways of computing PC and GM. In this work using the mechanism described in [14], PC and GM are computed. PC and GM extracts both content and structural information of the image.

The computation of FSI consists of two stages. In the first stage, the local similarity maps using PC and GM of the blocks of extracted and generated watermarks is calculated. In the second stage, these maps are pooled into a single similarity score. The similarity map based on PC is computed using equation below:

$$S_{PC} = \frac{2PC_1(b).PC_2(b)}{PC_1^2(b) + PC_2^2(b)} \tag{7}$$

where, $PC_1(b)$– Phase Congruency of generated first level DCLT coarse coefficient of block $b$, $PC_2(b)$– Phase Congruency of extracted first level DCLT coarse coefficient of block $b$. The similarity map based on GM is calculated using following equation:

$$S_{GM} = \frac{2GM_1(b).GM_2(b)}{GM_1^2(b) + GM_2^2(b)} \tag{8}$$

where, $GM_1(b)$ – Gradient Magnitude of generated first level DCLT coarse coefficient of block $b$, $GM_2(b)$ – Gradient Magnitude of extracted first level DCLT coarse coefficient of block $b$. FSI is computed by taking the product of $S_{PC}$ and $S_{GM}$ using the formula given below:

$$FSI = S_{PC}.S_{GM} \tag{9}$$

The tamper assessment is evaluated based on the value of *FSI* using the following equation. The threshold of 0.7 on FSI has been computed empirically, so that watermarked block is robust for most of the incidental attacks and fragile to the intentional attacks.

$$a(i) = \begin{cases} \text{tampered,} & FSI < 0.7 \\ \text{"not tampered",} & otherwise \end{cases} \tag{10}$$

where, $a(i)$ – Authentication status of each block $i$,

## 4    Experimental Results

We have created a corpus of different types of images for testing the proposed intelligent semi-fragile watermarking system. Most of the images are scanned document images like Cheque, Marks-cards, Identification and some images are taken from the standard image database USC-SIPI [25]. The existing method [1] has also been implemented. The results obtained from the existing method [1] and proposed method for a sample image of few example categories in the corpus are shown in Fig. 3. Each sample image is watermarked and subsequently tampered intentionally by performing malicious changes to the content. The blocks identified as tampered are marked as black patches for the existing method [1] and proposed method. The visual inspection of results shown in Fig. 3 reveals that the proposed method detects tampered regions more accurately.

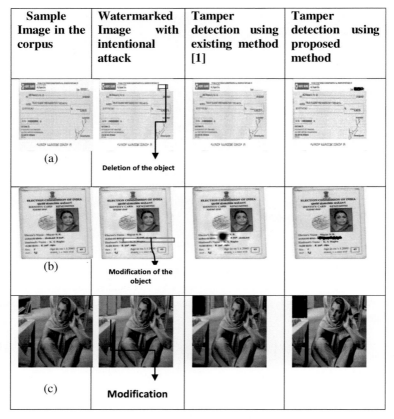

| Sample Image in the corpus | Watermarked Image with intentional attack | Tamper detection using existing method [1] | Tamper detection using proposed method |
|---|---|---|---|
| (a) | Deletion of the object | | |
| (b) | Modification of the object | | |
| (c) | Modification | | |

**Fig. 3** Results of tamper detection. (a) Cheque Image (b) Id (c) Sample Image from the USC-SIPI Image Database

# 5    Discussions

A comparative analysis of fragility and robustness of the existing method [1] and proposed method is carried out in detail in the following subsections.

## 5.1    Robustness Analysis

The robustness is evaluated using the parameter Normalization Correlation Coefficient (NCC) [20]. NCC is computed for the extracted and received first level DCLT coarse coefficients. The watermarked image is subjected to different incidental attacks. The NCC values are calculated for both existing method [1] and proposed method by varying the variance parameter of the noise. The graph is plotted for NCC values for different types of noise namely Salt and pepper, Speckle and Gaussian as shown in Fig. 4. In the plot shown in Fig. 4, it is observed that the

NCC values for the existing scheme [1] significantly drops as the variance increases. Consequently, robustness performance measured in terms of NCC also deteriorates with the increase in the variance of the noise. However, the NCC values for proposed method is stable for variance <= 0.1 and drops slightly for variance greater than 0.1. Thus, the proposed method exhibits a better and stable robustness compared to the existing method [1] for different types of noise.

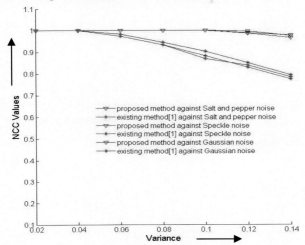

**Fig. 4** Robustness under Noise attacks

The watermarked image is also subjected to JPEG compression by varying the amount of compression. The amount of compression is varied from 30% to 5% in steps of 5% and NCC values are computed for both existing [1] and proposed methods. A plot depicting the robustness performance of the existing [1] and proposed schemes in terms of NCC values is shown in Fig. 5. It is observed from Fig. 5 that proposed scheme exhibits better NCC values than the existing scheme [1] for quality factors less than 15%.

The watermarked image is filtered using median filtering method for different size of the window. The size is varied from 3X3 to 11X11. For different sizes of the window, NCC values are computed for the existing [1] and proposed methods. The robustness performance of the existing [1] and proposed schemes in terms of NCC values is shown in Fig. 6. It is inferred that the NCC values for the existing method [1] drops substantially as the size of the window used in Median Filtering increases. The watermarked image is subjected to Histogram equalization. The number of bins in the Histogram Equalization is varied from 32 to 128. The plot showing robustness performance of the existing [1] and proposed methods in terms of NCC values is shown in Fig. 7. It is inferred from the NCC values in Fig. 7 that proposed method is highly robust than the existing method [1] against Histogram Equalization

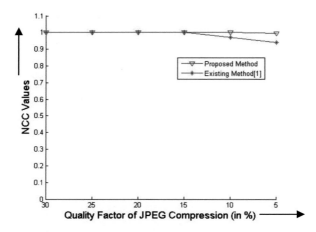

**Fig. 5** Robustness against JPEG compression

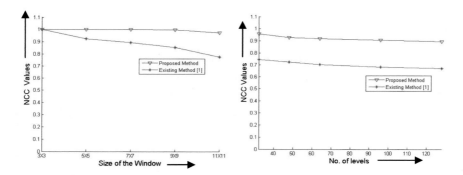

**Fig. 6** Robustness against Median Filtering    **Fig. 7** Robustness against Histogram Equalization

## 5.2    Fragility Analysis

The fragility is decided based on the tamper detection and localization of tampered regions of a semi-fragile watermarking scheme. We have tested all the images in the corpus for different types of tampering namely (i) introducing a new content (ii) deleting the existing content (iii) manipulating the existing content and (iv) performing multiple attacks in the same image.

The performance of detection of tampered regions for both existing method [1] and proposed method is measured in terms of parameters namely False Positive Rate (FPR) and False Negative Rate (FNR). FPR measures the number of bits incorrectly identified as tampered, whereas FNR measures the number of bits actually tampered but not identified. FPR and FNR are calculated as given below:

$$FPR = \frac{No.\, of\ \ bits\ incorrectly\ identified\ as\ tampered}{Total\ no.\, of\ bits\ tampered} \qquad (11)$$

$$FNR = \frac{No.\, of\ \ bits\ actaully\ tampered\ but\ not\ identified}{Total\ no.\, of\ bits\ tampered} \qquad (12)$$

The average values of FPR and FNR for both existing [1] and proposed method for all the images in the corpus against various intentional attacks are calculated and values are tabulated in Table 5. It is inferred from the average FPR and FNR values in Table 5 that, the proposed method results in better accuracy than the existing method [1]. The average accuracy is computed using the equation below and the average accuracy values for existing method [1] and proposed method is shown in Table 5.

$$Average\ Accuracy = 100 - (Average\ FPR\ or\ Average\ FNR) \qquad (13)$$

The average FPR or FNR in Equation (13) is calculated by taking average FPR or FNR values of all the attacks. It is evident from the average accuracy values shown in Table 5 that, we have achieved an average accuracy of more than 95% for different types of intentional attacks.  Thus the proposed approach performs better in detecting tampered regions than the existing method [1].

**Table 5** FPR and FNR analysis of tamper detection against various intentional attacks

| Attack | Average Accuracy of tamper detection (in %) | | | |
|---|---|---|---|---|
| | Existing Method [1] | | Proposed Method | |
| | FPR | FNR | FPR | FNR |
| Insertion | 15.11 | 13.67 | 4.51 | 3.97 |
| Deletion | 18.23 | 18.09 | 3.92 | 3.12 |
| Modification | 17.47 | 18.13 | 3.14 | 2.98 |
| Multiple Attacks | 24.16 | 20.98 | 5.01 | 4.21 |
| Avg. Accuracy | 81.26 | 82.29 | 95.86 | 96.43 |

# 6 Conclusions

An intelligent semi-fragile watermarking scheme based on DCLT has been proposed in this paper. This scheme intelligently embeds appropriate amount of watermark based on the energy contribution levels of the first level DCLT coarse coefficients. It does not require additional information to be communicated to the receiver. The proposed method is robust to most of the incidental attacks. In case of intentional attacks, the location of tampered regions is determined block-wise using FSI based on PC and GM features of the block. From the experimental results, it is observed that proposed method exhibits higher NCC values and thus, more robust than the existing method [1]. The proposed approach also leads to detection and localization of tampered regions with better accuracy in the watermarked image compared to the existing method [1]. The proposed work can be further enhanced by incorporating tamper recovery, which is considered as future work of the current study.

## References

1. Ghofrani, S., et al.: Image content authentication and tamper localization based on semi fragile watermarking by using the curvelet transform. In: TENCON 2012 - 2012 IEEE Region 10 Conference, Cebu, pp. 1–6 (2012)
2. Singh, P., et al.: A Survey of Digital Watermarking Techniques, Applications and Attacks. International Journal of Engineering and Innovative Technology (IJEIT) 2(9), 165–175 (2013)
3. Chang, W.H., Chang, L.W.: Semi-fragile watermarking for image authentication, localization, and recovery using tchebichef moments. In: International Symposium on Communications and Information Technologies (ISCIT), pp. 749–754 (2010)
4. Schirripa, G., Simonetti, C., Cozzella, L.: Fragile digital watermarking by synthetic holograms. In: Proc. of European Symposium on Optics/Fotonics in security & Defence, London, UK, pp. 173–182 (2004)
5. Dittmann, J., Ferri, L.C., Vielhauer, C.: Hologram watermarks for document authentications. In: Proceedings of IEEE International Conference on Information Technology, Las Vegas, pp. 60–64 (2001)
6. Aoki, Y.: Watermarking Technique Using Computer-Generated Holograms. Electronics and Communications in Japan, Part 3 84(1), 21–31 (2001)
7. Maeno, K., Sun, Q., Chang, S., Suto, M.: New semi-fragile image authentication watermarking techniques using random bias and nonuniform quantization. IEEE Trans. Multimedia 8(1), 32–45 (2006)
8. Gokhale, U.M., Joshi, Y.V.: A Semi Fragile Watermarking Algorithm Based on SVD-IWT for Image Authentication. International Journal of Advanced Research in Computer and Communication Engineering 1(4), 217–222 (2012)
9. Wu, X., Huang, J., Hu, J., Shi, Y.: Secure semi-fragile watermarking for Image authentication based on parameterized integer Wavelet'. Journal of Computers 17(2), 27–36 (2006)
10. Zou, D., et al.: A semi-fragile lossless digital watermarking scheme based on integer wavelet transform. In: IEEE 6th Workshop on Multimedia Signal Processing, pp. 195–198 (2004)
11. Wu, X., et al.: Reversible semi-fragile image authentication using zernike moments and integer wavelet transform. In: Digital Rights Management. Technologies, Issues, Challenges and Systems. Lecture Notes in Computer Science, vol. 3919, pp 135–145 (2006)

12. Kavadia, C., Shrivastava, V.: A Novel Digital Watermarking Technique based on Feature Attribute Selection using Integer Wavelet Transform Function and ID3 Algorithm. International Journal of Computer Applications **88**(16), 35–40 (2014)
13. Donoho, D.L., Duncan, M.R.: Digital curvelet transform: strategy, implementation and experiments. In: Proc. Society Optics and Photonics, vol. 4056, pp. 12–29 (2000)
14. Zhang, L., et al.: FSIM: A Feature Similarity Index for Image Quality Assessment. IEEE Transactions on Image Processing **20**(8), 2378–2386 (2011)
15. Starck, J.-L., Fadili, M.J.: Numerical issues when using wavelets. In: Meyers, R. (ed.) Encyclopedia of Complexity and Systems Science, vol. 14, pp. 6352–6368. Springer, New York (2009)
16. Chen, L., Lu, G., Zhang, D.S.: Effects of different gabor filter parameters on image retrieval by texture. In: Proc. of IEEE 10th International Conference on Multi-Media Modelling, Australia, pp. 273–278 (2004)
17. Candès, E.J., Demanet, L., Donoho, D.L., Ying, L.: Fast Discrete Curvelet Transforms. Multiscale Modeling and Simulation **5**, 861–899 (2005)
18. Candes, E., Donoho, D.: New tight frames of Curvelets and optimal representations of objects with C2 singularities. Comm. Pure Appl. Mathematics **57**(2), 219–266 (2004)
19. Dodis, Y., et al.: Threshold and proactive pseudo-random permutations. In: TCC 2006 Proceedings of the Third conference on Theory of Cryptography, Verlag Berlin, Heidelberg, pp. 542–560 (2006)
20. Lee, J., Won, Chee Sun: A Watermarking Sequence Using Parities of Error Control Coding For Image Authentication And Correction. IEEE Transactions on Consumer Electronics **46**(2), 313–317 (2000)
21. Morrone, M.C., Burr, D.C.: Feature detection in human vision: a phase-dependent energy model. Proc. R. Soc. Lond. B **235**(1280), 221–245 (1988)
22. Morrone, M.C., Ross, J., Burr, D.C., Owens, R.: Mach bands are phase dependent. Nature **324**(6049), 250–253 (1986)
23. Morrone, M.C., Owens, R.A.: Feature detection from local energy. Pattern Recognit. Letters **6**(5), 303–313 (1987)
24. Kovesi, P.: Image features from phase congruency. Videre: J. Comp. Vis. Res. **1**(3), 1–26 (1999)
25. http://sipi.usc.edu/database/

# Dental Image Retrieval Using Fused Local Binary Pattern & Scale Invariant Feature Transform

R. Suganya, S. Rajaram, S. Vishalini, R. Meena and T. Senthil Kumar

**Abstract**  In the field of dental biometrics, textural information plays a significant role very often in tissue characterization and gum diseases diagnosis, in addition to morphology and intensity. Failure to diagnose gum diseases in its early stages may leads to oral cancer. Dental biometrics has emerged as vital biometric information of human being due to its stability, invariant nature and uniqueness. The objective of this paper is to improve the classification accuracy based on fused LBP and SIFT textural features for the development of a computer assisted screening system. The swift expansion of dental images has enforced the requirement of efficient dental image retrieval system for retrieving images that are visually similar to query image. This paper implements a dental image retrieval system using fused LBP & SIFT features. The fused LBP & SIFT features identify the gum diseases from the epithelial layer in classifying normal dental images about 91.6% more accurately compared to other features.

**Keywords**  Dental image retrieval · Local binary pattern · Scale invariant feature transform · Fused LBP & SIFT · Gum diseases

R. Suganya(✉) · S. Vishalini · R. Meena
Department of Information Technology, Thiagarajar College of Engineering,
Madurai 625015, India
e-mail: rsuganya@tce.edu, vishalinisrithardeepi@gmail.com, rm201993@gmail.com

S. Rajaram
Department of Electronics and Communication Engineering,
Thiagarajar College of Engineering, Madurai 625015, India
e-mail: rajaram_siva@tce.edu

T. Senthil Kumar
Department of Computer Science and Engineering, Amrita School of Engineering,
Amrita Vishwa Vidyapeetham, Coimbatore, India
e-mail: t_senthilkumar@cb.amrita.edu

© Springer International Publishing Switzerland 2016
S.M. Thampi et al. (eds.), *Advances in Signal Processing and Intelligent Recognition Systems*,
Advances in Intelligent Systems and Computing 425,
DOI: 10.1007/978-3-319-28658-7_18

215

# 1   Introduction

Dental informatics is an emerging discipline applying computer and information science to dental practice, research, education and management. Dental radiograph and dental photograph are tools mostly used in biometrics as it provides information about teeth in detail. Teeth are parts of human organ that are not easily decayed and located inside mouth. It has its own characteristics based on a number of distinctive features for each individual tooth. Therefore,teeth based identification is one of reliable tools for human identification. An efficient CBIR system can help the doctors in retrieving similar medical image from the database to diagnose the disease efficiently. In CBIR system each image is stored in the database has its features extracted and compared to the features of the query image & similarity measurement techniques.

The main objective of this paper is to implements a dental image retrieval system using fused feature extraction methods-Local Binary Pattern (LBP) & Scale Invariant Feature Extraction (SIFT). The fused LBP and SIFT features identify the gum diseases associated in dental images. The paper is organized as follows: Section 2 presents literature survey;Section 3 outlines the proposed methodology; Section 4 discusses the experimental results. Finally, the paper is concluded in section 5.

# 2   Literature Survey

Several research works have been reported in the literature survey for analyzing the dental image retrieval for gum diseases. Many of these systems have utilized vectors to store and retrieve images,since comparison of vectors is relatively simple,but in this case,spatial relationship of images region is not considered. Distances metrics are now an important problem in information retrieval [6]. Ashish Oberoi, Deepak Sharma,Manpreet Singh provides the CBIR system for medical images based on various techniques for feature extraction and similarity measurement in bipartite graph partitioning and integrated minimum cost matching technique [9]. The combining features are giving the efficient retrieval of face Recognition in Labeled Faces in Wild (LFW) images efficiently achieving better retrieval rate [3]. Content based image retrieval for ultrasound kidney images by using LBP as a feature extraction method is used to retrieve the similar images from the database [7]. The performance of algorithms for data classification often depends heavily on the availability of a good metric. Understanding the relationship among different distance measures is helpful in choosing a proper one for a particular application. Texture can be defined as the spatial distribution of gray levels. Texture analysis able to extracts the texture features namely contrast,directionality,coarseness and busyness and it is applicable in computer vision, pattern recognition, segmentation and image retrieval [5].

## 3   Feature Extraction Methods

A texture is a manner in which the constituent parts are united. It is the structure or repeated patterns on an image. Texture in an image can be determined by if the neighboring pixels satisfy a specified criterion of similarity. Local binary pattern algorithm is used for texture feature extraction in dental image [6].

### 3.1   Local Binary Pattern (LBP)

LBP method provides a robust way for describing pure local binary patterns designed for extracting in a texture feature in dental images. The original $3 * 3$ neighborhoods threshold is identified by the value of the center pixel. The LBP operator represents the texture in the image by thresholding the neighborhood with the gray value of its center pixel and the results will be described as binary code format. The pixel-to-pixel comparison in the image produces the texture and the resulting image is in the form of texture histogram. Local binary pattern method is gray scale invariant and can be easily combined with a simple contrast measure by computing for each neighborhood the difference of the average gray level of those pixels which have the value 1 and those which have the value 0 respectively which is shown in fig 1. The resultant values are summed and then replaced the pixel value of center. The mathematical expression for LBP is given below.

$$LBP = \sum_{p=1}^{p} 2^{p-1} X_f (G_p - G_c) \tag{1}$$

$f(x) = \{1, x \geq 0$
$\quad\quad \{0, else$

LBP is a two-valued code. The LBP value is computed by comparing gray value of center pixel with its neighbors, using the above two equations,where

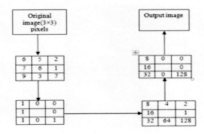

**Fig. 1**  Local Binary Pattern

$G_P$- gray value of its neighbors $G_C$-gray value of the center pixel P - number of neighbors R - radius of the neighborhood The dental disease like caries, Gum disease, Gingivitis and Peritonitis can be easily identified based on the texture of the teeth.

## 3.2 Scale Invariant Feature Transform (SIFT)

Scale invariant feature transform (SIFT) is an algorithm to detect and describe local features in the dental images. Important characteristic of these features is that the relative positions between them in the dental image shouldn't change from one image to another. SIFT can robustly identify objects even among clutter and under partial occlusion,because the SIFT feature descriptor is invariant to uniform scaling, orientation, and partially invariant to affine distortion and illumination changes. SIFT involves four different stages for extracting the features from the images is shown in the fig 2. The stages are Scale-space extreme detection, Key point localization, Orientation assignment, Key point descriptor.

$$L(X, Y, \sigma) = G(X, U, \sigma) * I(X, Y) \tag{2}$$

$$G(X, Y, \sigma) = 1/2\Pi\sigma * e^{(-(x^2+y^2))}/2s \tag{3}$$

$G(X, Y, \sigma)$ is a variable scale Gaussian,
$I(X, Y)$ is the input image,
$\sigma$ is a width of image.

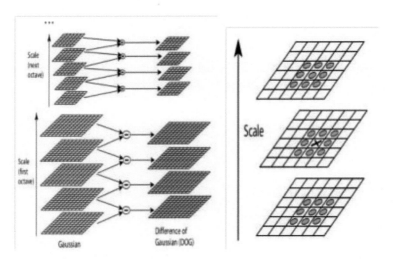

**Fig. 2**  Scale Invariant Feature Transform

SIFT extracted the shape features from the dental images. The dental image has multiple curves that can be detected from the image and matched that to database dental images and retrieved the similar dental images.

## 4 Proposed Methodology Fused LBP and SIFT

For dental image the texture and shape are very important features. The LBP only extract the texture features and the SIFT extract the shape features. The extracted features of LBP and SIFT has to be combined and the resultant feature is given to the similarity measurement technique. The block diagram for dental image retrieval by fused LBP & SFIT is shown in the fig 3. Combining of these features will produces a good result. Steps for fused LBP& SIFT for dental feature extraction:

1. For each pixel in the cell, compare it with other 8 neighboring pixels
2. Follow the pixel along a circle i.e. clockwise or anticlockwise
3. If the centre pixels value is greater than the neighbor, write 1 otherwise Write 0. This gives an bit binary number.
4. Compute the histogram over the cell, of the frequency of each number occurring.
5. Optionally normalize the histogram.
6. Concatenate normalized histogram of all cells
7. This gives the feature vector for the window. This can be used for classification (Eco-friendly computing and communication systems.

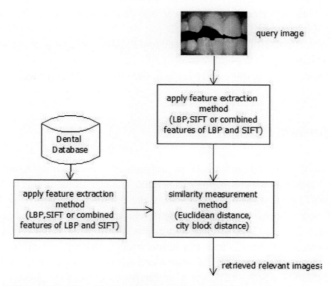

**Fig. 3** Block diagram for Dental Image Retrieval by fused LBP & SIFT

## 4.1  Image Retrieval Using Similarity Measures

Distance metric is the main tool for retrieving similar images from large dental databases. Euclidean distance and City block distance metrics give good performance in terms of precision using quantized histogram texture features in the DCT domain for compressed images [6]. In CBIR for Dental images, Euclidean distance and City block distances are used for the purpose of similarity comparison.

**Euclidean Distance (ED).** The Euclidean distance is calculated for every query image and the collection of images in the database to retrieve the similar images.

$$ED = \sqrt{\sum_{i=1}^{N} (f_q(i) - f_{db}(i))^2}$$

**City Block Distance (CBD).** City Block distance is the sum of absolute distance between the two vectors of query image p1 at (x1, y1) and database p2 at (x2, y2).

$$CB(p_1, p_2) = |X_1 - X_2| + |Y_1 - Y_2|$$

## 4.2  Dental Image Retrieval Process

The following steps are performed in the retrieval process:

1. Load dental database & input query image.
2. Read dental images one by one from the data base.
3. Extract the features from input query image & database images by using LBP, SIFT, fused LBP & SIFT.
4. Compare the features of the query image and database image using fused LBP & SIFT technique.
5. Store and sorting the result finally retrieve the similar dental images.

## 5  Experimental Results

The experimental is carried out with 2D dental dataset collected from Krishna Dental Care Centre, Madurai with anthropometrics data. The size of the image is 256 X 256 and it is in JPEG format. Some images from the database are shown in following fig. 4. The figure 4 shows the extracted features based on texture and shape. The dataset includes Caries, Gum disease, Gingivitis, Peritonitis, Teeth Fraction, Orthodontic, Trauma and Oral Cancer. For dental image retrieval, we extracted twenty three LBP features; thirty one SIFT features and twenty six fused LBP & SIFT features and fed them to a Support vector machine for automated diagnosis. 90 images (51 gum diseases and 39 normal images) were used for analysis. fused LBP & SIFT

**Fig. 4**   a. Original dental image, b. Texture feature extraction by LBP, c. Shape feature extraction by SIFT

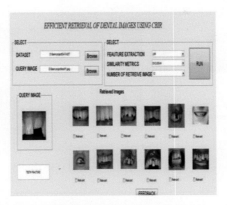

**Fig. 5**   Dental retrieval by using LBP with CBD For Teeth Fraction

**Fig. 6**   Dental retrieval by using LBP with ED For Teeth Fraction

feature provide a good precision of 91.6% and accuracy (91.5%). The experimental study conducted for 90 patients of dental diseases, The dental images for teeth fraction are retrieved by using feature extraction methods such as LBP, SIFT and fused LBP & SIFT which is shown in fig 5, 6, 7 & fig 8. Similarity measurement techniques are Euclidean Distance and City Block Distance. The feature extraction LBP with Euclidean distance produces 58.33% precision and 60.52% accuracy. The LBP with City Block Distance also produces 67.6% accuracy which does not

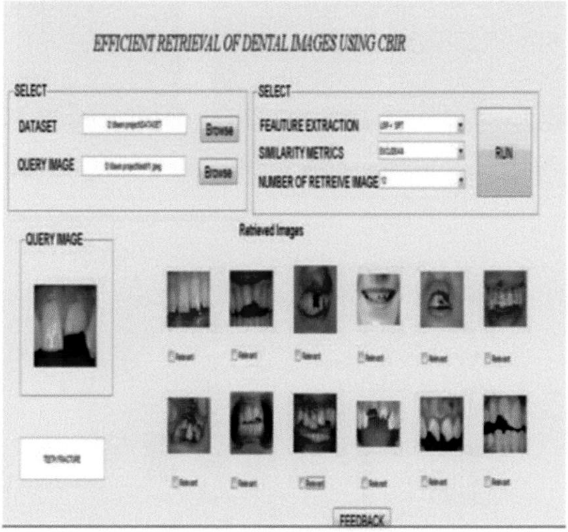

**Fig. 7** Dental retrieval by using fused LBP & SIFT with ED for Teeth Fraction

**Fig. 8** Dental retrieval by using fused LBP & SIFT with CBD for Teeth Fraction

support doctors decision. Similarly, by applying SIFT method alone for dental image retrieval produces 67% - 71% either by ED or CBD similarity metrics. Whereas, our proposed fused LBP & SIFT with City Block Distance gives more accurate results 91.6% accuracy compared to other features.

**Table 1** Comparison of Retrieval accuracy rate for LBP, SIFT and fused LBP & SIFT techniques by using similarity measures

| Feature extraction Method | Similarity measurement techniques | Precision (%) | Recall(%) | Accuracy (%) |
|---|---|---|---|---|
| LBP | ED | 58.33 | 41.17 | 60.52 |
| LBP | CBD | 66.67 | 53.33 | 67.64 |
| SIFT | ED | 66.67 | 53.33 | 67.64 |
| SIFT | CBD | 75 | 52.94 | 71.05 |
| Fused LBP& SIFT | ED | 84 | 22.94 | 81.05 |
| Fused LBP& SIFT | CBD | 91.6 | 13.45 | 91.57 |

## 6 Conclusion

The main novelty of our approach is to construct a similarity metric suited for dental image retrieval based on fused LBP & SIFT feature extraction. In this paper, we have developed an efficient content based image retrieval system based on various feature extraction methods and similarity measurement techniques. The dental images are retrieved by using feature extraction methods such as LBP, SIFT and fused LBP & SIFT with similarity measurement techniques like Euclidean Distance and City Block Distance. It has been observed from the experimental study that the fused LBP & SIFT helps the doctors to identify the gum diseases from the epithelial layer in classifying normal dental images about 91.6% precision, more accurately compare to other features.

**Acknowledgements** The authors convey their heartfelt thanks to Dr.K. Sudha, Dental Surgeon, Sri Krishna Dental care, Madurai for providing the dental image dataset used in this paper and continuous support for conducting this work and valuable suggestions at different stages of the work.

## References

1. Sable, D.R.G.: Teeth Feature Extraction and Matching for Human Identification Using Scale Invariant Feature Transform Algorithm. European Journal of Advances in Engineering and Technology (2015)
2. Velmurugan, K.: A Survey of Content-Based Image Retrieval Systems using Scale-Invariant Feature Transform (SIFT). International Journal of Advanced Re-search in Computer Science and Software Engineering **4**, January 2014
3. Tayade, Y.R., Bansode, S.M.: An Efficient Face Recognition and Retrieval Using LBP and SIFT. International Journal of Advanced Research in Computer and Communication Engineering **2**, April 2013
4. Senthil Kumar, R., Senthilmurugan, M.: Content-Based Image Retrieval System in Medical Applications. International Journal of Engineering Research & Technology **3**, March 2013
5. Ren, J., Jiang, X., Yuan, J.: Dynamic texture recognition using en-hanced LBP features. In: IEEE Int. Conf. Acoustics, Speech, and Signal Processing, May 2013

6. Malik, F., Baharudin, B.: Analysis of distance metrics in content-based im-age retrieval using statistical quantized histogram texture features in the DCT do-main. Journal of King Saud University-Computer and Information Sciences **25**, 207–218 (2013)
7. Hussain, S.A., Holambe, A.N., Shaikh, Z.: A Comparative Study Of Diffrent Transformation Techniques For Cbir. International Journal of Engineering Research and Technology **1**, October 2012
8. Oberoi, A., Singh, M.: Content Based Image Retrieval System for Medi-cal Databases (CBIR-MD) - Lucratively tested on Endoscopy, Dental and Skull Images. International Journal of Computer Science **9**, May 2012
9. Oberoi, A., Sharma, D., Singh, M.: CBIR-MD/BGP: CBIR-MD System based on Bipartite Graph Partitioning. International Journal of Computer Applications **52**, August 2012

# Block Based Variable Step Size LMS Adaptive Algorithm for Reducing Artifacts in the Telecadiology System

Thumbur Gowri and P. Rajesh Kumar

**Abstract** In this paper, an efficient Block Based Error Data Nonlinear Variable Step Size Least Mean Square (BBEDNVSSLMS) adaptive algorithm is used to reducing the noises present in the cardiogram signal. Now a day's Heart attack is main problem in the world. This problem is very important when patients are present far away from the medical diagnosis centre. So in these cases Telecardiology (ambulatory) system will help to patient for proper treatment within their area. When we are measuring the Electrocardiogram (ECG) signal from the patient it will undergo numerous noises. In this paper we proposed efficient BBEDNVSSLMS adaptive algorithms for reducing of Power Line Interference (PLI) noise and Electrode Motion (EM) artifact noise. Also, we derived sign based algorithms based on BBEDNVSSLMS algorithm, which will give less computational complexity. Finally these algorithms are applied to corrupted ECG signals. By analyzing the simulation values for different factors on, Signal to noise ratio, convergence characteristics and Mean square error values, these algorithm gives better elimination of noise in the ECG signal compared to LMS algorithm.

**Keywords** Adaptive filtering · Artifacts · Mean square error · Signal to noise ratio · Misadjustment ratio

## 1 Introduction

The exact interpretation of ECG signal is critical and extraction of PQTST values is also a very important in the clinical laboratories, so it is a challenging process [1].

T. Gowri(✉)
Department of ECE, GIT, GITAM University, Visakhapatnam 530045, AP, India
e-mail: gowri3478@gmail.com

P.R. Kumar
Department of ECE, AUCE, Andhra University, Visakhapatnam 530003, AP, India
e-mail: rajeshauce@gmail.com

© Springer International Publishing Switzerland 2016
S.M. Thampi et al. (eds.), *Advances in Signal Processing and Intelligent Recognition Systems*,
Advances in Intelligent Systems and Computing 425,
DOI: 10.1007/978-3-319-28658-7_19

225

When measuring ECG signal in the laboratory or ambulatory system and transmitting signal through telecardiology system, signal will undergo numerous artifacts. These artifacts, in general, are Electrode Motion (EM), PLI, Muscle Artifact (MA) noise, instrument noise [2-3] and while transmitting the signal for diagnosis  centre channel noise also add.  These noises may corrupt the original ECG signal characteristics and masks the tiny features of ECG signal. From the decades onwards there are number of approaches are used to remove the noise present in the signal.

We can use non adaptive [4] filters and adaptive filters to reduce the noise signal. Shazia Javed et.al [5] presented an adaptive noise cancelation model to remove baseline wander (BW) noise from mathematically modelled ECG signals. Gowri et al. [6] and Verulkar et al. [7] proposed filtering techniques for reduction of Power Line Interference in ECG signals. Pradeep et.al [8] used hybrid technique by combining of wavelet and empirical mode decomposition method to reduce the noise in ECG signal. There are various techniques are used to enhance the ECG signal quality in the literature [9-12] using adaptive filters. If the incoming signal is non- stationary then the adaptive filter gives good analyzation of the signal output, comparing with Weiner filter, with some known information from the signal. The variable step size LMS gives better convergence rate and less misadjustment factor compared to LMS algorithm [13-14].

In this paper, we proposed Block Based Error Data Normalized Variable Step Size LMS (BBEDNVSSLMS) algorithm. This algorithm gives less computational complexity than the EDNVSSLMS algorithm. Further, to reduce the computational complexity, we have derived three algorithms from the proposed BBEDNVSSLMS algorithm. These are Sign Regressor BBEDNVSSLMS, Sign BBEDNVSSLMS and Sign Sign BBEDNVSSLMS algorithms. The original ECG signal is taken from Physionet database for analysing of these algorithms.

## 2    Basic Least Mean Square (LMS) Adaptive Algorithm

For fast few decades onwards adaptive filter plays a very important role in the analysis of different signals. The basic adaptive filter structure is shown in Fig 1. Consider an adaptive filter of length K with desired signal $d(q)$. For the  input vector  $X(q)$, the system generates output signal  $y(q)$ is shown in  the following equation.

$$y(q) = W^T(q)X(q) \tag{1}$$

The  vector  representation  for  the  input  signal  $X(q)$  is  given  by $X(q) = [x(q), x(q-1),...., x(q-k+1)],$ and  the  tap  weight  vector  at  $q^{th}$ index  is $W(q) = [w(q), w(q-1),...., w(q-k+1)]^T.$

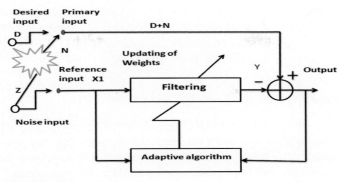

**Fig. 1** Adaptive filter structure

The weight updated vector for the LMS algorithm is given by the following equation.

$$W(q+1) = W(q) + \mu \, e(q)X(q) \, . \tag{2}$$

Where $\mu$ is the step size, it can be chosen as maximum eigen value of the auto correlation function of the input signal that is $\frac{1}{2\lambda_{max}}$ (or $\frac{1}{tr(R)}$), which controls the mean square error of the signal. Whereas $\lambda_{max}$ is the maximum eigen value.

## 3    Proposed Block Based Error Data Normalized VSSLMS Algorithm

The step size is fixed in the LMS algorithm; due to    this we cannot get fast convergence rate and also not get low steady state misadjustment when iterations increase. In the variable step size algorithm, step size controls these factors: which gives lower misadjustment ratio [10], and if we apply normalization to the step size it gives faster convergence rate [15] compared to the LMS algorithm. For reducing of the Excess Mean Square Error (EMSE), fast convergence and low misadjustment; we proposed an Error Data Normalized VSSLMS algorithm. To reduce the computational complexity, we adopt Block Based processing of normalized algorithms. The proposed EDNVSSLMS weight update equation is as follows.

$$W(q+1) = W(q) + \mu(q) \, e(q)X(q), \tag{3}$$

where $\mu(q)$ is the variable step size

$$\text{for ENVSS} \quad \mu(q) = \frac{\mu}{z + e^T(q)e(q)} \tag{4}$$

For EDNVSS,

$$\mu(q) = \frac{\mu}{\beta\left(\|E(q)\|^2\right) + (1-\beta)\left(\|X(q)\|^2\right)}.$$  (5)

where $\mu$ is defined as $\mu = \mu - \left(\frac{0.01}{q}\right)$ and chosen $\beta < 1$.

The variable step size $\mu(q)$ is high at the starting iterations of the signal so we will get fast convergence rate and when iterations increases then step size reduced, due to this steady state misadjustment error is reduced. For somehow, further fast convergence and error rate can be minimized with error data normalized VSSLMS. For less computational complexity and to reduce the gradient noise we applied Block Based Error Data Normalized VSSLMS algorithm (BBEDNVSSLMS). The weight updated equation for BBEDNVSSLMS algorithm is given below.

$$W(q+1) = W(q) + \frac{\mu(q)}{X_{Mj}^2 + \varepsilon} X(q)\, e(q),$$  (6)

where,   $X_{Mj} = \max\{|X_m|, m \in Z_j'\}$, $Z_j' = \{jk, jk+1, \ldots, jk+k-1\}$, $j \in Z$.   When $X_{Mj} \neq 0$   then choose $\varepsilon = 0$.

In Block based normalization, the maximum one value is chosen from that block for analysing the weight equation. For fast transmission of the signal in the telecardiology systems, we applied Signum function [16] to the Block based algorithms, which will further reduce the computational complexity of the signal. These algorithms are best suit in implementation of hardware design such as Application Specific Integrated Circuits (ASIC). The sign function is defined as follows.

$$\mathrm{sgn}\{e(q)\} = \begin{cases} 1: & e(\mathrm{q}) > 0 \\ 0: & e(\mathrm{q}) = 0 \\ -1: & e(\mathrm{q}) < 0 \end{cases}$$  (7)

The sign function tell that: when error value is greater than zero then assign target value as one, when its value is less than zero then target assigned negative one, otherwise when the incoming value is zero then assign target value also zero. This Sign function can be applied for only input data [15] function or apply only for error function   or we can also apply for both the error and data functions. For Sign Regressor BBEDNVSSLMS algorithm, the signum function is applied to data function. The updated weight vector equation is defined as follows.

$$W(q+1) = W(q) + \frac{\mu(q)}{X_{Mj}^2 + \varepsilon} \mathrm{sgn}\left(X(q)\right)\, e(q).$$  (8)

Comparing equation (6) and (8), the data is clipped in (8), i.e., $X(q)$ is replaced by $\mathrm{sgn}\big(X(q)\big)$ so the number of multiplication required is reduced. The signum can be applied to the error function in (6) and the resulting algorithm is called as Sign BBEDNVSSLMS algorithm. Due to clipping of error, the number of multiplications reduced. Its weight equation is as follows.

$$\mathrm{W}(q+1) = W(q) + \frac{\mu(q)}{X_{Mj}^2 + \varepsilon} X(q)\,\mathrm{sgn}\big(e(q)\big) \,. \tag{9}$$

Further, to reduce the computational complexity in (6), we applied sign function for both error and data. The weight for the resulting BBEDNVSSLMS algorithm as follows.

$$\mathrm{W}(q+1) = W(q) + \frac{\mu(q)}{X_{Mj}^2 + \varepsilon} \mathrm{sgn}(X(q))\,\mathrm{sgn}\big(e(q)\big) \,. \tag{10}$$

While analyzing (8) and (9) with (10), the output signal quality is reduced in the PLI reduction simulation results but in eliminating of EM, equation (10) gives better result. The mean square error verses number of iterations graph indicates the convergence characteristics of the signal. We have chosen step size is high in the initial    condition of the signal, for faster convergence rate and slow step size variation at middle of the iteration of the signal for less steady state error variations. The convergence characteristics for different algorithms are observed in fig 2. From the characteristics we analyzed that for data 105, the BBSREDNVSSLMS algorithms gives the better MSE and Power line interference that is  50/60 Hz noise elimination is compared to other algorithms as discussed above.

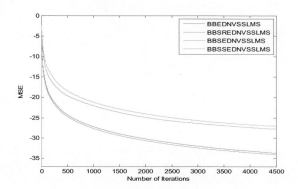

**Fig. 2** Convergence characteristics for different adaptive algorithms.

# 4    Simulation Analysis

In order to analyze the designed various adaptive algorithm characteristics are best, the original ECG signal is taken from the MIT-BIH physionet database [17]. These data base contains total 47 patients records. From these we have chosen 6 records: 100,101,102,103, 104 and 105 record numbers. The files are sampled at the rate of 360 Hz per channel with a resolution of 11 bits over 10mV range. For each record, we have collected 4500 samples. We plotted graph for the record number 105 and the values are tabulated for the remaining records. We have taken the number of samples x-axis and amplitude of the signal on y-axis for all graphs, expect except frequency spectrum graph.  We have taken he variance value as 0.001 and is added to the original noise. We consider the step size for BBEDN and BBSREDN algorithms as 0.2 and the step size for the remaining sign algorithms as 0.04. We have chose the order of the filter as 5.

From the MATLAB simulation results, the frequency spectrum graph for elimination of PLI noise is shown in fig 3. From this graph we can analyze that the marked line indicates 60Hz noise in fig 3(a), using different adaptive algorithms this 60Hz  main power  line noise is  reduced   is  shown in fig 3(b-e). The PLI noise is a non physiological noise i.e. whenever we connect instrument for measuring ECG signal from the person then this power line noise is automatically added; so removing of PLI noise is very important. The PLI noise eliminated using different algorithms is shown in fig (4) over the duration of 4500 samples. From this graph we can observe that the BBSEDN-VSSLMS algorithm has still some noise residue when compared with the remaining algorithms.

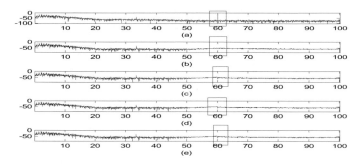

**Fig. 3** (a) Frequency spectrum of ECG with 60Hz power line inter ference (b) spetrum using BBEDNVSSLMS filtering (c)spectrum using BBSREDNVSSLMS filtering (d)spectrum using BBSEDNVSSLMS filtering  (e)spectrum using BBSSEDNVSSLMS filteriing

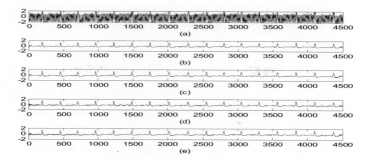

**Fig. 4** (a) Corrupted ECG  signal with PLI.    Recovered ECG signal after Appling of adaptive algorithms  (b)  BBEDNVSSLMS  (c) BBSREDNVSSLMS (d)BBSEDNVSSLMS and  (e)BBSSEDNVSSLMS.

**Table 1** Comparision of different adaptive algorithms  for reducing of PLI artifact. (in dB.)

| S. No. | Algorithm | Record No. | | | | | | Avg. SNR |
|---|---|---|---|---|---|---|---|---|
| | | 100 | 101 | 102 | 103 | 104 | 105 | |
| 1. | BBEDN VSSLMS | 13.128 | 12.658 | 11.937 | 12.787 | 12.953 | 12.562 | 12.670 |
| 2. | BBSREDN VSSLMS | 13.458 | 13.144 | 12.235 | 13.456 | 13.271 | 12.750 | 13.052 |
| 3. | BBSEDN VSSLMS | 8.4762 | 8.4534 | 7.8996 | 10.1385 | 9.4528 | 9.2345 | 8.9425 |
| 4. | BBSSEDN VSSLMS | 8.4161 | 8.1487 | 7.7647 | 9.9603 | 8.9127 | 9.6234 | 8.8043 |

The calculated Signal to noise ratio (SNR) values for different records using different adaptive algorithms are presented in Table 1. The average SNR for BBEDN-VSSLMS is 12.670 dB; Sign Regressor BB gets13.052dB; Sign BB gets 8.9425 dB and Sign Sign BB gets 8.8043 dB. From Table 1, it is clear that the BBSREDNVSSLMS algorithm gets high SNR value i.e. it eliminates more power line noise than other algorithms. The real non stationary or non physiological Electrode motion (EM) noise is collected from physionet noise stress database. For this noise we added random noise of 0.0001 because when we transmit the signal through telecardiology system the channel noise is added. The EM noise is occurring when we placing the electrode on the heart part. The reduction of EM noise using different adaptive algorithms is shown in fig (5). From the Table 2, we can observe that the BBSSEDNVSSLMS algorithm gets high SNR of 7.6899dB and the BBSEEDNVSSLMS algorithm gets low SNR of 3.0377dB. So for the EM elimination, the BB Sign Sign algorithm performs better with less computational complexity.

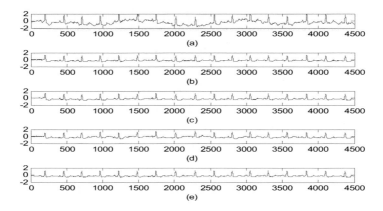

**Fig. 5** (a) ECG corrupted with EM. Elimination of EM using (b) BBEDNVSSLMS (c) BBSREDNVSSLMS (d) BBSEDNVSSLMS (e) BBSSEDNVSSLMS algorithms.

The Excess Mean Square Error (EMSE) characteristics for PLI reduction and EM reduction using different adaptive algorithms are shown in fig (6) and fig (7) respectively. The performance of EMSE and misadjustment ratio for different adaptive algorithms was presented in Table 3. The average misadjustment value and EMSE is less for BBEDNVSSLMS in the PLI elimination. For EM reduction the average EMSE is less in BBSSEDNVSS-LMS and misadjustment is less using BBEDNVSSLMS algorithm.

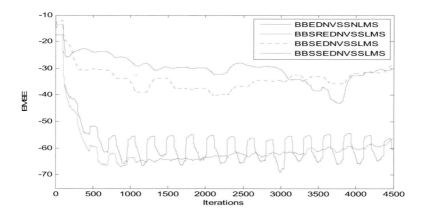

**Fig. 6** Excess Mean Square Error characteristics PLI using different adaptive algorithms.

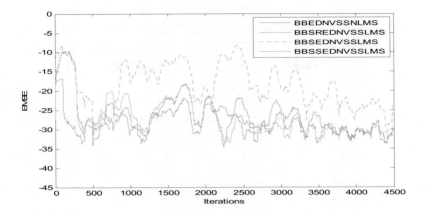

**Fig. 7** Excess Mean Square Error characteristics EM using different adaptive algorithms.

**Table 2** Comparision of different adaptive algorithms for reducing of EM artifact. ( in dB.)

| S. No. | Algorithm | Record No. | | | | | | Avg SNR |
|---|---|---|---|---|---|---|---|---|
| | | 100 | 101 | 102 | 103 | 104 | 105 | |
| 1. | BBEDN VSSLMS | 5.6778 | 6.8744 | 7.2499 | 4.8567 | 9.4566 | 6.5089 | 6.7707 |
| 2. | BBSREDN VSSLMS | 5.6342 | 6.3451 | 6.9872 | 3.2476 | 9.3453 | 6.9104 | 6.4116 |
| 3. | BBSEDN VSSLMS | 2.2834 | 3.1253 | 2.5261 | 3.3658 | 3.3434 | 3.5823 | 3.0377 |
| 4. | BBSSEDN VSSLMS | 7.6547 | 9.4563 | 9.2354 | 9.9547 | -0.1232 | 9.9617 | 7.6899 |

**Table 3** Comparision of different adaptive algorithms for reducing of PLI and EM artefact in terms of EMSE nad Misadjustment ratio (EMSE in dB.)

| Noise | Rec No | BBEDN VSSLMS | | BBSREDN VSSLMS | | BBSEDN VSSLMS | | BBSSEDN VSSLMS | |
|---|---|---|---|---|---|---|---|---|---|
| | | EMSE | $M_{ad}$ | EMSE | $M_{ad}$ | EMSE | $M_{ad}$ | EMSE | $M_{ad}$ |
| P L I | 100 | -54.187 | 2.9e-05 | -46.757 | 1.6e-04 | -23.985 | 0.031 | -22.946 | 0.030 |
| | 101 | -54.226 | 2.8e-05 | -47.254 | 1.4e-04 | -25.666 | 0.020 | -24.634 | 0.040 |
| | 102 | -51.385 | 9.0e-05 | -44.819 | 4.1e-04 | -36.424 | 0.003 | -35.424 | 0.003 |
| | 103 | -54.303 | 2.3e-05 | -46.789 | 1.3e-04 | -39.955 | 0.001 | -38.346 | 0.001 |
| | 104 | -52.963 | 3.9e-05 | -46.499 | 1.7e-04 | -39.690 | 0.001 | -38.755 | 0.001 |
| | 105 | -53.679 | 2.9e-05 | -47.259 | 1.3e-04 | -32.767 | 0.004 | -32.234 | 0.004 |

**Table 4** (*Continued*)

|   |   |   |   |   |   |   |   |   |   |
|---|---|---|---|---|---|---|---|---|---|
| | **Avg** | **-53.457** | **4.0e-05** | **-46.562** | **1.9e-04** | **-33.081** | **0.010** | **-32.056** | **0.013** |
| | 100 | -20.166 | 0.074 | -21.056 | 0.061 | -13.977 | 0.309 | -20.658 | 0.066 |
| | 101 | -23.026 | 0.037 | -22.276 | 0.044 | -15.834 | 0.195 | -24.772 | 0.025 |
| | 102 | -26.007 | 0.031 | -23.949 | 0.050 | -16.466 | 0.280 | -26.452 | 0.028 |
| E | 103 | -19.235 | 0.075 | -17.586 | 0.110 | -11.805 | 0.415 | -15.010 | 0.198 |
| M | 104 | -25.348 | 0.023 | -24.510 | 0.028 | -16.970 | 0.157 | -29.110 | 0.010 |
| | 105 | -25.025 | 0.022 | -25.257 | 0.021 | -15.881 | 0.177 | -27.060 | 0.014 |
| | **Avg** | **-23.134** | **0.043** | **-22.439** | **0.052** | **-15.155** | **0.255** | **-23.843** | **0.056** |

# 5    Conclusion

In this paper we have shown that reduction of PLI and EM noise, which occurs during recording of ECG signal from the patient. We developed Block Based Error Data Normalized VSSLMS based algorithms and their sign versions. These algorithms are applied to eliminate the PLI and EM noise in ECG signal. From the simulation analysis, we observed that PLI noise is better eliminated using BBEDN-VSSLMS algorithm and EM noise better eliminated using BBSSEDN-VSSLMS algorithms compared to other derived algorithms. We have also calculated different parameters such as EMSE, SNR and Misadjustment ratio for above algorithms.

# References

1. Yaqin, C., Li, G.: An improved algorithm of adaptive coherent model in the application of electrocardiograph. Signal Processing **18**, 244–248 (2002)
2. Limacher, R.: Removal of power line interference from the ECG signal by an adaptive digital filter. In: Proc. of European Tel. Conf, Garmisch-Part, pp. 300–309 (1996)
3. Rahman, M.Z.U., Shaik, R.A., Reddy, D.V.R.K.: Baseline wander and power line interference elimination from cardiac signals using error nonlinearity LMS algorithm. In: IEEE International Conference on Systems in Medicine and Biology, pp. 217–220 (2010)
4. Kabir, M., Shahnaz, C.: Denoising of ECG signals based on noise reduction algorithms in EMD and wavelet domains. Biomedical Signal Processing and Control **7**, 481–489 (2012)
5. Javed, S., Ahmad, N.A.: An adaptive noise cancelation model for removal of noise from modeled ECG signals. In: IEEE Region 10 Symposium, pp. 464–468 (2014)
6. Gowri, T., Sowmya, I., Rahman, M.Z.U., Reddy, D.V.R.K.: Adaptive power line interference removal from cardiac signals using leaky based normalized higher order filtering techniques. In: IEEE 1st Int. Conf. on Artificial Intelligence, Modeling & Simulation, pp. 259–263 (2013)
7. Verulkar, N.M., Zope, P.H., Suralkar, S.R.: Filtering Techniques for Reduction of Power Line Interference in Electrocardiogram Signals. International Journal of Research and Engineering Technology **1**(9), 1–7 (2012)

8. Pradeep Kumar, B., Balambigi, S., Asokan, R.: ECG denoising based on hybrid technique. In: Proc. of IEEE–ICAESM, pp. 285–290 (2012)
9. Bai, L., Yin, Q.: A modified NLMS algorithm for adaptive noise cancellation. In: IEEE on Intelligent Networks and Network Security, pp. 3726–3729 (2010)
10. Kwong, R.H., Johnston, E.W.: A variable step size LMS algorithm. IEEE Trans. Signal Processing 40(7), 1633–1642 (1992)
11. Li, N., Zhang, Y., Zhao, Y., Yanling, H.: An improved variable tap length LMS algorithm. Signal Processing 89, 908–912 (2009)
12. Haykin, S.: Adaptive filter theory, 4th edn. Pearson Education (2002)
13. Zhao, S., Manb, Z., Khoo, S., Hong, R.W.: Variable step-size LMS algorithm with a quotient form. Signal Processing 89, 67–76 (2009)
14. Aboulnasr, T., Mayyas, K.: A robust variable size LMS type algorithm: analysis and simulation. IEEE Trans. Signal Processing 45(3), 631–639 (1997)
15. Ahmed, I., Sulyman, S., Azzedine, Z.: Convergence and steady-state analysis of a variable step-size NLMS algorithm. Journal of Signal Processing 83, 1255–1273 (2003)
16. Rahman, M.Z.U., Shaik, R.A., Reddy, D.V.R.K.: An efficient noise cancellation technique to remove noise from the ECG signal using normalized signed regressor LMS algorithm. In: IEEE Int. Conf. on Bioinformatics and Biomedicine, pp. 257–260 (2009)
17. The-MIT-BIH-Arrhythmia-Database. http://physionet.org/physiobank/database/mitdb/

# Braille Tutor

## A Gift for the Blind

Anjana Joshi and Ajit Samasgikar

**Abstract** Education is the basic right of every human being. There are 37 million blind people in world and India has 15 million people among them. Visually impaired people use Braille language to write and read. Though there are many blind schools, the number of teachers who can effectively teach are very less. Moreover the blind children cannot read or write without the help of teacher. Thus, there is a need to develop a technology which helps these people to learn efficiently without anybody's help which will make them self-dependent. To meet this need the idea is to develop a product that can teach the blind children the basic alphabets and numbers. Further the product has data connectivity through Bluetooth link so that the teachers can use it to monitor what the students are learning. The result is the complete working device Braille tutor with the speakjet IC output through speaker and Bluetooth connectivity for monitoring each device.

**Keywords** Blind · Actuators · Arduino · Speakjet · Bluetooth · Braille

## 1 Introduction

The device called Braille Tutor is an automatic device which is capable of making the blind students learn alphabets in a Braille language without anybody's help. The basic idea is to use solenoid actuators for the purpose giving a feel of Braille script to the blind children. The actuators are controlled by atmega microcontroller (Arduino board). The device uses ten solenoid actuators for basic alphabets and numbers.

---

A. Joshi(✉)
School of Information Sciences, Manipal University, Manipal, India
anjana2492@gmail.com

A. Samasgikar
MMRFIC Systems PVT Limited, Bangalore, India
samajit@gmail.com

© Springer International Publishing Switzerland 2016
S.M. Thampi et al. (eds.), *Advances in Signal Processing and Intelligent Recognition Systems*,
Advances in Intelligent Systems and Computing 425,
DOI: 10.1007/978-3-319-28658-7_20

237

Further the device also has Bluetooth connectivity, the device can be connected to computers in the range of 300metres and this can be used by the teachers to monitor what the students are learning by user interface design on their respective computers. This device can be used to improve the literacy rate in the blind people. This also can give them a sense of independent learning without depending on any teachers.

## 2    Methodology

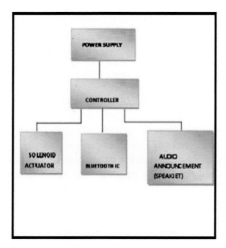

**Fig. 1** Block diagram of the device.

## 2.1    Description

Fig.1 shows the block diagram of the device Braille tutor. It takes the input from the user and gives output accordingly. There are three switches provided NEXT, PREVIOUS and a button to switch from alphabets to numbers and vice versa. The NEXT switch gives the combination of next letter/number and previous gives the combination of previous letter/number. For the output purpose LEDs and actuators are used. Actuators are used so that the user can sense the actuators which are high and learn the letters.

There are totally ten actuators used, six for letters and four for numbers. Also the audio announcement is given for the easy learning. LEDs are used as the reference for the teacher. Controller and actuator are powered up using 5v power supply. ULN 2003 is used for driving actuator .ATMEGA 128 (Arduino MEGA) is the heart of the device. This controls the overall working of the device.

## 2.1.1    Actuators

Fig. 2 shows the actuator used in the device. The actuators are high whenever the pin to which they are connected receives a high voltage. These are sensed by the user which tells them the letter present. The Operating Voltage is 5Volts and Stroke is 6mm.

**Fig. 2** Solenoid Actuator

## 2.1.2    ULN2003

ULN2003 is a monolithic IC consists of seven NPN Darlington transistor pairs with high voltage and current capability. It is commonly used for applications such as relay drivers, motor, display drivers, led lamp drivers, logic buffers, line drivers, hammer drivers and other high voltage current applications. It consists of common cathode clamp diodes for each NPN Darlington pair which makes this driver IC useful for switching inductive loads. Here in this case though the voltage of pin is sufficient to drive the actuator current is insufficient. Hence we are driving actuators through uln2003 which acts as current amplifier thus driving actuators.

**Fig. 3** ULN2003 IC which has seven inputs and seven outputs

### 2.1.3    Bluetooth Transceiver

We are using Bluetooth HC-05 transceiver to send the information about the things that user will be learning using Braille tutor device to teachers PC. Figure 4 shows the Bluetooth Transceiver. This transceiver is powered by 3.3 V and we make use of the TX, RX pins for the transmission. This acts as a generic serial COM port. Suppose there are ten blind students using the Braille tutor then all these are connected to a single PC that has user interface design for the teacher to monitor the learning of the student.

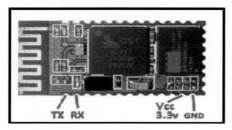

**Fig. 4** Bluetooth transceiver chip HC05

### 2.1.4    Speakjet

The Speakjet is a completely self-contained, single chip voice and complex sound synthesizer. It uses a mathematical sound algorithm to control an internal five channel sound synthesizer to generate on-the-fly, unlimited vocabulary speech synthesis and complex sounds.

The Speakjet is pre-configured with 72 speech elements (allophones), 43 sound effects, and 12 DTMF Touch Tones. Through the selection of these sounds, and in combination with the control of the pitch, rate, bend, and volume parameters, the user has the ability to produce unlimited phrases and sound effects, with Thousands of variations, at any time. In this device the speakjet is connected to arduino and is programmed to generate sounds of all the letters and numbers till ten.

The output of the speakjet IC after amplifying is given to the speaker which emits the sounds.

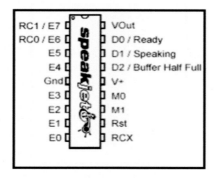

**Fig. 5** Speakjet IC

### 2.1.5 Arduino Mega- Atmega 128

The Arduino Mega 2560 is a microcontroller board based on the ATmega2560. It has 54 digital input/output pins (of which 15 can be used as PWM outputs), 16 analog inputs, 4 UARTs (hardware serial ports), a 16 MHz crystal oscillator, a USB connection, a power jack, an ICSP header, and a reset button. It contains everything needed to support the microcontroller; simply connect it to a computer with a USB cable or power it with a AC-to-DC adapter or battery to get started.

**Fig. 6** Arduino MEGA 2560

## 3 Implementation

The entire circuit is simulated in proteus software and results are obtained successfully. The mapping of PCB is done on eagle software. After integration of all the components the device is tested and errors are debugged. The speaker (100 ohm 5W) gives the voice output for the device. There is a provision on the device to use Headphones.

Arduino mega board has Atmega2560 which is programmed to control the Actuators, Bluetooth transceiver and speakjet IC. The program is in C language with some special keywords for functions. The Bluetooth transceiver can be tested using a software called teraterm, for which we need to pair the Bluetooth transceiver with the system that we will be using to monitor the Braille tutor, then enter the COM PORT of the Bluetooth transceiver in the tera term and the output will be displayed.

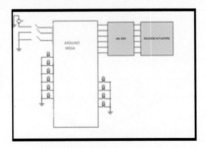

**Fig. 7** Shows how the Led's ,actuators and switches are connected to arduino. Led's are included in the device just for Reference of the onlooker.

This can be further improvised by developing a user interface by using Microsoft visual.

The simulation and the circuit diagram are done using Proteus software.LM386 IC (amplifier) is used to amplify the signal input to the speaker. A 10k potentiometer is used to adjust the volume. The volume adjustment can be provided on the front end of the device. A jack is used for connecting headphones.

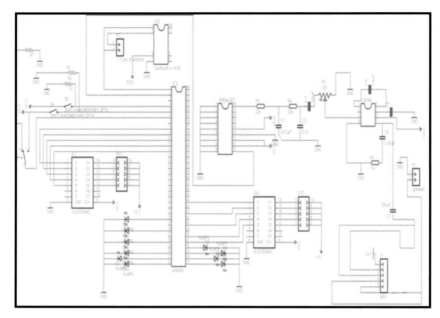

**Fig. 8** Circuit diagram

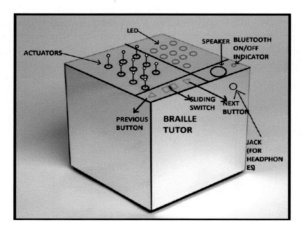

**Fig. 9** Prototype Product

# 4 Results and Evaluation

India is now home to the world's largest number of blind people. Of the 37 million people across the globe who are blind, over 15 million are from India. It is estimated that at least 200,000 children in India have severe visual impairment or blindness and approximately 15,000 are in schools for the blind. Although this represents a small percentage of the estimated 5 million blind in India, it is significant in terms of blind-years. Since this device helps a blind person learn independently, It will significantly increase the literacy rate among the blind.

The device was tested in Government Blind School, Hubli and we received a positive feedback. The blind children were able to learn the alphabets and the numbers on their own without the help of the teacher. They were able to sense the letters and the alphabets based on the output of the actuator.

# 5 Conclusion

This device "Braille tutor" makes visually impaired learns Braille without the help of a tutor and hence increases the literacy level in blind population. This was not possible with the Braille interpreter which only interpreted the written Braille script.

# References

1. Antonacopoulos, A., Bridson, D.: A robust braille recognition system. In: Marinai, S., Dengel, A. (eds.) DAS 2004. LNCS, vol. 3163, pp. 533–545 (2004)
2. Damit, A., et al.: Dual braille code translator: basic education tool for visually impaired children. In: 2014 International Conference on Computer, Communications, and Control Technology (I4CT), pp. 399–402 (2014)
3. Dias, M.B., et al.: An automatic braille writing tutor with multilingual exercises and educational games. In: IEEE Int. Conf. on Information and Communication Technologies and Development, p. 478 (2009)
4. Schroeder, F.: Literacy: The key to opportunity. Journal of Visual Impairment and Blindness, 290–293, June 1989
5. Halima, B.Z., Azlina, A.: Voice recognition system for virtually impaired. In: IEEE Int. Conf. on Information Technology (2008)
6. Shahbazkia, H.R., et al.: Automatic braille code translation system. In: Lazo, M., Sanfeliu, A. (eds.) CIARP 2005. LNCS, vol. 3773, pp. 233–241 (2005)
7. Yin, J., et al.: Mediated braille automatic recognition method. In: IEEE Int. Conf. on Frontier of Computer Science and Technology, pp. 618–624 (2010)
8. Chamot, J.: Electronic braille tutor teaches independence. National Science Foundation Press Release (2006)
9. Koedinger, K., Anderson, J., Hadley, W., Mark, M.: Intelligent tutoring goes to school in the big city. International Journal of Artificial Intelligence in Education **9**, 30–43 (1997)

10. Dias, M.F., et al.: Enhancing an automated braille writing tutor. In: Proceedings of the IEEE/RSJ International Conference on Intelligent Robots and Systems (IROS 2009), October 2009. https://www.ri.cmu.edu/pub_files/2009/10/Pper_IROS2009_Final_Submission_After_Review.pdf
11. Kalra, N., et al.: A braille writing tutor to combat illiteracy in developing communities. In: Artificial Intelligence in Information Communication Technology for Development Workshop at IJCAI 2007, Hyderabad, January 2007
12. Sanchez, B.R., et al.: Keyboard of automatic teller in braille code. In: IEEE Int. Conf. on Electronics, Communication and Conference 2006, p. 186974 (2006)

# Performance Evaluation of S-Golay and MA Filter on the Basis of White and Flicker Noise

Shivang Baijal, Shelvi Singh, Asha Rani and Shivangi Agarwal

**Abstract** EEG is one of the most effective diagnostic techniques to evaluate the electrical activity in brain. For that we use a variety of filters to reduce the artifacts existing in the EEG signal. The main objective of the paper is to compare the functioning of the two commonly used filters i.e. the moving average and S-Golay. This is done by creating a synthetic signal and then by adding white or flicker (pink) noise to it. The analysis is made on the basis of performance of filters on SNR of the final output while keeping the same number of reference points (frame length) and observing the distortion ratio. Furthermore the most important information of EEG signal lies in the peak of the signal so it becomes absolutely necessary to use a filter that not only filters out the noise better but simultaneously shows the ability to provide the least distortion from the original signal. The shape preserving characteristics of the filter are determined at different noise levels and peaks are detected.

**Keywords** Synthetic EEG · Savitzky-Golay · Moving average filter · Pink noise · White noise · S.N.R · Distortion ratio

## 1 Introduction

Our brain perform certain electrical activities Hence to monitor those activities EEG(electroencephalography)[1] is performed that enables us to check the electrical activities of human brain. This technique was developed by Hans Berger (1924, Jena). The electrical activities of the brain are due to the ionic current flowing within the neurons that is measured in terms of voltage fluctuations.

S. Baijal · S. Singh · A. Rani · S. Agarwal(✉)
Instrumentation and Control Engineering Division, NSIT, University of Delhi,
Sec-3, Dwarka, New Delhi, India
e-mail: {ishivangbaijal,shelvisingh,ashansit,agarwal.shivangi}@gmail.com

© Springer International Publishing Switzerland 2016
S.M. Thampi et al. (eds.), *Advances in Signal Processing and Intelligent Recognition Systems*,
Advances in Intelligent Systems and Computing 425,
DOI: 10.1007/978-3-319-28658-7_21

245

Any disorder in these electrical activities helps to diagnose diseases such as epilepsy, sleep disorder, tumors, coma encephalopathy and brain death [1]. Hence EEG is a valuable tool in the research and diagnosis of various brain activities and disorders. The amplitude of the EEG signal is around 100V in case of scalp measurement and is around 1-2 mV when we measure from the surface of the brain. The bandwidth of EEG signal varies from 1Hz to 50Hz. EEG recording setup consist of electrodes, amplifier, filter and the recording device. The commonly used electrode placement is 10-20 for a system [4]. This configuration provides coverage to all region of the brain and divides the skull with reference to skull landmarks [5]. The left side of the head is connected to the odd-numbered electrodes and the right side to the even-numbered electrode. The waveform from brain are classified into four categories[4] based on frequency range-$\beta(>13Hz)$, $\alpha(8\text{-}13Hz)$, $\Theta(4\text{-}8Hz)$ and the $\delta(0.5\text{-}4Hz)$. The most commonly studied signals are Alpha rhythm(induced during relaxation), Beta during wakefulness, Theta and Delta during the deep sleep. There [4] is a common observation at the time of EEG recording that there is introduction of the various artifacts. All the signals that appear in EEG recording which don't come from the brain are termed as the artifacts. The reasons of artifacts may be heart or muscle activity, blinking of eyes, eye ball movement, bad electrode positioning or electrode impedance. The most commonly encountered artifacts [4] are the power line interference that may even introduce artificial spikes in the EEG signal. To reduce and filter out the noise we use different filters. The filter choice depends on the ability to attenuate the noise in the EEG signal without affecting the significant information about the original signal.

It has been [2] observed that the digital filters used in EEG maybe time-domain filter or frequency domain filter. In this case [2] the raw data is directly applied to filter and the output is received, whereas in case of frequency domain the RMS value is computed and Fourier Transform of raw data is taken then filter is applied and finally the inverse Fourier transform of the data is evaluated. The Fourier transform bring delay and have more latency rate [3] as compared to time domain filter as it requires the full epoch to be acquired before the filtering.

In this paper a general emphasis is made on the two times domain digital filters namely S-Golay and moving average filter to filter out noise from a noisy signal. Certain studies [12] compared S-Golay and Moving average giving the inference that S-Golay is better in preserving the peak and the shape of the original signal. It has [14]also been observed that S-Golay filter is used to preserve the original property and it is used to develop a new method for baseline wander. The authors in [15]focus on S-Golay smoothing filter implementation for EEG application. The research work in [13] discusses the range and boundaries for application of digital filter(S-Golay). In the present work a synthetic EEG is created and known amount of noise is added to signal and finally the performance of the moving average and the S-Golay filter on the basis of SNR values as well as signal distortion are compared.

## 2 Synthesis of Synthetic EEG Signal

In order to evaluate the performance and comparison of filter, a synthetic EEG signal is created and then added to the known amount of white noise. In this paper EEG(synthetic) in nature, composed of different time segments is used. The time segments different in amplitude and frequency are applied for the same duration [11]. Equations

$0.5\cos(\Pi T1)+1.5\cos(4\Pi T1)+4\cos(5\Pi T1)$
$0.7\cos(\Pi T1)+2.1\cos(4\Pi T1)+5.6\cos(5\Pi T1)$
$1.5\cos(2\Pi T1)+4\cos(8\Pi T1)$
$1.5\cos(\Pi T1)+4\cos(4\Pi T1)$
$0.5\cos(\Pi T1)+1.7\cos(4\Pi T1)+3.7\cos(5\Pi T1)$
$2.3\cos(3\Pi T1)+7.8\cos(8\Pi T1)$
$0.8\cos(\Pi T1)+\cos(3\Pi T1)+3\cos(5\Pi T1)$

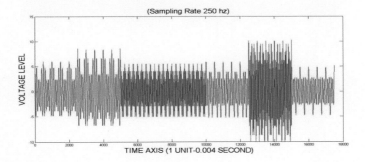

**Fig. 1** Synthetic EEG Signal

The above signal consist of all the possible states made from the above equations which ensures the signal consist of all states varied in terms of frequency, amplitude and phase. The data matrix for each segment is varied for equal interval of time and then finally merged to create the above signal which is free from any noise and baseline drift.

## 3 Addition of White and Pink Noise in the Synthetic EEG Signal

There are different kinds of noise that might get introduced into the EEG signal while recording the data. To validate the performance of the filter the paper uses two types of noises i.e. White and Pink noise. White noise is introduced in the synthetic signal as it affects all the frequency bands of the signal in equal amount [10]. All the components of the white noise constitute of samples that are independent and random(equal probability distribution normal distribution with zero mean)hence White Gaussian noise [8]. Colored noise [9] is different from the white noise as it affects the different components of frequency differently. Pink noise is a variant of white noise but it consist of signal with constant power in each octave band of frequency. The PSD of the pink noise is inversely proportional to the frequency of the signal used.

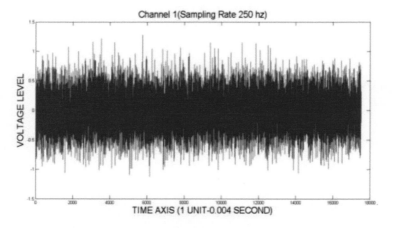

**Fig. 2** Plot of white noise

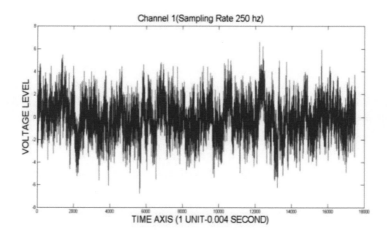

**Fig. 3** Plot of pink noise

The Figure 2 depicts the white noise and Figure 3 shows the pink noise respectively that are added in synthetic EEG signal. In the present work white noise to signal is applied to vary the SNR of signal at different levels whereas the creation of the pink noise is done through the random function and the noise is directly added to the signal.

## 4 Application of Savitzy-Golay Filter to the Noisy Signal

Savitzky and Golay filter is a digital LPF (low pass filter) which is applied on sample points smoothing purpose based on local least-squares polynomial approximation. Savitzky and Golay, in their "seminal" [7] paper [6], displayed that by fitting successive sub set of input samples by a low degree polynomial and then estimating the polynomial at a reference point in the calculated period. They substantiated that least squares smoothing minimizes artifact and at the same time maintains and preserves the morphology of waveform peaks. This S-G filters property is applied in the area of ECG and EEG signal processing. The Savitzky-Golay filtering method is often used with electroencephalogram (peak) data. The parameters of the S-Golay in this paper are chosen on the basis of genetic algorithm and 21 as the frame length and 8 as the polynomial order of the S-Golay filter is chosen. This filter is applied on EEG signals and this is how it works; the figure 4 is an EEG synthetic signal in which some known noise is added. Figure 4 is the SG filtered output.

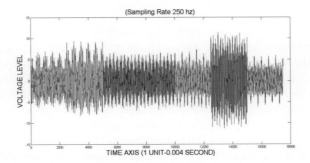

**Fig. 4** Noisy EEG signal with white noise

## 5 Moving Average Filter

Signals contain both desirable and undesirable information that may be one frequency or a range of frequencies. Henceforth, a circuit designed to perform this selection of information (or frequencies) is called a filter circuit, or simply a filter. For example,

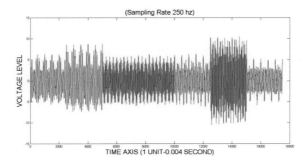

**Fig. 5** S Golay filtered EEG output

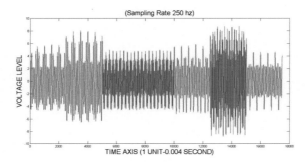

**Fig. 6** Moving average filtered output

a low-pass filter removes high frequency components so here we are concerned with the moving average filter in which all data point in the window is averaged (filtered) value. Since the moving average filter is easier to understand as compared to other filters hence it is most frequently operated upon. Therefore, it is a premier filter for time domain encoded signals and is very useful to perform ordinary tasks like minimizes random noise while upholding a sharp response. In figure 6, the moving average filter has a smoothening effect which attenuates the amplitude of the known noise, but also distorts the peaks. If the sharpness and shape of peaks is distorted then that can lead to a very situation which may also comprise relevant information in the EEG signal because peaks are the main source of information about the patients' health.

## 6   Signal to Noise Ratio(SNR) of EEG Signal

The SNR is calculated as the power of the signal upon the power of noise. This parameter help us to decide the amount of useful information a signal consist of with respect to the false information. We generally have many methods to evaluate the

**Table 1** SNR Comparison of S-Golay and MA filter using white noise

| Serial No. | White noise level(in dB) | SNR after S-Golay filter(dB) | SNR after MA filter(dB) |
|---|---|---|---|
| 1 | 1 | 13.98592 | 11.52761 |
| 2 | 3 | 15.88593 | 12.80929 |
| 3 | 5 | 17.8501 | 13.97887 |
| 4 | 10 | 22.932 | 16.034 |
| 5 | 15 | 27.64269 | 16.93539 |
| 6 | 20 | 32.28745 | 17.26933 |
| 7 | 25 | 36.41551 | 17.38351 |
| 8 | 30 | 39.4159 | 17.42223 |

**Table 2** SNR Comparison of S-Golay and MA filter using white noise

| Serial No. | White noise level(in dB) | SNR after S-Golay filter(dB) | SNR after MA filter(dB) |
|---|---|---|---|
| 1 | 1 | 13.98592 | 11.52761 |
| 2 | 3 | 15.88593 | 12.80929 |
| 3 | 5 | 17.8501 | 13.97887 |
| 4 | 10 | 22.932 | 16.034 |
| 5 | 15 | 27.64269 | 16.93539 |
| 6 | 20 | 32.28745 | 17.26933 |
| 7 | 25 | 36.41551 | 17.38351 |
| 8 | 30 | 39.4159 | 17.42223 |

SNR parameters that may include the calculation of PSD (power spectral density) [14] of the signal, calculation of the noise map or the standard deviation of the noise [15]. The method implied in the paper uses the subtraction method in which noise of the signal is evaluated by filtering the noisy signal using designed filter. The power of the signal and noise is evaluated for final value of SNR of the signal. The higher the value of the SNR the more relevant information the signal is said to possess. The table given below compares the performance of the two filters based on SNR performance.

## 7 Evaluation of Signal Distortion

Distortion of a signal is a characteristic of a signal that may be defined as the attenuation of the signal waveform from its original waveform. While filtering the noisy EEG signal we also need to reassure that the signal reconstructed by the filter is having the similar waveform as the synthetic EEG.

So in this paper the amount of distortion (in dB) is calculated that is introduced or still remains even after reconstructing our synthetic EEG signal by passing noisy EEG signal through the Moving average and S-Golay filter. The motive behind calculating the distortion is to compare the performances of the filter for the shape preservation as the EEG signal contains the information in the peak (voltage level) hence it becomes important to use a filter that preserves the shape of the EEG signal therefore bringing the least amount of distortion in the EEG signal.

## 8  Peak Detection

In this section the peak detection algorithm is applied that uses derivative method to evaluate the peaks in noisy signal and the original signal. So in the this section we have counted the number of peaks in the noise free signal and the filtered output keeping the threshold value of 8 and at the same time evaluation of the distortion ratio is done to compare the performance of these two filters. The main objective is to observe the effect of filtering on the peaks and compare the peak preservation performed by S-Golay and Moving average filter.

## 9  Application of Filters on Real EEG Signal

After the application of the filters on the Synthetic EEG we have collected the real EEG sample and collected the filtered output from our given two filters i.e. moving average and S-golay. Unlike the synthetic EEG the real EEG suffers from the baseline drift so we have applied the baseline correction first and then filtered the noise using the two filters.

## 10  Results and Discussion

In the present work comparative analysis between the S-golay and the moving average filter is made on the basis of SNR comparison. The SNR calculation is done using the subtraction method that indicates the amount of noise rejected or retained by a particular filter. Moreover this paper also focuses on the performance of the filters when amount of white noise is increased or decreased in synthetic EEG signal. Table 1 shows the performance of the filter. When the signal is made less noisy the SNR of S-Golay is increasing indicating that it is rejecting less amount of noise and producing the output signal with higher SNR as compared to the Moving Average filter. The SNR value according to Table 1 of the moving average is always less as compared to the S-Golay output SNR value showing that S-Golay is much more suitable then the moving average filter for the noise removal in EEG signal. It is observed from the table that using less noisy signal, the SNR of output of moving average filter is constant which infers that it is rejecting almost the same amount of artifactsin case of less corrupted signal whereas the S-Golay is rejecting almost the same amount of artifacts in case of less corrupted signal. The S-Golay is rejecting lesser noise and providing much higher value of SNR when the signal is made less corrupted. Table 2 evaluates the SNR value for the output when pink or flicker noise is added to the synthetic EEG. The SNR value in case of pink noise is better for the S-Golay as compared to the Moving Average. Table 3 provides the comparison of the filters on the basis of distortion value. It is noticed that the distortion value is tending towards zero as less noisy signal is used in case of S-Golay filter which infer that S-Golay is actually reconstructing the waveform of the output filtered signal whereas in case of the Moving average it is observed that the value is much higher than S-Golay so the filter is causing distortion in the signal. Therefore it is conclude that S-Golay is performing better as compared to moving average and is causing less distortion. Then peak detection is obtained keeping our threshold value of 8. It is observed that original signal has 22 peaks (Figure 7)after passing it through the S-Golay filter. We have 27 peaks(Figure 8) mainly due to the reason of addition of spikes due to the noise addition in the signal whereas in case of moving average as it can been seen in (Figure 9)that only 10 peaks are indicated in moving average filter. This filter increases the width of the peaks and decreases the height of the peak. Therefore it is inferred from the results that moving average fails to preserve the shape of the signal with respect to S-Golay filter.

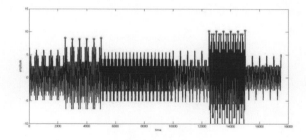

**Fig. 7** Peaks detected in original signal

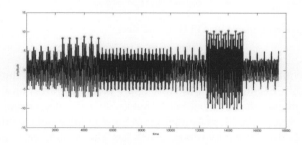

**Fig. 8** Peaks detected in S-Golay filtered output

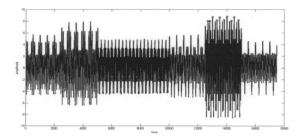

**Fig. 9** Peaks detected in Moving average filtered output

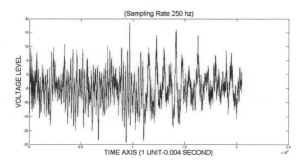

**Fig. 10** S Golay filtered Real EEG signal output

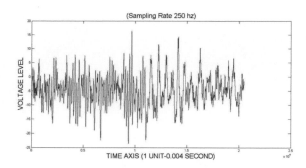

**Fig. 11** Moving average filtered Real EEG signal output

The last section describes the comparison of the performance of the filters on the basis of the real EEG signal on which we apply baseline correction first, and then the respective filters and the output infers that the shape is much better preserved in figure 10 that come as the filtered output from S-golay filter and the other output on figure 11 that is from moving average which has filtered the noise but still unable to preserve the shape of the EEG signal that makes S-golay filter better in real life applications also.

# 11 Conclusion

In the present work the S-Golay filter and Moving Average filter is applied on a synthetic EEG signal. The performance of the designed filters are evaluated on basis of SNR value, distortion value and peak preserving shape which further improves peak detection. It is concluded form the parameters obtained that S-Golay is a better filter for EEG signals as compared to Moving average filter.

# References

1. https://en.wikipedia.org/wiki/Electroencephalography
2. Nitschke, J.B., Miller, G.A., Cook, E.W.: Digital filtering in eeg/erp analysis: Some technical and empirical comparisons. Behavior Research Methods, Instruments, and Computers **30**(1), 4–67 (1998)
3. Babu, P.A., Prasad, K.V.S.V.R.: Removal of Ocular Artifacts from EEG Signals by Fast RLS Algorithm using Wavelet Transform. International Journal of Computer Applications **21**(4), 1–5 (2011)
4. Teplan, M.: Fundamentals of EEG Measurement. Measurement Science Review (2002)
5. Jasper, H.H.: The Ten-Twenty Electrode System of the International Federation. Electroencephalography and Clinical Neurophysiology **10**, 371–375 (1958)
6. Savitzky, A., Golay, M.J.E.: Smoothing and differentiation of data by simplified least squares procedures. Anal. Chem. **36**, 1627–1639 (1964)
7. Riordon, J., Zubritsky, E., Newman, A.: Top 10 articles. Anal. Chem. **72**(9), 324–329 (2000)
8. ENSC327 Communication System(22: Gaussian processes and White Noise). http://www2.ensc.sfu.ca/people/faculty/ho/ENSC327/Pre22white.pdf
9. Hargittai, S.: Savitzky-Golay least-squares polynomial filters in ECG signal processing. In: Proc. Comput. Cardiol., Budapest, Hungary, pp. 763–766 (2005)
10. Chen, K., Zhang, H., Wei, H., Li, Y.: Improved Savitzky-Golay-method-based fluorescence subtraction algorithm for rapid recovery of Raman spectra. Applied Optics **53**, 5559–5569 (2014)
11. Azami, H., Mohammadi, M., Saraf, K.: A new signal segmentation approach based on singular value decomposition and intelligent Savitzky-Golay filter. In: Comm. in Comp. and Information Sci., vol. 427, pp. 212–224. Springer (2014)
12. Guinon, J.L., Ortega, E., Anton, J.G., Herranz, V.P.: Moving average and Savitzki-Golay smoothing filters using mathcad. In: International Conference on Engineering Education, ICEE (2007)
13. Bromba, M.U.A., Ziegler, H.: Application hint for Savitsky-Golay digital smoothing filters. Anal. Chem. **53**, 1583–1586 (1981)
14. Gharaibeh, K.M., Gard, K.G., Steer, M.B.: Accurate Estimation of Digital Communication System Metrics-SNR, EVM and $\rho$ in a Nonlinear Amplifier Environment
15. http://massive-data.org/methods/snr-calculation.html

# An Efficient Multi Object Image Retrieval System Using Multiple Features and SVM

Nizampatnam Neelima and E. Sreenivasa Reddy

**Abstract** Retrieval of images containing multiple objects has been an active research area in recent years. This paper presents an efficient retrieval system for images containing multiple objects using shape, texture, edge histogram features and SVM (Support Vector Machine). The process starts with preparing the knowledge database. In this, all the images in dataset are segmented and shape, texture, edge histogram features are extracted. A combined feature vector is generated by combining these multiple features. These features are trained by using SVM and all the images in database are classified into different classes. This information is stored as knowledge data base. When the user enters his choice of query image, it is segmented and features are extracted. The combined feature vector is generated by using all these features. SVM is used for matching of the query image feature vector class with the classes in knowledge database. The similarity distance is calculated between the feature vector of the query image and the images in the corresponding class. The images are sorted and displayed according to the ascending order of the similarity distance. Experimental results demonstrate better retrieval efficiency.

**Keywords** Retrieval of multiple object images · Adaptive k-means · Edge histogram · Shape features · Support Vector Machine

## 1 Introduction

User requirements to retrieve the images of their choice have become endless as the digital media becomes globalized. Sometimes user interest for retrieval can be

N. Neelima(✉)
Department of Electronics and Communications,
Jain University, Bangalore, Karnataka, India
e-mail: neelima.niz@gmail.com

E.S. Reddy
Department of Computer Science and Engineering,
Acharya Nagarjuna University, Guntur, Andhra Pradesh, India

© Springer International Publishing Switzerland 2016
S.M. Thampi et al. (eds.), *Advances in Signal Processing and Intelligent Recognition Systems*,
Advances in Intelligent Systems and Computing 425,
DOI: 10.1007/978-3-319-28658-7_22

257

like "An airplane on runway" or "A palace with blue sky". It is the scenario where foreground and background are considered as two different objects. But if the user choice of retrieval is "Birthday cake with candles" (or) "Watermelon and Apple", in such cases both the objects are to be considered for retrieval. Another example is in crime investigations few times the choice can be "Retrieval of a record with person A as investigator and person B as criminal". Also In wild life photography the images with multiple objects are to be retrieved. For example, A lion and deer (or) a group of different animals in the images. In such cases where the user's interest is multiple objects together in the image, such image retrieval is complex because the size of the objects and their orientation are different. This motivates our research into image retrieval system for retrieving images containing multiple objects. Several methods were proposed where the retrieval is based on considering foreground and background as two different objects. But the proposed method is focused on retrieving of multiple objects in the image.

Humans can perceive most of the information through pictures rather than words. At any instant images can convey more information to the user rather than the words especially in medical and scientific applications. So instead of keywords the proposed method uses image as query. Image retrieval not only retrieves images from a huge dataset according to the user interest, also searching and indexing is also performed in some of its applications. Retrieval by image, known as Content- based image retrieval (CBIR) has gained much recognition in the past years and is the extensive research topic in   many research fields. These techniques are used to retrieve the semantically relevant images from the database based on the user's request.

As the growth develops exponentially in theoretical research and systems, several techniques for CBIR systems were developed and are widely used in all the visual content based systems. The indexing of the images is the major challenge when the images consist of multiple objects. In earlier days, text-based retrieval techniques were used, in which searching is based on the text entered manually by the user. The main drawback of this type of systems is they will not consider the amount and depth of   information present in the images. Content-based images retrieval (CBIR) was introduced to minimize the drawbacks of text-based retrieval. An improved image retrieval system for multiple objects in terms of retrieval efficiency is proposed and discussed in this paper.

## 2    Proposed Methodology

This method takes a user interested multiple object image as a query image, and retrieves relevant images from an image database. At the first segmentation is the most significant task to be performed to identify the objects in the given image. Segmentation is the process which divides the image into multiple segments. Several techniques were proposed for better segmentation. In this paper, adaptive k-means technique is used for segmenting the images because it is fast and easy to understand. It determines automatically the number of clusters and the center of clusters iteratively.

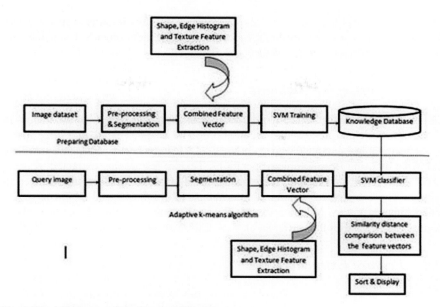

**Fig. 1** Block Diagram of the Proposed System

After segmenting, shape, texture, edge histogram features are extracted. These features are stored and trained by using SVM. SVM classifies these features into different classes. When the user enters the query image, it is segmented and features are extracted. Segmentation improves the signal to noise ratio and provides better approximation for feature description [5]. This combined feature vector is compared with the feature vectors of knowledge database. According to the similarity distance the images are sorted in ascending order and displayed. Feature Extraction is also another principal task, which is discussed in next sections. The proposed system architecture for retrieving multiple objects is shown in Fig.1.

The structural flow of the proposed method is organized as

step 1: All the images in database are pre-processed and segmented.
step 2: Shape, texture and edge histogram features are extracted and a combined feature vector is formed as per equation (3).
step 3: All these features are trained & classified into different classes by using SVM and stored as Knowledge data base.
step 4: The user provides a query image to retrieve similar images, first the image is segmented.
step 5: Features are extracted from the query image [8] and combined feature vector is generated according to equation (3).
step 6: This combined feature vector is compared with combined feature vectors in knowledge database by using SVM classifier. Feature matching is performed at this stage.

step 7: Based on the similarity distance between the feature vector of the query image and the feature vectors of images [8] in knowledge data base, images are sorted in ascending order of distance and displayed.

## 2.1    Preparing the Database

This is the preliminary step of the proposed system. In this phase, all the images in the database are pre-processed to segment properly. Shape, Texture and Edge Histogram features are extracted and combined as a feature vector. Shape features are extracted by using morphological operations. Texture features are extracted by using DWT and Gabor convolution. Edge histogram is obtained by using canny edge detection. These extracted features are combined into a vector and stored in a file. Then Support Vector Machine (SVM) is used to train the images based on these features and classifies into different classes. This is known as knowledge data base.

## 2.2    Feature Extraction

Feature extraction is the inevitable asset of this system. It obtains the useful information that can describe the image with its content. There are three types of features can be extracted for retrieving the images, they are shape, texture and Edge Histogram features.

### 2.2.1    Shape Features

In Image retrieval, discrimination between the images which are closely separated is the principal task. For this, shape of the objects is considered and the features are used. The low-level visual features can be extracted directly from the images. These features gives the outer line configuration of an object which provides the curves and edges. Morphological operations are used to pre-process the image [6] and to extract the shape features. The shape features which are extracted are perimeter-area ratio [6], form factor, axis ratio [6], roundness, compactness, bounding box and density [6].

### 2.2.2    Texture Features

Calibrating the texture content of an image is another important task in content description. Texture provides the knowledge about the repetitive patterns in the image (or) arrangement of intensities in an image. In this work, texture features are extracted by using DWT and Gabor filter. The images are decomposed into co-variance matrices LL, LH, HL and HH by using DWT [4]. Out of these four, LL coefficients are selected which represent low-level co-variance matrix. Low level images give the appropriate co-efficient matrix and other gives detailed co-efficients i.e. LH, HL and HH bands. The sum of LL coefficients gives the texture content.

Another texture feature is obtained by using gabor filter. Gabor filters can describe the feature intensity levels both in time and frequency domain. The expression for texture feature extraction is given by equation 1.

$$G(w) = \frac{e^{-\left(\log\left(\frac{\omega}{\omega_0}\right)\right)^2}}{2*\log\left(\frac{f}{\omega_0}\right)^2} \qquad (1)$$

The sensitivity and perception of gabor filters [2] is almost similar to human visual system. These filters are suitable for texture description and differentiation. The filter generates real and imaginary coefficients which represents orthogonal directions. These two can be used individually or a complex form can be formed by combining them. The complex form is given in equation 2.

$$g(x, y; \lambda, \theta, \varphi, \sigma, \gamma) = \exp\left(-\frac{x^2 + \gamma^2 y^2}{2\sigma^2}\right) \exp\left(i\left(2\pi\frac{x'}{\lambda} + \varphi\right)\right) \qquad (2)$$

where

$$x = x\cos\theta + y\sin\theta \quad \text{and} \quad y = -x\sin\theta + y\cos\theta$$

In the above equation, the wavelength of the sinusoidal factor is given by $\lambda$, $\theta$ represents the orientation, $\varphi$ is the phase offset, $\sigma$ is the standard deviation of the Gaussian envelope and $\gamma$ is the spatial aspect ratio [11].

### 2.2.3    Edge Histogram Descriptor

Edge detection is a process of locating the edges of image. Detection of edges is essential while understanding the features of image. Most of the relevant features and indicative information is pertained by edges. Edges are the pixels where the intensity levels are different form their neighborhood pixels; by considering only these pixels the image size can be reduced. The edge histogram descriptor defines five different edges for the images [3]. Four directional edges (vertical, horizontal, $45^0, 135^0$ diagonal edges and arbitrary edges without any directionality. We used the Edge Histogram descriptor from [3].

[combined feature vector] = [shapefeatures texturefeatures edge histogram descriptor].                                                                          (3)

## 3      Support Vector Machine

Support Vector Machines (SVM's) are supervised learning machines used for classification and regression. These are used as classifier in the proposed method. These machines implement statistical based learning means the data decides the algorithm. SVM's are based on kernel trick where a kernel function is used to compute similarity in the given feature space. Several linear and non-linear kernels provide the flexibility for the user to choose the kernel function based on application. There are four basic types of SVM are available which are linear, polynomial, radial basis function and sigmoid. For a given training vector set xi SVM allows mapping into the infinite dimensional space by the parameter φ. The general kernel function [7] is given by equation 4.

$$K(X_i, X_j) = \phi(X_i)\phi(X_j) \qquad (4)$$

The Gaussian RBF kernel [1] used for the nonlinear data in this work is given by equation 5.

$$K(X_i, X_j) = e^{-\gamma\|x_i - x_j\|^2/2} \qquad (5)$$

The non-linear machines trained on non-separable data. A separating hyper plane is considered which can classify effectively the feature space. The lagrangain multipliers for each inequality conditions are given in equation 6. The optimal dividing hyper plane [7] is given by equation 6.

$$w^* = \sum_{i \in sv} h_i Y_i X_i \qquad (6)$$

The maximum [7] [10] is given by equation 7.

$$W(\alpha) = \sum_{i=1}^{l} \alpha_i - \frac{1}{2}\sum_{i=1}^{l} y_i y_j \alpha_i \alpha_j \, K(x_i, x_j) \qquad (7)$$

The features from (3) are trained by SVM and classify the features into number of classes. When the user defines the query image SVM maps the feature vector of the query image into the corresponding class and according to the similarity distance the similar images are retrieved and displayed.

$$b^* = \frac{1}{|sv|} \sum_{i \in sv}(y_i - \sum_{j=1}^{N} h_j y_j x_j^T x_i) \qquad (8)$$

The output vector for the classification is given in equation 8. Where sv is the support vector, x is the list of training vectors, hj are lagrangain coefficients and yj are their respective labels.

## 4    Results

To implement the proposed method COREL Image dataset is used. The dataset consists 210 images with seven categories each with 30images. The dataset consists of "bonsai &shoji", "gun &bullet", "two horses", "flowers", "blue bus & roadway", "blue sky & red bus", "pyramid & blue sky" [9]. Consider a query image which is shown in Figure 2. It has two horses; the retrieval of such similar images is very difficult. The Top 10 retrieved results are shown in Figure 6. Another two such examples are shown in Figure 3 & Figure 4. The top 10 retrieved results for these query images are shown in Figure 7 & 8 simultaneously. The retrieval efficiency is above 80% in all these examples. The precision is the best even though the number of retrieved images is increased, which is shown in Table 1. The results are good when compared to the system used in [9].

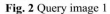

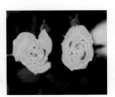

**Fig. 2** Query image 1          **Fig. 3** Query image 2          **Fig. 4** Query image 3

**Table 1** Retrieval efficiency for different query images

| No. of images to be retrieved as per user | horses | bon-sai+shoji | flowers | blue bus + roadway | red bus + blue sky | pyramid + blue sky | gun + bullet |
|---|---|---|---|---|---|---|---|
| | | | Retrieval Accuracy in percentage | | | | |
| 5 | 100 | 100 | 100 | 100 | 100 | 100 | 100 |
| 10 | 100 | 100 | 100 | 100 | 100 | 100 | 100 |
| 15 | 100 | 100 | 90 | 100 | 100 | 100 | 100 |
| 20 | 95 | 95 | 80 | 90 | 95 | 95 | 95 |
| 25 | 95 | 95 | 80 | 84 | 88 | 84 | 80 |

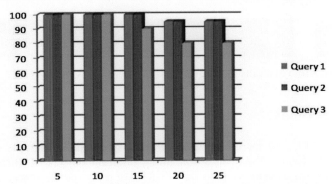

**Fig. 5** Precision comparison of different query images for the proposed system

**Fig. 6** Top ten retrieved results for the query image 1

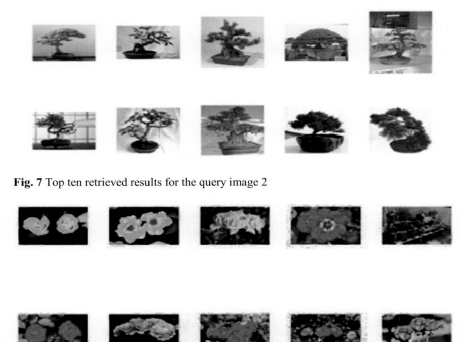

**Fig. 7** Top ten retrieved results for the query image 2

**Fig. 8** Top ten retrieved results for the query image 3

## 5    Discussions

The Proposed Method for the retrieval of images containing multiple objects is effective and efficient for images containing different objects of different size, shape and texture. Several systems for CBIR were proposed for object based Image retrieval, but they focused on foreground and back ground objects. Wei-bang CHEN [9] proposed multiple object retrieval based on color of the objects. Color may not be a single attribute to be considered in image retrieval. In COREL dataset or real time images color based retrieval may not be sufficient to retrieve the images. For example, if the query image is "Birthday cake with candles", "Bonsai and Shoji", in such cases instead of color, shape and texture features are more relevant to retrieve. Vigilant choice of features, shape and texture and edge histogram features in the combined feature vector reduces the time to retrieve. This is an added advantage for the proposed system. Improper segmentation is one of the drawbacks in CBIR; by using adaptive k- means the proper segmentation is achieved. Training and classification using SVM provides accuracy in retrieving. Experimental results shown in Table 1 proved that the precision is greater than 80% for the entire query images even the number of retrieved images is increased.

# References

1. Muandet, K., Fukumizu, K., Dinuzzo, F., Scholkopf, B.: Learning from distributions via support measure machines. In: Advances in Neural Information Processing Systems 25, pp. 10–18 (2012)
2. He, K., Sun, J., Tang, X.: Guided Image Filtering. IEEE Transactions on Pattern Analysis and Machine Intelligence 35(6) (2013)
3. Won, C.S.: Feature extraction and evaluation using edge histogram descriptor in MPEG-7. In: Lecture Notes in Computer Science, vol. 3333, pp. 583–590. Springer (2005)
4. Wang, C., Zhang, X., Shan, R., Zhou, X.: Gradient Image Retrieval based on DCT and DWT Compressed Domains Using Low Level Features. Journal of Communications 10(1) (2015)
5. Arandjelovic, R., Zisserman, A.: Smooth object retrieval using a bag of boundaries. In: Proceedings of IEEE International Conference on Computer Vision, pp. 375–382 (2011)
6. Nunes, J.F., Moreira, P.M., Tavares, J.M.R.S.: Shape based image retrieval and classification. In: Iberian Conference on Information Systems and Technologies (CISTI), p. 433 (2010)
7. Burges, C.J.C.: A Tutorial on Support Vector Machines for Pattern Recognition, pp. 1–43. Kluwer Academic Publishers, Boston
8. Eini, S., Chalechale, A.: Shape Retrieval using Smallest Rectangle Centroid Distance. International Journal of Signal Processing, Image Processing & Pattern Recognition 6(4), 61–68 (2013)
9. Chen, W.-B., Zhang, C.: Multiple object retrieval in image databases using hierarchical segmentation tree. In: Proceedings of 19th ACM International Conference on Multimedia, pp. 881–882 (2011)
10. Ramesh Babu Durai, C., Balaji, V., Duraisamy, V.: Improved Content Based Image Retrieval using SMO and SVM Classification Technique. European Journal of Scientific Research 69(4), 560–564 (2012)
11. Vaithiyanathan, V.: Variational level set segmentation for the human motion capture analysis. In: Proceedings of International Conference on Computer Communication and Informatics (2012)
12. Zhang, H., Gao, W., Chen, X., Zhao, D.: Object detection using spatial histogram features. Image and Vision Computing, 327–341 (2006)

# Blood Cells Counting by Dynamic Area-Averaging Using Morphological Operations to SEM Images of Cancerous Blood Cells

Kanik Palodhi, Dhrubajyoti Dawn and Amiya Halder

**Abstract** In this paper, we describe a novel method of determining the complete blood cell count using region-based segmentation method of highly magnified images obtained from SEM. It provides an efficient way of measuring blood cell counts and it also be used to diagnose the presence of many life-threatening diseases such as leukemia, allergies etc. In many cases, mainly, due to diseases and cell longevity, the actual sizes of micron-order blood cells appear to change their sizes. Therefore, this proposed method uses dynamical averaging of red blood cell (RBC) and white blood cell (WBC) areas so that no a-priori area (fixed area of blood cell) is required. This is a significant step forward since no fixed area cut-offs are used to separate WBC and RBC. The SEM images are preprocessed with a global threshold and then, significant noises are removed using morphological processing. Finally, the areas are separated using dynamical averaging and separately, WBC and RBC counts are computed. Experimental data are provided for different real samples of cancer patients which show encouraging results.

**Keywords** Morphological operation · Image segmentation · RBC · WBC

## 1 Introduction

One of the most challenging research areas in image processing has been content based image indexing and information retrieval. It has huge applications in the field of biomedical images where complex cell or cell clusters need to be identified from the noisy images. A particular application in this field is to determine the complete blood count (CBC) that is used as a screening test for a number of diseases. Diseases

K. Palodhi(✉) · D. Dawn · A. Halder
St. Thomas College of Engineering and Technology, 4, D. H. Road Kidderpore,
Kolkata 23, West Bengal, India
e-mail: {kanikpalodhi,dawndhruba,amiya.halder77}@gmail.com

© Springer International Publishing Switzerland 2016       267
S.M. Thampi et al. (eds.), *Advances in Signal Processing and Intelligent Recognition Systems*,
Advances in Intelligent Systems and Computing 425,
DOI: 10.1007/978-3-319-28658-7_23

such as infections in different body parts, allergies and critical health issues related to heart and lungs need precise CBC for proper diagnosis of diseases and disorders. Currently, for determination of CBC of blood sample a number of tests are performed to calculate the constituent particles of the sample, namely, RBC and WBC. By far the most common technique is to image the stained blood sample with transmission light microscope and then, manually counting the particles identifying and deciding on the deviations due to disorders [1-4]. This arduous task takes prolonged time and is prone to human error. Not only that, trained personnel need to be employed which escalates the cost of this basic test to a large extent, considering this test is employed widely for the first diagnosis. The other automated techniques include laser based flow-cytometer or sceptre counter both of which are rather costly and not employed normally, since they can destroy the cells of the samples if due care is not taken [5-8]. Therefore, there is a need for a software-based image processing approach which needs minimal human intervention. Among the very firsts in this field of automated blood slide sample analysis was Bentley and Lewis and later, Rowan presented the first fully automated system [8,9]. There are numbers of literatures available on CBC using segmentation, watershed and other feature extraction algorithms, which predominantly uses microscope images with normal (not diseased) blood cells [9-13]. Our endeavour is to propose a technique with blood samples obtained from diseased body (having leukaemia) where the normal blood cells sizes vary significantly. In this article, instead of normal microscope images, scanning electron microscope (SEM) images were investigated that provide large magnification of the order of 1000 that provides better resolution of the blood cells in gray scale images.

The article is organized as follows: The proposed algorithm is presented in section 2. Experimental results describes in section 3 and conclusions are drawn in section 4.

## 2   Proposed Algorithm

In this article, an averaging method termed as area-based approach is employed for separating WBC and RBC according to their respective sizes with appropriate use of denoising and filtering. The proposed algorithm is presented in subsequent sections.

### 2.1   *Image Acquisition and Binarization*

A scanning electron microscope (SEM) produces images of a sample by scanning it with a focused beam of electrons. The electrons interact with atoms in the sample, producing various signals that contain information about the sample's surface topography and composition. In our case, beam voltage of the SEM is kept constant at 15 KV for the images under investigation that provides lateral resolution of the order of $0.1 \mu m$ (Figure 1(a) and 2(a)). These images obtained from leukaemia patient shows different sizes of RBC and WBC, with WBCs characterised by spiky edges, a characteristic of diseased sample. Using the method of thresholding, the image f(x,y)is converted to a binary image g(x,y) that belongs to a set A in real space to get rid of the grey-variation within the contours of RBC and WBC within the images, with the threshold level kept at 20% of the maximum gray value i.e. $g(x, y) = 1$ for $f(x, y) \geq 0.2$ and $g(x, y) = 0$ for $f(x, y) < 0.2$. The threshold value was arrived at after considering a number of SEM images.

## 2.2 Morphological Processing and Filtering

Morphological image processing is a collection of non-linear operations related to the shape or morphology of features of an image. Morphological operations rely only on the relative ordering of pixel values, therefore, are particularly, suited to the processing of binary images. In this case, based on the connected component concept, the area of the structuring elements (RBCs and WBCs) with 8-connectivity of g(x,y)are determined. Along with the RBCs and WBCs there were several smaller area-components which have been introduced at the time of binarization. These areas are removed considering an area threshold of $5 - 7.5 \mu m^2$ pixels. Considering g(x,y) belongs to set A and one of the connected components is known in A, then, standard way of representing extracting connected components and region filling can be represented as,

$$X_k = (X_{k-1} \oplus B) \cap A^c, \{k = 1, 2, 3...\} \tag{1}$$

where, B is symmetric structuring element, $A^c$ means complemented A and the method ends at the final step with $X_k = X_{k-1}$ for a particular k i.e. when region filling is completed [14].

## 2.3 Re-filtering with Area-Based Separation and Counting

In this step we have labelled all the structuring elements in the binary image with an index value. The reason for doing this is to identify each element and access them individually. More precisely, considering a no. of SEM images, manually, the area in terms of pixels are computed and then, WBC and RBC areas are separated (similar process to Eq. 1). This separation,essentially, is a process of assigning a label to every pixel in an image such that pixels with the same label share certain visual characteristics. In this case, area property is used to separate WBCs and RBCs where with each labelling an average area is calculated dynamically. This is an improvement over previous methods of assuming a-priori areas of RBC and WBC sizes. From the area variation, we have observed that in these SEM images, the RBCs are larger than the WBCs. Therefore, a filtering mechanism is used to first eliminate the WBCs from the binary image setting a cutoff area of $7.5\mu m^2$ pixels and only calculate RBC using blob counting. Later, by subtracting the image having only the RBCs from the binary image with both RBCs & WBCs, RBCs can be eliminated and WBCs can be counted using blob counting.

## 3  Experimental Results

There were several images which were subjected to the algorithm coded both in MATLAB and C programming languages and here an illustrative example is shown in Figure 1 and Figure 2. The different steps of the operations on the image, suitably chosen, are presented in this section. The acquired image (Figure 1(a)) is first subjected to global threshold which produces grainy noises (Figure 1(b)) across the images, removed using a morphological area cut-off (Figure 1(c)). SEM images are, then, filled and labelled depending upon their column-wise positions. Applying manual consideration gained through processing a number of each of the filled area is re-filtered using dynamical area averaging. In this process, initially, an RBC area with bounds are mentioned but with each new RBC added, the total area is averaged to provide a more accurate area for each particular image. The same process is followed for WBC and both of them are separated to count CBC, accurately.

Although the method is advantageous on multiple counts, two main drawbacks of these methods are extracting information on overlapping areas and separating artefact from actual blood cells. Currently, efforts are going on to use edge-linking methods together with the above algorithm for providing a better result.

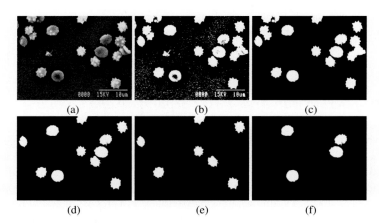

(a)                              (b)                              (c)

(d)                              (e)                              (f)

**Fig. 1** Different steps of the method:(a) Sample SEM image (b) binary image (c) denoising using higher area-based cut-off, (d) Morphological processing without overlapping (e) area-based separation of WBC and (f) area-based separation of RBC.

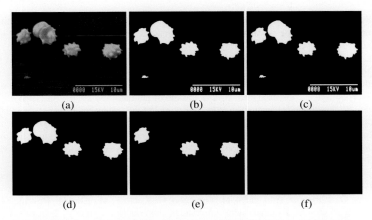

(a)             (b)             (c)

(d)             (e)             (f)

**Fig. 2** Different steps shown in another sample:(a) Sample SEM image (b) binary image (c) de-noising using higher area-based cut-off, (d) Morphological processing without overlapping (e) area-based separation of WBC and (f) area-based separation of RBC.

## 4 Conclusions

This dynamic area-based separation of blood cell counting is fast and accurate method, though, in the overlapping region area-based approach needs to be modeled carefully. The dynamical averaging provides significant improvement over previous method since this can be used to count cells of blood cells, sizes of which deviates significantly, from the normal blood cells. With further experimentation, this method can potentially be used for blood samples obtained from different image acquisition methods such as optical microscope image.

**Acknowledgement** We hereby acknowledge the support provided by St. Thomas College Engineering and Technology. We also accord our thanks to University Science Instrumentation Centre, University of Burdwan for the scanning electron microscope images provided for this method.

## References

1. Lewis, S.M., Bain, B.J., Bates, I., Lewis, D.: Blood cell morphology in health and disease practical haematology, 10th edn. Churchill Livingstone Elsevier, Philadelphia (2006)
2. Ramoser, H., Laurain, V., Bischof, H., Ecker, R.: Leukocyte segmentation and classification in blood smear images. In: Engineering in Medicine and Biology Society (2005)
3. Abbott Diagnostics. http://www.abbott.com/products/diagnostics.htm/
4. Beckman Coulter. http://www.coulter.com/coulter/Hematology/
5. Ongun, G., Halici, U., Leblebicioglu, K., Atalay, V., Beksac, M., Beksac, S.: An automated differential blood count system. In: Proceedings of the 23rd Annual International Conference of the IEEE Engineering in Medicine and Biology Society, vol. 3, pp. 2583–2586 (2001)

6. De Roover, C., Herbulot, A., Gouze, A., Debreuve, E., Barlaud, M., Macq, B.: Multimodal segmentation combining active contours and watersheds. In: Proceedings of the 13th European Signal Processing Conference (2005)
7. Kass, M., Witkin, A., Terzopoulos, D.: Snakes: Active contour models. International Journal of Computer Vision **4**, 321–331 (1988)
8. Bentley, S., Lewis, S.: The use of an image analyzing computer for the quantification of red cell morphological characteristics. British Journal of Hematology **29**, 81–88 (1975)
9. Rowan, R.: Automated examination of the peripheral blood smear. In: Automation and Quality Assurance in Hematology, pp. 129–177. Blackwell Scientific, Oxford (1986)
10. Di Rubeto, C., Dempster, A., Khan, S., Jarra, B.: Segmentation of blood images using morphological operators. In: 15th International Conference on Pattern Recognition, vol. 3, pp. 397–400 (2000)
11. Gauch, J.M.: Image segmentation and analysis via multi scale gradient watershed hierarchies. IEEE Transactions on Image Processing **8**(1), 69–79 (1999)
12. Leznray, O., Elmoataz, A., Cardot, H., Gougeon, G., Lecluse, M., Elie, H., Revenu, M.: Segmentation of cytological images using color and mathematical morphology. European Congress of Stereology **18**(1), 1–14 (1999)
13. Beucher, S.: Watersheds of functions and picture segmentation, acoustics, speech, and signal processing. In: IEEE International Conference on ICASSP, vol. 7, pp. 1928–1931 (1982)
14. Gonzalez, R.C., Woods, R.E.: Digital Image processing. Pearson Education (2002)

# Texture Guided Active Contour for Object Segmentation in Natural Images

Glaxy George and M. Sreeraj

**Abstract** Object segmentation based on active contour approach, uses an energy minimizing spline to localize object boundary. The major limitations of parametric and nonparametric active contour approaches are sensitivity to noise, high computation cost and poor convergence towards boundary. To address these problems, this paper proposes a novel approach utilizing the textural characteristics of the image for detecting the boundary of the salient object. From the textural patterns of the image a saliency map is generated and convolved with the user-defined vector field kernel. The performance of proposed approach is analyzed on three datasets such as Weizmann single object dataset, MSRA-500 dataset and Brodatz texture images. The experimental results demonstrates that proposed method can effectively converges towards the boundary of the salient object.

## 1 Introduction

Object segmentation is an important area of research in image processing due to its numerous applications such as remote sensing [1], object tracking [2] and robotics [3]. Several approaches has been proposed for segmenting objects and one of the popular approach is active contour models. Kass et al. [4] proposed a spline that minimizes energy using both internal and external forces. The major limitations of this model are user initialization, poor and slow convergence towards object boundary and parameter dependencies. To overcome these limitations, many approaches has been proposed [5]-[19] based on original active contour model.

Active contour approches can be categorized into two:parametric and nonparametric. The problems associated with parametric active contour models are sensitive to noise, parameter sensitivity, slow convergence rate, ineffective stopping criteria, and user initialization. Nonparametric approaches are able to handle high

G. George(✉) · M. Sreeraj
FISAT, Ernakulam, India
e-mail: {glaxygeorge,sreerajtkzy}@gmail.com

© Springer International Publishing Switzerland 2016
S.M. Thampi et al. (eds.), *Advances in Signal Processing and Intelligent Recognition Systems*,
Advances in Intelligent Systems and Computing 425,
DOI: 10.1007/978-3-319-28658-7_24

273

curvature regions and automatically initialize the contour of the salient object. A recent approach proposed in [19], utilizes both structural and textural distinctiveness for contour initialization. The major limitation of this approach is high compuation cost and also the usage of dimensionality reduction.

The objective of this paper is to present a robust object segmentation method using active contour that automatically do contour initialization and convergence towards object boundary. To accomplish this, texture variations are employed. The sensitivity to noise is get reduced by incorporating variations to vector field convolution. The proposed non-parametric approach can handle the limitations of parametric model by utilizing the textural chracterics of the image.

The remainder of the paper is discussed as follows: Section 2 reviews related works regarding active contours. The proposed method is explained in Section 3. Experimental evaluation with results and comparisons with state-of-art methods is presented in Section 4. Finally, conclusion in Section 5.

## 2  Related Works

Active contour approach has a wide variety of applications such as image segmentation and motion tracking. This approach is based upon the usage of deformable contours or snakes which adapts to object shapes and motions. Active contour model, also called snakes is guided by both internal and external forces. External force pulls the snake towards object boundary and internal force controls the deformation of the snake. Internal force is defined by the chracterics of the image.

Active contour models can be categorized into two types: parametric [5]-[7] and non- parametric [8]-[10]. Fig. 1 shows the taxonomy of active contour approaches. Parametric active contour model minimizes an energy function to determine object boundaries. The external forces have been proposed in recent past can enhance the capture range of the original snake approach. But these approaches fail to locate the boundary of the object. In [11], Xu et al. proposed Gradient vector Flow (GVF) external force, has improved capture range, insensitive to initialization and better

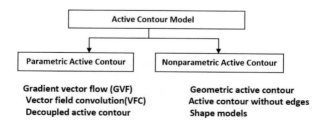

**Fig. 1** Taxonomy of active contour approach

convergence to boundary. But this method is noise and parameter sensitive and has high computation cost.

Another parametric model named, Vector field convolution (VFC) [12] is insensitive to noise and has less computation cost. However, limitations are dependence on edge map and fail to converge in cluttered backgrounds. A recent model, decoupled active contour (DAC) [13] is less sensitive to parameter and noise, better convergence to boundary but doesn't automatically initialize the contour.

Nonparametric methods [8]-[10] are capable with high curvature boundaries. In [9] proposes a new model based on curve evolution, MumfordShah model can handle particular segmentation of the image since the stopping term is independant of the gradient of the image. In [10], proposes that shape modeling can be applied to shapes with significant protrusions, and no priori information about the objects. A single instance of the model, when subjected to an image having multiple region of interest(ROI), has the ability to split freely to represent each ROI. But all these models of nonparametric fail in the presence of broken edges, are slow to converge and exhibit high noise sensitivity.

Texture is one of the image characteristic having significant role in object segmentation. In the presence of high texture patterns, high intensity variations or object with weak edges, all aforementioned methods fail to locate the boundary. The textural characteristics can handle such situations by taking advantage of image transformations, intensity and statistical models [14]. A new approach, Gabor-based texture segmentation [15], transforms image to Gabor feature space and can express texture changes accurately. In [16], Savelonas et al. proposed local binary pattern (LBP) textures representation. This method is constrained by its patch size ie smaller one is unable to distinguish large textures. LBP has high computation cost due to the iterative calculations associated with histograms.

The main challenges of active contours are noise sensitivity and limited capture range for the contour. Gradient Vector Field(GVF) and Vector Field Convolution(VFC) approaches have increased capture range. The drawbacks of GVF are computationally expensive and sensitive to noise. VFC handled limitations of GVF to a great extent, but does not take full advantage of the structural property of the image boundary. The proposed method is formulated by adapting the VFC kernel using the textural information of an image. This results in more stable field in the presence of noise and thus provides improved active contours.

## 3 Proposed Methodology

Segmentation based on texture analysis, one recent approach, has acheived better performance in exploring the boundaries of objects. It is a difficult task on images having cluttered background due to the presence of numerous texture patterns. Fig. 2 depicts the architecture of the proposed method. Texture guided region clustering identifies pixels having same texture features and texture discreteness computes saliency map. The external force is convolved with variation map obtained from

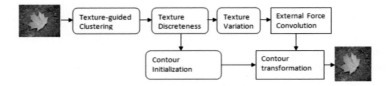

**Fig. 2** Overall architecture of the proposed methodology

discreteness. The intial contour is localized to the boundary through several contour transformations. Fig. 3 shows the image deformation on each section of the proposed method. The detailed description is given in the following sections.

## 3.1 Texture Guided Region Clustering

Identification of region of interest is the first task in segmentation process. In case of images with cluttered background, textural variations hepls to discrete objects and background. Usually, region of interest have high texture intensity value compared to background. In proposed approach, a texture model is created which depicts the underlying texture patterns present in the image. Using this model, a textural variation map is calculated between different texture regions. Pixels having same texture belongs to same region. The textural feature representation is defined as follows:Let I be the M N image and $\lambda$ be a neighborhood centered at pixel location x in I. The local textural representation $l_c(x)$ for each color channel c, defines a neighborhood around the pixel of interest. The statistic for that neighborhood is calculated, and use that value as the value of the pixel of interest in the output image. The vector representation for each pixel comprised of three columns such as its own structural variation value $s_c(x)$, a filter value $f_c(x)$ with respect to the neighborhood and its local randomness $r_c(x)$. The filter value of each pixel is the difference between the highest and lowest one in the neighborhood set.

$$f_c(x) = max(s_c(x)) - min(s_c(x)) \tag{1}$$

Input Image          Texture region clustering          Texture discreteness          Contour Initialization          Final Contour

**Fig. 3** The resultant image on each section in the proposed method.

For p neighbors in the nxn patch, compute $r_c(x)$ as

$$r_c(x) = \sum_p s_c(x) \cdot log s_c(x) \qquad (2)$$

The local textural representation $l_c(x)$ of a pixel, for each color channel c, is $[s_c(x)\ f_c(x)\ r_c(x)]$.

$$l_c(x) = [s_c(x)\ \ f_c(x)\ \ r_c(x)] \qquad (3)$$

This feature model can be used to identify the different texture patterns in the image by means of clustering. The different texture regions forms a texture model of the image.

Experiments demonstrate that the 5x5 pixel neighborhood has high performance in texture pattern identification as compared to 3x3 and 9x9 patch size. Fig. 4 shows the local textural representation of each texture region in the image.

**Fig. 4** Local textural representation of each texture region in the Weizmann dataset image.

## 3.2 *Texture Pattern Discreteness*

Texture pattern discreteness is defined as the relationship between different texture regions. ie, to what extent a region t1 differs from region t2. Given the texture model $T=t_1,t_2,...t_{mxn}$, texture pattern discreteness between $t_1$ and $t_2$ is defined as:

$$\alpha(t_1, t_2) = \sum_k (t_{1k} - t_{2k})^2 / (\sigma^2_{(t1,t2)}) \qquad (4)$$

where $\sigma_{(t1,t2)}$ represents the $L_2$-norm between regions t1 and t2. Texture pattern discreteness depicts the statistical relationship between different texture regions in

**Fig. 5** Texture discreteness and resultant image after convolution

the image. Fig. 5 shows texture discreteness of the input image and also the resultant image after applying vector convolution.

## 3.3  Saliency Map Computation

Saliency map projects the edges of salient objects and darkens the background. Two possibilities regarding salient object are, more visible than any other in the image and closeness to the center of the image. Based on these possibilities, a texture variation map is generated.

The texture discreteness $D(m, n)$ of image $I(m, n)$ is computed as:

$$D(m, n) = \lambda_i, if \ (m, n) \ \varepsilon \ R_i \tag{5}$$

where $\lambda_i$ is texture value of region $R_i$ and this value is assigned to each pixel belonging to the region $R_i$. The texture value is unique for each texture region and is calculated from texture pattern discreteness and spatial variance of pixels.

$$\lambda_i = \omega_i \cdot v_i, if \ (m, n) \ \varepsilon \ R_i \tag{6}$$

where i=1,2...r texture regions. The texture discreteness map $\omega_i$ of each region $R_i$ is

$$\omega_i = P(t_j) \cdot \alpha_{(t_i, t_j)} \tag{7}$$

The discreteness map is the expected statistical measure of texture region $t_i$ with respect to other regions and implements the first possibilty. $P(t_j)$ is the size or total pixels of each region. Second possibility is based on how far the pixels in each region away from the center of the image.

$$v_i = \psi_i / n_{t_i} \tag{8}$$

where $\psi_i$ and $n_{t_i}$ represent the spatial distance and the number of pixels of texture region $t_i$ respectively. The spatial distance $\psi_i$ is calculated as:

$$\psi_i = \sum_{(m,n)\varepsilon R_i} ((m, n) - (m_c, n_c))^2 / \sigma_c^2 \tag{9}$$

where $\sigma_c$ is the spatial distance variance between the pixels and center of the image.

## 3.4 *External Force*

The external force proposed by Li et al. in VFC [12] is used in the proposed architecture. The gradient based external force can localize the object boundary of an image with homogeneous background. For images with inhomogeneous backgrounds, additional features are more adequate and powerful. The external force,

$$E = D(m, n) * k(m, n) \tag{10}$$

where k(m,n) is the vector field kernel in [12]. The VFC [12] approach convolves a vector field kernel k(m,n) with the structural variation map v(m,n)of an image I(m,n).

$$f(m, n) = v(m, n) * k(m, n) \tag{11}$$

The kernel is defined in [12] as

$$k(m, n) = w(m, n) \cdot u(m, n) \tag{12}$$

where w and u is the magnitude of the vector field kernel and the unit vector pointing to the origin of the kernel respectively.

$$w(m, n) = (d + \eta)^{-\gamma} \tag{13}$$

$$d = (m^2 + n^2)^{-1/2} \tag{14}$$

where d is the distance to the origin of kernel and $\eta$, $\gamma$ are positive parameters. These two are used to control variation of the magnitude. For better performance,$\eta$ =0 and $\gamma$=2 is used.

The unit vector,

$$u(m, n) = [-m/d, -n/d], if \ (m, n) \neq (0, 0) \tag{15}$$

otherwise u(m,n)=(0,0).

**Fig. 6** Initial(red) and final contour(green) of Weizmann dataset image.

## 3.5 *Contour Initialization*

Some of the parametric active contour approaches provide user interactive initialization. In some situations this may leads to ineffective performance. In proposed approach, texture model guides the automatic initialization of contour. The texture discreteness of the image determines the region of interest and an adaptive thresholding method proposed in [17] do localization of ROI. The set of points $\rho$ belonging to the object is selected as follows:

$$\rho = (p, q)\varepsilon I(p, q), where D(p, q) > \top \qquad (16)$$

where $\top$ is the threshold proposed in [17]. The connected regions $\mathfrak{R}$ of different sizes are obtained on applying thresholding. These regions may be foreground or background of the image. The region having largest area than all other regions is selected as the region of interest, due to the fact that ROI usually have more points. A convex hull around the object is computed using the method proposed in [18]. Fig. 6 shows intial and final contour based on proposed method. The Fig. 7 shows convex hull and circle initialization of the image based on the groundtruth.

## 4 Experimental Results and Analysis

This section presents the experiments conducted to evaluate the performance of the proposed approach. The experiments was tested in MATLAB 2012a. Performance comparison was performed against four approaches such as gradient vector flow (GVF) [11], vector field convolution (VFC) [12], decoupled active contour (DAC) [13] and non-local active contours (NLAC) [20]. VFC and GVF implementation was provided by Li et al. DAC and NLAC code was also provided by respective authors. The performance comparison was performed using Precision, Recall and F1-measure. The execution times was also compared based on common initializations.

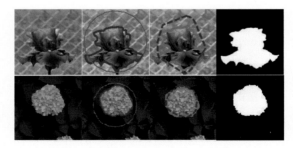

**Fig. 7** Top(MSRA image) and bottom(Weizmann dataset image) rows shows input image, circle initialization, iniatialization based on proposed method and ground truth

**Table 1** 3x3,5x5 and 9x9 patch size comparison on datasets.

|         | 3x3   | 5x5   | 9x9   |
|---------|-------|-------|-------|
| Weizmann |      |       |       |
| Rec     | 82.77 | 83.36 | 70.64 |
| Pre     | 64.06 | 69.24 | 61.97 |
| F1      | 72.22 | 75.65 | 66.02 |
| MSRA    |       |       |       |
| Rec     | 70.04 | 69.63 | 69.81 |
| Pre     | 67.34 | 70.94 | 62.58 |
| F1      | 68.66 | 70.28 | 65.99 |
| Brodatz |       |       |       |
| Rec     | 78.45 | 64.09 | 67.03 |
| Pre     | 69.03 | 70.1  | 71.3  |
| F1      | 73.44 | 66.47 | 69.09 |

## 4.1 Data Sets

The experiments was conducted using three datasets which contains images with uniform and highly textured background, weak edges and high intensity variation. Experiments was conducted with gray scale images also.

1. Weizmann single object dataset: This dataset includes 100 images with single object and all was used for conducting experiments. Some images have complex background and others have homogeneous background.
2. Brodatz texture images: The gray scale images in this dataset was tested and better result was obtained.
3. MSRA-500 dataset:It consists of 500 images, and is widely used for performance evaluation of segmentation approaches.

## 4.2 Parameter Selection

The two parameters include:patch size or neighborhood pixel size and number of texture regions or clusters(k). Experiments were conducted in 3x3,5x5 and 9x9 patch size. For Brodatz images, 9x9 patch size is suitable to obtain accurate result. Experiments shows that 5x5 patch acheived the better performance and the precisionand recall evaluation is shown in Table 1. The performance of the proposed approach also depends on number of texture regions. As the number of clusters increases, more accurate result was obtained. The performance evaluation based on number of cluster regions(k) is shown in Table 2. For MSRA dataset images, k=3 was appropriate choice for texture representation while k=5 strikes the better result for Weizmann and Brodatz images.

**Table 2** Comparison based on number of texture regions on datasets.

|          | k=2   | k=3   | k=5   |
|----------|-------|-------|-------|
| Weizmann |       |       |       |
| Rec      | 72.70 | 63.06 | 60.16 |
| Pre      | 69.06 | 70.24 | 71.97 |
| F1       | 70.83 | 66.45 | 65.03 |
| MSRA     |       |       |       |
| Rec      | 75.64 | 69.63 | 69.81 |
| Pre      | 61.04 | 70.94 | 62.58 |
| F1       | 67.56 | 70.28 | 70.27 |
| Brodatz  |       |       |       |
| Rec      | 73.55 | 63.79 | 62.73 |
| Pre      | 65.13 | 67.71 | 72.31 |
| F1       | 69.08 | 65.69 | 67.18 |

**Table 3** The average Precision, Recall and F1-Measure results for automatic initialized approaches.

|          | GVF   | VFC   | DAC   | NLAC  | Proposed method |
|----------|-------|-------|-------|-------|-----------------|
| Weizmann |       |       |       |       |                 |
| Rec      | 81.6  | 94.9  | 71.9  | 33.7  | 60.09           |
| Pre      | 73.0  | 52.1  | 59.5  | 74.2  | 76.70           |
| F1       | 80.6  | 65.2  | 65.3  | 25.2  | 67.38           |
| MSRA     |       |       |       |       |                 |
| Rec      | 76.6  | 89.6  | 61.8  | 43.7  | 65.40           |
| Pre      | 57.0  | 59.1  | 72.5  | 64.2  | 72.70           |
| F1       | 65.36 | 71.21 | 66.72 | 52.40 | 68.63           |
| Brodatz  |       |       |       |       |                 |
| Rec      | 98.60 | 99.8  | 87.2  | 73.2  | 62.70           |
| Pre      | 49.3  | 44.1  | 56.3  | 18.2  | 59.19           |
| F1       | 65.8  | 61.2  | 68.4  | 29.9  | 60.89           |

## 4.3   Contour Initialization

In this experiment, comparison of the contour initialization schemes was performed. Some approaches like VFC, DAC, NLAC and GVF doesn't provide automatic initialization. Two initialization schemes was used- circle and texture guided initialization. In circle initialization, using the groundtruth a circle is generated and covers the object. The radius of the circle was choosen as R=30. Texture discreteness provided the automatic contour initialization. Table 3 demonstrates that proposed method provides effective performance on Weizmann dataset compared to other approaches. It was investigated that GVF has high recall value but low precision value compared with the proposed method for the three datasets.

**Table 4** Average execution times of various approaches in sec. All methods are implemented in MATLAB

|         | GVF   | VFC   | DAC   | NLAC   | Proposed method |
|---------|-------|-------|-------|--------|-----------------|
| Time(s) | 14.95 | 16.24 | 42.55 | 497.56 | 15.89           |

## *4.4 Computional Cost*

The average running time of proposed approach against the other methods on datasets Weizmann, Brodatz and MSRA datasets was evaluated. The execution times is tested on an Intel i5 with 16 GB RAM. For this configuration, proposed approach consumes 80% on texture representation, 0.016 % on computing texture discreteness, 0.026 % on texture map,0.024 % for initializing the active contour,0.005% on external force convolution and 0.019% on contour transformation. Table 4 shows comparison regarding the execution time. NLAC has longest time and GVF has the lowest time. VFC and proposed method has medium computation cost.

## 5 Conclusion

In this paper, an active contour approach for natural image segmention based on the texture discreteness was proposed. Experimental results on datasets demonstrated the effectiveness of the proposed methodology. Experiments on gray scale images depicts that the approach is not limited to color. The automatic contour initialization leads to accurate segmentation. Future work involves extension to medical image segmentation, video data and remote sensing.

## References

1. Yang, Y., Gao, X.: Remote sensing image registration via active contour model. Int. J. Electron. Commun. **63**, 227–234 (2009)
2. Leymarie, F., Levine, M.: Tracking Deformable Objects in the Plane Using an Active Contour Model. IEEE Trans. Pattern Analysis and Machine Intelligence **15**(6), 617–634 (1993)
3. Mata, M., Armingol, J.M., Fernández, J., De La Escalera, A.: Object learning and detection using evolutionary deformable models for mobile robot navigation. Robotica **26**, 99–107 (2008)
4. Kass, M., Witkin, A., Terzopoulos, D.: Snakes: active contour models. Int. J. Comput. Vis. **1**, 321–331 (1988)
5. Li, B., Acton, S.T.: Active contour external force using vector field convolution forimage segmentation. IEEE Trans. Image Process. **16**(8), 2096–2106 (2007)
6. Xu, C., Prince, J.L.: Snakes, shapes, and gradient vector ow. IEEE Trans. Image Process. **7**(3), 359–369 (1998)

7. Mishra, A.K., Fieguth, P.W., Clausi, D.A.: Decoupled active contour (DAC) for boundary detection. IEEE Trans. Pattern Anal. Mach. Intell. **33**(2), 310–324 (2011)
8. Caselles, V., Kimmel, R., Sapiro, G.: Geodesic active contours. Int. J. Comput. Vis. **10**(10), 1467–1475 (1997)
9. Chan, T., Vese, L.: Active contours without edges. IEEE Trans. Image Process. **10**(2), 266–277 (2001)
10. Jalba, A.C., Wilkinson, M.H.F., Roerdink, J.B.T.M.: CPM: A deformable model for shape recovery and segmentation based on charged particles. IEEE Trans. Pattern Anal. Mach. Intell. **26**(10), 1320–1335 (2004)
11. Xu, C., Prince, J.: Gradient vector flow: a new external force for snakes. In: IEEE Computer Society Conference on Computer Vision and Pattern Recognition, pp. 66–71 (1997)
12. Li, B., Acton, S.: Vector field convolution for image segmentation using snakes. In: IEEE International Conference on Image Processing, pp. 1637–1640 (2006)
13. Derraz, F., Taleb-Ahmed, A., Peyrodie, L., Pinti, A., Chikh, A., Bereksi-Reguig, F.: Active contours based Battachryya gradient flow for texture segmentation. In: International Congress on Image and Signal Processing, pp. 1–6. IEEE (2009)
14. Sagiv, C., Sochen, N., Zeevi, Y.: Integrated active contours for texture segmentation. IEEE Trans. Image Process. **1**, 1–19 (2006)
15. Savelonas, M.A., Iakovidis, D.K., Maroulis, D.: LBP-guided active contours. Pattern Recogn. Lett. **29**, 1404–1415 (2008)
16. Otsu, N.: A threshold selection method from gray-level histograms. IEEE Trans. Syst., Man Cybernet. **9**, 62–66 (1979)
17. Barber, C.B., Dobkin, D.P., Huhdanpaa, H.: The quickhull algorithm for convex hulls. ACM Trans. Math. Softw. **22**, 469–483 (1996)
18. Jung, M., Peyré, G., Cohen, L.D.: Texture segmentation via non-local nonparametric active contours. In: Energy Minimization Methods Comput. Vis. Pattern Recogn., pp. 74–88. Springer (2011)
19. Lui, D., Scharfenberger, C., Fergani, K., Wong, A., Clausi, D.A.: Enhanced decoupled active contour using structural and textural variation energy functionals. IEEE Trans. Image Process. **23**(2), 855–869 (2014)

# Development of Tomographic Imaging Algorithms for Sonar and Radar

Ashish Roy, Supriya Chakraborty and Chinmoy Bhattacharya

**Abstract** Tomographic principle is now widely known for medical imaging using CT-scan machines. Whereas CT-scan uses absorption tomography by X-rays new areas of applications such as sonar and radar imaging are recently explored for reflection tomography. In this paper, we have shown that same backprojection principle can be utilized for imaging under water objects by sonar and ground objects by inverse synthetic aperture radar (ISAR). Analysis of results of image reconstruction from data collected in experiments are shown in the paper.

## 1 Introduction

Tomographic imaging is now a standard nonevasive method for medical imaging of cross-sections of human body with varied applications such as diagnosis of tumors, bulging, ulcers, etc. The method in computer-aided tomography (CT) involves collections of projection data from an object either by transmission of rays such as in X-ray tomography or by reflection or backscattering of rays such as in case of seismic tomography for geophysical studies [1], or non-destructive testing and analysis of structures and materials [2]. Recent applications of reflection tomography include imaging of underwater scenes by using acoustic sensors like sonar, and imaging ground footprints by microwave sensors such as radar [3], [4]. In reflection tomography reconstruction of a scene is done by backprojections unlike optical

A. Roy(✉)
Savitribai Phule Pune University, Pune 411007, India
e-mail: ashish_roy2006@yahoo.co.in

S. Chakraborty
Indian Institute of Technology, Delhi 110016, India
e-mail: supriya.chkrbty@gmail.com

C. Bhattacharya
ARDE, Pune 411021, India
e-mail: cbhat0@ieee.org

© Springer International Publishing Switzerland 2016
S.M. Thampi et al. (eds.), *Advances in Signal Processing and Intelligent Recognition Systems*,
Advances in Intelligent Systems and Computing 425,
DOI: 10.1007/978-3-319-28658-7_25

imaging that is a forward projection or mapping of illumination intensity from an object scene. Reflections from a three-dimensional (3D) object are collected as several one-dimensional (1D) projection views taken at different aspect angles. Backprojection is implemented by creating a two-dimensional (2D) cross-section view from superposition of these 1D backscatterings or projections at different aspect angles. This imaging technique has become popular in ocean acoustic tomogrpahy (OAT) as by backprojections it is possible to reconstruct high resolution images of underwater objects such as sea mines at safe standoff distances (100–200m). The sonar transducer is moved from point to point around the entire circumference of a circle so as to view the object from all possible aspect angles, thus compiling a complete set of projections over 360° of aspect angles [5]. The image represents a 2D magnitude profile of the acoustic reflectivity functions when backprojected on the image plane from the object. Similarly, in case of inverse synthetic aperture radar (ISAR) 1D projection slices of a rotating object are acquired as microwave radar backscattering. By backprojections of these 1D backscatterings obtained at different look angles of radar observation the 2D image of the rotating object can be reconstructed at far off distances [6], [7].

In this paper, we demonstrate that same backprojection principle can be utilized in the case of OAT or ISAR to obtain high resolution cross-section images of either underwater objects or rotating objects on ground. The same imaging principle can be applied either to underwater acoustic wave propagation or to free space microwave propagation for obtaining high resolution images of cross-section at each aspect angle of viewing. The algorithms for reconstructing images from 1D projections are based either on filtered backprojections (FBP) or by 2D inverse Fourier transform of stacks of 1D projections. We show here working of the FBP algorthim with a spline filter specifically designed for reconstruction of images of underwater objects such as mines and buoys. For ISAR imaging we obtain actual radar backscattering at X-band collected from rotating corner reflectors (CR) and utlize a 2D inverse Fourier transform algorithm to reconstruct images of the CRs with high azimuth resolution. Analytical results of the quality of reconstructed images from experimental received data from radar and sonar are presented in the paper.

## 2  Principle of Backprojection for Under Water Imaging by Sonar

The sonar-object geometry for recording OAT data is shown in Fig. 1. Here, $x$-$y$ plane represents a Cartesian coordinate system with origin $O$ located at the center of the scene including the object to be imaged by the sonar. The transmitter and receiver are colocated such that direction of the backscattered energy follows the same path opposite to the direction of transmission. Assuming sonar transmission to be rectilinear $f(x, y)$ denotes the 2D distribution of the signal backscattered from the object when reconstructed from all aspect angles of viewing. The backscattered signal at an aspect angle $\theta$ is the 1D projection of the object function $f(x, y)$. This 1D projection operation is known as Radon transform and is denoted as $s(l, \theta)$ in

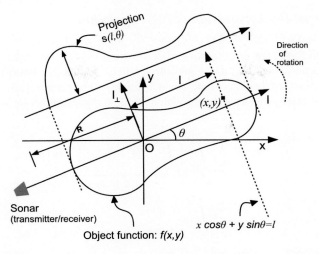

**Fig. 1** Sonar-object geometry for a tomographic imaging sonar and the projection operation result $s(l, \theta)$ for an aspect angle $\theta$.

Fig. 1. It is defined as the line integral of $f(x, y)$ along a line $x \cos \theta + y \sin \theta = l$, inclined at an angle $\theta$ from the $x$-axis at a distance $l$ from the origin. In the rotated coordinate system $s(l, \theta)$ is written as

$$s(l, \theta) = \mathscr{R} f(x, y) = \int f(x, y) dl_\perp = \int f(l \cos \theta - l_\perp \sin \theta, l \sin \theta + l_\perp \cos \theta) dl_\perp \quad (1)$$

where $-\infty < l < \infty, 0 < \theta < \pi$, $\mathscr{R}$ is the Radon transform or 1D projection operator. The 1D spatial Fourier transform of $s(l, \theta)$ is

$$S(k, \theta) = \int s(l, \theta) e^{-j2\pi l k} dl \quad (2)$$

Here, $k = \sqrt{k_x^2 + k_y^2}$ represents the spatial frquency variable. Reconstruction of the object function requires backprojection of the recorded projections taken at all angles $\theta$. Backprojection operation is the superposition of all reflected wavefront integrals from the point at $(x, y)$ at different positions of $\theta$. Using the result of 2D spatial inverse Fourier transform in the polar coordinate system the backprojected object function is expressed as

$$\hat{f}(x, y) = \frac{1}{4\pi^2} \int_0^\pi \int_{-\infty}^\infty S(k, \theta) e^{j2\pi l k} |k| dk d\theta \quad (3)$$

## 2.1   Filtered Backprojection (FBP) Algorithm

According to the result of projection slice theorem [8] the 1D Fourier transform
of the projection, i.e., $S(k, \theta)$ is a central "slice" of the 2D Fourier transform
$F(k \cos \theta, k \sin \theta)$ of the object function at an angle $\theta$. The backprojection operation
of (3) can be rearranged as a superposition of filtered projections in the following
manner,

$$\hat{f}(x, y) = \frac{1}{4\pi^2} \int_0^\pi \underbrace{\left\{ \int_{-\infty}^\infty S(k, \theta)|k|e^{j2\pi lk} dk \right\}}_{Filtered\ projections} d\theta$$

$$= \frac{1}{4\pi^2} \int_0^\pi \hat{s}(l, \theta) d\theta \qquad (4)$$

Also the term $|k| = \sqrt{k_x^2 + k_y^2}$ represents the effect of a spatial filter applied to
each of the projections at various aspect angle $\theta$. Since the projection data $s(l, \theta)$
is sampled by the sonar receiver at discrete incremental angles the FBP algorithm
of (4) is implemented in a discrete manner.

$$\tilde{f}(x, y) \triangleq \Delta \sum_{n=0}^{N-1} \hat{s}(x \cos n\Delta + y \sin n\Delta, n\Delta) \qquad (5)$$

where $N$ is the total number of projections at discrete incremental angular steps
$\Delta = \pi/N$, and $0 \leq n \leq N - 1$. Each projection data has to be sampled at least at
Nyquist interval $d \leq 1/2k_o$ where $k_o$ is the highest spatial frequency of interest of
the object.

The modulus function $|k|$ accentuates noise in output image at high-spatial fre-
quencies. Since practical reconstructed images by OAT show a low signal to noise
ratio (SNR) at high frequencies, use of this unmodified filter function degrades the
SNR even further. Generally underwater objects are space-limited and employing a
bandlimiting filter with sharp cut-off frequency response is not very suitable especially in presence of noise. Hence, a class of filters given by the modified response
function $H(k) = |k|W(k)$ is generally used where $|k|$ is a ramp like filter function.
A bandlimiting window function $W(k)$ is selected such that a better trade-off between the filter bandwidth and SNR is achieved. Various filters such as Ram-Lak,
Shepp-Logan are utilized for X-ray tomography to improve reconstrcution $\tilde{f}(x, y)$
[9]. Applications of spline functions are shown in simulation with standard phantom models for backprojection with FBP [10]. We utilize spline interpolation filter
for implementing the window function $W(k)$ in the FBP reconstruction to improve
the SNR for reconstuction of underwater scenes in OAT. In the Fourier domain a
symmetric B-spline function of degree $\alpha$ is given as [11]

$$W(k) = \frac{|1 - e^{jk}|^{\alpha+1}}{|jk|^{\alpha+1}} \tag{6}$$

where $k$ is a spatial frequency variable. For degree $\alpha = m$ where $m$ is an integer the Fourier domain spline interpolating filter spectrum is derived as

$$|H(e^{jk})| = |\frac{k}{2\pi}| \frac{1}{\sum_{l \in Z} |sinc(\frac{k}{2\pi} + l)|^{m+1}} \tag{7}$$

The steps for implementing the FBP algorithm are summarized below:

- Computation of the 1D Fourier transform of $s(l, \theta)$ received at each angle $\theta$.
- Multiplication of the 1D transform of the projections with the spline filter response function $|H(e^{jk})|$ in the frequency domain.
- Computation of the inverse Fourier transform of the above results that gives the filtered projections $\hat{s}(\theta)$.
- Summation of discrete projections $\hat{s}(\theta)$ for all $\theta$ to deliver $\tilde{f}(x, y)$.

The operational steps of FBP algorithm using the spline interpolation filter are shown in Fig. 2 below.

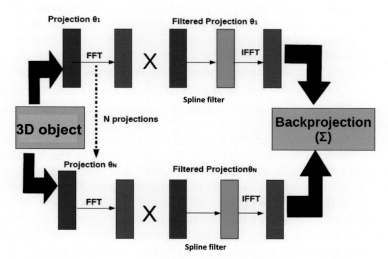

**Fig. 2** Flowchart of FBP algorithm utilizing spline interpolation filter for reconstruction of $\tilde{f}(x, y)$ by sonar.

## 3  Principle of Backprojection for ISAR Imaging of Rotating CR

In the case of ISAR tomographic image reconstruction the colocated radar transmitter-receiver is kept stationary. Samples of backscattered microwave signal are collected from a rotating object for each incremental angle of rotation. The same geometry of Fig. 1 can also be considered for ISAR imaging scenario. Here projection slices are collected with the source-receiver system in a fixed position and the object is rotating at incremental angle with uniform angular speed. The source is a frequency modulated continuous wave (FMCW) transmitter at X-band with center frequency of 9.75GHz. The spatial projection data contained in the baseband time signal is available at the output of receiver as a beat signal and is represented as [12], [13]

$$\bar{r}_\theta(t) = \int_{-L}^{L} s(l, \theta) e^{-j\frac{2}{c}(\omega_o + 2\alpha t - 2\tau_o)l} dl \tag{8}$$

where $\omega_o$ is the carrier frequency, $\alpha = \frac{B}{T}$ is the FM sweep rate, $B$ is the transmitted sweep bandwidth in time $T$. The time delay due to two-way path of travel $R$ between the radar and object is $\tau_o = \frac{2R}{c}$ and $c$ is speed of the electromagnetic wave. As seen in (8) the received beat signal is represented as the Fourier transform of $s(l, \theta)$ over kernel $\frac{2}{c}(\omega_o + 2\alpha t - 2\tau_o)$. The projection function $s(l, \theta)$ represents complex reflectivity of the object function $f(x, y) = |f(x, y)| e^{\angle f(x,y)}$ measured at all aspect angle positions of the rotating object.

The transform is bound in spatial domain by the dimensions of the object $-L \leq l \leq L$. Therefore $\bar{r}_\theta(t) = S\left[\frac{2}{c}(\omega_o + 2\alpha t - 2\tau_o), \theta\right]$ is a Fourier transform of radial projection at an angle $\theta$ representing backscattered data collected on a polar raster grid.

The experimental setup for collecting radar backscattering in ISAR mode for a standard trihedral CR object using FMCW radar is shown in Fig. 3. The setup is configured inside a semi-anechoic chamber. The CR is positioned on a rotating pedestal for recording the backscatter for 180° angular rotation. The resolution in the range direction (along the lines of backscattering) is $\frac{c}{2B}$ and for the present case is approximately 1m. Resolution in the cross range direction is $\frac{\lambda}{2\theta_{rot}}$ where $\lambda$ is operating wavelength of the radar and $\theta_{rot}$ is the total angle of viewing over which the ISAR beat signal is measured. Beat signal containing the projection information is recorded by the receiver and stored in digital memory.

The reconstruction of the 2D object function $\tilde{f}(x, y)$ from the 1D transform of the projections $S\left[\frac{2}{c}(\omega_o + 2\alpha t - 2\tau_o), \theta\right]$ is achieved using a 2D inverse Fourier transform

$$\tilde{f}(x, y) = \mathscr{F}_2^{-1}[S\left(\frac{2}{c}(\omega_o + 2\alpha t - 2\tau_o), \theta\right)] \tag{9}$$

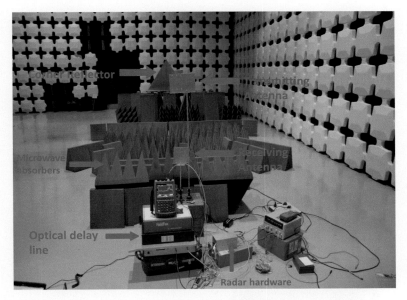

**Fig. 3** Radar data collection for rotating CR in ISAR mode of image reconstruction.

The projection data record for angular measurement of $180°$ in steps of $\theta = 1°$ is sequentially populated in a matrix. Rows of the data matrix constitute the sampled projection at each incremental $\theta$. For performing a 2D inverse Fourier transform the data is converted from a polar to cartesian raster through linear interpolation along the rows and columns as the total viewing angle $\theta_{rot}$ is small.

## 4 Results of Backprojection on OAT and ISAR Data

The results of FBP algorithm for underwater objects MK56 mine and Manta buoy are shown in Sect. 4.1. Reconstruction of images using Shepp-Logan filter and spline interpolating filter of order 4, 8 and 24 are shown here. Sect. 4.2 shows the results of reconstruction of the CR images from projections obtained in the ISAR experiment.

### 4.1 Results of Backprojections for Underwater Objects

We present simulation results in laboratory for collection of backprojection data over actual underwater objects such as MK56 mine and Manta buoy. In practical scenario a number of underwater sonar transducers will be placed in a circumference to collect the backprojection data at fine angular resolution. The object MK56 mine and

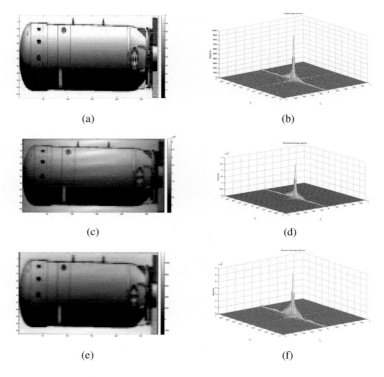

(a)                                                                    (b)

(c)                                                                    (d)

(e)                                                                    (f)

**Fig. 4** Results of backprojetion of actual underwater object (MK56 mine) showing (a) image of original object and (b) its frequency response. Reconstructed object (c) using Shepp-Logan filter and (d) its frequency response. Reconstructed object (e) using spline of order 24 and (f) its frequency response.

its frequency response is shown in Fig. 4(a) and 4(b) respectively. In the simulations projections over $180°$ are taken at $\Delta = 1°$ for reconstructing the 2D image of the object. The backprojected image of the object using Shepp-Logan filter and spline filter of order 24 are shown in Fig. 4(c) and 4(e) respectively. Their frequency responses are shown in Fig. 4(d) and 4(f) respectively. Reconstructed image of Fig. 4(c) shows artifacts with a visible pattern of intensity variation that is due to the response of Shepp-Logan filter. It can be seen in Fig. 4(e) that no such artifacts are observed and reconstruction using spline filter of order 24 gives a better image. Similar results are shown for the object Manta buoy in Fig. 5.

The error in reconstruction is measured by two norms. First, the normalized cross correlation gives the degree of similarity between the image function and backprojected images and is defined as

$$\sigma_N = \frac{1}{N} \sum_1^N \frac{(f - \overline{f})(t - \overline{t})}{s_f s_t} \tag{10}$$

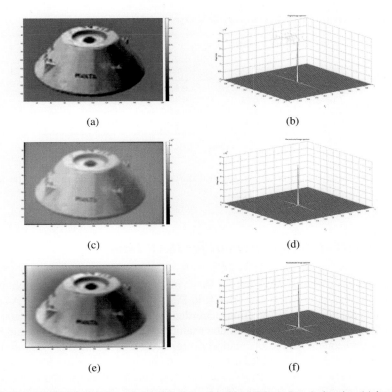

(a)          (b)

(c)          (d)

(e)          (f)

**Fig. 5** Results of backprojetion of actual underwater object (Manta buoy) showing (a) image of original object and (b) its frequency response. Reconstructed object (c) using Shepp-Logan filter and (d) its frequency response, reconstructed object (e) using spline of order 24 and (f) its frequency response.

where $f$ represents the actual image function and $t$ is the backprojected image both having $N$ pixels, $\overline{f}$ and $\overline{t}$ are the mean values of their pixels and $s_f$ and $s_t$ are their standard deviations respectively. Second, root mean square error (RMSE) distance is used for the comaprison of performance of the Shepp-Logan and spline filters. The normalized RMSE is given as

$$RMSE(p1, p2) = \sqrt{E\left\{(p1 - p2)^2\right\}} \qquad (11)$$

where $p1$, $p2$ are vectors containing normalized pixel values of the original and reconstructed images respectively and $E$ is the expectation of the squares of their errors. The calculated RMS error and the cross correlation between the image function and its reconstruction using Shepp-Logan and spline filters are given in Table 1. It is seen that compared to the Shepp-Logan filter RMS error is less in case of the reconstructed images of both the objects using spline filter and decreases with in-

**Table 1** Comparative analysis of errors in the backprojected OAT images

| Object | Error Calculation | Shepp-Logan filter | Spline of different orders | | |
|--------|-------------------|--------------------|----------|--|--|
|        |                   |                    | Order 4 | Order 8 | Order 24 |
| MK 56  | RMS error         | 0.45               | 0.41    | 0.35    | 0.33 |
|        | Correlation coeff.| 0.94               | 0.94    | 0.95    | 0.96 |
| Manta  | RMS error         | 0.68               | 0.64    | 0.61    | 0.58 |
|        | Correlation coeff.| 0.90               | 0.91    | 0.93    | 0.95 |

creasing order of the filter. Also, the normalized correlation is higher in the latter case.

## 4.2 Results of Backprojections for ISAR Data

The radar cross section (RCS) plot for ISAR measurement over 180° rotation of the turntable is shown for 50cm and 1m CR in Fig. 6(a) and 6(b) respectively. As seen in the plots the number of projections recorded for 180° rotation of the turntable are oversampled. The projection data is downsampled in the cross range direction such that each subsequent degree of rotation is represented by a single projection. Data within $\pm 6°$ and $\pm 12°$ are selected for image reconstruction for the 50cm and 1m CRs respectively. A linear interpolation is performed in the range and cross range direction by padding zeros to the ISAR data matrix. A Taylor window is applied for reducing the sidelobe levels along both the direction. The reconstructed images using 2D IFFT are shown in Fig. 7(a) and 7(b) for 50cm and 1m CR respectively. The 3D magnitude impulse response functions (IRFs) of the 50cm and 1m CR are shown in Fig. 7(c) and 7(d) respectively. The spatial dimensions of a pixel in the 2D image of 1m CR is $7.32 \times 0.13 \text{cm}^2$ and that of the 50cm CR is $7.32 \times 0.06 \text{cm}^2$. Ths 2D images contain total 8092 pixels in the range direction and 512 pixels in the cross range direction. It is seen in the 2D images of the CRs that the response of

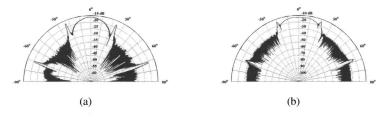

(a)                                                  (b)

**Fig. 6** Raw data collection for ISAR imaging showing RCS polar plot for (a) 50cm CR and (b) 1m CR.

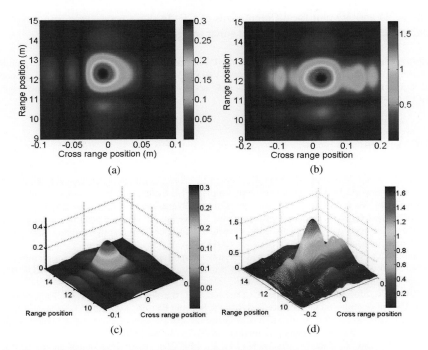

**Fig. 7** Results of ISAR data processing showing backprojcted 2D images of (a) 50cm CR and (b) 1m CR, 3D impulse responses of (c) 50cm CR and (d) 1m CR.

1m CR occupies larger number of pixels in the cross range direction. The peak RCS value of 1m CR is larger in magnitude by 7.4dB in comparsion to the 50cm CR as seen in the 3D IRFs. The 3dB range resolution measured for both the CRs is 1.2m. The cross range resolution obtained from the response of 50cm CR is 8cm and that of the 1m CR is 4cm.

## Conclusion

In this paper we have shown backprojections of cross section views of objects obtained using both sonar and radar principle. A filtered backprojection algorithm is implemented for the reconstruction of images of underwater objects and its working is shown utilizing Shepp-Logan and spline filters. From the comparative analysis of error it is shown that spline filters of higher order are capable of producing smoother backprojected images than Shepp-Logan filters. We have demonstrated the reconstruction of 2D ISAR images of standard rotating CRs of different sizes from backscattered radar data. The resolution of the reconstructed CR image is measured both as 2D and 3D plot for different sizes of CR. In these ways same tomographic principle can be utilized for radar and sonar image formation.

**Acknowledgment** We acknowledge Director, NPOL, Kochi, India for awarding the sonar project to the third author. The cooperation of Director VRDE, Ahmednagar, India for providing support and logistics for the ISAR measurements is duly acknowledged. The first author acknowledges UGC, India for granting research scholarship.

# References

1. Shi, L., Song, W.Z., Xu, M., Xiao, Q., Lees, J.M., Xing, G.: Imaging seismic tomography in sensor network. In: Proceedings of 10th Annual IEEE Communications Society Conference on Sensor, Mesh and Ad Hoc Communications and Networks (SECON), pp. 327–335 (2013)
2. Athanasios, B., Nick, P., Laurent, P., Antonios, V., Vassilis, K.: Damage identification in carbon fiber reinforced polymer plates using electrical resistance tomography mapping. Journal of Composite Materials **47**(26), 3285–3301 (2013)
3. Sabo, T.O., Hansen, R.E., Austeng, A.: Synthetic aperture sonar tomography: a preliminary study. In: Proceedings of 10th European Conference on Synthetic Aperture Radar (EUSAR), pp. 1–4, June 2014
4. Zhu, X.X., Montazeri, S., Gisinger, C., Hanssen, R.F., Bamler, R.: Geodetic SAR Tomography. IEEE Transactions on Geoscience and Remote Sensing **PP**(99), 1–18 (2015)
5. Brian, G.F., Ron, J.W.: Application of acoustic reflection tomography to sonar imaging. The Jornal of the Acoustic Society of America **117**(5), 2915–2928 (2005)
6. Zheng, J., Su, T., Zhu, W., Zhang, L., Liu, Z., Liu, Q.H.: ISAR Imaging of Nonuniformly Rotating Target Based on a Fast Parameter Estimation Algorithm of Cubic Phase Signal. IEEE Transactions on Geoscience and Remote Sensing **53**(9), 4727–4740 (2015)
7. Sego, D.J., Griffiths, H., Wicks, M.C.: Radar tomography using Doppler-based projections. In: Proceedings of IEEE Radar Conference, pp. 403–408, May 2011
8. Kak, A.C., Slaney, M.: Principles of Computerized Tomographic Imaging. Society of Industrial and Applied Mathematics (2001)
9. Jain, A.K.: Fundamentals of Digital Image Processing. Prentice Hall, Eaglewood Cliffs (1989)
10. Horbelt, S., Liebling, M., Unser, M.: Filter design for filtered back-projection guided by the interpolation model. In: Proceedings of SPIE 4684, Medical Imaging 2002: Image Processing, p. 806, May 2002
11. Unser, M.: Splines: a perfect fit for signal and image processing. IEEE Signal Processing Magazine **16**(6), 22–38 (1999)
12. Munson Jr., D.C., O'Brien, J.D., Jenkins, W.: A tomographic formulation of spotlight-mode synthetic aperture radar. Proceedings of the IEEE **71**(8), 917–925 (1983)
13. Carrara, W.G., Goodman, R.S., Majewski, R.M.: Spotlight Synthetic Aperture Radar signal processing Algorithms. Artech House (1995)

# Part III
# Signal and Speech Processing

# Design and Development of an Innovative Biomedical Engineering Application Toolkit (B.E.A.T. ®) for m-Health Applications

**Abhinav, Avval Gupta, Shona Saseendran, Abhijith Bailur, Rajnish Juneja and Balram Bhargava**

**Abstract** Obtaining on-the-go medical grade reliable data in noisy environment is a challenge faced by developers across the globe. While transporting patients, mobile phones connected wirelessly to the sensors have tremendous lifesaving applications. Recent trends in research have shown promises in the use of continuous biomedical signals for Biometric applications. However, practical applications of such systems will be possible only if they can generate real time medical grade data in environment having radical noise. This paper details the design and development of B.E.A.T. ® - an innovative system which captures medical grade physiological signals using 2 leads with or without the use of pre-gelled electrodes over Android OS.

## 1 Introduction

Healthcare research and services are of prime importance and is estimated that the Indian market for diagnostics is worth US$1.1 billion accounting for up to 4% of

Abhinav(✉) · A. Bailur
Cardea Labs, New Delhi 110067, India
e-mail: abhinav@cardeabiomedical.com, abhijit_bailur@yahoo.co.in

A. Gupta
Computer Science Department, IIT Delhi, Delhi 110016, India
e-mail: avval.07@gmail.com

S. Saseendran
M.B.A., IMT Dubai, Dubai, UAE
e-mail: shona.027@gmail.com

R. Juneja · B. Bhargava
Department of Cardiology, AIIMS, New Delhi 110019, India
e-mail: {rjuneja2,balrambhargava}@gmail.com

© Springer International Publishing Switzerland 2016
S.M. Thampi et al. (eds.), *Advances in Signal Processing and Intelligent Recognition Systems*,
Advances in Intelligent Systems and Computing 425,
DOI: 10.1007/978-3-319-28658-7_26

299

the total healthcare delivery market (PwC, India). Because widespread screening is often tempered by high costs, there is an urgent need for methods to increase screening throughput while maintaining diagnostic reliability and reducing cost.

The ubiquitous presence of mobile phone holds great potential for improving the quality of life for the masses. There are more than approximately 6 billion mobile phone subscribers worldwide, out of which 76% are in developing countries such as India [1]. With rapid innovation fueled by competition, the mobile phone is getting smarter by the second with innovative applications designed to provide health related information on the go. In recent years, mHealth has shown promising growth mainly because it can alter how healthcare is delivered, the quality of patient experience, and the cost of health care. The processed report over internet can also be automatically received and viewed on the mobile phone as required. With advancement in technology, mobile phones are connected wirelessly to wearable portable sensors, to enable medical measurement and monitoring for various stakeholders [2][3][4].

Among the various Bio-medical signals, ECG is one of the vital signals and its monitoring is very crucial during pre-operative and post-operative period which needs regular and in several cases continuous monitoring [5]. Also during patient transport, monitoring of ECG plays a crucial role. With deep penetration of mobile phones and mobile phone networks in developing countries, ECG monitoring on-the-go is the need of the hour and this need can be met if ECG signals could be plotted on all affordable smart phones. Such systems can find major applications in various areas including alternatives for event recorders [6]. Recently scientists, researchers and developers have migrated from sporadic analytics such as fingerprint, iris scan, face recognition etc. to continuous authentication systems involving biomedical signals such as ECG, EEG among others. Because ECG can be comparatively easy to obtain, it is almost considered to be a new gold-standard for continuous biometric identification. [7] But from practical view point, the applications of such systems will be heavily dependent on the signal to noise ratio of the data acquired from the hardware. For environments with sporadic noise, the systems are bound to fail or give erroneous results at worse.

Developing a smart, user-friendly system which can capture bio-signals with minimum leads, wirelessly transmit it over smart phones and yet consume minimalistic power to plot and save the data over smart phones of medical grade can go a long way in development of portable devices for monitoring for both m-health and Internet of Things application [1]. The challenge, however, lies in getting reliable medical grade data immune to noisy environment, and handling real time data while performing signal processing on affordable phones.

This paper details the design and development of a system which allows its users to capture medical grade signals by mere touch over Android OS wirelessly. The associated mobile phone application not only plots the data in real-time, but it also processes the data to give relevant output while saving the data on user request. The system performs real time signal processing of raw data at both hardware level as well as on the Android Phone. The mobile phone algorithm is cleverly developed, allowing the application to run on affordable handsets having low speed processors. The hardware can be easily customized to obtain medical grade single lead ECG, EOG and EMG signals. The entire system works on 3.3 Volts making the system completely portable.

## 2    Hardware Design

Designing of B.E.A.T. ® hardware was performed in 2 stages. In the first stage, independent modules for Power Section, Analog Section and Digital Section were designed and separately tested. After stability of the modules, the modules were interconnected to check for compatibility. Once the data obtained was found to be satisfactory and of medical standards, a final board was fabricated comprising all the different modules.

### 2.1   Power Supply Module

Power board Fig. 2(a) provides a stable 3.3 Volts rail for powering the other boards. The input voltage source can be a commonly available 3.7 Volts mobile phone battery a 9 V commercially available battery. The voltage is regulated to 3.3 Volts by using LT1117 voltage regulator. A low voltage indicator though a comparator circuit is provided.

### 2.2   Analog Module

The Analog board has AD8236 at its kernel as shown in Fig. 1. It is the lowest power Instrumentation Amplifier in the Industry and can operate on voltages as low as 1.8V. Because of its extremely small size consuming a maximum current of 40 μA, it is an excellent choice for battery-powered applications.

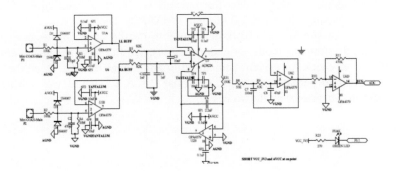

**Fig. 1** Schematic of Analog module

The system is a single lead ECG designed to work on both 2 as well as 3 electrodes connected to Left Arm, Right Arm and Left Leg to get lead I, lead II or lead III data as required by the user. It can be replaced with AD620 instrumentation amplifier as well. Fabricated Analog Board is shown in Fig.2 (b)

The signals from the body are fed through a buffer circuit which is primarily a voltage follower which is then given to a pre-amplifier. The pre-amplifier was designed using a quad op-amp (OPA4379). The Analog signals from the two electrodes are fed to AD8236. The output raw signal is fed to an active high pass filter and the processed output is fed to the reference pin of the instrumentation amplifier. This helps in efficient removal of baseline wander and also allows the system to be driven on just 2 electrodes.

The optional 3rd lead is connected to the system's ground and may be used in extremely noisy environment. The raw data is fed to a 4$^{th}$ order band pass Sallen-key filter of band 0.5 Hz – 40 Hz (for ECG and EOG) or 0.5 Hz- 200 Hz for EMG. This provides a fairly clean data for feeding to the Digital Board and for wireless transmission.

# 3    Design of Digital Board

Digital board Fig. 2 (C) is designed by selectively choosing the components for minimum power consumption. The board is designed using ULP (Ultra Low Power) Microcontroller PIC24. The Analog signal is fed to the analog input port for Analog to Digital Conversion. The data is sampled at 12 bits. The digital data is sent to a sub-program where the data is convoluted with a filter coefficients designed for a 100th order Hamming Window based filter.

This allows the implementation of a real time digital filter. The digital filter is designed using MATLAB where the filter co-efficient for 100th order filter for a frequency band of 0.5Hz to 40 Hz (for ECG and EOG) and 0.5 Hz to 200 Hz (for EMG) is obtained. The incoming data undergoes convolution with the filter co-efficient to obtain a noise free data for transmission. Data transmission is facilitated wirelessly using Bluetooth 2.0 Module LMX9838. The

Bluetooth is programmed to work in Serial Port Profile which allows easy pairing and data transmission with peripheral devices. All the three designs were interconnected to check for system compatibility and signal fidelity. Once the system was found to perform perfectly, the final B.E.A.T.® board was fabricated having all the subsystems as shown in Fig. 3.

**(a) Power Board**

**(b) Analog Board**

**(c) Digital Board**

**Fig. 2**

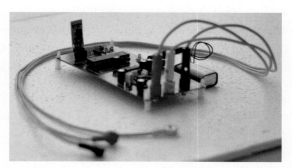

**Fig. 3** Final B.E.A.T.® Board

# 4    Mobile Phone Application

The B.E.A.T.® Android application is built to interact with the B.E.A.T hardware via Bluetooth and plot the captured signals on the smart phone in real time. The real time data plotting is accomplished by means of a shared memory buffer and producer-consumer threads running on it. The producer thread writes data points received from the B.E.A.T. ® hardware on to the memory buffer. The consumer thread reads these data points and plots them on to the UI. In addition to real-time signal plotting, the App features the following functionalities:

   a)  Event recording and data saving,
   b)  PDF report generation,
   c)  Heart-rate (based on Pan Tompkins Algorithm),
   d)  Trigger pause-resume & filtering on-off signals from hardware.

   Major sections include:

## 4.1   Bluetooth Service

The Bluetooth Service interacts with the external B.E.A.T.® hardware and is responsible for:

- Connection Establishment
- Reading data from input stream and supplying to UI thread
- Writing signals into output stream based on user requests

   The Bluetooth service achieves these responsibilities by the following threads.

- *Connect thread* – creates a bluetooth RFCOMM socket with the remote Bluetooth device, recognized by the Bluetooth MAC address, and attempts to connect to it using a Serial Port Profile. On connection establishment, it initiates the connected thread.
- *Connected thread* – Once connection is established, this thread fetches the input and output streams of the socket, and uses them to exchange data.

   The socket input stream receives data points from the B.E.A.T. ® hardware. It is the responsibility of the connected thread to write these data points on to a shared memory buffer for the plotting thread to consume, thereby acting as producer. The connected thread is also responsible for sending START, STOP, FILTER_ON and FILTER_OFF signals to the BEAT hardware using the output stream of the socket connection on demand.

## 4.2   Pan Tompkins Algorithm

A modified Pan Tompkins algorithm performs real-time beat detection from ECG signals. It is achieved in the following steps:

- Band Pass Filter: QRS energy is concentrated approximately in the 5-15 Hz band. This band pass filter removed and eliminates unnecessary muscle noise, 50Hz interference, baseline wandering and T wave. The high pass filter is designed by subtracting the output of first order low-pass filter from an all-pass filter with a delay [7]. The critical frequency of the high pass filter has a delay of around 80ms.
- First differentiation: The signal from the Band-pass is then differentiated to calculate the slope where it uses a five-point derivation.
- Squaring: After the previous stage, the signal is then squared. The squaring operation is done to obtain only positive values and thereby gives a non-linear amplification of the output.
- Integration: Moving-window integration is performed to the data after the previous stage. The output consists of multiple peaks within duration of QRS complex. A smoothing function, smoothens the output by moving-window integration.
- Thresholding: An Adaptive threshold which adapts to changes in the ECG signal by estimates of signal and noise peaks eventually helps in finding the R peaks.

## *4.3   Real-time Plotting*

For plotting real time signals received from the B.E.A.T.® hardware, we use a Drawing class that extends the Android SurfaceView and implements onDraw (android.graphics.Canvas) method which is responsible for redesigning the surface. A secondary thread contains a SurfaceHolder for this SurfaceView and triggers rendering of data points into the surface. This thread acts as consumer of the shared memory buffer which contains the data points produced by the Bluetooth service, and passes these data points to the surface view's draw method.

Based on Pan Tompkins algorithm, for ECG data, a real time heart rate value is also displayed as shown in Fig. 4 (a). For EMG this loop is disabled Fig. 4 (b)

## 5    Data Saving and Event Recorder

The Android application supports saving of data captured up to 90 seconds in raw as well as in PDF format as shown in *Fig. 6*. The B.E.A.T data files can be opened directly by this app [*Fig. 5 (a)]* to re-view the data collected with scrolling and zoom [*Fig. 5 (b)]*. The app uses the *AChartEngine* library for plotting data from the saved files. The PDF report saves an accurate 25mm/s recording of the data, and is helpful to be shared and used among users and systems that do not have the B.E.A.T App.

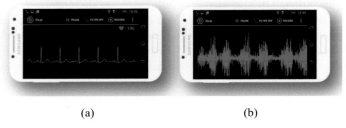

(a)                                                    (b)

**Fig. 4** (a) Real Time ECG plot with Heart rate (b) Real Time EMG plot

The mobile application also features a unique Event Recording functionality which supports data saving from past in addition to real-time data. This is facilitated by allowing the data to be temporarily saved in a memory for up to 15 seconds before being overwritten by new data values. Thus when a user triggers the Event button by pressing the Record button twice in quick succession, previous data from buffer is stored along with the continuous real time data. This data can be saved in a .pdf format and sent to a doctor for consultation through internet.

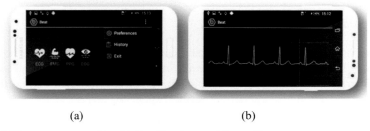

(a)                                                    (b)

**Fig. 5** (a) History to access files (b) Scroll and zoom history files

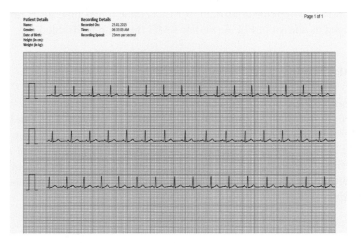

**Fig. 6** PDF File Generated Automatically

# 6 Desktop Application

A desktop application was developed using National Instrument's Lab View. Data from B.E.A.T. ® hardware is streamed wirelessly through Bluetooth and is captured through a virtual Serial Port Profile (SPP) for plotting & saving on the screen in real-time as shown in Fig. 7.

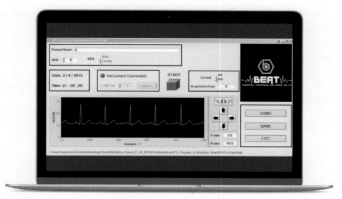

**Fig. 7** Desktop Application of B.E.A.T. ®

# 7 Results and Discussions

As shown in Fig. 6 (a), Fig. 6(b), Fig. 7(b) and Fig. 8, this system plots a medical grade real-time ECG on all mobile phones working on Android OS without any delay. The system developed was put on 10 healthy volunteers who did not have any known cardiac disorder. The volunteers were all of the age group of 20 to 30 years. The data was simultaneously captured from this device and MP 150 from Biopac (gold standard) for comparative analysis. The PQRST waveforms were compared and found to be matching within acceptable limits.

The purpose of developing this novel platform technology was to make health monitoring possible in resource limited settings making it a perfect candidate for m-health applications. Since the system works perfectly in extremely noisy environment it finds itself useful n Biometric Identification. It will be useful during patient transport of sick patients from the periphery to secondary/tertiary care centers. Sick patients need monitoring even within a hospital not only for transport but also during procedures like CT Scans, MRI etc. Most rural hospitals/secondary care hospitals cannot afford costly ECG monitors and our device will be a perfect solution for such places.

**Acknowledgment** This work was carried out by the support of Department of Scientific and Industrial Research (DSIR), Government of India.

# References

1. Kansal, A., Gupta, A., Paul, K., Prasad, S.: mDROID - an affordable android based mHealth system. In: HEALTHINF, pp.109–116 (2014)
2. Cisco Visual Networking Index: Global Mobile Data Traffic Forecast Update (2014–2019). (accessed February 3, 2015)
3. Fitbit study: UK adults find mobile health tracking, not public messaging. Mobile Health News
4. Schuman, A.J.: Improving patient care: Smartphones and mobile medical devices. Contemporary Pediatrics, June 2013
5. Secerbegovic, A., Mujcic, A., Suljanovic, N., Nurkic, M., Tasic, J.: The research mHealth platform for ECG monitoring. In: 11th International Conference on Telecommunications (ConTEL), pp.103–108 (2011)
6. Bonacina, S., Masseroli, M.: A web application for managing data of cardiovascular risk patients. In: EMBS 2006, pp. 6324–6327. IEEE (2006)
7. Istepanian, R. H., Laxminarayan, S., Pattichis, C.S.: M-Health: Emerging Mobile Health Systems. Springer (2006)

# A VMD Based Approach for Speech Enhancement

B. Ganga Gowri, S. Sachin Kumar, Neethu Mohan and K.P. Soman

**Abstract** This paper proposes a Variational Mode Decomposition (VMD) based approach for enhancement of speech signals distorted by white Gaussian noise. VMD is a data adaptive method which decomposes the signal into intrinsic mode functions (IMFs) by using the Alternating Direction Method of Multipliers (ADMM). Each IMF or mode will contain a center frequency and its harmonics. This paper tries to explore VMD as a Speech enhancement technique. In the proposed method, the noisy speech signal is decomposed into IMFs using VMD. The noisy IMFs are enhanced using two methods; VMD based wavelet shrinkage (VMD-WS) and VMD based MMSE log STSA (VMD-MMSE). The speech signal distorted with different noise levels are enhanced using the VMD based methods. The level of noise reduction and speech signal quality are measured using the objective quality measures.

## 1 Introduction

Speech signal information is distorted by the addition of noise either during the recording or during transmission. The presence of noise reduces the perceptual quality and intelligibility of the speech signal. Speech enhancement aims to reduce the noise level present in the speech signal and enhances the quality and intelligibility. In a noisy speech signal if the noise variation over time is slow compared to speech, then the noise is considered to be stationary such as car noise and if the variation is fast the noise is non-stationary such as white noise, restaurant noise, etc. The enhancement of speech signals distorted with non-stationary noise is more challenging than speech signals distorted with stationary noise [22]. The different types of

B.G. Gowri(✉) · S.S. Kumar · N. Mohan · K.P. Soman
Center for Excellence in Computational Engineering and Networking,
Amrita Vishwa Vidyapeetham, Coimbatore, India
e-mail: {gangab.90,sachinnme,neethumohan.ndkm}@gmail.com, kp_soman@amrita.edu

© Springer International Publishing Switzerland 2016
S.M. Thampi et al. (eds.), *Advances in Signal Processing and Intelligent Recognition Systems*,
Advances in Intelligent Systems and Computing 425,
DOI: 10.1007/978-3-319-28658-7_27

noise that affect the speech signal are white noise, colour noise, reverberation noise, fan noise, etc. Speech enhancement techniques are applied in the filed of forensics, VoIP, mobile phones, pre-processing step in speech recognition systems, telephone conversations, teleconferencing systems, hearing aids, etc. Several articles discuss speech enhancement techniques such as spectral subtraction, filtering based methods, modelling based methods and decomposition based methods.

Spectral subtraction is one of the earliest methods proposed for speech enhancement, in which an enhanced speech is obtained by subtracting the spectrum of the noise estimate from the spectrum of the noisy speech [21]. A modified spectral subtraction using cascaded-median gives a continuous noise update for pre-processing stage in real time speech recognition systems [28]. The modification in spectral subtraction reduces the computational complexity and memory requirement of the recognition system. Filtering based techniques such as Wiener filter uses a linear time invariant filter to reduce noise and produces a linear estimate of the speech signal [4]. In adaptive Wiener filter the transfer function of the filter is adapted based on the input speech statistics [10]. Kalman filter which tracks the state of a dynamic system is used to estimate the state of the varying noise structure (non-stationary) from the noisy speech signal [22]. Statistical model based method such as Minimum Mean Square Error (MMSE) estimator estimates the amplitude spectrum of the speech from the noisy speech signal by additional use of *a priori* speech information [3, 11]. If there is no *a priori* speech information and only noisy signal is available, Steins Unbiased Risk Estimate of Mean Square Error (SURE-MSE) can be used. Spectral subtraction and MMSE estimator techniques are based on Discrete Fourier Transform. Similarly Discrete Cosine Transform (DCT) can be used in speech enhancement to produce state of art results [26]. In DCT combined with SURE-MSE, the enhanced signal is obtained by thresholding the DCT coefficients and SURE-MSE provides the thresholding function parameters [31]. Probabilistic model based techniques such as Hidden Markov Model (HMM) is used to develop the MMSE and Maximum *a posteriori* estimators for clean speech and noise [13]. Autoregressive HMM gives a better model of the speech compared to HMM [18]. Sparse HMM adds an additional regularization term to the objective function of the sparse autoregressive HMM [6], here sparsity improves noise estimation. In MMSE, Gaussian mixture model (GMM) can be used to model the clean speech [20]. In deep neural network (DNN) based speech recognition systems, DNN is trained with multi-conditioned acoustic models [24]. This method provides state-of-art performance even without explicitly providing any noise compensation.

Decomposition based techniques such as subspace approach decomposes the noisy signal into a noisy subspace and signal-plus-noise subspace using singular value decomposition (SVD) [14, 16, 23]. The enhanced signal is obtained by removing the noisy subspace and estimating the speech signal from the remaining subspace. In SVD combined with Genetic algorithm, SVD performs the noise removal using singular values and singular vectors and the Genetic algorithm optimizes the control parameters for it [30]. Empirical mode decomposition (EMD) is a data driven approach where the speech signal is decomposed into IMFs by *sifting* process [1]. The noisy IMFs are enhanced using the soft thresholding and Savitzky Golay filter. In bivariate EMD a reference signal is constructed based on the noisy speech signal and a portion of Gaussian noise [15]. A complex signal is constructed with the speech and reference signals as real and imaginary parts and is decomposed using EMD. The noisy IMFs are denoised using adaptive soft thresholding. In DCT combined with EMD, the noisy sub bands are denoised using soft thresholding method in DCT domain [5]. The thresholded signal is further enhanced using EMD based soft thresholding method. Ensemble EMD (EEMD) avoids the mode mixing problem in EMD by making use of white noise characterisitcs during decomposition. The enhanced speech is the ensemble mean of IMFs from several decompositions [29]. In multivariate EMD the signal is projected in multiple directions on hyperspheres. Denoising is done by analysing and eliminating the modes related to noise [25]. However the lack of proper formulation restricts the efficient application of EMD in real time systems [5].

Variational mode decomposition (VMD) is a non recursive method for signal decomposition based on calculus of variation [9]. It decomposes the signal concurrently into its IMFs while reproducing the input. In VMD the optimization is carried out using ADMM. In this work we have considered the speech signal distortion due to white Gaussian noise. The noisy speech signal is decomposed into IMFs using VMD. The proposed work includes two methods for enhancing the noisy IMFs. In the first method, VMD is combined with the wavelet shrinkage (VMD-WS) and in the second, VMD is combined with minimum mean square error (MMSE) STSA estimator (VMD-MMSE) [8, 12]. The objective measures are calculated for the enhanced output.

## 2 VMD Algorithm

Variational Mode Decomposition (VMD) decomposes a signal synchronously into a series of band-limited IMFs. The objective of VMD is to find the IMF with a centre frequency under the condition that the IMF has a minimum bandwidth. The given variational problem is solved using a modern powerful optimization algorithm, Alternating Direction Method of Multipliers (ADMM) [2]. Here the bandwidth of the IMF is minimized while reproducing the input. The IMFs are modelled as amplitude modulated frequency modulated (AM-FM) signals and is given by,

$$u_k(t) = A_k(t)\cos(\phi_k(t)) \tag{1}$$

where $u_{(k)}(t)$ is a pure harmonic signal with amplitude $A_{(k)}(t)$ and the instantaneous frequency $\phi'_{(k)}(t)$. Eq (1) represents the IMF for mode $k$ with the mean frequency $\omega_k$. In general, the bandwidth of the AM-FM signal depends on both, maximum deviation and rate of change of the instantaneous frequency. The bandwidth of the IMF $u_{(k)}(t)$ is measured by creating a new function using Hilbert transform. The Hilbert transform of $u_{(k)}(t)$ is $u_k^H(t)$ from which an analytical signal $(u_k(t) + ju_k^H(t))$ is created. The analytical signal has a positive frequency spectrum (unilateral) from which the original signal $u_{(k)}(t)$ can be directly retrieved by taking the real part alone. The unilateral frequency spectrum of the analytical signal is then shifted to the baseband by multiplying it with the exponential signal $e^{-j\omega_k t}$.

$$u_k^M(t) = \left(u_k(t) + j u_k^H(t)\right) e^{-j\omega_k t} \tag{2}$$

Eq (2) represents the analytical signal which is centred at the origin. It can be also expressed as

$$u_k^M(t) = \left(\delta(t) + \frac{j}{\pi t}\right) * u_k(t) \tag{3}$$

Now the bandwidth is estimated by finding the time derivative of Eq (3) and taking the L2-norm for it as in Eq ( 4)

$$\Delta\omega_k = \left\| \partial_t \left[ \left(\delta(t) + \frac{j}{\pi t}\right) * u_k(t) \right] \right\|_2^2 \tag{4}$$

where $\Delta\omega_k$ represents the bandwidth of the mode $u_{(k)}(t)$. The variational formulation is defined as minimization of the bandwidth of a mode function $u_{(k)}(t)$

$$\min_{u_k,\omega_k} \left\{ \sum_k \left\| \partial_t \left[ \left(\left(\delta(t) + \frac{j}{\pi t}\right) * u_k(t)\right) e^{-j\omega_k t} \right] \right\|_2^2 \right\}$$

$$s.t. \sum_k u_k = f \tag{5}$$

where $f$ is the original signal. The constraint enforces the reproducibility of the input. The constrained problem is converted to an unconstrained problem using an augmented Lagrangian multiplier as follows

$$L(u_k, w_k, \lambda) = \alpha \sum_k \left\| \partial_t \left[ \left(\left(\delta(t) + \frac{j}{\pi t}\right) * u_k(t)\right) e^{-j\omega_k t} \right] \right\|_2^2 + \left\| f - \sum_k u_k \right\|_2^2$$

$$+ \left\langle \lambda, f - \sum_k u_k \right\rangle \tag{6}$$

The Lagrangian multiplier enforces the constraint and the quadratic penalty (second term) enforces the convergence. Optimization is done using ADMM in which estimation of a parameter is obtained assuming all others are known. The optimization takes place in the Fourier domain and the sub-signals are obtained by taking the inverse Fourier transform over the optimized analytical signals. The updation for the mode function $u_k$ in the Fourier domain is given by

$$\hat{u}_k^{n+1} = \left(\hat{f} - \sum_{i \neq k} \hat{u}_i + \frac{\hat{\lambda}}{2}\right) \frac{1}{\left(1 + 2(\omega - \omega_k)^2\right)} \tag{7}$$

The update given in Eq (7) can be obtained by finding the residual of the corresponding mode function $u_k$ and applying Wiener filtering with a signal prior $1/(\omega - \omega_k)^2$. The update equation for centre frequency $\omega_k$ in the Fourier domain is given by

$$\omega_k^{n+1} = \underset{\omega_k}{\arg \min} \left\| \partial_t \left[ \left(\left(\delta(t) + \frac{j}{\pi t}\right) * u_k(t)\right) e^{-j\omega_k t} \right] \right\|_2^2 \tag{8}$$

and $\lambda$

$$\lambda^{n+1} \leftarrow \lambda^n + \tau \left(f - u_k^{n+1}(t)\right) \tag{9}$$

In case of denoising the noise content has to removed during reconstruction. Here exact reconstruction of the input signal is not needed and the Lagrangian multiplier is dropped. Fig. 1 represents the IMFs obtained through VMD decomposition of a clean speech signal. Here Fig. 1(a) corresponds to the IMF having lowest central frequency and Fig. 1(j) represents the IMF with highest frequency. On decomposition the first mode has the lowest frequency component and the last mode has the highest frequency component.

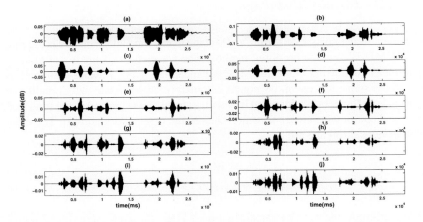

**Fig. 1** (a) - (j) Represents the IMFs obtained through VMD decomposition of a clean speech signal

## 3    Proposed Method

The speech signal $x(t)$ is corrupted by white Gaussian noise $n(t)$ giving a noisy speech signal $y(t)$ represented as,

$$y(t) = x(t) + n(t) \tag{10}$$

The noisy speech signal is decomposed into IMFs using VMD

$$y(t) = \sum_{k=1}^{M} u_k(t) \tag{11}$$

where $u_k(t)$ represents the noisy IMFs. In the proposed method the noisy IMFs are processed using the methods: wavelet shrinkage, MMSE separately to get the enhanced speech signal. The enhanced speech signal is represented as

$$\widehat{x(t)} = \sum_{k=1}^{M} \widehat{u}_k(t) \tag{12}$$

where $\widehat{u}_k(t)$ represents the enhanced IMF.

### 3.1    VMD - Wavelet Shrinkage

Enhancement of the noisy IMFs from VMD stage using wavelet shrinkage is the first strategy (VMD-WS) in the proposed method. The wavelet shrinkage is a nonlinear, non-parametric denoising algorithm. The enhancement using wavelet shrinkage is given in three steps; first the noisy signal is transformed on to wavelet domain, the transformed coefficients are denoised based on a shrinkage value and in the final step the denoised coefficients are transformed back to the original domain [8]. The main steps in wavelet shrinkage are

$$U_k = W(u_k) \tag{13}$$
$$Z_k = T(U_k, \lambda) \tag{14}$$

where $W$ represents the wavelet transformation, $\lambda$ is the shrinkage value and $Z$ represents the coefficients obtained after the shrinkage process. The shrinkage value can be obtained using shrinkage rule such as rigrsure, heursure, minimax and sqtwolog. The noisy IMFs are enhanced by soft thresholding using the heursure threshold value [27]. The thresholding is represented as

$$T(d_k, \lambda) = \begin{cases} (d - \text{sgn}(d_k)\lambda), & (|d_k| > \lambda) \\ 0 & , & (|d_k| \leq \lambda) \end{cases} \tag{15}$$

where $d_k$ is the detailed coefficient corresponding to the $U_k$ in wavelet domain

$$\widehat{u_k} = W_{-1}(Z) \tag{16}$$

where $W_{-1}$ represents the inverse wavelet transformation and $\widehat{u_k}$ is the $k^{th}$ enhanced IMF. All the noisy IMFs are enhanced using the above method separately and the enhanced speech signal is obtained by combining the enhanced IMFs.

## 3.2 VMD - MMSE

Enhancement of the noisy IMFs from the VMD stage using the minimized mean square error log short time spectral amplitude estimator (MMSE log-STSA) is the second strategy (VMD-MMSE) in the proposed method. The MMSE log STSA method is a statistical approach with *a priori* information which models the uncertainty of the speech information in the noisy observation. The enhancement is carried out in the Fourier domain where the spectral amplitude of the speech signal is estimated by minimising the error. Speech enhancement is achieved in three stages. In the first stage the noisy observation is transformed in to the Fourier domain. The error between the estimated amplitude and the speech is minimised in the second stage. The additional priori SNR information improves the estimation accuracy. In other words MMSE is used to derive an optimal STSA estimator from the noisy observation. In the last stage the estimated spectral amplitude is combined with the phase information from the noisy observation to get the enhanced signal [12]. The Fourier coefficients of the IMFs corresponding to clean signal and the noisy speech signal is given as

$$U_k = A_k e^{j\alpha k} \tag{17}$$

$$\widehat{U_k} = R_k e^{j\vartheta k} \tag{18}$$

The estimator is obtained by minimizing the mean square error, $E\{(\log A_k - \log \widehat{A_k})^2\}$ and is given by,

$$\widehat{A_k} = \frac{\xi_k}{1 + \xi_k} \exp\left\{ \frac{1}{2} \int_{v_k}^{\infty} \frac{e^{-t}}{t} dt \right\} R_k \tag{19}$$

where $\xi_k$ is the *a priori* probability SNR. The above expression gives the $k^{th}$ estimated amplitude. The estimator can be obtained using a multiplicative nonlinear gain function which depends only on the *a priori* and *a posteriori* SNR values is given by,

**Table 1** Comparison of SNRseg at different noise levels

| SNRseg | 10dB | 15dB | 20dB | 25dB | 30dB |
|---|---|---|---|---|---|
| Noisy speech | 2.2 | 5.78 | 10 | 15 | 19.8 |
| VMD-WS | 4.5 | 8.1 | 12.2 | 16.6 | 20.7 |
| VMD-MMSE | 7 | 9.8 | 12.6 | 15.1 | 16.1 |

$$G\left(\xi_k, \gamma_k\right) \triangleq \frac{\widehat{A}_k}{R_k} \tag{20}$$

where $\gamma_k$ is the *a posteriori* SNR value. The enhanced IMFs are estimated seperately from the corresponding noisy IMFs. The enhanced speech signal is reconstructed by combining the enhanced IMFs.

## 4   Results and Discussion

The VMD based enhancement methods are applied on self generated speech signals distorted with white Gaussian noise in the range of 10dB-30dB. The speech quality of the enhanced signals are estimated using the following objective measures: segmental signal-to-noise ratio ($SNR_{seg}$), perceptual evaluation of speech quality (PESQ), signal distortion ($C_{sig}$), background noise distortion ($C_{bak}$), overall quality ($C_{ovl}$) [17, 19]. All these measures are scaled from 1 to 5 and the quality of the enhanced signal increases with increase in this measure. $SNR_{seg}$ is the measurement of signal to noise ratio over speech frames.

$$SNR_{seg} = \frac{10}{M} \times \sum_{m=0}^{M-1} \frac{\sum_{j=1}^{K} W(j,m) \log_{10} \frac{|X(j,m)|^2}{(X(j,m) - |\widehat{X}(j,m)|)^2}}{\sum_{j=1}^{K} W(j,m)} \tag{21}$$

where $W(j, m)$ is the weight of the $j^{th}$ frequency band, $M$ is the total number of frames in the signal, $K$ is the number of bands. $X(j, m)$ and $\widehat{X}(j, m)$ are the weighted clean signal spectrum and weighted enhanced signal spectrum respectively in the $j^{th}$ frequency band at the $m^{th}$ frame. Since speech is a non-stationary signal $SNR_{seg}$ gives better ratio compared to SNR. Comparison of $SNR_{seg}$ values of a speech signal distorted with white noise and the enhanced speech outputs from VMD-WS and VMD-MMSE are given in Table 1 and the improvement in $SNR_{seg}$ is shown in Fig. 2. From the two proposed strategies VMD-WS and VMD-MMSE, the improvement in $SNR_{seg}$ is high in VMD-WS with comparatively better noise reduction.

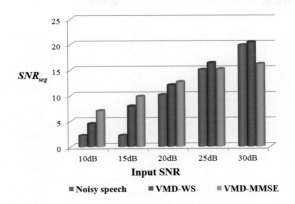

**Fig. 2** Comparison of SNRseg of noisy speech signal, VMD-WS signal and VMD-MMSE signal at different noise levels

**Table 2** Comparison of PESQ at different noise levels

| PESQ | 10dB | 15dB | 20dB | 25dB | 30dB |
|------|------|------|------|------|------|
| Noisy speech | 2.09 | 2.44 | 2.79 | 3.12 | 3.42 |
| VMD-WS | 2.63 | 3.01 | 3.26 | 3.53 | 3.76 |
| VMD-MMSE | 2.83 | 3.01 | 3.24 | 3.45 | 3.67 |

The perceptual evaluation of speech quality (PESQ) is an important measure which estimates the perceptual quality of the enhanced speech signal.

$$PESQ = a_0 + a_1 D_{ind} + a_2 A_{ind} \qquad (22)$$

where $D_{ind}$ is the average disturbance value and $A_{ind}$ is the average asymmetrical disturbance value. Comparison of PESQ values of VMD-WS and VMD-MMSE enhanced outputs are made in Table 2 and their corresponding plot is shown in Fig. 3(a). Both methods show improvement in the PESQ measure; however VMD-WS gives relatively higher value compared to VMD-MMSE. Higher values of PESQ shows the improvement in the perceptual quality of the speech signal. $C_{sig}$, $C_{bak}$ and $C_{ovl}$ denotes reduction in the level of signal distortion, reduction in the noise level and overall quality of the enhanced signal.

$$C_{sig} = 3.093 - 1.029LLR + 0.603PESQ - 0.009WSS \qquad (23)$$

$$C_{bak} = 1.634 + 0.478PESQ - 0.007WSS + 0.063SNRseg \qquad (24)$$

**Table 3** Comparison of $C_{sig}$, $C_{bak}$ and $C_{ovl}$ at different noise level

| $C_{sig}$ | 10dB | 15dB | 20dB | 25dB | 30dB |
|---|---|---|---|---|---|
| Noisy speech | 2.02 | 2.71 | 3.38 | 3.95 | 4.4 |
| VMD-WS | 2.16 | 3.77 | 4.24 | 4.64 | 4.93 |
| VMD-MMSE | 2.73 | 3.28 | 3.75 | 4.16 | 4.54 |
| $C_{bak}$ | 10dB | 15dB | 20dB | 25dB | 30dB |
| Noisy speech | 2.56 | 2.99 | 3.47 | 3.96 | 4.4 |
| VMD-WS | 2.9 | 3.38 | 3.8 | 4.24 | 4.6 |
| VMD-MMSE | 3.1 | 3.47 | 3.8 | 4.09 | 4.34 |
| $C_{ovl}$ | 10dB | 15dB | 20dB | 25dB | 30dB |
| Noisy speech | 2.04 | 2.57 | 3.1 | 3.56 | 3.93 |
| VMD-WS | 2.4 | 3.37 | 3.76 | 4.09 | 4.35 |
| VMD-MMSE | 2.68 | 3.13 | 3.5 | 3.8 | 4.13 |

$$C_{ovl} = 1.594 + 0.805\text{PESQ} - 0.512\text{LLR} - 0.007\text{WSS} \qquad (25)$$

where LLR is the Log Likelihood ratio and WSS is the Weighted Spectral Slope. In the proposed strategies, the VMD-WS retains the signal quality better compared to VMD-MMSE. Comparison of $C_{ovl}$, $C_{sig}$ and $C_{bak}$ measures for the distorted speech signal and enhanced signals are made in Table 3 and the corresponding plots are shown in Fig. 3(b), Fig. 4(a) and Fig. 4(b). The clean speech signal (Fig. 5(a)) distorted by white Gaussian noise is enhanced using the the proposed techniques: VMD-WS and VMD-MMSE. The distorted signal is shown in Fig. 5(b) and the

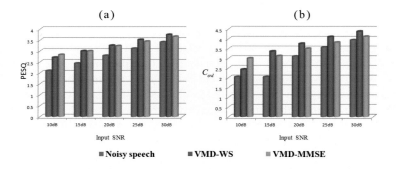

**Fig. 3** Comparison of PESQ and $C_{ovl}$ at different noise levels in (a) and (b)

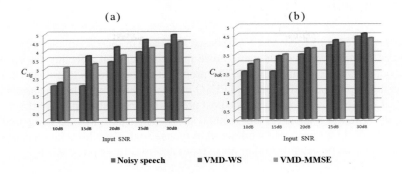

**Fig. 4** Comparison of $C_{sig}$ and $C_{bak}$ in (a) and (b) at different noise levels

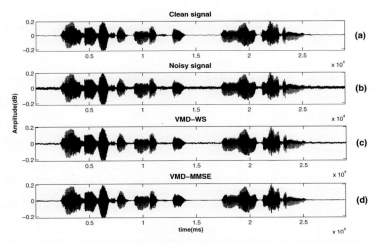

**Fig. 5** Performance of the proposed method (a) clean signal, (b) noisy signal, (c) VMD-WS enhanced signal and (d) VMD-MMSE enhanced signal

corresponding enhanced signals are shown in Fig. 5(c) and Fig. 5(d). Here the signal structure is better preserved by using VMD-MMSE compared to the VMD-WS method. Comparing the proposed method with other techniques based on non-stationary signal analysis shows that the proposed method gives a good and competitive performance. The results are competitive with those of EMD based speech enhancement techniques [5, 15].

## 5   Conclusion

In this work we have proposed a VMD based approach in speech enhancement. The enhanced output shows considerable noise reduction without affecting the speech signal quality and perceptual quality. The proposed method shows improvement not only in the segmental SNR value and PESQ but the other composite measures such as $C_{sig}$, $C_{bak}$ and $C_{ovl}$. Comparing the post processing methods applied to the noisy IMFs, the VMD-WS enhancement method shows better performance compared to VMD-MMSE. In the proposed work we have focused only the white Gaussian noise. In future this can be extended other noise types and for higher noise levels.

## References

1. Boudraa, A.O., Cexus, J.C., et al.: Denoising via empirical mode decomposition. In: Proc. IEEE ISCCSP, p. 4 (2006)
2. Boyd, S., Parikh, N., et al.: Distributed optimization and statistical learning via the alternating direction method of multipliers. Foundations and Trends in Machine Learning 3(1), 1–122 (2011)
3. Breithaupt, C., Martin, R., et al.: MMSE estimation of magnitude-squared DFT coefficients with supergaussian priors. In: IEEE International Conference on Acoustics, Speech, and Signal Processing. Proceedings (ICASSP 2003), vol. 1, pp. I-896. IEEE (2003)
4. Chirtmay, S., Tahernezhadi, M., et al.: Speech enhancement using Wiener filtering. Acoustics Letters 21, 110–115 (1997)
5. Deger, E., Molla, M.K.I., Hirose, K., Minematsu, N., Hasan, M.K., et al.: Subband DCT and EMD Based Hybrid Soft Thresholding for Speech Enhancement. Advances in Acoustics and Vibration (2014)
6. Deng, F., Bao, C.C., Kleijn, W.B., et al.: Sparse HMM-based speech enhancement method for stationary and non-stationary noise environments. In: 2015 IEEE International Conference on IEEE Acoustics, Speech and Signal Processing (ICASSP), pp. 5073–5077 (2015)
7. Donoho, D.L., Johnstone, J.M.: Ideal spatial adaptation by wavelet shrinkage. Biometrika 81(3), 425–455 (1994)
8. Donoho, D.L., Johnstone, I.M., et al.: Adapting to unknown smoothness via wavelet shrinkage. Journal of the American Statistical Association 90(432), 1200–1224 (1995)
9. Dragomiretskiy, K., Zosso, D.: Variational mode decomposition. IEEE Transactions Signal Processing 62(3), 531–544 (2014)
10. El-Fattah, M.A.A., Dessouky, M.I., et al.: Speech enhancement with an adaptive Wiener filter. International Journal of Speech Technology 17(1), 53–64 (2014)
11. Ephraim, Y., Malah, D., et al.: Speech enhancement using a minimum-mean square error short-time spectral amplitude estimator. IEEE Transactions Acoustics, Speech and Signal Processing (ICASSP) 32(6), 1109–1121 (1984)
12. Ephraim, Y., Malah, D.: Speech enhancement using a minimum mean-square error log-spectral amplitude estimator. IEEE Transactions Acoustics, Speech and Signal Processing (ICASSP) 33(2), 443–445 (1985)
13. Ephraim, Y.: A Bayesian estimation approach for speech enhancement using hidden Markov models. IEEE Transactions Signal Processing 40(4), 725–735 (1992)
14. Ephraim, Y., Van Trees, H.L.: A signal subspace approach for speech enhancement. IEEE Transactions Speech and Audio Processing 3(4), 251–266 (1995)
15. Hamid, M.E., Molla, M.K.I., et al.: Single channel speech enhancement using adaptive soft-thresholding with bivariate EMD. ISRN Signal Processing (2013)

16. Hu, Y., Loizou, P.C.: A generalized subspace approach for enhancing speech corrupted by colored noise. IEEE Transactions Speech and Audio Processing **11**(4), 334–341 (2003)
17. Hu, Y., Loizou, P.C.: Evaluation of objective quality measures for speech enhancement. IEEE Transactions Audio, Speech, and Language Processing **16**(1), 229–238 (2008)
18. Juang, B.H., Rabiner, L.R.: Mixture autoregressive hidden Markov models for speech signals. IEEE Transactions Acoustics, Speech and Signal Processing (ICASSP) **33**(6), 1404–1413 (1985)
19. Krishnamoorthy, P.: An overview of subjective and objective quality measures for noisy speech enhancement algorithms. IETE technical review **28**(4), 292–301 (2011)
20. Kundu, A., Chatterjee, S., et al.: GMM based Bayesian approach to speech enhancement in signal/transform domain. In: IEEE International Conference on IEEE Acoustics, Speech and Signal Processing (ICASSP), pp. 4893–4896 (2008)
21. Loizou, P.C.: Speech enhancement: theory and practice. CRC Press (2013)
22. Mathe, M., Nandyala, S.P., et al.: Speech enhancement using Kalman Filter for white, random and color noise. In: 2012 International Conference on IEEE Devices, Circuits and Systems (ICDCS), pp. 195–198 (2012)
23. Rezayee, A., Gazor, S.: An adaptive KLT approach for speech enhancement. IEEE Transactions Speech and Audio Processing **9**(2), 87–95 (2001)
24. Seltzer, M.L., Yu, D., et al.: An investigation of deep neural networks for noise robust speech recognition. In: 2013 International Conference on IEEE Acoustics, Speech and Signal Processing (ICASSP), pp. 7398–7402 (2013)
25. Sole-Casals, J., Gallego-Jutgla, E., et al.: Speech Enhancement: a multivariate empirical mode decomposition approach. In: Advances in Nonlinear Speech Processing, pp. 192–199. Springer, Heidelberg (2013)
26. Soon, Y., Koh, S.N., et al.: Noisy speech enhancement using discrete cosine transform. Speech Communication **24**(3), 249–257 (1998)
27. Vidakovic, B.: Nonlinear wavelet shrinkage with Bayes rules and Bayes factors. Journal of the American Statistical Association **93**(441), 173–179 (1998)
28. Waddi, S.K., Pandey, P.C., et al.: Speech enhancement using spectral subtraction and cascaded-median based noise estimation for hearing impaired listeners. In: 2013 National Conference on Communications (NCC), pp. 1–5. IEEE (2013)
29. Wu, Z., Huang, N.E.: Ensemble empirical mode decomposition: a noise-assisted data analysis method. Advances in Adaptive Data Analysis **1**(1), 1–41 (2009)
30. Zehtabian, A., Hassanpour, H., et al.: A novel speech enhancement approach based on singular value decomposition and genetic algorithm. In: 2010 International Conference of IEEE Soft Computing and Pattern Recognition (SoCPaR), pp. 430–435 (2010)
31. Zheng, N., Li, X., et al.: SURE-MSE speech enhancement for robust speech recognition. In: 2010 7th International Symposium on IEEE Chinese Spoken Language Processing (ISCSLP), pp. 271–274 (2010)

# Opportunistic Routing with Virtual Coordinates to Handle Communication Voids in Mobile Ad hoc Networks

Varun G. Menon and P.M. Joe Prathap

**Abstract** Communication voids or the unreachability problem has been one of the major issues in mobile ad hoc networks. With highly mobile and unreliable nodes, they can occur very often and may lead to loss of data packets in the network. Most of the existing methods that are available to handle this issue try to find out a route around the communication void. Due to this the data packets may have to travel more number of hops in reaching the destination and thus causing more delay in data delivery in the network. Such techniques do not allow us to fully utilize the advantages brought by geographic routing and opportunistic forwarding. In this paper we propose a method, Opportunistic Routing with Virtual Coordinates (ORVC) that would improve the potential forwarding area for a packet by sending the packets to multiple locations near the destination. The analysis and simulations show that using this method a data packet takes less number of hops to reach the destination compared to the existing methods.

## 1 Introduction

Mobile Ad hoc networks have gained wide popularity over these years due to their numerous advantages and applications. With decentralized control and infrastructure less transmission, the nodes always have the choice to enter or leave the network at any point of time, leading to the networks dynamic topology. This networking concept originated from the needs of battlefield communications has

V.G. Menon(✉)
Sathyabama University, Chennai, India
e-mail: varungmenon46@gmail.com

P.M.J. Prathap
RMD Engineering College, Chennai, India
e-mail: joeprathappm@rediffmail.com

© Springer International Publishing Switzerland 2016
S.M. Thampi et al. (eds.), *Advances in Signal Processing and Intelligent Recognition Systems*,
Advances in Intelligent Systems and Computing 425,
DOI: 10.1007/978-3-319-28658-7_28

323

been a major area of research and study and have led to the development of its major applications in disaster relief operations, emergency situations and community networks.

Due to the mobility, random movements of nodes and frequent failures of wireless links, routing of data packets from the source to destination has been a major issue in MANETs. Over these years a number of routing protocols has been proposed to handle this issue and to guarantee the required quality of service in the network. But it is often seen that the protocols like DSR [11], DSDV [19], AODV [20] and TORA [26, 27] which depend on predetermined routes suffer from the rapid changes in the network topology [1, 2, 18, 28, 5], and give very low performance. Geographic routing [16, 25] which uses the location information to route packets from one hop to another gives better performance in this type of network. But even this routing technique has a major drawback that it is sensitive to inaccuracy of location information [14, 23, 24, 25]. So the opportunistic forwarding technique was developed to route the data packets in mobile ad hoc networks and to provide better performance [2, 17]. The major advantage of opportunistic routing is that it does not maintain predetermined routes. Here the sender node broadcasts its data packet in to the wireless channel. The nodes within its transmission range would receive the data packet. Based on opportunistic forwarding, the node that is nearest to the destination would be chosen as the next forwarding node. Thus opportunistic routing helps in making the data packet reach the destination node with less number of hops and less delay [21, 22, 31].

Both geographic and opportunistic routing mechanisms use a simple technique to forward the data packets. The data packets are forwarded to the neighbor node that is nearest to the destination with a maximum positive progress to the destination node. But in some cases the source node may not find any suitable forwarder nearer to the destination to forward the data.Also the source node may suffer from the absence of neighbouring nodes in its transmission area that are located in the direction of the destination. This is called a communication void or routing void or communication gaps and this problem is often referred as the unreachability problem [7, 8]. Figure 1 shows an example of a communication void in the wireless network. In our paper we would refer this area as a communication gap because we are trying to remove the gap or delay caused by such empty spaces in the network. We would also refer the nodes as mobile devices as it gives more meaning to the practical application of mobile ad hoc networks. In Figure1 the source node S does not have any neighbour node that has positive progress towards the destination. So in order to send a data packet to the destination D, the sender node S would have to find a route S-W-X-Y-Z-D around the gap to deliver the packet. This would result in increased delay in delivering the data packets. With highly mobile nodes, communication gaps can occur very often in ad hoc networks and this may lead to loss of data packets. This important issue needs to be addressed for better data delivery and performance in the network.

In our paper we first analyze the various techniques that are used in managing the communication gaps in ad hoc networks. Various disadvantages and issues with each of the methods are discussed in detail in this section. We then present our proposed method Opportunistic Routing with Virtual Coordinates (ORVC) and explain its working in detail. The various advantages of using this method are discussed and a performance comparison is made using simulations with the existing methods. Finally, we present the conclusion and possible advancements that can be made further in this area.

## 2    Problem Description

The two modern routing techniques, Geographic routing and opportunistic routing uses greedy forwarding [14, 23] to forward the data packets to the next hop in mobile ad hoc networks. In order to get the position of every mobile device in the network, these methods make use of some services given in [9, 6, 15, 29]. Once the position of the nodes and the destination is determined, the location information is attached to the header of the transmitted data packet. When a source node wants to send a packet to a particular destination, using this routing it sends the packet to the next hop node that has maximum progress to the destination. This greedy forwarding strategy is used till that particular data packet reaches the end device. When a particular device is unable to forward a packet due to the unavailability of a neighboring node with maximum progress towards the destination device, the void handling strategy is used [3]. Here a new path is developed around the empty space of communication and the data packet is forwarded along that path.

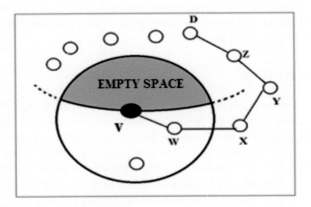

**Fig. 1** Illustration of a Communication gap

In Figure 1 we can see that node V is unable to find next hop node to forward the packet. So it routes the packet around the empty space with V-W-X-Y-Z-D. Here we have node V as the void node and the region in front of the node with no nodes available for communication is called the communication gap area. Communication gaps are a serious challenge in designing routing protocols and ensuring better data delivery in the ad hoc networks with less delay. Due to the dynamic nature of the nodes in the mobile ad hoc networks it is quite unpredictable to know in advance about the position of these communication gaps. If an effective mechanism is not used to handle these gaps, long delays may occur in the network and many data packets may get lost leading to very low performance of routing in the network. Most of the existing communication gap handling techniques cause long delays in the network and also does not guarantee an optimal path to the destination.

# 3    Different Techniques Used to Handle Communication Gaps

In this section we analyse some of the popular techniques used to handle communication gaps and see the problems and issues associated with each method.

## 3.1    Perimeter Routing

This method of managing communication gaps was introduced with one of the most famous geographic routing protocol, GPSR [12, 13, 14]. This technique works with the help of a planar traversal algorithm [3, 4]. Using this method, when a communication gap occurs at a mobile device, this algorithm is used to find out a path around that empty space. The data packet is then forwarded along the newly computed path. Information about the location of the mobile device facing this problem is added to the header of the data packet. This information helps all the communicating nodes to make routing decisions locally. The main disadvantage of this method is that additional delay is incurred in routing the data packet around the empty space and also there is no guarantee that this method would find optimal paths in the network.

## 3.2    Face2 Routing

The Face2 algorithm [3] is another to handle communication gaps. This method has a lot of similarity with the perimeter routing strategy and also uses a planar graph traversal algorithm to effectively manage communication gaps. In case of communication gaps, the algorithm finds new routes by walking along the faces of the planar graph and then proceeding along the line that connects the sender and receiver devices. This technique also suffers from the drawback that an additional delay is incurred in setting up the new route for the data packets.

## 3.3    Geometric Methods to Handle Communication Gaps with BOUNDHOLE Technique

Geometric technique is one of the popular methods in managing communication gaps. This method makes use of the geometric properties of the empty space and that of the mobile devices in the network. The main aim of this method is to find out empty spaces in the network where there are no nodes for communication.

The advantage of this method compared to face2 and Perimeter routing strategies is that storing information regarding communication gaps in only required for the regions in the network that encounters any difficulty or a problem. To find out a communication gap around a gap node a TENT rule [7, 8] is used. One of the major drawbacks of this algorithm is that this method would put a lot of strain on the devices that are located on the boundaries of these empty spaces. Also if the destination device is inside the empty space, this method cannot be used.

## 3.4  One Hop Flooding

This is one of the gap handling technique that exploits the advantages of the flooding the data packet in the network to route around the empty spaces. This is a partial or restricted flooding technique [3, 7, 8] and is used at the node experiencing a communication gap. Here the packet that was not forwarded due to communication gap is broadcasted to the neighboring mobile devices that are one hop away. After broadcasting the packets to all its neighbors, using a cache, the void node stores the identifier value of that particular packet and thus refuses to accept the copy of the same data packet from its neighbors. As soon this packet reaches the neighboring mobile devices, each device forwards the packet to the device that has maximum progress to the destination. Full flooding techniques utilizes large amount of resources and increases the delay in the network. Although one hop flooding technique utilizes fewer resources, optimal path to forward the data packet cannot be guaranteed.

## 3.5  Hybrid Gap Handling: BOUNDHOLE Plus Restricted Flooding (BRF)

This method involves a combination of BOUNDHOLE technique with restricted flooding. BOUNDHOLE method cannot be used when the destination mobile device is located inside the communication hole. This problem is solved by restricted flooding technique which allows to broadcast the data packet from the node experiencing the communication gap to the mobile devices that are located just a hop away. Although this hybrid routing technique gives better performance

**Table 1** Analysis of Performance Parameters of the Gap handling Methods

| Gap handling technique | Optimal Path | Complexity | Overhead | Scalability | Delay |
|---|---|---|---|---|---|
| Perimeter | No | Medium | Medium | Yes | High |
| Face2 | No | Medium | High | Yes | High |
| BOUNDHOLE | No | High | Medium | Yes | High |
| One-hop flooding | No | Low | High | No | Medium |
| BRF | No | High | High | No | Medium |

compared to other methods, it cannot guarantee the optimal path to the destination. Table 1 shows a comparison of the most important performance parameters of these five gap handling techniques. We can see that most of these techniques suffer from many issues and drawbacks.

# 4    Opportunistic Routing with Virtual Coordinates

Most of the existing techniques used to handle communication gaps try to find a route around the empty space resulting in more delay in the network. The advantages brought by greedy forwarding and opportunistic routing cannot be achieved under these methods. One of the major reasons for the low performance of these methods is that these schemes are only introduced at the mobile device that experiences the communication gap. In our proposed method we improve on these techniques by setting trigger nodes that are one hop away from a potential node near the empty space. A message called void alert is send from the mobile device that experiences the communication gap to the devices that are one hop away. This would inform the trigger node about the device that is experiencing the communication gap and only one hop away. As soon as this information is received the trigger node would send data packets to virtual coordinates towards the direction of the void node (that is near to the destination node). This would significantly improve the potential forwarding area, thus resulting in improved performance in data delivery.

**Packet Forwarding Algorithm**

NNList: List of neighboring nodes
$N_V$: Node that is experiencing the communication gap
$N_T$: Trigger Node
$N_R$: Destination Node or Receiver Node
$V_{MSG}$: Gap alert message
$V_C$: Virtual Coordinate

Node $N_T$ receives a packet for $N_R$
if find (NNList, $N_R$) then
fwd-node $\leftarrow N_R$
return
elseif $V_{MSG}$ is equal to 1
ListN.remove ($N_V$)
ListN.update
set $V_C$ in the direction of $N_R$
forward data packet to $V_C$
else forward the packet to next hop
end if

In Fig. 2 we can see that the mobile device V which experiences the communication gap is unable to proceed with the data packet. To alert about this problem we send a message, void alert (set to 1 in case of a communication gap) from the device V to the neighbouring device T which is just one hop away. This would alert the node T about the communication gap and would then send the data packets to virtual destinations V`, V1`, V2` etc. that are located near the destination. This would improve the potential forwarding area for the nodes. These virtual coordinates are calculated based on the transmission range of the mobile device T. As the data packet reaches these virtual coordinates, devices in this region receiving the packet would forward it to the destination and thus data delivery is guaranteed with minimum delay.

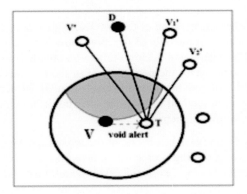

**Fig. 2** Improving the Forwarding Area with Virtual Coordinates

## 5    Performance Evaluation

We evaluate the performance of Opportunistic Routing with Virtual Coordinates (ORVC) using the ns-2 simulator [10]. For this we integrate the ORVC method with opportunistic routing OR protocol. We set up a mobile network with 100 nodes with a communication hole as shown in Figure 3.The shaded portion in the figure is set as the communication hole with no mobiles distributed in the region. The source node S and destination node D are fixed at the two ends. The two-ray Ground propagation model is used for the simulation. Mobility is introduced in the network with the Random Way Point [30] mobility model without pausing. We then send CBR traffic into the network and vary the speed of the mobile nodes and analyse the performance of the ORVC and GPSR communication gap handling mechanisms.

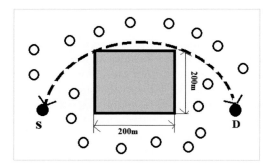

**Fig. 3** Network Topology with Communication Hole

The analysis of the performance of the two techniques used to manage communication gaps are done with the help of the three most important performance metrics for mobile ad hoc networks. Using the results obtained from the simulation, graphs are created to compare the performance of the two mechanisms. We have taken the communication gap handling mechanism in GPSR because it is the most popular and widely used method in handling communication gaps. Comparison is made based on the following metrics.

- Packet Delivery Ratio (PDR): It is one of the most important metric in deciding the performance of a routing protocol in a network. It is defined as the ratio of data packets received at the destination(s) to the number of data packets sent by the source(s).
- Average end-to-end delay: The average delay in receiving an acknowledgement for a delivered data packet.
- Path length: It is the average end-to-end (node to node) path length for successful packet delivery.

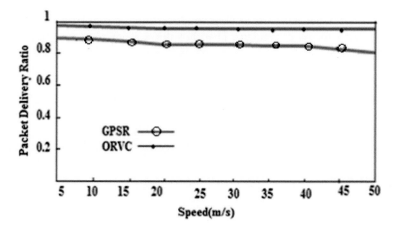

**Fig. 4** PDR vs Speed.

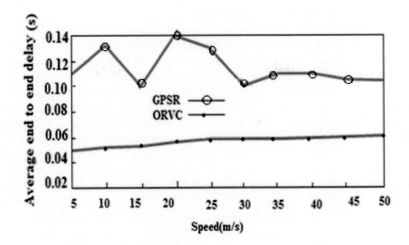

**Fig. 5** Average end to end delay vs Speed.

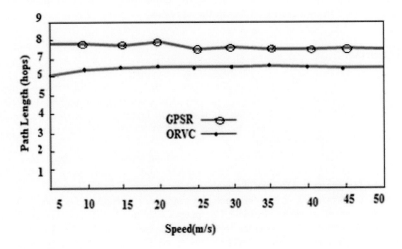

**Fig. 6** Path Length vs Speed.

Fig. 4 shows that the packet delivery fraction for ORCV is much better compared to GPSR communication gap handling technique especially when the mobility increases. Fig. 5 shows that average end to end delay in transmitting the data packets in ORCV is much lower than GPSR protocol. Fig. 6 shows that the number of hops taken by ORCV is much less compared to GPSR scheme. As the mobility increases ORCV maintains a steady performance while the performance of GPSR mechanism comes down. Thus the simulation results clearly show that the Opportunistic Routing with Virtual Coordinates (ORCV) handles communication gaps much better than geographic routing handling techniques.

## 6    Conclusion and Future Work

In this paper we addressed the issue of handling communication gaps in mobile ad hoc networks. Most of the existing methods try to find a route around the communication gap to deliver the data packet at the destination. Also the communication gap handling mechanism is only triggered at the void node that results in additional delay in the network. These methods are often unable to guarantee an optimal path for the transfer of the data packet towards the destination. The proposed solution Opportunistic Routing with Virtual Coordinates (ORVC) sends a void alert message from the node that is experiencing a communication gap to any neighboring node that is one hop away. This neighboring node is the trigger node which would send the incoming data packets towards virtual destinations that are situated in the direction of the gap node near to the destination. This enhances the potential forwarding area for the nodes. Simulation results show that ORVC guaranteed a very high rate of data delivered in mobile ad hoc networks. Further research can be done to develop communication gap handling techniques that would consume very less memory and energy of the nodes in the network. It would be very promising to find out certain methods that would prevent or minimize the occurrence of communication holes in the network.

## References

1. Biswas, S., Morris, R.: ExOR: opportunistic multi-hop routing for wireless networks. In: Proc. of ACM SIGCOMM, pp. 133–144 (2005). doi:10.1145/1080091.1080108
2. Chachulski, S., Jennings, M., Katti, S., Katabi, D.: Trading structure for randomness in wireless opportunistic routing. In: Proc. of ACM SIGCOMM, pp. 169–180 (2007). doi:10.1145/1282380.1282400
3. Chen, D., Varshney, P.: A survey of void handling techniques for geographic routing in wireless networks. IEEE Comm. Surveys and Tutorials 9(1), 50–67 (2007). doi:10.1109/comst.2007.358971
4. Chen, D., Varshney, P.K.: On demand geographic forwarding for data delivery in wireless sensor networks. Elsevier Computer Commun. J., special issue on Network Coverage and Routing Schemes for Wireless Sensor Networks, 2954–2967 (2006). doi:10.1016/j.comcom.2007.05.022
5. Chlamtac, I., Conti, M., Liu, J.N.: Mobile ad hoc networking: imperatives and challenges. Ad hoc Networks 1(1), 13–64 (2003). doi:10.1016/s1570-8705(03)00013-1
6. Das, S., Pucha, H., Hu, Y.: Performance comparison of scalable location services for geographic ad hoc routing. In: Proc. IEEE INFOCOM, vol. 2, pp. 1228–1239 (2005). doi:10.1109/infcom.2005.1498349
7. Chen, D., Deng, J., Varshney, P.: Selection of a forwarding area for contention-based geographic forwarding in wireless multi-hop networks. IEEE Transactions on Vehicular Technology 56(5), 3111–3122 (2007). doi:10.1109/tvt.2007.900371
8. Chen, D., Varshney, P.K.: Geographic routing in wireless ad hoc networks. In: Part Of Series Computer Communications And Networks in Guide to Wireless Ad Hoc Networks, pp. 151–188. Springer (2009)

9. Hsiao, P.: Geographical region summary service for geographical routing. ACM MC2R **5**(4), 25–39 (2002). doi:10.1145/509506.509515
10. Ivanov, S., Herms, A., Lukas, G.: Experimental validation of the ns-2 wireless model using simulation, emulation, and real network. In: 2007 ITG-GI Conference on Communication in Distributed Systems (KiVS), pp. 1–12 (2007)
11. Johnson, D.B., Maltz, D.A., Broch, J.: DSR: the dynamic source routing protocol for multihop wireless ad hoc networks. Ad Hoc Networking (2001)
12. Karp, B.: Challenges in geographic routing: sparse networks, obstacles, and traffic provisioning. In: The DIMACS Workshop on Pervasive Networking, Piscataway, NJ (2001)
13. Karp, B., Kung, H.T.: GPSR: greedy perimeter stateless routing for wireless networks. In: Proc. ACM MobiCom, pp. 243–254 (2000). doi:10.1145/345910.345953
14. Kim, Y., Lee, J., Helmy, A.: Impact of location inconsistencies on geographic routing in wireless networks. In: Proc. ACM MSWIM 2003, pp. 124–127 (2003). doi:10.1145/940991.941013
15. Li, M., Lee, W.C., Sivasubramaniam, A.: Efficient peer-to-peer information sharing over mobile ad hoc networks. In: MobEA 2004 (2004)
16. Mauve, M., Widmer, A., Hartenstein, H.: A survey on position-based routing in mobile ad hoc networks. IEEE Network **15**(6), 30–39 (2001). doi:10.1109/65.967595
17. Nikolov, M., Haas, Z.J.: Towards optimal broadcastin wireless networks. IEEE Transactions On Mobile Computing **14**(7), 1530–1544 (2015). doi:10.1109/TMC.2014.2356466
18. Xiao, M., Jie, W., Huang, L.: Community-aware opportunistic routing in mobile social networks. IEEE Transactions on Computers **63**(7), 1682–1695 (2014). doi:10.1109/TC.2013.55
19. Perkins, C.E., Bhagwat, P.: Highly dynamic destination-sequenced distance-vector routing (DSDV) for mobile computers. In: Proc. of ACM SIGCOMM, pp. 234–235 (1994). doi:10.1145/190314.190336
20. Perkins, C.E., Royer, E.M.: Ad hoc on-demand distance vector routing. In: Proc. of the 2nd IEEE Workshop on Mobile Computing Systems and Applications, pp. 90–100 (1999). doi:10.1109/MCSA.1999.749281
21. Rozner, E., Seshadri, J., Mehta, Y., Qiu, L.: SOAR: simple opportunistic adaptive routing protocol for wireless mesh networks. IEEE Transactions on Mobile Computing **8**(12), 1622–1635 (2009). doi:10.1109/tmc.2009.82
22. Yang, S., Yeo, C.K., Lee, B.S.: Towards reliable data delivery for highly dynamic mobile ad hoc networks. IEEE Transactions on Mobile Computing **11**(1), 111–124 (2012). doi:10.1109/tmc.2011.55
23. Son, D., Helmy, A., Krishnamachari, B.: The effect of mobility induced location errors on geographic routing in mobile ad hoc sensor networks: analysis and improvement using mobility prediction. IEEE Transactions on Mobile Computing **3**(3), 233–245 (2004). doi:10.1109/tmc.2004.28
24. Menon, V.G., Joe Prathap, P.M.: Performance of various routing protocols in mobile ad hoc networks-a survey. Research Journal of Applied Sciences, Engineering and Technology **6**(22), 4181–4185 (2013)
25. Menon, V., Joe Prathap, P.M.: Performance analysis of geographic routing protocols in highly mobile ad hoc network. Journal of Theoretical and Applied Information Technology **54**(1), 127–133 (2013)

26. Park, V.D., Scott Corson, M.: Temporally-Ordered Routing Algorithm (TORA) version 4: Functional specification. Internet-Draft, draft-ietfmanet-TORA-spec- 04.txt (2001)
27. Park, V.D., Scott Corson, M.: A performance comparison of TORA and ideal link State routing. In: Proceedings of IEEE Symposium on Computers and Communication, pp. 592–598 (1998). doi:10.1109/ISCC.1998.702600
28. Shin, W.-Y., Chung, S.-Y., Lee, Y.H.: Parallel opportunistic routing in wireless networks. IEEE Transactions on Information Theory 59(10), 6290–6300 (2013). doi:10.1109/tit.2013.2272884
29. Xue, Y., Li, B., Nahrstedt, K.: A scalable location management scheme in mobile ad-hoc networks. In: Proc. of IEEE, LCN, pp. 102–111 (2001). doi:10.1109/lcn.2001.990775
30. Yoon, J., Liu, M., Noble, B.: Random waypoint considered harmful. In: Proc. IEEE INFOCOM, pp. 1312–1321 (2003). doi:10.1109/infcom.2003.1208967
31. Yuan, Y., Yuan, H., Wong, S.H., Lu, S., Arbaugh, W.: ROMER: resilient opportunistic mesh routing for wireless mesh networks. In: Proc. of IEEE WiMESH (2005)

# Acoustic Echo Cancellation Technique for VoIP

**Balasubramanian Nathan, Yung-Wey Chong and Sabri M. Hansi**

**Abstract** Acoustic Echo Canceller (AEC) is a crucial element to eliminate acoustic echo in a Voice over Internet Protocol (VoIP) communication. In the past, AEC have been implemented on Digital Signal Processing (DSP) platforms. This work uses software to perform the very demanding real-time signal processing that stack up against traditional implementation of AEC. Several issues are taken into consideration in the design of the AEC software. First, the AEC software is able to handle a mismatch in sampling rate between A/D and D/A conversion. Second, the playback sampling rate of playing CD-quality music with voice chat is considered. Third, the AEC has to overcome the major challenge of separating the reference signal in the presence of noise. Finally, the computational complexity and convergence problem has to be minimal. This work presents an AEC method that enable optimal audio quality in VoIP.

**Keywords** Acoustic Echo Canceller · Voice over Internet Protocol · Sampling rate · Signal processing · Noise canceller

## 1 Introduction

In hands-free duplex mode communication system, acoustic echo happens when signal from loudspeaker is reflected, picked up by microphone and transmitted back to the sender. It is very annoying and disrupting due to the significant roundtrip delay making it difficult to hear the other party. For years, researchers turned to DSPs and other hardware solutions to stop the loudspeaker-to-microphone feedback by implementing AEC. The AEC removes unwanted far-end

B. Nathan(✉) · Y.-W. Chong · S.M. Hansi
National Advanced IPv6 Centre, Universiti Sains Malaysia, Malaysia, Penang, Malaysia
e-mail: balasubramanian@nav6.usm.my, chong@usm.my

S.M. Hansi(✉)
Seiyun Community College, Hadhramount, Yemen
e-mail: sabri@nav6.usm.my

© Springer International Publishing Switzerland 2016
S.M. Thampi et al. (eds.), *Advances in Signal Processing and Intelligent Recognition Systems*,
Advances in Intelligent Systems and Computing 425,
DOI: 10.1007/978-3-319-28658-7_29

echo signal that is captured by the near-end microphone. However the increase of computation power in personal computers (PCs) has created the path that enables the implementation of real-time signal processing in PC environment [1].

Although AEC can be implemented in PCs, special care has to be taken into consideration in the design. The performance of AEC can be degraded drastically due to manufacturing error that causes sampling frequency offset between A/D and D/A interfaces [2]. AEC will not work as expected because there will be a nonlinear time-varying disturbances of the effective echo path as well as buffer overflow and underflow, causing it unable to estimate the echo path correctly [3]. Besides sampling frequency offset, mismatch can also happened in a scenario that require the playback sampling rate to be different than the capture sampling rate [4]. For example, a PC will capture the microphone signal at 44.1 kHz, and the echo from any source played by the computer such as DVD audio stream is sampled at 48 kHz. This scenario requires a sampling rate conversion (SRC) that is able to match the playback and capture sampling rates. Moreover the sampling rate for the signal capture from the microphone and the signal played through the speakers may not be identical [17]. In a VoIP system, the speech signal is usually sampled at 8 kHz or 16 kHz sampling frequency, which is sufficient for intelligible speech [5]. Since bandwidth is a precious resource in telecommunication, sampling a speech signal at 44.1 kHz may not be optimal especially in capturing the information present in the speech. Thus it is necessary to perform sampling rate conversion to match the playback and capture sampling rates and provide an optimal sampling frequency for speech signal.

During a VoIP session, the near-end signal may be corrupted not only by far-end speaker echo signal, but also ambient noise. The level of ambient noise can affect the performance of AEC. To avoid ambient noise, user can try to find a quiet location to make VoIP call but this option may not always be possible. Alternatively, the ambient noise can also be removed by the AEC to increase signal-to-noise ratio. Although this feature is highly desirable, it is not an easy task to subtract noise from a received signal. It could result in an increase in output noise power if the adaptive noise cancellation is not designed properly [6].

The remainder of this paper is organised as follows: In Section 2, we relate our work to other AEC methods. It is followed by the presentation of proposed AEC. Section 4 discusses the performance evaluation of the proposed system. Section 5 concludes the proposal and ideas for future work.

## 2    Background

Adaptive filter is the core in an acoustic echo canceller. Generally, the estimated echo, $\hat{d}(n)$, is subtracted from the near-end signal input signal, $y(n)$ plus the echo resulting transmitted near-end signal of

$$z(n) = y(n) + d(n) - \hat{d}(n) \tag{1}$$

The residual error signal, $e(n)$ will be piggybacked to the adaptive filter so that it can self-adjust the transfer function to achieve optimum performance [19].

$$e(n) = \hat{d}(n) - d(n) \tag{2}$$

Several adaptive algorithms has been proposed and used in the past namely, Least Mean Square (LMS), Normalised Least Mean Squares (NLMS), Recursive Least Square (RLS), Affine Projection Algorithm (APA), Sub band Adaptive Filtering, and Frequency Domain Adaptive Filter (FDAF). Although a wide variety of recursive algorithms have been developed in the past, the choice of one algorithm over another is determined by many factors such as convergence rate, robustness etc. In an AEC application, the adaptive filter must be able to filter real time input signals, thus the convergence rate must be fast.

Although many adaptive algorithms have been proposed, conventional adaptive filter such as LMS, NLMS, APA do not solve the effect of different sampling rate between A/D and D/A converter of low quality PC audio hardware. Pawig et al. [3] found that the sampling frequency offset causes the NLMS algorithm to fail in identifying the effective acoustic echo. When the frequency offset, $\Delta f$, between the loudspeaker signal, $x(n)$, and microphone signal $y(n)$ is 6 kHz, the maximum of the estimated impulse response moves with time to compensate for the time-varying delay of the effective echo path. Time jumps due to the finite buffer length, $L_B$ also caused the performance of the AEC to degrade. Pawig et al. [3] further proposed a Least Mean Square (LMS)-based framework to estimate the offset by recursively estimates the sampling rate offset and uses an external resampling filter to correct the offset. Miyabe et al. [7] proposed a blind compensation of sampling frequency mismatches between channels of asynchronous microphone array. It estimated and corrected sampling rate mismatch in the frequency domain based on the maximum likelihood estimation of linear phase shift. Abe et al. [8] addresses the problem in the frequency domain by recursively estimates the sampling rate offset through frame-step control and phase rotation. The estimation and correction are carried out in a single feedback loop without an external resampling filter. Stokes and Malvar [4] addressed the effects of different sampling rate between microphone and playback audio signal of CD-quality in the PC using frequency domain interpolation, which allows AEC to be performed while a 44.1 kHz signal is played through the speakers.

Another problem to be solved is to increase the signal-to-noise ratio. In the past, it has been recognized that AEC and Acoustic Noise Canceler (ANC) can be integrated together. Gupta et al. [9] proposed a combined adaptive noise canceler and blind source separation (ANC-BSS) system to remove noise and far-end echo signal from near-end signal. Gustafsson et. a. [10] combined acoustic echo control and noise reduction using a conventional NLMS-based adaptive echo canceler and a frequency domain postfilter in the sending path to attenuate both the residual echo remaining after an imperfect echo cancellation and the ambient noise. The postfilter approach helps to reduce the computational complexity of the adaptive echo canceler and more robust in the presence of loud background noise. On the other hand, Mahbub et al. [11] employ two-stage scheme by feeding the output of the AEC block to an ANC block where a spectral subtraction-based algorithm with adaptive spectral floor estimation is used.

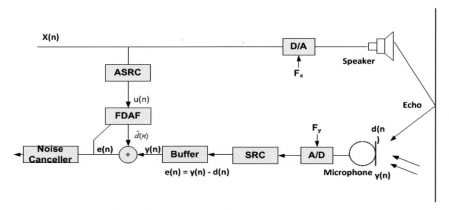

**Fig. 1** Acoustic Echo Cancellation System

# 3    Acoustic Echo Canceller

Fig. 1 shows a block diagram of the system design of the AEC. In this figure, $x(n)$ represents the far-end signal, $d(n)$ is the undesired echo, $y(n)$ is the near-end signal and $\hat{d}(n)$ is the replica of the echo. The AEC system consists of four components; the sampling rate conversion (SRC), arbitrary sampling rate conversion (ASRC), adaptive filter, and noise canceller. The AEC is designed to ensure that the sampling rate mismatch issue is minimized and noise cancellation is taken into consideration. The next subsection presents overview of these modules:

## 3.1    Sampling Rate Conversion (SRC)

Sampling rate conversion component uses cubic interpolation to converts the discrete-time signal $x(k)$ at a rate $F_x$ to another signal $y(k)$ sampled at a rate $F_y$ [12]. Finite Impulse Response (FIR) is integrated as part of the SRC component because it is suitable for linear-phase designs. We first assume that sampling rate converter starts with a set of $M$ interpolation filters that calculate output values spaced every $1/M$ samples. The output signal is calculated from the 4 nearest $X$ values using the interpolation filters. It will then fit a cubic polynomial to these 4 values. Desired output time will be evaluated and the integer and fractional portion is separated. For each output time sample, the fractional time offset, $t_c$, will be monitored using the input and output sampling rate where

$$\tau_\chi = M \ (\tau_{out} - T_M) \tag{3}$$

It will then calculate the weighting coefficients $w_a$, $w_b$, $w_c$ and $w_d$ where

$$w_a = -\frac{1}{6}t_c^3 + \frac{1}{2}t_c^2 - \frac{1}{3}t_c$$

$$w_b = \frac{1}{2}t_c^3 - t_c^2 - \frac{1}{2}t_c + 1$$

$$w_c = -\frac{1}{2}t_c^3 + \frac{1}{2}t_c^2 + t_c$$

$$w_d = \frac{1}{6}t_c^3 - \frac{1}{6} \tag{4}$$

The filter $h_{net}[i]$ is computed as a weighted sum of FIR filters.

$$h_{net}[i] = w_a h_k[i] + w_b h_{k+1}[i] + w_c h_{k+2}[i] + w_d h_{k+3}[i] \tag{5}$$

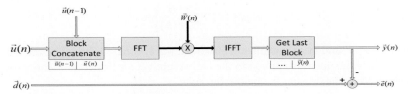

**Fig. 2** Calculating output and error signals

This technique separates the variation in the time-varying filter's design from the adaptive filter. It solved the issue where there is sampling rate mismatch in a multichannel environment such as signal capture from the microphone and the signal played through the speakers [13].

## 3.2 Adaptive Filter

The fast block LMS (FBLMS) algorithm, also known as overlap-save Frequency Domain Adaptive Filter (FDAF), is a derivative of LMS algorithm designed as adaptive filter algorithm in AEC. In FBLMS, the desired signal and input signal are transformed into discrete frequency domain using fast Fourier transform (FFT) before the adaptive processing [14]. It updates the filter coefficients in the frequency domain so that it can achieve fast convergence rate and low computational resources. Instead of linear convolution and correlation that are performed in LMS and NLMS-based adaptive filter, circular operation is performed in frequency domain on a block-by-block rule.

According to Haykins [15], in FBLMS algorithm, the $M$ tap weights of the filter are padded with an equal number of zeros and an $N$-point Fast Fourier Transform (FFT) is used for the computation. The frequency domain weight vector, $\hat{W}(n)$, is initialized with a $M$-by-1 null vector. For each new block of $M$ input samples, the FBLMS algorithm will compute the filtering, error estimation, and tap weight adaptation. As shown in Fig. 2, the current input signal block will be concatenates to the previous block. The tap input vector, $u(n)$, are transformed using FFT resulting

$$U(k) = diag\{FFT[u(kM - M), \ldots, u(kM - 1), u(kM), \ldots u(kM + M - 1)]^T\} \quad (6)$$

The input signal blocks are then multiplied by the frequency-domain weight vector, $\widehat{W}(k)$, and an inverse FFT (IFFT) is performed on the multiplication result. The last block from the results is retrieved as the output signal vector, $z^T(k)$.

$$z^T(k) = [z(kM), z(kM + 1), \ldots, z(kM + M - 1)]$$
$$= last\ M\ elements\ of\ IFFT\ [U(k)\widehat{W}(k)] \quad (7)$$

The error signal vector is produced by comparing the filter output against the desired response, $d(k)$, of the filter and it can be transformed into frequency domain as follows

$$E(k) = FFT \begin{bmatrix} 0 \\ e(k) \end{bmatrix} \quad (8)$$

After output and error signals are obtained, the FBLMS will update the weight vector. The adjustment is applied to the tap weight vector from one iteration of the algorithm to the next using the below formula:

$$\begin{pmatrix} Adjustment\ to\ the \\ weight\ vector \end{pmatrix} = \begin{pmatrix} step-size \\ parameter \end{pmatrix} X \begin{pmatrix} tap\ input \\ vector \end{pmatrix} X\ (error\ signal) \quad (9)$$

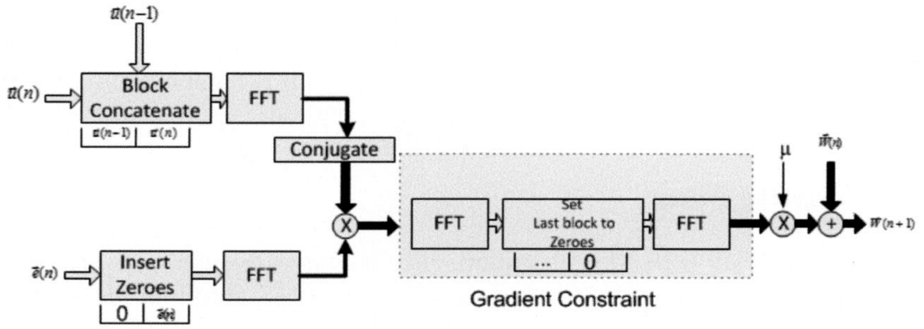

**Fig. 3** Updating filter coefficients

As shown in Fig. 3, zeroes is inserted before the error signal vector. This step is necessary to ensure that the length of the error signal vector is the same as concatenated input signal blocks. FFT is performed on the error signal blocks and the results will be multiplies by the complex conjugate of the FFT. An IFFT is then implemented on the multiplication results and the value of the last block of the IFFT is set to zeroes before performing FFT again.

$$\phi(k) = first\ M\ elements\ of\ IFFT[D(k)U(k)E(k)] \quad (10)$$

The FFT result is multiplied by step size, μ. For the $k$th block, we can sum the product of $u(kL+i)e(kL+i)$ for all the possible values of $i$, that is $i=0,1,...,L-1$. The weight vector $\widehat{W}(k)$ is then added to the multiplication results:

$$\widehat{W}(k+1) = \widehat{W}(k) + \mu \sum_{i=0}^{L-1} U(kL+i)e(kL+i) \qquad (11)$$

The update equation for the tap weight vector can also be summarized as follows:

$$\widehat{W}(k+1) = \widehat{W}(k) + \alpha FFT\begin{bmatrix} \phi(k) \\ 0 \end{bmatrix} \qquad (12)$$

## 3.3   Adaptive Arbitrary Sample Rate Conversion (ASRC)

Before the system can correct a frequency offset, it is necessary to estimate the sampling rate offset during the course of the conversation. This can be done by comparing the waveforms of the near end signal y(n) and estimated echo signal $\hat{d}(n)$ [3]. From the estimation, adaptive resampling is used to correct the frequency offset, $\Delta f$. In the resampling process, signal is upsampled by an integer factor $U$ by a digital upsampling stage and a digital interpolation filter $h_d(n)$. The upsampling increases the accuracy of the next interpolation stage. To reach the desired rate of conversion between upsample and downsample integer factor, $D$, the upsampled signal is interpolated using Lagrange interpolation. Langrage interpolation with $L+1$ samples of $\hat{x}(i)$ uses polynomials to approximate the signal where the coefficients $l_n(\Delta k)$ of the interpolation filter is [3, 16]

**Fig. 4** ASRC implementation

$$l_2(\Delta k) = -\frac{1}{6}(\Delta k + 1)\,\Delta k(\Delta k - 1) \qquad (13)$$

The resample far end input signal, $x'(k)$ with current ratio $\hat{a}(n)$ will be used by the adaptive filter to create the input vector as well as estimate the echo.

$$x'(k) = ASRC(x(k), \hat{a}(k)) \qquad (14)$$

## 3.4   Noise Canceller

During a VoIP communication, a moderate background noise levels can cause low SNRs. In order to reduce the noise, we included a noise canceller before sending to far-end speaker, as illustrated in Fig. 1. The noise canceller is implemented by using auditory masking. For a mix of speech and noise, the noise components which lie below masked threshold are inaudible, reducing the effects of quantization noise. In the noise canceller, a rough estimation of the speech spectrum is obtained using a preliminary spectral weighting. From the preliminary spectrum, the masked threshold is estimated. A conventional weighting rule is modified to attenuate the signal only at those frequencies where the noise is not completely masked by the speech [11]. The weighting coefficient is approximated by

$$H(\Omega) = \min\left(1, \sqrt{\frac{R_{TT}(\Omega)}{R_{bb}(\Omega) + R_{nn}(\Omega)}} + \frac{\zeta_n R_{nn}(\Omega) + \zeta_b R_{bb}(\Omega)}{R_{nn}(\Omega) + R_{bb}(\Omega)}\right) \tag{15}$$

where $\zeta_n$ and $\zeta_b$ are attenuation factor for noise and residual echo respectively. $R_{TT}$ represents the masked threshold whereas $R_{bb}$ and $R_{nn}$ are the final noise estimation and residual echo estimation respectively.

## 4    Performance Evaluation

We now evaluate the performance of the new AEC system using real-time VoIP signal. The setup of the evaluation is an office with two clients (PCs) directly connected through local area network. The playback and capture sampling rate is set at 44.1 kHz. The audio stream of $y(n)$, and the filtered output speech signal are recorded [16]. We noticed that when the audio stream has not been processed by the AEC, acoustic echo and background noise is noticeable as shown in Fig. 5. The echo may be caused by the synchronization problem of the input and output audio streams of the soundcard. By including SRC and ASRC in the system, the performance of AEC in the presence of a sampling rate mismatch increased dramatically as compared to pure FDAF algorithm. The signal-to-noise ratio improved as well when noise canceller is added to the system. The background noise that is present in the original VoIP system was eliminated providing clearly voice signal and improved user experience.

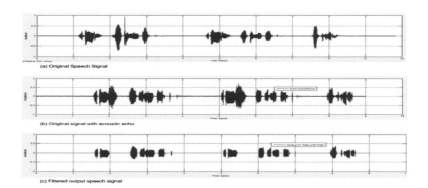

(a) Original Speech Signal

(b) Original signal with acoustic echo

(c) Filtered output speech signal

**Fig. 5** Real-time VoIP results proposed for AEC system

# 5    Conclusion

In this paper, we proposed a novel method of acoustic echo canceler that handles background noise and sampling rate mismatch between A/D and D/A conversion as well as mismatch in a multichannel environment without using a hardware based solution. The results indicate that the echo of real-time VoIP communication dramatically reduced, providing a better user experience. Although the complexity increased, causing slight delay in the VoIP communication, this can further improved by using high power computers for real-time VoIP communication.

**Acknowledgments** The authors would like to thank the Ministry of Science, Technology and Innovation (MOSTI) Malaysia for providing ScienceFund grant titled Development of Acoustic Echo Cancellation Software for Personal Computers (01-01-05-SF0638) to fund this research project.

# References

1. Fischer, V., Gansler, T., Diethron, E.J., Benesty, J.: A software stereo acoustic echo canceler for Microsoft Windows. In: Proc. Int. Workshop on Acoustic Echo and Noise Control, Darmstadt, Germany, pp. 87–90, September 2001
2. Robledo-Arnuncio, E., Wada, T.S., Juang, B.H.: On dealing with sampling rate mismatches in blind source separation and acoustic echo cancellation. In: Proc. WASPAA, pp. 34–37 (2007)
3. Pawig, M., Enzner, G., Vary, P.: Adaptive sampling rate correction for acoustic echo control in Voice-over-IP. IEEE Transactions on Signal Processing **58**(1), 189–199 (2010)
4. Stokes, J.W., Malvar, H.S.: Acoustic echo cancellation with arbitrary playback sampling rate. In: Proc. IEEE Int. Conf. Acoust., Speech, Signal Process. (ICASSP 2004), vol. 4, pp. iv-153–iv-156 (2004)
5. Sakshat Virtual Labs, Sampling Frequency and Bit Resolution for Speech Signal Processing. http://iitg.vlab.co.in/?sub=59&brch=164&sim=474&cnt=1
6. Widrow, B., Glover, J.R.J., McCool, J.M., Kaunitz, J., Williams, C.S., Hearn, R.H., Zeidler, J.R., Dong, J.E., Goodlin, R.C.: Adaptive noise cancelling: principles and applications. Proc. IEEE **63**(12), 1692–1716 (1975)
7. Miyabe, S., Ono, N., Makino, S.: Blind compensation of inter-channel sampling frequency mismatch with maximum likelihood estimation in STFT domain. In: IEEE International Conference on Acoustics, Speech and Signal Processing (ICASSP), pp. 674–678 (2013)
8. Abe, M., Nishiguchi, M.: Frequency domain acoustic echo canceller that handles asynchronous A/D and D/A clocks. In: IEEE International Conference on Acoustics, Speech and Signal Processing (ICASSP), pp. 5924–5928 (2014)
9. Gupta, V., Chandra, M., Sharan, S.: Acoustic Echo and Noise Cancellation System for Hand-Free Telecommunication using Variable Step Size Algorithms. Radioengineering, 200–207 (2013)

10. Gustafsson, S., Martin, R., Jax, P., Vary, P.: A psychoacoustic approach to combined acoustic echo cancellation and noise reduction. IEEE Trans. Speech and Audio Processing **10**(5), 245–256 (2002)
11. Mahbub, U., Fattah, S.A., Zhu, W.P., Ahmad, M.O.: Single-channel acoustic echo cancellation in noise based on gradient-based adaptive filtering. EURASIP JASMP **1** (2014)
12. Beckmann, P., Stilson, T.: An efficient asynchronous sampling-rate conversion algorithm for multi-channel audio applications. In: Audio Engineering Society Convention 119 (2005)
13. Shynk, J.J.: Frequency-domain and multirate adaptive filtering. IEEE Signal Processing Magazine **9**, 14–37 (1992)
14. Haykin, S.: Adaptive Filter Theory, 5th edn. Pearson Education Limited, Upper Saddle River (2014)
15. Hanshi, S.M., Chong, Y.-W., Naeem, A.N.: Review of Acoustic Echo Cancellation Techniques for Voice Over IP. Journal of Theoretical & Applied Information Technology, 77 (2015)
16. Hanshi, S.M., Chong, Y.-W., Ramadass, S., Naeem, A.N., Ooi, K.-C.: Efficient acoustic echo cancellation joint with noise reduction framework. In: 2014 International Conference on Computer, Communications, and Control Technology (I4CT), pp. 116–119, September 2–4, 2014. doi:10.1109/I4CT.2014.6914158

# MRAC for a Launch Vehicle Actuation System

Elizabath Rajan, Baby Sebastian and M.M. Shinu

**Abstract** Actuators are used to provide Thrust Vector Control (TVC) for lifting the launch vehicle and its payload to reach its space orbits. The TVC comprises actuation and power components for controlling the engines direction and thrust to steer the vehicle. Here the actuator considered is a linear Electromechanical Actuator (EMA) system. The design is based on the Model Reference Adaptive Controller (MRAC) technique. It calculates an error value as the difference between actual system and desired model and attempts to minimize the error by adjusting the controller parameters. Adaptive control laws are established with the gradient method. The experimental result shows that the MRAC has better tracking performance than that with the traditional compensated system when both controllers are subjected to the same parameter variations. The simulation is done using MATLAB/SIMULINK software.

**Keywords** EMA · Compensator · MRAC · MIT rule

## NOMENCLATURE

$B_e$    Viscous damping coefficient of engine
$B_m$    Viscous damping coefficient of torque motor
$J_e$    Engine Moment of inertia (MI)
$J_m$    MI of torque motor rotating assembly
$K_{cf}$    Feedback gain of current loop
$K_l$    Actuator mounting structure stiffness

E. Rajan(✉) · M.M. Shinu
Department of Electronics and Instrumentation, Vimal Jyothi Engineering College,
Chemperi, Kannur, Kerala, India
e-mail: {elizabath91,shinu.mohanan.m}@gmail.com

B. Sebastian
Scientist/Engineer, CECG, VSSC, ISRO, Trivandrum, Kerala, India
e-mail: baby_sebastian@vssc.gov.in

© Springer International Publishing Switzerland 2016
S.M. Thampi et al. (eds.), *Advances in Signal Processing and Intelligent Recognition Systems*,
Advances in Intelligent Systems and Computing 425,
DOI: 10.1007/978-3-319-28658-7_30

345

Kp    LVDT scale factor
$K_T$    Torque sensitivity of motor
lm    Actuator lever arm length
nb    Roller screw ratio
$N_{ch}$    Number of operating channels in torque motor
$\omega_d$    Demodulator frequency
$\xi d$    Demodulator damping factor

# 1    Introduction

The linear actuation system plays a major role in the thrust vector control of spacecraft system. Linear actuators are divided into two types: Electrohydraulic Actuators (EHA) and Electromechanical Actuators (EMA). Nowadays EMAs [1] are widely used in the field of Thrust Vector control (TVC) applications. Replacing a hydraulic actuator with a high power electromechanical actuator can result in the overall system being lighter, cleaner, more efficient, easier to integrate both mechanically and electrically into the aircraft, and easily maintainable [2].The actuator under consideration is a linear EMA system used for moving the nozzle of engine for TVC [3]. It is based on a Brushless Direct Current (BLDC) torque motor.

Servo system parameters frequently vary due to changing operating conditions and resulting in altered dynamic behavior of the system .The control algorithm should follow the same system dynamic behavior in the case of parameter variations [4-7]. In this paper, the design and implementation of MRAC is presented. In this scheme, the system response is forced to track the response of a model output regardless of process parameter variations.MRAC is used in the feedback loop to improve the performance of system. Here the gradient approach of adaptation technique based on the minimization of a chosen loss function (J) is applied. MRAC is also known as Model Reference Adaptive System (MRAS).

The paper is organized as follows. A brief description of the system, modelling and compensator scheme is presented in section 2. Section 3 deals with the implementation of the Model Reference Adaptive Controller design. Simulation results are given in section 4. Finally, concluding remarks are presented in the final section.

# 2    Electromechanical Actuation System

## 2.1    System Description

The configuration of actuator placement with respect to the engine [8] is shown in figure 1. The engine is gimbaled at a point and actuators help in moving the engine about the pivot point to achieve thrust vector control (TVC). Generally for an engine, there would be two actuators, mounted on mutually perpendicular axes, one in yaw plane and the other in pitch plane. In the launch vehicle [12], the actuator is mounted to the engine at its one end and the stage is attached to the other end. The servo system should provide closed loop position control by tracking the commands issued from the autopilot loop.

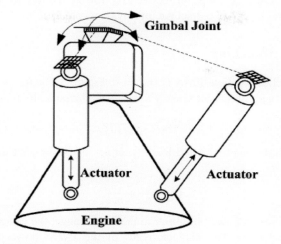

**Fig. 1** Actuator placement with respect to the engine.

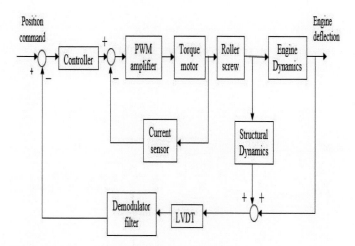

**Fig. 2** Block diagram of a electromechanical actuator system.

Block diagram of typical electro mechanical actuator system is shown in figure 2. The LVDT (Linear Variable Differential Transformer) sense the angular position of the nozzle which is compared with the command input and error signal is generated. This error voltage is processed by means of controller and then fed to the brushless DC torque motor which drives the roller screw/engine. The torque motor produces the necessary torque to nullify the error. The required power amplification [16] of the control signal is given by (Pulse Width Modulated) PWM power amplifier working inside a high gain current loop. This current loop ensures steady current being fed to the motor. The resulting torque gives the angular deflection of the nozzle. Roller screw is used for converting rotary motion to linear motion. Demodulator filter is used to extract the information from the LVDT.

## 2.2   System Modelling

The linear model of a typical EMA is shown in Figure 3.The torque motor can be approximated as a second order system which is mathematically represented as

$$G_M = \frac{1}{J_m.s^2 + B_m.s} \tag{1}$$

The second order approximated load dynamics equation used in the actuation system is given by

$$G_E = \frac{1}{J_e.s^2 + B_e.s} \tag{2}$$

For the operation of power amplifier in the linear range, the current loop should have larger bandwidth. The characteristics of the PWM power amplifier and current loop dynamics are used to determine the torque motor current and it is given by

$$i_{mc} = V_{in} \times K_A \tag{3}$$

where $V_{in}$ is the input voltage of power amplifier and $K_A$ is the net power amplifier gain.

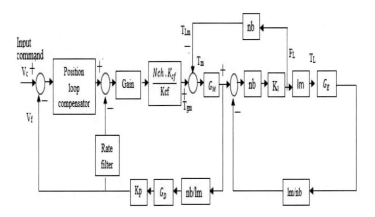

**Fig. 3** Linear model of the EMA.

The torque motor is a quadruplex motor, consisting of four identical three phase coils. So the total current is given by

$$i_m = i_{mc} \times N_{ch} \tag{4}$$

Torque developed by motor ($T_{gm}$) is given by

$$T_{gm} = K_T \times i_m \tag{5}$$

where $K_T$ is the torque sensitivity of motor.
The load torque transmitted to the motor side ($T_{Lm}$) is given by

$$T_{Lm} = F_L \times nb \tag{6}$$

The load force ($F_L$) transmitted to motor side is given by

$$F_L = \left( \theta_m - \delta e \times \frac{lm}{nb} \right) \times nb \times K_l \tag{7}$$

where $\theta_m$ and $\delta e$ denotes the motor angular position and the nozzle deflection respectively. Driving torque of the nozzle/load dynamics ($T_L$) given by

$$T_L = F_L \times lm \tag{8}$$

Demodulator filter with dynamics equivalent to that of a second order Butterworth filter is used to extract the signal from the LVDT output. LVDT output is related to motor output as

$$V_{LVDT} = \theta_m \times \frac{nb}{lm} \times K_p \tag{9}$$

where $K_p$ is the LVDT scale factor.
Demodulator dynamic equation is given by

$$G_D = \frac{\omega_d{}^2}{s^2 + 2\xi_d \omega_d s + \omega_d{}^2} \tag{10}$$

where $\xi_d$ and $\omega_d$ are the demodulator damping factor and frequency respectively.

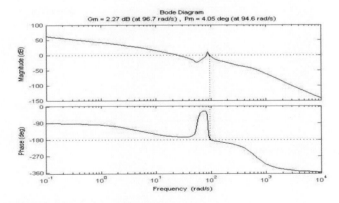

**Fig. 4** Frequency response of the open loop system.

The open loop frequency response of the plant indicates a phase margin of 4.05° and gain margin of 2.27 dB. From the phase margin it is observed that even though it is positive, its value is very less so the system does not meet the required specifications. Therefore in order to meet all the required specifications, suitable compensation scheme is used.

## 2.3   Compensation Scheme

The compensation scheme was designed based on the requirements of the system. This classical controller comprises a position loop and pseudo rate loop, as shown in the figure 5. The position loop compensator consists of a lag compensator, notch filter, and an appropriate gain. To increase the low frequency gain, lag compensator [9] is used in the forward path .Thus it reduces the steady state error. It also increases the phase margin of the system, but there is an unnecessary increase in the gain of the system at high frequency. This will decrease the stability margin. So the system can be closed using a rate filter, which have the same function as that of a lead compensator without increasing the gain. In the actuator nozzle system, velocity is not directly available as a signal to be fed back. This pseudo rate filter derived from the LVDT output is provided to increase the relative stability of the system. To suppress the resonant frequency of the system, a notch filter is included in the forward path of the position loop. The transfer function of the rate filter, lag compensator and notch filter are given below.

$$G_r = \frac{4.2s}{s+125} \tag{11}$$

$$G_{lag} = \frac{s+9.7}{s+6.59} \tag{12}$$

$$G_{nh} = \frac{s^2+1.772s+7849}{s^2+17.72s+7849} \tag{13}$$

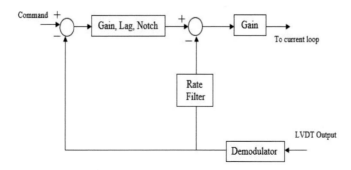

**Fig. 5** Compensation Scheme.

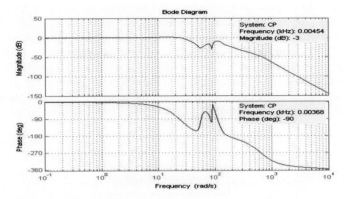

**Fig. 6** Frequency response of the closed loop compensated system.

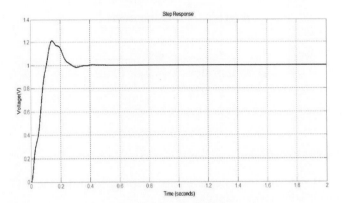

**Fig. 7** Step response of the closed loop compensated system.

The frequency response of the compensated system is given in figure 6.The step response is given in the figure 7.The system is tracking the command input. The parameters of both step and frequency response are within the required range. The bandwidth of the system is obtained as 4.5Hz.

# 3    Proposed Work

## 3.1    *Model Reference Adaptive Controller*

The influence of minor parameter variations on the actuation system may be satisfactorily compensated using classical compensator design methodology. However, the deleterious effects of substantial parameter changes can no longer be effectively compensated by this method. In order to provide a solution to this problem, adaptive controller is chosen. The adaptive controller   is a controller that automatically and continuously measures the dynamic behavior of process. The block diagram [10] of the system is shown in Figure 8.The controller has two loops. Inner loop is a normal feedback loop consisting of the process and the controller, outer loop which adjusts the controller parameters. The model represents the ideal response of the process to a command signal. The model output is compared to the system output and the controller parameters are adjusted based on the error. The methods for adjusting the parameters are the gradient and lyapunov method. Here the gradient method (MIT Rule) is used to develop the adaptation mechanism.

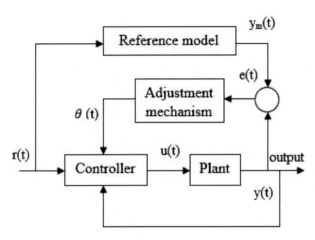

**Fig. 8** Block diagram of MRAC.

The original approach to model reference adaptive control is the MIT rule. The loss function chosen for the adaptation is

$$J(\theta) \ = \ \frac{1}{2}e^2 \tag{14}$$

$\theta$ is the adjustable parameter and e is the error. According to the MIT rule, the time rate of change of $\theta$ is proportional to the negative gradient of J, that is

$$\frac{d\theta}{dt} \ = \ -\gamma\frac{\partial J}{\partial \theta} = -\gamma\frac{\partial e}{\partial \theta} \tag{15}$$

$\frac{\partial e}{\partial \theta}$ is called sensitivity derivative of the system. The parameter $\gamma$ is the adaptation gain, which is tuned to get desired response. The MIT rule is a gradient method that aims to minimize the squared model cost function. The advantage of this gradient method is its simplicity of implementation.

## 3.2   Controller Design

Consider a system described by the model,

$$\frac{dy}{dt} \ = \ -ay + bu \tag{16}$$

where u is the control variable and y is the measured output. Assume we want to obtain a closed loop system described by

$$\frac{dy_m}{dt} = -a_m y_m + b_m u_c \tag{17}$$

Let controller be given by,

$$u(t) = \theta_1 u_c - \theta_2 y \tag{18}$$

Controller parameters are chosen to be

$$\theta_1 = \frac{b_m}{b} \tag{19}$$

$$\theta_2 = \frac{a_m - a}{b} \tag{20}$$

The input output relations of both the system and the model are same. Then it is called the perfect model following. i.e.

$$\frac{dy}{dt} = \frac{dy_m}{dt}$$

$$\frac{dy}{dt} = -ay + b(\theta_1 u_c - \theta_2 y)$$

$$= -a_m y + b_m u_c \tag{21}$$

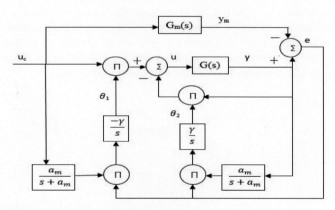

**Fig. 9** Block diagram of MRAC based on MIT rule.

Let $\frac{dy}{dt} = Py$ , where P is the differential operator.

To apply the MIT rule, introduce the error, $e = y - y_m$

$$e = \frac{b\theta_1}{p + a + b\theta_2} u_c - \frac{b_m}{p + a_m} u_c \tag{22}$$

Sensitivity derivative is obtained by taking the partial derivative w.r.to controller parameters $\theta_1$ and $\theta_2$.

$$\frac{\partial e}{\partial \theta_1} = \frac{b\theta_1}{p + a + b\theta_2} u_c \tag{23}$$

$$\frac{\partial e}{\partial \theta_2} = \frac{-by}{p + a + b\theta_2} \tag{24}$$

Approximations are required because the process parameters a and b are not known. So we use the approximation.

$$p + a + b\theta_2 = p + a_m \tag{25}$$

Therefore the equations for updating the controller parameters are obtained as

$$\frac{d\theta_1}{dt} = -\gamma \frac{a_m u_c}{p + a_m} e \tag{26}$$

$$\frac{d\theta_2}{dt} = \gamma \frac{a_m y}{p + a_m} e \tag{27}$$

## 4    Simulation Results

The model reference adaptive control scheme is applied to the electromechanical actuation system by using MIT rule. The model is simulated in MATLAB which is shown in figure 10.The input signal is a sine wave with unit amplitude. Here, the adaptation gain is adjusted in order to achieve a good tracking response. The value of $\gamma$ is chosen as 1.8.Results of comparing model reference method and traditional compensator design methodology by varying the parameter values of $K_P$ (LVDT scale factor) and $K_T$ (Torque sensitivity of motor) are shown in figure 11 and figure 12 respectively. It is inferred that the MRAC gives better tracking performance. It can also cope up with the parameter variations.

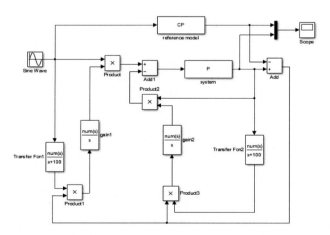

**Fig. 10** Simulink model for the MIT rule.

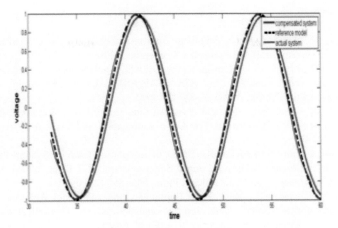

**Fig. 11** Parameter variation of $K_P$.

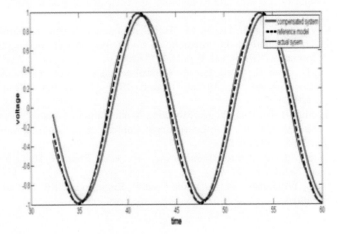

**Fig. 12** Parameter variation of $K_T$.

## 5    Conclusion

The experimental results verify the gradient method of the MRAC approach for the EMA system. MRAC is designed for the stability of a closed loop system. In this scheme, the controller parameters are adjusted automatically to give a desired closed loop performance. The controller maintains constant performance in the presence of parameter variations. The MIT rule of the model reference adaptive controller provides a good tracking performance and it can eliminate the parameter variations in the output when compared to the classical controller (compensated system).MRAC is applicable only for low frequency signals and is the major drawback associated with this MIT rule.

# References

1. Balaban, E., Saxsena, A.: A diagnostic approach for electro mechanical actuators in aerospace system. In: IEEE Aerospace Conference (2009)
2. Cowan, J.R., Myers, W.N.: Design and test of a high power electromechanical actuator for thrust vector control. In: AIAA/SAE/ASME/ASEE 28th Joint Propulsion Conference and EXhibit, AIAA 92-3851 (1992)
3. Sumathi, R.,Usha, M.: Pitch and Yaw Attitude Control of a Rocket Engine Using Hybrid Fuzzy- PID Controller. The Open Automation and Control Systems Journal, 29–39 (2014)
4. Swarnkar, P, Jain, S, Nema, R.K.: Effect of Adaptation Gain on System Performance for Model Reference Adaptive Control Scheme Using MIT Rule. World Academy of Science, Engineering and Technology, Paris, 545–550 (2010)
5. Joseph, A., Isaac, J.S.: Real Time Implementation of Model Reference Adaptive Controller for a Conical Tank. International Journal on Theoretical and Applied Research in Mechanical Engineering 2(1), 57–62 (2013)
6. Jain, P., Nigam, M.J.: Real time control of ball and beam system with model reference adaptive control strategy using MIT rule. In: IEEE International Conference on Computational Intelligence and Computing Research (ICCIC), Madurai, pp. 305–308 (2013)
7. Duka, A.-V., Oltean, S.E., Dulău, M.: Model reference adaptive vs. learning control for the inverted pendulum. In: International Conference on Control Engineering and Applied Informatics (CEAI), vol. 9, pp. 67–75 (2007)
8. Li, Y., Lu, H., Tian, S., et al.: Posture Control of Electromechanical-Actuator-Based Thrust Vector System for Aircraft Engine. IEEE Trans. Ind. Electron. 59(9), 3561–3571 (2012)
9. Oh, C.-S., Sun, B.-C., Park, Y.-K.: Modeling & simulation of a launch vehicle thrust vehicle control system. In: 12th International Conference on Control, Automation & System, pp. 2088–2092 (2012)
10. Ogatta, K.: Modern control Engineering, 4th edn. Prentice Hall, New Jersey publications (2002)
11. Astrom, K.J., Wittenmark, B.: Adaptive Control, Englewood Cliff, 2nd edn. Prentice Hall, NJ (2000)

# Analysis of Inspiratory Muscle of Respiration in COPD Patients

Archana B. Kanwade and Vinayak Bairagi

**Abstract** Chronic Obstructive Pulmonary Disease (COPD) is a disease of lungs, in this airway becomes narrow. To perform excess work of respiration necessary and accessory muscles such as sternomastoid (SMM) muscle has to work. In this paper correlation among obstruction level of airways and activity of SMM has invented with time and frequency domain features of electromyography (EMG). Spirometric data and EMG signal of SMM is collected for ten COPD patients. Features used for the analysis are root mean square (RMS), peak to peak voltage, integration and number of peaks in time and frequency domain. Features of EMG and spirometric data are correlated. Significant correlation has found between peak to peak amplitude and number of peaks in time and frequency domain.

**Keywords** Sternomastoid · Electromyography · Time · Frequency

## 1 Introduction

Chronic Obstructive pulmonary disease (COPD) is disease of lungs in that airways become narrow and shortness of breath which is progressive, can lead to death. A Common risk factor for COPD is tobacco smoking, biomass fuel consumption, and diesel exhaust, occupational exposure to organic or inorganic dust. In COPD there are two abnormalities chronic bronchitis and Emphysema. Emphysema involves the gradual destruction of the air sacs (alveoli) in the lungs, making it even harder for a person to breathe. Chronic bronchitis causes a person's lungs to become very inflamed. It commonly affects the windpipe and passageways in the lungs.

A.B. Kanwade(✉)
Sinhgad Institute of Technology and Science, Pune, India
e-mail: archana_kanwade@yahoo.com

V. Bairagi
AISSMS Institute of Technology and Science, Pune, India
e-mail: vbairagi@yahoo.co.in

© Springer International Publishing Switzerland 2016
S.M. Thampi et al. (eds.), *Advances in Signal Processing and Intelligent Recognition Systems*,
Advances in Intelligent Systems and Computing 425,
DOI: 10.1007/978-3-319-28658-7_31

COPD is a major cause of mortality and morbidity all over the world [1]. COPD burden is progressive and increasing year by year. It has been estimated that by the year 2030, COPD will become the third biggest cause of death [14]. COPD is relatively unknown or ignored by the public as well as public health and government officials [1], [13].

COPD is confirmed by a simple diagnostic test called spirometry, that measures how much air a person can inhale and exhale, and how fast air can move into and out of the lungs [13]. In spirometry two major factors of lungs are studied. They are Force Expiratory Volume (FEV) and Force Vital Capacity (FVC) [1]. COPD is a systemic disease; Spirometry does not give overall idea about disease status. Limitations of spirometry are, it is effort dependent, alter broncho-motortone, difficult to perform on morbid patients, subjective, time consuming, require patient co-operation and specific respiratory maneuvers[2],[3]. There is a need of easy to perform but physiologically accurate method to access pulmonary mechanics in COPD patients.

## 1.1    Muscles of Respiration

Each respiratory cycle consists of inspiration followed by expiration. Diaphragm and intercostals muscles are inspiratory muscles, active during quite breathing, although SMM and scalene are accessory muscles of respiration and recruited during rigorous activity [4]. Patients with COPD increasingly need to use the inspiratory accessory muscles like SMM even during quite breathing. Therefore we may find severity of COPD by analyzing work done by accessory muscle of respiration. We can correlate severity of COPD with EMG of Sternomastoid muscle. Sternomastoid muscle is selected as it is easily accessible and EMG data can be collected using invasive method.

## 1.2    Electromyography

To study activity of muscle Electromyography (EMG) is best well known technique. EMG signal is electrical manifestation of the neuromuscular activity of muscle. It is complicated and affected by anatomy, physiology of muscle, peripheral nervous system and technique of detection. Shape of EMG depends on orientation of recording electrode contacts with respect to active fibers. Amplitude of action potential depends on diameter of muscle fiber and distance between electrodes and recording site. Duration of action potential will be inversely proportional to conduction velocity [5]. Muscle fiber conduction velocity (v) is the propagation velocity of the depolarization along the membrane of the muscle fiber.

According to Lindstrom model, power spectral density (PSD)[6] of EMG signal is

$$S(f) = \frac{1}{v^2} G\left(\frac{fd}{v}\right) \tag{1}$$

Where:   f            is frequency.

   D            is the distance between the bipolar electrodes.

   $G\left(\dfrac{fd}{v}\right)$        indicates, spectrum shape.

Power Spectral density of EMG signal undergoes shift in frequency because of a shift in conduction velocity. Conduction velocity is associated with muscle activity and localized muscle fatigue. As muscle fatigue, frequency of firing decrease (mean and median frequency decreases)[6]. Alterations in amplitude and Frequency are found depending on level of muscle contraction. EMG can be analyzed in time and frequency domain. There are two basic techniques to acquire EMG, non-invasive and invasive. EMG accesses muscle function, by using surface electrode to detect electrical activity of muscle.

Miguel Angel Mañanas, has done an analysis of EMG, vibromyographic (VMG) of SMM muscle to detect the severity of COPD [7]. The Activity of SMM was found by means of several indexes: root-mean-square (rms) values, mean frequency, median frequencies and ratio between high and low-frequency they have used some parametric and non- parametric methods to estimate power spectral density. Results vary with spectral estimator technique. Dornelas de Andrade evaluated the activity of diaphragm and the SMM muscle with a 30% Threshold load as found correlation with obstruction level about r=-0.5370. Analysis is done using RMS value [8].

## 2    Methodology

Ten COPD (6 Male, 4 Female) patients are selected with 56.35± 20 age, surface EMG of SMM muscle is collected for maintained protocol using, RMS SALUS (2/4 Channel Portable EMG) EMG machine. Selected electrode site was between center and tedious areas of muscle. Data was collected while patient was appliedwith constant load on respiration. Distance between center of electrodes or

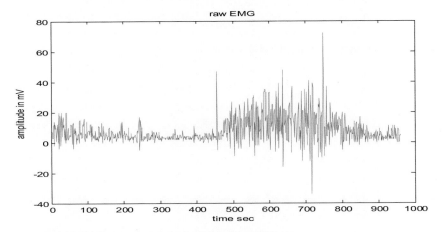

**Fig. 1** Raw EMG of SMM muscle at maintained protocol.

detecting surfaces should be only 1-2 cm. Patient information such as height, weight, body mass index and spirometric data is collected for correlation with existing system. EMG analysis is done in time and frequency domain for linear envelope, peak to peak value, root mean square value, mean rectified voltage, number of turns, number of peaks. Figure 1 shows the raw EMG of the SMM muscle for maintained load at respiration in time and frequency domain.

## 2.1   Linear Envelope

It represents profile of myoelectric activity. Full wave rectifier and low pass filter of 20 Hz is used to get linear envelope. Finite impulse response filter using Gaussian window wit order of 20 is used to design a low pass filter. Signal processing tool box is used for design of filter and obtaining the linear envelope. Refer figure 2 and 3 for rectified output and linear envelope of EMG signal. Results vary with filter used and window length.  Linear envelope rises with muscle activity [13]. This feature provides information about instantaneous activity of muscle, onset and duration and pattern of muscle contraction, inspiration and expiration activity. SMM muscle becomes active while doing inspiration. Low pass filter d.

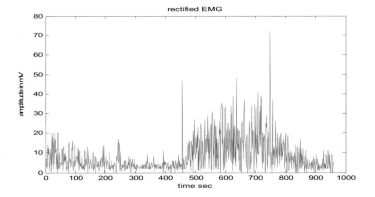

**Fig. 2** Rectified output of EMG of SMM muscle

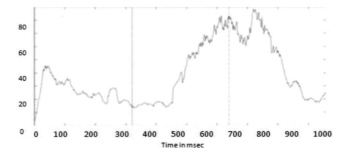

**Fig. 3** Linear envelope of rectified EMG

## 2.2   Root Mean Square

RMS voltage is the effective value of an alternating current, It is similar to linear envelope but with positive time indicator. It gives an instantaneous measure of power output. RMS depends on i) number of motor units firing ii) area of motor units iii) motor unit duration iv) propagation velocity electrode configuration vi) instrumentation control [5],[12]. It represents motor unit behavior during muscle contraction.

$$V_{rms} = \frac{\sqrt{x1^2 + x2^2 + \cdots + xn^2}}{n}$$

## 2.3   Integration

Integration gives total amount of muscle activity during given time interval and it is represented by area under the curve. Integration value depends on amplitude, duration and frequency of action potentials.

## 2.4   Zero Crossings

Number of times EMG crosses the baseline is called zero crossing. As muscle activity increases zero crossing increases for low level of activity. But same is not true for high level of activity [10]. Here zero crossings is calculated for non-rectified signals.

## 2.5   Spike Counting

Numbers of peaks in time and frequency domain are found out for COPD patients. Number of peaks increases with activity of muscle. Positive and negative spikes or peaks are counted, spikes of low amplitude and high amplitude are considered equally. Linear relationship is there among spikes and contraction force up to 70% load [13]. This method is suitable at low force level.

## 2.6   EMG Features in Frequency Domain

Frequency domain representation decomposes the EMG signal into sinusoidal components of different frequency. It gives energy distribution of signal with respect to frequency. Spectrum analysis is insensitive to interference between motor unit contributions and responds to single motor unit. It provides possibility to relate myoelectric signal to physiologic events. It measures recruitment of motor unit but firing rates are not shown. Limitation of this method is frequently repeated motor unit dominate spectrum. Figure 4 gives frequency spectrum of COPD patients. Power line interference is observed at 50 Hz. We have used notch filter of 50 Hz to remove power line interference. We can clearly observe that frequencies are dominant only up to 20 Hz.

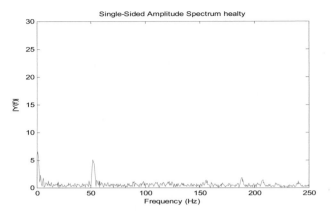

**Fig. 4** Frequency domain representation of EMG of SMM muscle

It detects amplitude of common frequency signal. Area under power spectral curve gives the total power. As force of contraction increases PSD shifts to upper side, it occurs only at low level of tension after 50% of maximum vital capacity there is no increase in frequency value. Spectrum is affected by changes in duration and shape of motor units. Firing rate and amplitude does not affect spectrum shape. We can do analysis in frequency domain using Fast Fourier transform function (FFT). FFT gives distribution of energy with respect to frequency. We have done frequency analysis with features mean frequency, median frequency and number of peaks.

## 3    Results and Conclusions

Features of EMG in time domain and frequency domain are compared with spirometric data FEV1 and FEV1/FVC. Figure 5 shows the plot of peak to peak amplitude verses FEV1 of COPD patients.  Correlation coefficient is used for validating results. It measures the strength and direction of the linear relationship between two variables. Correlation coefficient is calculated by taking ratio of covariance of variable and product of standard deviation.  Correlation coefficient varies between -1 to 1. We can observe that as force expiratory volume (FEV1/FVC) decreases, excess work on respiration increases, results in activity of SMM and increase in peak to peak voltage of EMG, with significant correlation coefficient (r= -0.63212), (refer table 1). Significant relationship has also found between level COPD and number of peaks in frequency domain with (r=-0.53553). Body mass index of the patients gives the information about Force Vital Capacity (FVC) with correlation coefficient of -0.53953.

**Table 1** Features of Surface-EMG and FEV1, FVC and FEV1/FVC.

| Peak to Peak Amplitude(μV) | RMS Voltage | No. of peaks in freq domain | No. of peaks in time domain | BMI | FVC | FEV1 | FEV1/FVC |
|---|---|---|---|---|---|---|---|
| 276 | 30 | 222 | 172 | 17.6 | 1.64 | 0.78 | 47.9 |
| 187 | 19 | 216 | 150 | 15.75 | 1.96 | 1.68 | 85.7 |
| 145 | 12 | 173 | 165 | 26.4 | 1.07 | 0.65 | 60.01 |
| 123 | 11 | 210 | 144 | 17.6 | 1.45 | 1.11 | 76.2 |
| 233 | 23 | 233 | 166 | 16.7 | 1.56 | 0.71 | 45.3 |
| 251 | 27 | 239 | 154 | 22.76 | 1.46 | 0.73 | 50 |
| 168 | 15 | 192 | 159 | 19.2 | 1.34 | 0.78 | 58.21 |
| 205 | 21 | 213 | 167 | 16.8 | 1.04 | 0.54 | 51.92 |
| 192 | 19 | 210 | 173 | 20.6 | 1.25 | 0.88 | 70.4 |
| 147 | 12 | 143 | 149 | 19.6 | 1.46 | 0.93 | 63.7 |

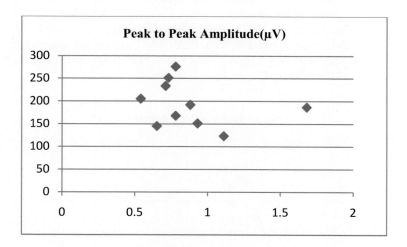

**Fig. 5** Plot of peak to peak amplitude verses of FEV1 of COPD patients.

When root mean square value is compared with FEV1 and Ratio between FEV1/FVC same relationship has observed. Refer figure 6 for comparison of RMS voltage with FEV1.

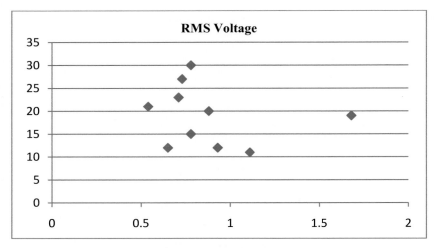

**Fig. 6** Plot of rms voltage verses of FEV1 of COPD patients.

When number of peaks in time and frequency domain is compared with FEV1 parameter, we discover that, relationship is approximately constant for number of peaks in time domain. While number of peaks are increasing in frequency domain for decrease in value of FEV1 parameter with correlation coefficient of (r=-0.63527) and when correlated with FEV1/FVC significant correlation has found with (r=-0.53553). Refer figure 7 and 8 for comparison of number of peaks in frequency and time domain.

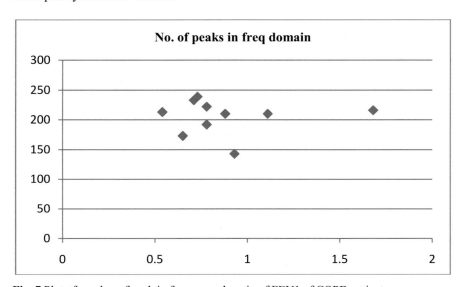

**Fig. 7** Plot of number of peak in frequency domain of FEV1 of COPD patients.

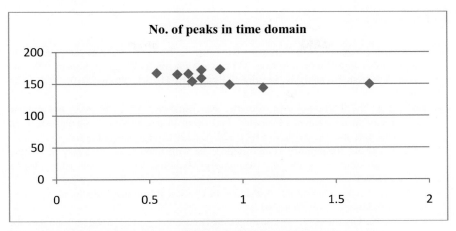

**Fig. 8** Plot of number of peak in time domain of FEV1 of COPD patients.

From the above result we can conclude that, features of EMG such as peak to peak amplitude, RMS voltage, number of peaks in frequency domain shows a significant relationship with severity of COPD (FEV1/FVC). But number of peaks in time domain are approximately constant with respect to severity of COPD(r=-0.23732). We can also conclude that body mass index of patient shows definite correlation with FVC with correlation coefficient of r=-0.53953. Therefore, greater body mass index, possibility of increase in the vital capacity from a maximally forced expiratory effort increases. We conclude that as severity increases, muscle activity of SMM increases.

## 4    Discussion

Features of EMG in time domain and frequency domain are compared among COPD patients with different level of severity (FEV1/FVC). We have observed that, significant correlation with negative correlation factor has found between features of EMG and severity of COPD. Further when compared with number of peaks in time and frequency domain, frequency domain peaks can be used to correlate with severity of COPD. Electromagnetic interference and interference due to muscles near to SMM should be reduced. Large sample of COPD patient needs to be studied to find definite relation between different stages of COPD and activity of sternomastoid muscle.

# References

1. Lozano, R., Naghavi, M., Foreman, K., Lim, S., Shibuya, K., Aboyans, V., et al: Global and Regional mortality from 235 causes of death for 20 age groups in 1990 and 2010: A systematic analysis for global Burdon of Disease Study 2010, 2095–2128 (2012)
2. El-Naggar, T., Mansour, M., Mounir, N., Mukhtar, M.: The role of impulse oscillometry in assessment of airway obstruction in smokers and ex-smokers. J. of Chest Diseases and Tuberculosis 61(4), 323–328 (2012). Egyptia, Elsevier
3. Crim, C., et al.: Respiratory system impedance with impulse oscillometry in healthy and COPD subjects. Trans. on Respiratory Medicine 105(7), 1069–1078 (2011). Science Direct
4. Spiro, S.G., et al: Respiratory mechanics, in Clinical Respiratory Medicine, 4th edn., pp. 19–28. Saunders, Elsevier (2012)
5. Koutsos, E., Georgiou, P.: An analogue instantaneous median frequency tracker for EMG fatigue monitoring. In: IEEE Int. Sym. on Circuits and Systems, pp. 1388–1391 (2014)
6. Geddes, L.A.: Electrodes and the Measurement of Bioelectric Events, p. 364. Wiley, New York (1972)
7. Mañanas, M.A., et al.: Study of Myographic Signals from Sternomastoid Muscle in Patients with Chronic Obstructive Pulmonary Disease. IEEE Trans. Biomed. Eng 47(5), 674–681 (2000)
8. Georgakis, A., et al.: Fatigue Analysis of the Surface EMG Signal in Isometric Constant Force Contraction Using the Averaged Instantaneous Frequency. IEEE Trans. on Biomedical Engineering 50(2), 262–266 (2003)
9. Oskoei, M.A., Hu, H.: Myoelectric control systems- A survey. Science Direct Trans. on Biomedical Signal Processing and Control 2, 275–295 (2007)
10. Raez, M.B.I., Hussain, M.S., Mohd-Yasin, F.: Techniques of EMG signal analysis: detection, processing, classification and applications. Biological Procedures Online, pp. 11–36. Springer (2006)
11. De Luca, C.J.: Physiology and Mathematics of Myoelectric Signals. IEEE Trans. Biomed. Eng. BME-26, 313–325 (1979)
12. Kamen, G.: Electromyographic kinesiology. In: Robertson, D.G.E., et al. Research Methods in Biomechanics, pp. 179–223. Human Kinetics Publ., Champaign (2004)
13. Leveau, B., Andersson, G.B.J.: Output Forms: Data Analysis and Applications, chapter 5, pp. 70–102 (1992)
14. World health organisation for respiratory disease Chronic Obstructive Pulmonary Disease, January 15, 2015. http://www.who.int/Frespiratory/copd/burden/en

# Interconversion of Emotions in Speech Using TD-PSOLA

B. Akanksh, Susmitha Vekkot and Shikha Tripathi

**Abstract** Emotional speech plays a major role in identifying the mental state of the speaker. The presented work demonstrates the significance of incorporating dynamic variations in prosodic parameters in speech for the effective conversion of one emotion to another. Most existing methods focus on the incorporation of emotions in synthetically generated neutral speech. The proposed model discusses the usage of TD-PSOLA as an important tool in converting one emotion to another for speech samples in English language. It analyses the significance of various prosodic parameters in distinct emotions and explores how prosodic modifications to one emotion can result in different perceived emotions. The paper explores a technique for interconversion of emotions in synthesised speech and demonstrates a methodology for prosodic modifications on an English database.

## 1  Introduction

In natural human interaction, a person conveys critical non-linguistic information to the listener along with the message through body reactions, gestures, facial expressions or through speech. This establishes certain factors such as the mental state and the emotional state of the speaker. In general, these aspects comprise the humane element of communication. Considering the communication between a robot and a human being, the interaction is considered effective not only when the robot is able to understand the message and emotional state of the human, but also when the robot is able to convey an appropriate response in the required tone and emotion. This makes conversation more natural and human-like. Emotional Speech Synthesis (ESS) or Expressive Speech Synthesis deals with incorporation or generation of emotional

B. Akanksh · S. Vekkot(✉) · S. Tripathi
Amrita Vishwa Vidyapeetham, Bangalore, India
e-mail: akanksh.b@tamu.edu, susmitha@blr.amrita.edu, shikha.eee@gmail.com

© Springer International Publishing Switzerland 2016
S.M. Thampi et al. (eds.), *Advances in Signal Processing and Intelligent Recognition Systems*,
Advances in Intelligent Systems and Computing 425,
DOI: 10.1007/978-3-319-28658-7_32

content in synthesized speech [1]. In ESS, the speech may either be recorded in a neutral tone or synthesized by the conventional neutral text-to-speech systems and then converted to emotional speech by incorporating the expressive information related to the target emotion. The three stages involved in neutral to expressive speech conversion are analysis, estimation and incorporation of expression-specific speech parameters into neutral speech. In the analysis stage, parameters that significantly affect the emotional content are identified. The values of these parameters are then estimated from a speech database in the estimation stage. Finally, methods to incorporate these expressive parameters are devised and implemented so as to obtain expressive speech. Prosodic parameters such as Pitch, Duration and Intensity greatly affect the emotional content portrayed by a speaker. Thus prosody modification, which is the process of modifying these parameters without affecting the perceptual quality of the speech, plays an important role in the incorporation of emotions. Pitch modification can be done both in time and frequency domains. In time domain, a popular method of pitch modification is the Time Domain Pitch Synchronous Overlap Add (TD-PSOLA) method [11], which allows to scale the pitch and time scale independently. This paper demonstrates the significance of incorporating dynamic variations in prosodic parameters in speech for effective inter-emotion conversion.

The background theory and existing literature in the area of emotion synthesis is described in section 2 while the proposed algorithm and implementation process is detailed in sections 3 and 4. Objective and subjective analysis of results obtained through the experiments are discussed in section 5. Finally, the paper is concluded in section 6 with insights into future progression of the work done.

## 2 Background Theory

Over the past decade, extensive research has been directed towards synthesis of speech rich in emotion. Literature suggests that early developments of expressive speech systems were built on top of the formant speech synthesis systems [2]. Perceptual experiments were conducted to find the relevant acoustic features for each emotion by varying the acoustic parameters and finding optimum values of each of these features for emotional speech synthesis [3]. As a result of these experiments, supra-segmental features like mean pitch and pitch range, speech rate, and voice quality parameters like phonation and vowel precisions, are found to be significant for effectively synthesizing emotions [2]. Tao et al. achieved expressive speech conversion by prosody (pitch and duration) modification of neutral speech for Mandarin language [4]. Typical prosodic features used for expressive analysis are parameters of F0 contour, duration (sentence duration, syllable duration etc.) and intensity [5], [6], [12]. Tao et al. used F0avg, F0topline, F0baseline, syllable duration and intensity as the prosodic parameters. A multi-level prosody and spectrum conversion method as applied to emotional speech synthesis is described in literature [7]. Pitch contour at the beginning and ending in a sentence, along with pitch frequency form significant factors in deciding the emotional content of the sentence [8].

Based on studies on F0 characteristics of expressions, it is found that the features relevant for emotion conversion are F0range (absolute difference between F0min value and the F0max value), F0avg, overall F0 inflections (variations in F0 values) and slope of F0 contour. The effect of prosodic parameters on actor simulated emotional expressions such as neutral, sorrow, angry and fear has been explored further. The prosodic parameters considered for the study are F0med (median of F0 values), F0range and speech rate. From the results of this study, it is concluded that anger expressions show increased F0med and F0range while sorrow depicts reduced F0med and F0range [8].

The role of epochs in Expressive Speech Synthesis (ESS) has been extensively researched and documented. In ESS by explicit control, emotional speech is obtained by modifying synthesized speech (which is essentially neutral) according to prosodic transformation scales derived from an emotional speech database [12].

Most of the work in this field is based on data collected locally and is publicly unavailable. The most common public database is the Berlin emotional speech database [9]. These expressive databases differ by the language, type of expressions considered, type of text materials used, number of speakers etc [2]. Majority of work is concentrated on German and Mandarin languages, with very few databases in English language.

The expression specific F0, duration and intensity parameters can be incorporated by prosody modification algorithms. There are several methods discussed in literature for prosody modification [10],[11]. The approaches like Pitch Synchronous Over Lap Add (PSOLA) operate directly on the speech waveform for prosodic parameter modification. Development of PSOLA allowed both time scale and pitch scale modification by using pitch marks as the pivotal points [11].

Most of the existing work deals with conversion of a synthetically generated neutral speech derived from a text-to-speech synthesiser into expressive speech. This paper proposes and implements a systematic approach for the inter-conversion of emotions in speech taking anger and sadness as test emotions. Also, the proposed algorithm is tested on English database which in turn makes the perception experiment much simpler and meaningful.

# 3   Process Flow of Proposed Work

The proposed approach involves multiple modules and steps as shown in Fig1 to obtain the output speech samples.

The Database or Corpus used plays an important role in the determination of parameters to be considered for incorporation of emotion in speech. This work utilizes the eNTERFACE05 Emotional Database [13], an English database of emotional speech recordings. Pre-processing of the database is required to make the database uniform for better accuracy and results. After database pre-processing, the values of the prosodic parameters of the corpus files are determined. PRAAT software is applied to obtain the prosodic data and expresses it using suitable statistical measures to generate a transform scale. Based on the transform scales, prosodic changes are implemented in the speech files using the Pitch-Synchronous Overlap Add (PSOLA) method. PSOLA enables the modification of prosodic parameters independently i.e. change in one parameter does not affect other parameters significantly. Intensity scaling is also implemented as specified by the transform scales.

## 3.1   PRAAT

PRAAT is an open source software package for the analysis of speech [14]. The software displays the speech waveform and a broadband spectrogram showing the spectral energy of the sound over time. In addition, markings that represent formants, pitch contour and intensity can also be viewed. In this work, the Voice Report feature of PRAAT is used to obtain a summary of all the prosodic and vocal parameters of the given speech file.

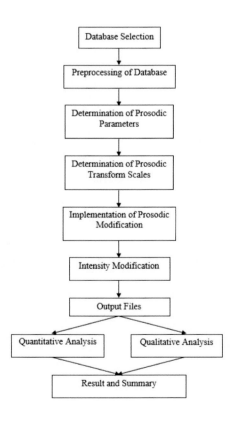

**Fig. 1** Process Flow Block Diagram

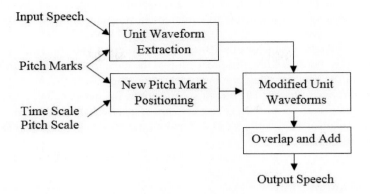

**Fig. 2** Implementation of TD-PSOLA

## 3.2 TD-PSOLA

Pitch and time-scale modifications to the speech waveform are carried out by the time-domain technique called Time Domain Pitch Synchronous Overlap Add (TD-PSOLA) which provides high quality output synthetic speech. The pitch and time scale modification system has three major segments: Pitch Estimation, Pitch Marking and TD-PSOLA as shown in Fig 2 [15]. Pitch estimation is achieved through the use of autocorrelation function (ACF) measurement in time-domain.

To implement TD-PSOLA, pitch marks or epochs corresponding to glottal openings are placed at each pitch cycle. Unit waveforms are then extracted from the speech signal using a Hanning window of length two pitch periods. Based on pitch and time scaling requirements, new synthetic pitch mark positions are determined. When a pitch scale is lowered using the TD-PSOLA, there is perceptual distortion in speech. To overcome this, a spectrum reconstruction method [16] is applied to reconstruct the low-band F0 harmonic components. Unit waveforms are then placed at the target pitch marks and overlap added to produce an output signal with desired pitch and duration.

## 3.3 Corpus Used

The eNTERFACE05 Emotional Database is an audio-visual emotion reference database for testing and evaluating emotion recognition and synthesis/conversion algorithms [13]. The final version of database contains speech recordings of 42 subjects from 14 different nationalities. Among these subjects, 81% were men, while the remaining 19% were women. The database contains speech recordings in 5 emotions- Anger, Sadness, Disgust, Fear and Surprise. This work focusses on anger and sadness as target emotions as these are opposite in nature and provide high perceptual distinguish-ability due to variations in prosodic parameters like pitch, duration and intensity.

The eNTERFACE database consists of video recordings of actors simulating various emotions. For this work, pre-processing of the database involved extraction of audio from the database files as mono track audio clips using the Aiseesoft Total Video Converter software. The output specifications for the sampling rate and the output bit-rate were set as 48KHz and 256kbps respectively. Therefore, the corpus used for implementation consisted of 410 audio speech samples pertaining to Anger and Sadness emotions.

# 4   Implementation

The implementation process is divided into 5 modules to produce the final output files required.

## 4.1   PRAAT Scripting and Extraction of Data from Voice Reports

PRAAT Software is used to generate voice reports of each of the audio files in the corpus to obtain prosodic data such as Mean F0, Median F0, Maximum F0, Minimum F0 and Intensity. Due to a large corpus of 410 speech files, scripting in PRAAT is used to automate the process of generation of voice reports for each file.

## 4.2   Statistical Calculation of Prosodic Parameters of Target Emotion

The data extracted from the voice reports of all speech files pertaining to anger and sadness are expressed statistically so that analysis on the prosodic parameter changes can be conducted for the entire database. The Mean, Median and Standard Deviation of each prosodic parameter provides a statistical summary of the distribution of these features in the entire database. Each of the prosodic parameters obtained are given in Table 1.

## 4.3   Transformation Scales

From Table 1, the difference in the prosodic parameters for anger and sadness can be expressed as a percentage change. For analysis, this parameter is chosen by taking

**Table 1** Statistical values of Prosodic parameters of database used.

| Parameters | Statistical method | Sadness | Anger |
|---|---|---|---|
| Mean F0 (Hz) | Mean | 179.9301 | 202.7764 |
| | Median | 153.9470 | 195.5680 |
| | Std. Deviation | 50.6658 | 50.4701 |
| Median F0 (Hz) | Mean | 163.0710 | 196.8649 |
| | Median | 148.8280 | 192.4690 |
| | Std. Deviation | 53.5325 | 52.7619 |
| Maximum F0(Hz) | Mean | 278.6062 | 358.2432 |
| | Median | 231.3940 | 330.7120 |
| | Std. Deviation | 137.8168 | 129.4954 |
| Minimum F0(Hz) | Mean | 92.2916 | 103.8823 |
| | Median | 89.8652 | 91.9774 |
| | Std. Deviation | 26.4397 | 28.6196 |
| Intensity E (dB) | Mean | 82.12 | 95.26 |
| | Median | 80.14 | 92.96 |
| | Std. Deviation | 9.7 | 10.7 |

**Table 2** Prosodic Transformation Scales for inter-conversion between Anger and Sadness

| Parameters | Sadness to Anger | Anger to Sadness |
|---|---|---|
| Mean F0 (Hz) | +27.04% | -21.28% |
| Median F0 (Hz) | +29.32% | -22.67% |
| Max F0 (Hz) | +42.92% | -30.03% |
| Min F0 (Hz) | +2.35% | -2.30% |
| Intensity E (dB) | +16.01% | -13.79% |

the median value of each prosodic parameter in the database. Median is chosen in place of mean to overcome the changes in mean value due to certain irregularities arising from ineffective portrayal of emotion in few speech samples.

## 4.4 Implementation of TD-PSOLA

The PSOLA algorithm is used to implement the Pitch and Time scaling of the input speech samples block diagram of which is given in Fig 3[15]. Pitch Scaling factors that are provided as input arguments to the PSOLA algorithm are obtained from Table 2 and are normalized between zero and one. Time scaling factors for the inter-conversion are derived from [2]. The resultant files are stored as separate output converted files of same sampling rate and bit rate.

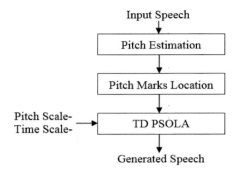

**Fig. 3** Flowchart of TD-PSOLA

## 4.5  *Intensity Scaling*

Intensity scaling of the resultant files are carried out in both time and frequency domain. The time domain intensity of the signal waveform is performed by enhancing or dampening the waveform as per the transformation scale table. The resultant waveform is then passed through an FIR filter that increases the intensity at the frequency corresponding to the Median F0 value as a Gaussian distribution with standard deviation 10 Hz and peak intensity +3dB. The output speech waveforms are stored as .wav files as with implementation of PSOLA.

## 5  Results

From the eNTERFACE Database consisting of 205 Files for sadness and angry emotional speech, a test corpus of 50 files was selected at random for both sadness and angry. The implementation of prosodic features was performed to obtain a resultant database. To determine the new prosodic values of the obtained speech samples, the Praat script designed to obtain voice reports was executed again.

## 5.1  *Quantitative*

The mean, median and standard deviation of each of these parameters was calculated again. Table 3 shows the Prosodic values of the generated files.

Table 4 shows the standard deviation of the prosodic parameters of the generated and original database values. As this deviation is less, the conversion using TD-PSOLA has been successful. The time domain waveforms of one of the speech files before and after prosodic modification from sadness to anger are shown in Figures 4 and 5.

**Table 3** Statistical values of Prosodic parameters of converted speech files.

| Parameters | Statistical Method | Anger to Sadness | Sadness to Anger |
|---|---|---|---|
| Mean F0 (Hz) | Mean | 175.2347 | 208.7462 |
| | Median | 160.1272 | 192.3510 |
| | Std. Deviation | 92.4015 | 89.6581 |
| Median F0 (Hz) | Mean | 158.2492 | 199.0710 |
| | Median | 147.4291 | 189.2720 |
| | Std. Deviation | 94.1976 | 92.2525 |
| Maximum F0 (Hz) | Mean | 247.7313 | 394.2860 |
| | Median | 225.1254 | 367.2313 |
| | Std. Deviation | 209.4234 | 210.3668 |
| Minimum F0 (Hz) | Mean | 93.1273 | 105.9229 |
| | Median | 84.2743 | 102.7295 |
| | Std. Deviation | 57.3469 | 54.4245 |

**Table 4** Standard Deviation of Prosodic Values of original and converted speech files

| Parameters | Anger to Sadness | Sadness to Anger |
|---|---|---|
| Mean F0 (Hz) | 4.38 | 2.28 |
| Median F0 (Hz) | 0.99 | 2.26 |
| Max F0 (Hz) | 4.43 | 25.82 |
| Min F0 (Hz) | 3.95 | 7.60 |

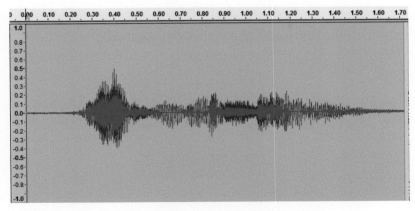

**Fig. 4** Speech waveform corresponding to sadness - before modification

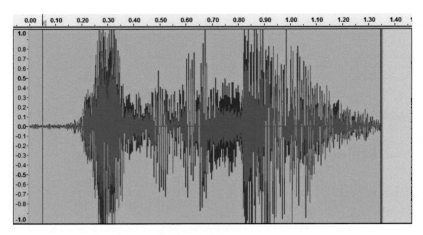

**Fig. 5** Speech waveform after modification - converted to anger

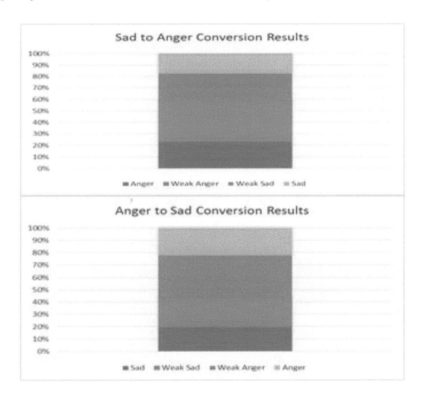

**Fig. 6** Results of (a) Sadness to Anger Conversion (b) Anger to Sadness Conversion.

**Table 5** Accuracy Tabulation of Original Emotion Perception

| Emotion in Speech Samples | Accuracy |
|---|---|
| Original Angry | 89.31% |
| Original Sadness | 86.37% |

## 5.2 Perception Test

The typical way of evaluating the quality of emotional speech synthesised is through a forced choice perception test including the emotion categories modelled in the process. Since speech and the emotion it portrays are subjective in nature, a perception test shows the effectiveness of the conversion of emotions. The perception test involves a focus group to whom a random order of input speech samples and output speech samples derived from this work has been played back and they were to identify the perceived emotion from each audio sample. Perception test focus group consisted of 15 members of current students, alumni and professors with mean and median age of 31.8 years and 22 years respectively. The group consisted of 10 males and 5 females. 40 speech samples consisting of 20 original files and 20 output files were played and they were given choices of strong and weak for each emotion.

Figure 6 shows the depiction of the perception test for converted speech samples. From Table 5, we see that the perception test showed a high accuracy in differentiating anger and sadness. After conversion, the proportion of perception test results that indicate a change in emotion portrayed by the speech is significantly high. However, the proportion of results that imply otherwise can depend on factors such as the linguistic content of speech affecting the objectivity of listeners.

The objective and subjective evaluations show that the extent to which the actual emotion is correctly perceived in each case is much higher. Also, the technique uses an English database and demonstrates that inter-conversion between different emotions can be carried out successfully thereby eliminating the need for a synthesised neutral speech at the source.

## 6 Conclusion

The paper proposes an effective approach for inter-conversion of emotions in speech. From the results, it is seen that prosodic modifications to speech signals using TD-PSOLA is highly effective in this aspect. Accuracy of prosodic changes depends on the corpus used and algorithm used to define pitch marks accurately. Perception test showed a significant accuracy in converted speech samples which was again reflected in the quantitative analysis. Further work can be carried out by using more accurate emotional databases as well as considering prosodic parameters as dynamic.

The technique discussed can be used for inter-conversion of variety of emotions whose results need to be evaluated with existing neutral speech based expressive systems. Vocal tract resonant frequencies, jitter, shimmer are some other parameters that may increase the perceptual quality of converted emotions.

# References

1. Schroder, M.: Emotional speech synthesis- a review. In: Proc. Eurospeech, pp. 561–564 (2001)
2. Burkhardt, F., Sendilmeier, W.F.: Verification of acoustical correlates of emotional speech using formant synthesis. In: Proc. ISCA Workshop on Speech & Emotion, pp. 151–156 (2000)
3. Schroder, M.: Expressive speech synthesis: past, present and possible futures. In: Affective Information Processing, vol. 2, pp. 111–126. Springer (2009)
4. Tao, J., Kang, Y., Li, A.: Prosody conversion from neutral speech to emotional speech. IEEE Trans. Audio, Speech, and Language Processing **14**, 1145–1154 (2006)
5. Bulut, M., Narayanan, S.: On the robustness of overall f0 only modifications to the perception of emotions in speech. J. Acoust. Soc. Am. **123**, 4547–4558 (2008)
6. Fairbanks, G., Hoaglin, L.W.: An experimental study of pitch characteristics of voice during the expression of emotion. Speech Monographs **6**, 87–104 (1939)
7. Wang, Z., Yu, T.: Multi-level prosody and spectrum conversion for emotional speech synthesis. In: ICSP 2014 Proceedings (2014)
8. Anil, M.C., Shirbahadurkar, S.D.: Speech modification for prosody conversion in expressive marathi text-to-speech synthesis. In: Proceedings of IEEE International Conference on Signal Processing and Integrated Networks (SPIN), pp. 56–58 (2014)
9. Burkhardt, F., Paeschke, A., Weiss, B.: A database of German Emotional Speech. In: Proceedings of Interspeech, Lisbon, Portugal (2005)
10. Rao, K.S., Yegnanarayana, B.: Prosody modification using instants of significant excitation. IEEE Trans. Audio, Speech and Language Processing **14**, 972–980 (2006)
11. Mourlines, E., Laroche, J.: Non-parametric techniques for pitch-scale and time-scale modification of speech. Speech Commun. **16**, 175–205 (1995)
12. Govind, D., Prasanna, S.R.M.: Expressive Speech Synthesis: A Review. International Journal of Speech Technology **16**, 237–260 (2013)
13. Martin, O., Kotsia, I., Macq, B., Pitas, I.: The eNTERFACE 2005 Audio-Visual Emotion Database (2005)
14. Praat: Doing Phonetics by Computer. (n.d.). (retrieved from). www.praat.org. http://www.fon.hum.uva.nl/praat/
15. Dung, N.T., Hai, N.D.: Time-Domain Pitch and Time Scale Modification of Speech Signal, School of Computing, National University of Singapore (2008)
16. Mochizuki, R., Kobayashi, T.: A low-band spectrum envelope modeling for high quality pitch modification. In: Proceedings of International Conference on Acoustics, Speech, and Signal Processing (2004)

# Part IV
# Pattern Recognition/Machine
# Learning/Knowledge-Based Systems

# A Technique of Analog Circuits Testing and Diagnosis Based on Neuromorphic Classifier

Sergey Mosin

**Abstract** The technique of functional testing the analog integrated circuits based on neuromorphic classifier (NC) has been proposed. The structure of NC providing detection both catastrophic and parametric faults taking into account the tolerance on parameters of internal components has been described. The NC ensures the associative fault detection reducing a time on diagnosis in comparison with parametric tables. The approach to selection of essential characteristics used for the NC training has been represented. The wavelet transform of transient responses, Monte Carlo method and statistical processing are used for the essential characteristics selection with maximum distance between faulty and fault-free conditions. The experimental results for the active filter demonstrating high fault coverage and low likelihood of alpha and beta errors at diagnosis have been shown.

## 1 Introduction

The testing plays essential role at design and manufacturing the electronic circuits. The actions on the integrated circuits test take about 40-60 % of total time and 40-70 % of total costs required for development. The testing and diagnosis of analog circuits are more complex and expensive in contrast to digital ones. The main reasons of such complexity are dealt with the component tolerances, nonlinearities and poor fault models, etc. As rule, the component tolerances make the parameters of circuit elements uncertain; together with nonlinearities this provides the complexity of computational equations. There are two main types of faults for analog circuits – catastrophic and parametric. The set of catastrophic faults determined by the opens and shorts is finite, and set of parametric faults is infinite.

S. Mosin(✉)

Vladimir State University (VSU), Gorky Str., 87, Vladimir 600000, Russia
e-mail: smosin@ieee.org

© Springer International Publishing Switzerland 2016
S.M. Thampi et al. (eds.), *Advances in Signal Processing and Intelligent Recognition Systems*,
Advances in Intelligent Systems and Computing 425,
DOI: 10.1007/978-3-319-28658-7_33

Thus, the functional testing and diagnosis are widely used for analog circuits. The functional approach consists in the homologation of output characteristics to specification. The comparison analysis of circuit under test responses with golden responses of fault-free circuit is performed during testing, and with responses of the circuit with considered faults is performed during diagnosis for all test signals.

The methods based on fault dictionary have been proposed for functional testing and diagnosis of analog circuits [1, 2]. The fault dictionary here is considered as parametrical table each row of which corresponds to fault-free and faulty circuit conditions, the columns represent the output responses, each cell contains the value or range values of the controlled output characteristic. Such methods provide testing and diagnosis only that faults the output responses for which have been estimated and included into the fault dictionary. The main problems of methods based on a fault dictionary are the following: 1) high dimension of parametrical table; 2) the limited number of considered faults and faulty conditions included into the table. The first problem requires a huge memory for the table storage and leads to high time cost for the consecutive comparison of circuit under test (CUT) response with responses at all rows. The second one does not ensure diagnosis of faults which have not been included in the table.

The alternative methods based on artificial neural networks (ANN) allow eliminate many disadvantages of methods using a fault dictionary [3-11]. An ANN with fixed structure can be used for representing the different quantity of considered fault-free and faulty conditions, thus only the limited small number of coefficients describing synaptic links should be stored. The ANN provides essential time reduction on the responses comparison due to associative operation, i.e. the indicator of test passing is generated immediately at the output layer after application of CUT's output response to neurons in input layer.

Due to the component tolerances and nonlinearities, the output responses of fault-free and faulty analog circuits can be similar and provide wrong test and diagnosis result. This paper proposes the technique of analog circuits testing and diagnosis using NC where the preprocessing of transient responses is stipulated for extraction of essential characteristics. The selection of essential characteristics with maximum distance between faulty and fault-free conditions provides increasing the distinction and consequently improves accuracy of testing and diagnosis. The neuromorphic classifier trained by patterns with such essential characteristics ensures functional testing and diagnosis of linear and nonlinear analog circuits.

The paper is organized as following. Section 2 describes the conceptual overview of proposed technique. Structure of neuromorphic classifier is given in Section 3. The approach to extraction the essential characteristics based on wavelet transform, Monte Carlo method and statistical processing is represented in Section 4. The procedure of construction and training the NC is proposed in Section 5. The experimental results for the Sallen-Key bandpass active filter are given in Section 6.

## 2    Decomposition of a Technique

The proposed technique of functional testing the analog integrated circuits based on neuromorphic classifier (NC) consists of several procedures (Fig. 1).

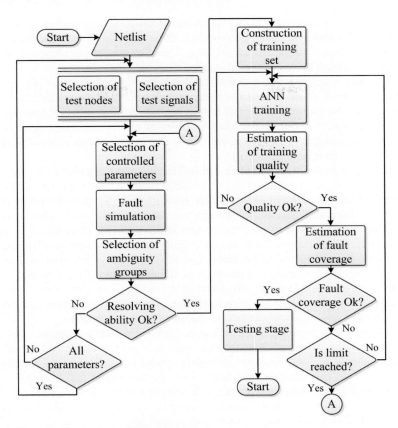

**Fig. 1** Conceptual overview of the technique

The circuit under test (CUT) is represented as Spice-like *netlist*, which contains the list of used components, the nodes of connection, nominal values of internal parameters and tolerances. The results of a CUT simulation in frequency and time domains are used for the procedures execution.

*Selection of test nodes* and *test signals* can be implemented in parallel. The target function in both cases is the choice of minimum set of internal nodes and possible input stimuli, which provides detection the maximum number of considered faults in the circuit. The sensitivity analysis in frequency domain or testability analysis is used as math background for these two procedures.

The *selection of controlled parameters* ensures the consideration for testing purpose only minimum subset of possible physical characteristics measured in test nodes. Usually the circuit behavior is estimated by the voltage values in accessible test nodes. Since the voltage signal in each output node is a time-dependent function, there is necessity to extract its essential characteristics for compact representation of the circuit behavior. There are different strategies for this purpose, which are based on methods of simple average or/and geometric mean calculation. In this case the information about dynamic characteristics of investigated circuit is lost as rule. Therefore for circuits, behavior dynamics of which is very important parameter (for example non-linear analogue circuits), other approaches of extracting essential characteristics of output signals are used. These approaches are based on signal transformation from the time domain into the frequency domain.

The Fourier transformation and wavelet decomposition are the most perspective approaches, which allow to save the dynamic behavior of analogue circuit during extracting essential characteristics of output signals. At that, wavelet transform is more suitable than Fourier transformation for cases of non-stationary signals. According to wavelet theory any signal with finite energy can be represented by linear combination of wavelet-signals [12].

Procedure of *fault simulation* allows to collect the output responses for fault-free and faulty circuit for determined list of possible faults. These responses show the influence of faults on circuit serviceability and difference in controlled output parameters. In this case the catastrophic faults are simulated by casual inclusion in the circuit the resistor instead of used component with small (for short effect) or large (for open effect) resistance. The parametric faults are simulated by casual inclusion the compensating elements in the analog circuit. The Monte Carlo method ensures taking into account component tolerances at simulation both fault-free and faulty conditions of the analog circuit. The Monte Carlo method provides calculation the most realistic region of possible responses using statistical distribution function of the component parameters which as rule described by the normal law.

All faults which have the same influence on CUT's output characteristics and ensure the similar values of controlled parameters at all test nodes and for all applied test signals are combined into ambiguity group. The *selection of the ambiguity groups* allows revealing all such faults from the considered list. Meanwhile the diagnostics of CUT is become possible with accuracy up to ambiguous groups. The iterative selection of additional controlled parameters, test nodes and/or test signals can be required if such resolving ability does not satisfy.

The adaptation and minimization of data set obtained during fault simulation and providing required resolving ability are performed at *construction of training set*. The main aim of this procedure is related with increasing the quality of ANN training due to using essential characteristics and reduction the total data set. *Training of ANN* is performed until the mean-square error becomes less a specified limit or the quantity of training epochs exceeds the limited number. *Estimation of fault coverage* is realized if training quality satisfies; otherwise the retraining of ANN is initiated. Calculated fault coverage shows quantitative efficiency of trained neural network in testing and diagnostics of CUT. Obtained ANN can be used as neuromorphic classifier in *testing stage* if the fault coverage is acceptable; otherwise the retraining of ANN is needed or even changing controlled parameters and training set can be required before retraining.

## 3    The Structure of Neuromorphic Classifier

According to Kolmogorov's theorem the multilayer neural networks can be used for function approximation. Any continuous function of $n$ variables at unit segment

$f : [0,1]^n \to [0,1]$ can be approximated using three-layer neural network with $n$ input neurons, $2n+1$ hidden neurons and one output neuron.

Geometrically it means that neural network splits the space of possible solutions onto the set of hyperplanes. The number of hidden layers defines the features of splitting the input space on subspaces of small dimension. For instance, two-layer neural network with one layer of nonlinear neurons splits input space on classes by hyperplane. The three-layer neural network with neurons in two last layers described by nonlinear transfer functions allows generating any convex domains in the space of solutions. So, the neuromorphic classifier based on three-layer neural network is proposed for testing and diagnosis of analog circuits (Fig. 2).

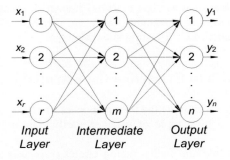

**Fig. 2** The structure of neuromorphic classifier

The input layer executes the distribution functions. The output layer of neurons provides generating and delivery the result. Intermediate layer ensures data processing. The neurons' output from a previous layer is connected to inputs of all neurons in next layer by the synaptic links. So, the topology of neuromorphic classifier is homogeneous and regular.

The hyperbolic tangent (tan-sigmoid) is used as transfer function for neurons in input and output layers and logistic sigmoid (log-sigmoid) is used as transfer function for intermediate neurons.

The proposed architecture of neuromorphic classifier with used transfer functions allows generating the convex domains in the space of solutions even with small numbers of neurons in the layers.

The number of neurons in the input layer is equal to the quantity of controlled parameters or essential characteristics for CUT output responses. The number of neurons in the output layer is selected as $n$ for reflection all condition from fault dictionary. Generally, the number of neurons in the intermediate layer is estimated by (1) and (2)

$$N = N_w / (N_x + N_y),\qquad(1)$$

$$\frac{N_y N_p}{1 + \log_2(N_p)} \leq N_w \leq N_y \left( \frac{N_p}{N_x} + 1 \right)(N_x + N_y + 1) + N_y,\qquad(2)$$

where $N$ is the number of neurons in the intermediate layer, $N_w$ is the number of synaptic links in neuromorphic classifier, $N_y$ is the length of output indicator, $N_p$ is the cardinality of training set, $N_x$ is the length of input pattern.

The training of the neuromorphic classifier for faults diagnosis is carried out by $p$ patterns reflecting the dependences input-output. Each training pattern is represented by pair $(x_i, y_i)$, where $x_i$ is $i$-th element of input vector, and associated vector $y_i$ is defined as (3)

$$y_i(k)=\begin{cases} 0, \text{if component } k \text{ is fault-free at } i\text{-th measurement;} \\ 1, \text{if component } k \text{ is faulty at } i\text{-th measurement;} \end{cases} \tag{3}$$

The problem of large dataset has essential impact on neuromorphic classifier overfitting and convergence at training stage, and consequently on increasing the training time and quality of CUT testing and diagnosis.

## 4    An Approach to Extraction the Essential Characteristics

According to proposed technique the time responses of analog circuits for each test signal are represented by a tuple of wavelet coefficients using discrete wavelet transform.

Two sets of fault-free and faulty analog circuit responses are required for the neuromorphic classifier training. The set of training patterns is generated based on simulation of fault-free and faulty analog circuit using Monte Carlo method and performing wavelet transform of output responses in the test nodes.

Multiple iterative calculations of output responses at random deviation of the component parameters at each step provides generation of the data sample $\mathbf{Y}$ ( $y_k \in \mathbf{Y}$ , where $k$ is the number of Monte Carlo iteration). The responses from $\mathbf{Y}$ reflect the most realistic behavior of the controlled output parameters at the test nodes especially when $k$ is a large value. Wavelet transform of all responses from the sample $\mathbf{Y}$ leads to generation the matrix $\mathbf{X} \subseteq \Re(s, r)$. Each row of $\mathbf{X}$ corresponds to $s$-th measurement. The elements of the row represent the $r$ coefficients of discrete wavelet transform of corresponding response. These $r$ coefficients can be used as input vector $x$ of training pattern for NC training. However, the maximum distinction between coefficients of fault-free and faulty conditions of analog circuit should be provided in order to ensure efficient classification. Moreover, an exclusion of ambiguous inputs from the structure of neuromorphic classifier improves the problem of possible overfitting and convergence.

The statistical characteristics such as mean and standard deviation are calculated for each coefficient in the rows of matrix $\mathbf{X}$ separately for fault-free and each faulty condition.

The criterion of the coefficient inclusion into final set of coefficient for testing purpose is the maximum of normalized Euclidean metric which is calculated for each coefficient as (4)

$$D_i = \sqrt{\frac{1}{q}\sum_{k=1}^{q}\left(\frac{m_i^{f_k} - m_i^{ff}}{m_i^{ff}}\right)^2}\,,$$ (4)

where $D_i$ is normalized Euclidean metric for $i$-th coefficient; $q$ is the quantity of fault-free and faulty conditions in the fault dictionary; $m_i^{f_k}$ is the mean of $i$-th coefficient sample for fault $f_k$; $m_i^{ff}$ is the mean of $i$-th coefficient sample for fault-free condition.

So, the final set of coefficients includes only such coefficients which provide essential distinction between fault-free and faulty conditions, i.e.

$$Coef = \{i\}\,,\ \forall i = 1..r\,,\ \exists D_i \geq min\_rate\,,$$ (5)

where $min\_rate$ is the minimal rate of difference.

The selection of essential wavelet coefficients for solving both tasks testing and diagnostics is performed based on 3D distance matrix (**DM**). Elements of distance matrix are calculated using Euclidean metric similarly to (4) but for all pairs of considered conditions (fault-free and faulty) [13].

The distance matrix is used for selection such wavelet coefficients that provide maximum distances between any pairs of circuit conditions and consequently increasing the distinction for all conditions. The numbers of maximum coefficients obtained according to (6) are included into subset of coefficients which will be used for training and test pattern generation. The cardinality of the *Coef* subset is less or equal to total number of coefficients obtained after wavelet transform of transient response.

$$Coef = \bigcup_{x=2}^{q}\bigcup_{y=x+1}^{q}\arg\max_{z}\left(\mathbf{DM}_{x,y,z}\right),$$ (6)

$$Coef \in \{1,2,...,r\}\,,\ \|Coef\| \leq r\,.$$

So, only coefficients with values from *Coef* subset are used from each row of matrix **X** for training the ANN. The same subset of coefficients is used also for testing the CUT.

The correctness of neuromorphic classifier training is estimated by the ability to distinguish between input patterns corresponding to fault-free and faulty conditions.

The correctness of training for the fault-free behavior (*FF_Corr*) is calculated using the trained classifier and the training pattern as input vectors as ratio of the number of correctly detected fault-free conditions to the total number of fault-free conditions in the training pattern (7)

$$FF\_Corr = \frac{N\_CorrFF}{\|\mathbf{TS_{FF}}\|}100\%\,,\ \mathbf{TS_{FF}} \subset \mathbf{TS}\,,$$ (7)

where **TS** is the set of training patterns; $\mathbf{TS_{FF}}$ is the subset of training patterns corresponding to fault-free condition; $N\_CorrFF$ is the number of correctly detected the *fault-free* condition.

The correctness of training for the faulty behavior ($F\_Corr$) is calculated using the trained classifier and the training pattern as input vectors as ratio of the number of correctly detected faulty conditions to the total number of faulty conditions in the training pattern (8)

$$F\_Corr = \frac{CorrF}{\|\mathbf{TS_F}\|}100\%, \; \mathbf{TS_F} \subset \mathbf{TS}, \; \mathbf{TS_F} \cup \mathbf{TS_{FF}} = \mathbf{TS} \tag{8}$$

where $\mathbf{TS_F}$ is the subset of training patterns corresponding to faulty conditions; $N\_CorrF$ is the number of correctly detected the faulty conditions.

## 5 The Procedure of Construction and Training the Neuromorphic Classifier

The test process using neuromorphic classifier can be presented by two main stages: training stage and testing stage (Fig. 3.). The training stage is realized during circuit design. Selection of neural network structure and its training are performed for an analog circuit during this stage. The following algorithm describes the process of neural network training.

1.  Selection of test frequencies and test nodes.
2.  $i = 0$.
3.  Simulation of fault-free circuit behavior by Monte Carlo method. $i = i + 1$.
4.  Performing discrete wavelet transform of transient output responses in the test nodes.
5.  Construction of matrix **X** and associated vector $y$. The coefficients of wavelet transform obtained on previous step are stored in $i$-th row of **X**. Vector $y_i$ is set up to the code of fault-free condition.
6.  If $i \leq N_{ff}$ then go to step 3, otherwise go to step 7. $N_{ff}$ is the limiting number of Monte Carlo iterations for fault-free circuit.
7.  Simulation of faulty circuit behavior by Monte Carlo method using casual inclusion of faults (parametric, catastrophic, single and multiple). $i = i + 1$.
8.  Performing discrete wavelet transform of faulty circuit's transient output responses in the test nodes.
9.  Construction of matrix **X** and associated vector $y$. The coefficients of wavelet transform obtained on previous step are stored in $i$-th row of **X**. Vector $y_i$ is set up to the code corresponding to the fault considered on this step.
10. If $i \leq N_{ff} + N_f$ then go to step 7, otherwise go to step 11. $N_f$ is the limiting number of Monte Carlo iterations for faulty circuit.
11. Repeat steps from 3 to 10 for each test frequencies.

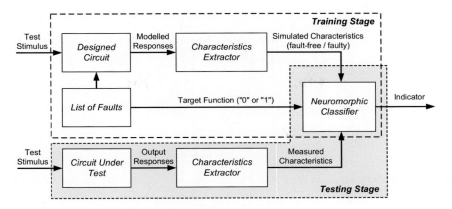

**Fig. 3** Scenario of functional testing and diagnosis

12. Calculate the mean for each circuit condition and column of matrix **X**. Estimate the Euclidean metrics using (4). Construction of distance matrix **DM**. Select and include in set *Coef* only such coefficients which satisfied to (6).
13. Select the number of neurons for input and output layers according to the cardinality of set *Coef* and the length of indication code respectively. Select the number of neurons in the intermediate layer using (1) and (2).
14. Training of neuromorphic classifier by $N_{ff} + N_f$ input vectors using only those columns of matrix **X**, which were included in set *Coef*.
15. Estimation the correctness of detection fault-free and faulty conditions. If the obtained correctness is reasonable then go to step 14 and retraining the classifier, otherwise the neuromorphic classifier has been successfully trained and ready to be used.

The neuromorphic classifier obtained in result is used further for testing the circuit. The circuit under test investigation is performed on the testing stage. Here the input pattern is generated on the base of wavelet transform of CUT measured output responses in test nodes. The indicator generated in the output layer of the neuromorphic classifier is the reaction on the applied input pattern which reflects the status of classification (fault-free or code of fault).

## 6    Experimental Results

The bandpass Sallen-Key filter (Fig. 4) has been used for practical consideration of the proposed technique. The gain and phase frequency characteristics are shown in Fig. 5.

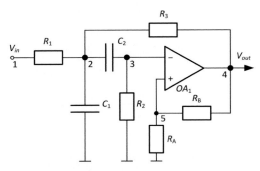

**Fig. 4** Sallen-Key bandpass filter $R_1$ = 10k, $R_2$ = 20k, $R_3$ = 10k, $R_4$ = 5k, $R_B$=10k, $C_1$ = 220n, $C_2$ = 220n

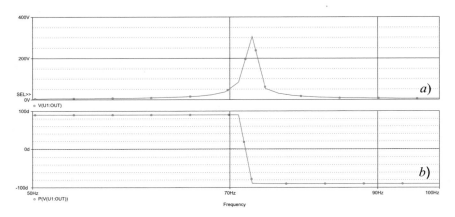

**Fig. 5** Frequency characteristics of the Sallen-Key bandpass filter: gain (*a*) and phase (*b*)

The sine wave with amplitude 1 Volt and frequency 72 Hz is used as the input test signal. The output time response is measured in the test node 4.

The fourth order Daubechies wavelet (db4) is used as mother wavelet. The duration of output response which is delivered to wavelet transform is equal to a period of the input test signal 14 ms. Each response is represented in result by 74 wavelet coefficients.

One fault-free and 28 faulty conditions have been included into the fault dictionary. The faulty conditions are comprised of 14 catastrophic faults represented by short and open effects for each circuit component and 14 parametric faults represented by deviations + 50 % и – 50 % from nominal of each circuit component. The number of iterations for Monte Carlo simulation consists of 5 000. The tolerances of the circuit components are 5 % for resistors and 10% for capacitors. 22 ambiguity groups were selected in result with cardinality from 1 up to 3.

The training set consists of 1 550 test patterns, 500 of which correspond to fault-free condition and 50 to other ambiguity groups.

The three-layer neural network with 30 neurons in the input layer, 50 neurons in the intermediate layer and 29 neurons in the output layer has been selected as NC.

The following characteristics have been obtained in result of training the NC: Training time = 12573.1 s; Mean-square error = 0.0002; FF_Corr = 100 %; F_Corr = 100 %. The simulation has been done on SMP system with four-core processor Intel® Core™ i7 920 @ 2.67 GHz 2.67 GHz, 64 RAM 24 Gb.

The performance estimation of the trained neuromorphic classifier has been fulfilled by 160 000 test patterns where 20 000 correspond to fault-free circuit, 70 000 to catastrophic faults and 70 000 to parametric faults. These test patterns are independent of patterns from the training set. The results of testing and diagnosis of the Sallen-Key filter using trained classifier are represented in the Table 1.

**Table 1** The Efficiency of Using the Trained Neuromorphic Classifier for the Filter Testing and Diagnosis

| Correctness of detection the fault-free condition, % | | | | | 99.03 |
|---|---|---|---|---|---|
| Correctness of detection the faulty conditions, % | | | | | |
| Fault | Testing | Diagnosis | Fault | Testing | Diagnosis |
| $R_1$_short* | 0 | 0 | $R_1$+50% | 89.20 | 80.80 |
| $R_1$_open | 100 | 98.06 | $R_1$−50% | 100 | 99.08 |
| $R_2$_short | 100 | 99.99 | $R_2$+50% | 99.94 | 88.92 |
| $R_2$_open | 100 | 100 | $R_2$−50% | 100 | 87.42 |
| $R_3$_short | 100 | 100 | $R_3$+50% | 95.38 | 77.38 |
| $R_3$_open | 100 | 98.52 | $R_3$−50% | 100 | 97.80 |
| $R_A$_short | 100 | 100 | $R_A$+50% | 99.56 | 85.60 |
| $R_A$_open | 100 | 83.14 | $R_A$−50% | 99.98 | 87.62 |
| $R_B$_short | 100 | 86.22 | $R_B$+50% | 100 | 88.64 |
| $R_B$_open | 100 | 100 | $R_B$−50% | 100 | 97.54 |
| $C_1$_short | 100 | 100 | $C_1$+50% | 99.34 | 98.72 |
| $C_1$_open | 100 | 91.12 | $C_1$−50% | 100 | 96.88 |
| $C_2$_short | 100 | 99.98 | $C_2$+50% | 99.34 | 99.12 |
| $C_2$_open | 100 | 100 | $C_2$−50% | 100 | 97.75 |

* Fault $R_1$_short is included into the same ambiguity group with fault-free condition.

The NC generated according to proposed technique has provided the correct detection of fault-free condition at 99.03%, detection at testing of catastrophic faults – 92.86 % (including fault $R_1$_short) and parametric faults – 98.76%. Diagnosis of catastrophic faults consists of 89.78%, and parametrical faults – 91.66%. The results for Sallen-Key bandpass active filter demonstrate the high fault coverage and low likelihood of alpha and beta errors at diagnosis. The training time of neuromorphic classifier using only 30 selected essential wavelet coefficients in 46 times less in comparison with a case of consideration all 74 wavelet coefficients.

# 7 Conclusion

The neural network technology has become the research hotspot in the field of analog circuit fault test and diagnosis. The technique of analog circuits testing and diagnosis based on neuromorphic classifier has been proposed. The NC provides fault classification in associative mode in contrast to methods based on parametric tables where fault detection is realized by consecutive comparing with each row.

The paper demonstrates what only coefficients providing the distinction between fault-free and faulty conditions are reasonable to be used for NC training and subsequent analog circuits testing. The experimental results show high efficiency of faults testing and diagnosis using NC. The fault coverage at testing is about 100 % (excluding the masking fault $R_1$_short). The diagnosis of parametric faults provides more 91 % and catastrophic about 90 %. These results are not inferior to results obtained by parametric tables, but ensure the reduction of time on testing and diagnosis. The binary code of the output indicator allows combine this technique with methods of digital circuits testing and diagnosis for the mixed-signal applications.

The ongoing work is oriented to adapting the proposed technique to application for mixed-signal integrated circuits.

# References

1. Slamani, M., Kaminska, B.: Analog circuit fault diagnosis based on sensitivity computation and functional testing. IEEE Design & Test of Computers **9**(1), 30–39 (1992). doi:10.1109/54.124515
2. Spina, R., Upadhyaya, S.: Linear circuit fault diagnosis using neuromorphic analyzers. IEEE Trans. Circuits Syst. II **44**(3), 188–196 (1997). doi:10.1109/82.558453
3. Cherubal, S., Chatterjee, A.: Parametric fault diagnosis for analog systems using functional mapping. In: Proc. of the DATE Conference (1999). doi:10.1145/307418.307489
4. Aminian, M., Aminian, F.: Neural-network based analog-circuit fault diagnosis using wavelet transform as preprocessor. IEEE Trans. Circuits Syst. II **47**(2), 151–156 (2000). doi:10.1109/82.823545
5. Barua, A., Kabisatpathy, P., Sinha, S.: A method to diagnose faults in analog integrated circuits using artificial neural networks with pseudorandom noise as stimulus. In: Proc. of 10th IEEE Int. Conf. on Electronics Circuits and Syst. (2003). doi:10.1109/ICECS.2003.1302050
6. Mosin, S.: Neural network-based technique for detecting catastrophic and parametric faults in analog circuits. In: Proc. of IEEE 18th Int. Conf. on Syst. Eng. (2005). doi:10.1109/ICSENG.2005.58
7. Stopjakova, V., Malosek, P., Matej, M., Nagy, V., Margala, M.: Defect detection in analog and mixed circuits by neural networks using wavelet analysis. IEEE Trans. on Reliability **54**(3), 441–448 (2005). doi:10.1109/TR.2005.853041
8. Wang, L., Liu, Y., Li, X., Guan, J., Song, Q.: Analog circuit fault diagnosis based on distributed neural network. J. of Comp. **5**(11), 1747–1754 (2010). doi:10.4304/jcp.5.11.1747-1754
9. Yuan, L., Yigang, H., Huang, J., Sun, Y.: A new neural-network-based fault diagnosis approach for analog circuits by using kurtosis and entropy as a preprocessor. IEEE Trans. on Instr. and Measurement **59**(3), 586–595 (2010). doi:10.1109/TIM.2009.2025068

10. Li, X., Zhang, Y., Wang, S., Zhai, G.: A method for analog circuits fault diagnosis by neural network and virtual instruments. In: Proc. of 3rd Int. Workshop on Intelligent Syst. and App. (2011). doi:10.1109/ISA.2011.5873270
11. Zhou, S.G., Li, G.J., Luo, Z.F., Zheng, Y.: Analog circuit fault diagnosis based on LVQ neural network. J. App. Mechanics and Materials **380–384**, 828–832 (2013). doi:10.4028/www.scientific.net/amm.380-384.828
12. Chui, C.K.: An introduction to wavelets. Academic Press (1992)
13. Mosin, S.: An approach to construction the neuromorphic classifier for analog fault testing and diagnosis. In: Proc. of 4th Mediterranean Conf. on Embedded Computing, pp. 258–261 (2015). doi:10.1109/MECO.2015.7181917

# A Genetic PSO Algorithm with QoS-Aware Cluster Cloud Service Composition

Mohammed Nisar Faruk, G. Lakshmi Vara Prasad and Govind Divya

**Abstract** The QoS-aware cloud service composition is a significantly crucial concern in dynamic cloud environment. There is multi-nature services are clustered together and integrated with multiple domains over the internet. Because of increasing number private and public cloud sources and predominantly all cloud services offers similar services. However this differs in their functionalities depend on the QoS constraints. This drags more complexity in choosing a clustered cloud services with optimal QoS concert, an enhanced Genetic Particle Swarm Optimization (GPSO) Algorithm is anticipated to crack this crisis. With the intention to construct the QoS-aware cloud composition algorithm, all the parameters to be redefined such as price, position, response time and reputation. The Adaptive Non-Uniform Mutation (ANUM) approach is proposed to attain the best particle globally to boost the population assortment on the motivation of conquering the prematurity level of GPSO algorithm. This strategy also matched with other similar techniques to acquire the convergence intensity. The efficiency of the anticipated algorithm for QoS-aware cloud service composition is exemplified and evaluated with a Modified Genetic Algorithm (MGA), GN_S_Net, and PSOA the outcomes of investigational assessment signifies that our model extensively achieves than the existing approaches by means of execution time with improved QoS performance parameters.

**Keywords** Cloud composition · QoS-aware · GPSO · ANUM · Cluster services

M.N. Faruk(✉)
Department of CSE, QIS College of Engineering and Technology, Ongole, India
e-mail: justdoit.fff@gmail.com

G.L.V. Prasad
Department of CSE, Bharath University, Chennai, India

G. Divya
Dhanalakshmi College of Engineering, Chennai, India

© Springer International Publishing Switzerland 2016
S.M. Thampi et al. (eds.), *Advances in Signal Processing and Intelligent Recognition Systems*,
Advances in Intelligent Systems and Computing 425,
DOI: 10.1007/978-3-319-28658-7_34

395

# 1    Introduction

Cloud computing as a world-shattering technology is shifting the entire IT environment, and the entire aspects of our lives. It carries not only the technological transformation, but also reflective influence on venture trade platforms and business processes. Platforms are distributed as services via the Internet in the cloud model [1]. At current situation, consumers are gradually more habituated to utilizing the Internet to expand tangible software sources by means of web services. The hosted web services are self-manipulating software models that offers assured processes autonomously from essential implementation mechanisms and technologies [2,3]. In the course of service composition process, the loosely-coupled services which are self-determining from each other that preserve to be incorporated into complex and significant composited services. This will be situated only when every component service's edge pattern is subject to regular protocols.

# 2    Related Works

The QoS of cloud services denotes to numerous nonfunctional features for example availability, response time, reliability, and throughput [4]. Assumed that the intellectual illustration of a composition model request, an amount of sub services which deliver the similar function however this vary in QoS parameters. The composition concept tend to select the preeminent sub service from a group of functionally identical services based on the QoS factors. The presented work in [5,6] practices combinational prototype model to discover the optimum choice of component services. This article uses linear programming method that is best suitable for small scale issues. However with the growing scale of difficulties, the complication of this method grows exponentially. Subsequently in the work of Mahammad Alrifai et al. mentions the issues by merging global optimization through local selection methods [7,8]. By disintegrating the optimization complications into small sub-problems, in particular this tactic is capable to solve the problem in a distributed mode. In another work author [9] extends the approaches above and presented a approach to further condense the search space through scrutinizing only the subsets of sub services meanwhile the amount of sub services for a configuration may be enormous too. Correspondingly the mentioned articles do not resolve the composition issue in the cloud background. Therefore, the distributed system is not deliberated in these articles. The Cloud computing is a progressively prevalent computing paradigm. It consumes a consistent substance for a wide collection of industry and end-user platforms [10]. Tao et al. examine the composition problem of several cloud resource containing software and hardware resources with various purposes and restrictions [11]. With the growing amount of cloud service consumers around the globe, the major cloud service providers positioning and functioning geographically detached datacenters to assist the internationally distributed cloud consumers. Simultaneously, cloud services endure to cultivate quickly. The resource volume of a datacenter is inadequate, so dispensing the load to worldwide datacenters will be reliable in offering stable facilities [12,13]. Increasing number of web services are organized on geographically distributed cloud datacenters and are obtainable all over the globe.

## 3    QoS-Aware Cloud Service Selection Model

The QoS could comprise a quantity of nonfunctional characteristics that are tend to define the superiority of Cloud Services (CS's) and manifested by Cloud Service Providers (CSP's), such as availability, price, response time, reputation and so on [16]. Nevertheless, to provide easiness, four characteristics: availability, service price, response time, and reputation remain deliberated on situations that do not disturb the main outcomes of this proposed work. The descriptions and the computation technique as assumed as follows.

**Cloud Service Price** $CS_P$. The charge that a cloud service consumers must pay to the Cloud Service Provider (CSP) for the service solicitation. The Fee is calculated by dollars (\$). $CS_P$ is categorized into two fragments: $CS_t$ cost for sending the request which is disables in practical approach, and the other $CS_s$ cost of services which is consumed.

$$CS_p = CS_t + CS_s \tag{1}$$

**Response Time** $CS_{RT}$. The anticipated delay among the prompted time when a request is directed and the stipulated time when the outcome is attained. Response time can be calculated by seconds. $CS_{RT}$ is characteristically encompasses the Execution Time ($ET_{Trans}$) other Data Transmission Time ($DT_{Trans}$). This can be computed by,

$$CS_{RT} = ET_{Trans} + DT_{Trans} \tag{2}$$

$$CS_{RT} = \sum ET_{Trans} + DT_{Trans} / N \tag{3}$$

Normally, is well-defined by average of times.

**Availability** $CS_A$. The possibility that the service process is manageable. Availability is in the defined range of [0, 1]. $CS_A$ is described by the ratio of service's readily obtainable and the total successful request times to the request times ($T_{Req}$) in assured interval.

$$CS_A = T_{Req} / N \tag{4}$$

Where $T_{Req}$ represents the over-all requested times.

**Reputation** $CS_{Reput}$. This metric defines the service invocation dependability. It is distinct as the proportion among the total number of service solicitations that comply the negotiated the total number of QOS service solicitations. It largely relies on cloud consumers' experiences of consuming services. Diverse cloud consumers might have dissimilar attitudes on the equivalent service. The worth of reputation is well-defined through average ranking consideration to the defined service by cloud consumers.

$$CS_{Reput} = \sum CS^i_{Reput} / N \qquad (5)$$

Such that $CS^i_{Reput}$ signifies the value specified by the $i^{th}$ cloud consumers, the range between 0 and 1. $CS_{Reput}$ with the range [0, 1].

## 4    QoS-Aware Computation on Cluster Cloud Services

*Definition 1.* **Task**: The fundamental operational unit of clustered service, each task comprises four elements: Name, function, respective input & output, which defines an abstract class of sub services.

*Definition 2.* **Sub Service**: The sub service is the cloud services perceptible by multiple CSP's (Cloud Service Providers), in-turn it is used to accomplish precise tasks. The sub services may have diverse QoS parameter values.

*Definition 3.* **Cluster Services (CSs)** are essentially a service-dependent workflow with several atomic functions, and the QoS elements of Cluster services are described by the QoS parameters and the Clustering models. There are usually four elementary Cluster models for cloud clustering services: 1. Sequence (Fig. 1a) this specifies that tasks are accomplished in ordering manner, 2. Parallel (Fig. 1b) this directs that all divisions of tasks would run concurrently, provided all divisions must be ended before the next node initiates. , 3. Branch (Fig. 1c) this specifies that subsequent task could initiate after any other division is finished, that have the similar function. 4. Loop (Fig. 1d) which demonstrates that performing of the same task for many times in ordering format.

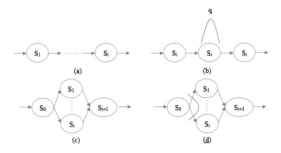

**Fig. 1** Basic Clustered Service of Cloud model

Towards the cloud cluster Service, the QoS elements can be observed by collective method since it delivers service as a group. This will relies on definition of QoS elements; this work originates aggregative properties of a cloud cluster service by its principal workflow outlines. The combined rules of every feature are described in Table 1. In this article, we emphasis on the chronological configuration model.

**Table 1** Aggregate rules for clustered services with QoS notations

| Model | Availability | Price | Response Time | Reputation |
|---|---|---|---|---|
| Sequence | $\prod\limits_{i=1}^{j} CP_A(i)$ | $\sum\limits_{i=1}^{j} CS_P(i)$ | $\sum\limits_{i=1}^{j} CS_{RT}(i)$ | $\prod\limits_{i=1}^{j} CP_{Reput}(i)$ |
| Parallel | $\prod\limits_{i=1}^{j} CP_A(i)$ | $\sum\limits_{i=1}^{j} CS_P(i)$ | $\max\{CS_{RT}(i)\}$ | $\prod\limits_{i=1}^{j} CP_{Reput}(i)$ |
| Branch | $CP_A(i)$ | $CP_P(i)$ | $CP_{Reput}(i)$ | $CP_{Reput}(i)$ |
| Loop | $CS_A^q$ | $q.CS_P$ | $q.CS_{Reput}$ | $CS_{Reput}^q$ |

# 5    The Proposed Genetic Particle Swarm Optimization (GPSO)

The GPSO is used for discover optimum regions from complex search spaces via the collaboration of individuals in a crowd of particles. The iterative process of GPSO brand to fit to resolve NP-hard optimization complications [18]. The GPSO preserves a swarm particles then each distinct comprises three factors, which are the Current Position $CP_i$ , the Previous optimal position $CP^i_{p\_best}$ , and finally velocity

$\sigma_i$ . The optimal solution originated by each element is stored in $CP^i_{p\_best}$ up to now; on other wing the optimality of the swarm $CP^i_{g\_best}$ refers new locations of particles are finalized by the current positions and velocity level. In spirit, the route of each particle is informed conferring to its own swarming knowledge in addition to the global best element in the swarming. These considerations of each particles apprises as follows:

$$\sigma_{i+1} = W_i\sigma_i + CW_1R_1(CP^i_{p\_best} - CP_i) + CW_2R_2(CP^i_{g\_best} - CP_i) \qquad (6)$$

$$CP_{i+1} = CP_i + \sigma_{i+1} \qquad (7)$$

Such that $W_i$ refers inertia weight, $CW_1$ defines cognition weight and $CW_2$ general social weight; $R_1$ and $R_1$ and are dual random values consistently disseminated with the range of [0, 1]. The $CP^i_{p\_best}$ and $CP^i_{g\_best}$ could be attained by computing $Fun(CP_i)$, this refers fitness function on calculating better target positions of the particles. The GPSO is typically tend to resolve multi-objective problems in series space since the advantages resembles perception, easy to control, ease of distribution of execution, and efficacy. Decent effects are attained through multiple optimization techniques with multiple constrained by GPSO, however in the isolated domain, the required results might not obtained properly. Additionally the swarm congregates quickly within the provisional neighborhood region of the group finest. Nevertheless, such high conjunction swiftness often consequences in: 1. the loss in multiplicity. 2. Precipitate conjunction if the global finest resembles to native targets, this will tend to worst quality of resolution.

## 6    ANUM Strategy on Genetic Particle Swarm Optimization (GPSO)

In the technique of GPSO, all additional particles explore along with the path of the global optimal one excluding itself. This development direction is visionless, so the optimal solutions would fade out and results in prematurity. In [17], the transformation impression in GPSO is familiarized to the association of the global best particle with the intention to discover more new resolutions in solution search space. Nevertheless, the likelihood of the transformation is highly sensitive. Complex transformation prospect increases the search space and brands it conceivable to discovery fine solutions, however it decreases the conjunction velocity at the same period. Inferior transformation likelihood makes the algorithm congregate quicker, but the chance of deteriorating into native goals will upsurge. In this segment, the non-uniform transformation process is considered depend on our earlier work [17]. The progression has the universally diminished features, but with arbitrary variation in the execution procedure. The function is specified as follows:

$$f(q,R) = \delta * \left[ 1 - R^{(1-\frac{q}{Q})^h} \right] \tag{8}$$

Where $Q$, $\delta$ and are constants coefficients. $q$ and $R$ are variables of the present group and random number in the range (0, 1) respectively. The mutation function $f(q,R)$ relies on generation number $q$. obviously; generation number $q$ is a distinct variable. Without losing of generalization, we presume presence of continuous variable in (0, Q), perceptibly, $f(q,R)$ is differential on $q$. simultaneously, the random variable $R$ is essentially autonomous with given generation number $q$. So the dualistic mutation function $f(q,R)$ is as follows,

$$\{(q,R) \mid q \in (0,Q) R \in (0,1)\}$$

Its fractional derivative represented as,

$$\frac{\partial f(q,R)}{\partial Q} = \delta R^{(1-\frac{q}{Q})^h}(1-\frac{q}{Q})^{h-1}\frac{h\ln R}{Q} < 0 \tag{9}$$

$$\lim_{q \to Q^-} f(q,R) = 0 \tag{10}$$

It could be resembles as mutation prospect "tediously" reduces and approaches to zero assuming $R$ being constant coefficient. Nevertheless, the independent along with uniform random variable $(0, 1)$ and the derived on $R$ of $f(q, R)$ as follows,

$$\frac{\partial f(q,R)}{\partial R} = -\delta * (1-\frac{q}{Q})^h * R^{(1-\frac{q}{Q})^{h-1}} < 0 \tag{11}$$

At this point the mutation is not reduces monotonously. However it decreases and diminishes to zero by means of probability. Consequently this method associates to decrease the probability of premature conjunction and directs the exploration to the promising realm. Fig. 2 brightens the fluctuating arrangement of non-uniform mutation against the generation number with these limitations,

$$[\delta = 0.5, h = 6, Q = 3000]$$

From Fig. 2, the mutation prospect reduces in the general with arbitrary fluctuation over the generations; finally the mutation probability is close to 0 that is decent for the conjunction after the $1500^{th}$ generation.

### a.  Procedure Description for GPSO

**Step 1:**  Initialization: The particle swarm size denoted as T, utmost generation number Q. Engendering current position $CP`$ and velocity $\sigma`$ for every particle arbitrarily, by examining the probability of particle's position (such that the position duly meet QoS limits).

**Step 2:** Repetition control parameter: if $(q==Q)$, Then goto Step 4; or else Continue;

**Step 3:** Repetition:

**Step 3 (a):** Computing fitness values by $Fun(CP_i)$ for every particle;

$$if\,(Fun(CP_i) > Fun(CP^i_{p\_best}))CP^i_{p\_best} = CP_i;$$
$$if\,(Fun(CP^i_{p\_best}) > Fun(CP_{g\_best}))CP_{g\_best} = CP_{p\_best};$$

**Step 3 (b):** Updating the current position $CP_i$ and velocity $\sigma_i$ consistent with equation (10) and (11);

**Step 3 (c):** ANUM process for discovery of global best particle, engendering a new position of it;

**Step 4** Output: collecting composition structure and the equivalent fitness values.
     **End.**

### b.  Fitness Function

The Fitness values are accustomed to assess the concert of current positions (CP) of particles to direct other particles to trail the global best particle. In the QoS-aware cloud service selection scope, the aim of it is to choose one clustering service from sub services of each individual task, therefore the accumulated QoS concert of Cluster Service can be explored as well as the limitations are fulfilled. Subsequently conflicts might occur among various objectives, that consequences the solution by sets. Hence a performance value which have an agreement with the trade-off between goals which need to be design, this is also required for the fitness assessment in GPSO. Finally the service composition, the fitness function requisite to redirect the global QoS-aware of a cluster service. In this article, the weighted sum of every characteristic for the QoS –aware cluster service is espoused.

$$f(x) = \sum_{i=1}^{4} W_i S_i \tag{12}$$

At this point $W_i$ signifies the equivalent weight for a assured attribute of QoS, and $\sum_{i=1}^{4} W_i = 1$. $S_1, S_2, S_3, S_4$ symbolize respectively. However the attribute scores is not on the similar level and have dissimilar units, such as service execution time mentioned as *ms* (millisecond) and every service price mentioned as $ (U.S dollars). If the values of $S_i$ is computed openly by the equation (12), it will unavoidably tend to unfair assessment. In other angle diverse attributes will tend to different properties for the integer calculation. For example, the service price plus service execution time may consume negative impressions on assessment, those values are the higher and poorer. Nevertheless service reliability plus service availability consume positive impressions, whose values are the higher the improved. Consequently, $S_i$ need to be standardized by means of proposed method,

$$S_i^{'} = \left[ \begin{array}{ll} \dfrac{S_i - S_i^{min}}{S_i^{max} - S_i^{min}}, & S_i^{max} - S_i^{min} \neq 0 \\ 1, & S_i^{max} - S_i^{min} = 0 \end{array} \right] \tag{13}$$

$$S_i^{'} = \left[ \begin{array}{ll} \dfrac{S_i^{max} - S_i}{S_i^{max} - S_i^{min}}, & S_i^{max} - S_i^{min} \neq 0 \\ 1, & S_i^{max} - S_i^{min} = 0 \end{array} \right] \tag{14}$$

Where $Q_i^{min}, Q_i^{max}$ signifies the lowest and highest of $i^{th}$ attributes of QoS for all the cloud sub services. Therefore the fitness function can be articulated as:

$$f^{'}(x) = \sum_{i=1}^{4} W_i S_i^{'} \tag{15}$$

$$\sum_{i=1}^{4} W_i = 1,\ 0 < W_i < 1,\ 0 \leq S_i^{'} \leq 1$$

In general, the cloud consumers typically assert their function necessities as well as global restrictions. Thus, the Cluster structure will be forbidden initially if the QoS violates the limitation.

# 7    Simulation Configuration

The proposed model compared with other models to define the optimal efficiency. All iteration were accompanied on numerous PCs with equivalent configuration of Intel Dual Core 1.2 GHz processor, 2 GB of RAM, and OS-Windows XP with the network speed of 15Mbps. The evaluations are reserved in the similar cloud clustered service request, along with workflows of diverse dimensions i.e. the cloud tasks were 10, 20, 30, and 45. The numbers of sub services for every individual task were arbitrarily attained from 10 to 20 and The QoS aware attributes of every cloud service. All cloud service signified by ID number.

### a.    Experiments on ANUM Strategy for GPSO

The GPSO with ANUM strategy has connected with other comparable schemes. The above crafted graph resembles the efficacy and equated with identical techniques (GN_S_Net [17], PSOA [18], and MGA [19]). All the algorithm given with same input date to validate the optimal performance which includes population of particles along with standard parameters. Initially the fitness function with the population size 25 particles and set of tasks were 15 respectively. The all defined algorithms iterated for 50 times to verify the optimality.

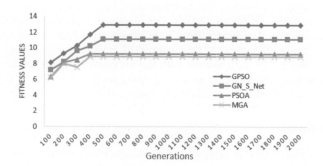

**Fig. 2** Fitness Comparison on ANUM

The GPSO congregates rapidly but scopes shows worst fitness significance. Since the particles gathered fleetly to attain the global best, however the random travel tends to prematurity. Correspondingly a static mutation proportion is familiarized to attain global best particle. Moreover the mutation is fair in attaining the optimal solution in search space. Nevertheless, the larger static mutation probability may cause longer time in execution and sluggish convergence. Since the ANUM strategy can broaden the particles searching space in prior and make them gathered fleetly gradually.

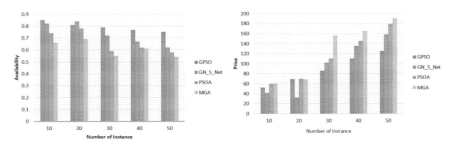

**Fig. 3(a).** Feasible Rate – Availability          **Fig. 3(b).** Feasible Rate – Price

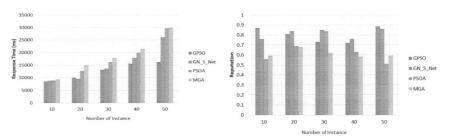

**Fig. 3(c).** Feasible Rate – Response Time (ms) **Fig. 3(d).** Feasible Rate – Reputation

The cloud cluster service is achievable only if its QoS fulfills the QoS limitations of given SLA. The legitimacy of a methodology can be formatted via viable rate. In this configuration, we computed the achievable rate of our method. For this determination, we formed 40 circumstances. Initially 20 scenarios have been adopted, the quantity of feasible services per cluster was set and allotted a constant figure 350 and the total of group of services ranges from 10 to 30. In the final session we have taken 20 scenarios, the quantity of feasible services ranges from 40 to 800 with a rate of 50 per instance with the assigned constant services were 15. We calculated the viable rate by executing three different methods: GA_S_NET, PSOA, and MGA. The Fig. 4. briefs the QoS data along with SLA constraints.

## 8    Conclusion

The choice of cloud service with global QoS-aware constraints is hot pursuits of current investigators. It is significant to identify which cloud services could be clustered conferring by the consumers with respect to QoS parameters In this article, we deliberate a unique heuristic method to resolve the QoS-aware cloud service selection by means of an enhanced GPSO which is broadly utilized to crack hefty scale optimization issues, In this paper includes the unique contribution as follows, Firstly the QoS-aware cloud service selection scheme, that could be transmuted into a general controlled optimization scope, provided we reconstructed the factors and operational mechanisms to pursuit GPSO along with QoS-aware cloud service selection. Secondly we deliver ANUM stratagem to attain best particle globally to decrease the prematurity on GPSO. This proposed method is also equated with other schemes to treasure the stability on QoS constraints.

# References

1. Armbrust, M., et al.: A view of cloud computing. Commun. ACM **53**(4), 50–58 (2010)
2. Bichier, M., et al.: Service-oriented computing. Computer **39**(3), 99–101 (2006)
3. Hatzi, O., et al.: An integrated approach to automated semantic web service composition through planning. IEEE Trans. Serv. Comput. **5**(3), 319–332 (2012)
4. Zheng, Z., et al.: Qos-aware web service recommendation by collaborative filtering. IEEE Trans. Serv. Comput. **4**(2), 140–152 (2011)
5. Qi, L., et al.: Combining local optimization and enumeration for qos-aware web service composition. In: Proceedings of the 2010 IEEE International Conference on Web Services, pp. 34–41. IEEE (2010)
6. Ardagna, D., et al.: Adaptive service composition in flexible processes. IEEE Trans. Software Eng. **33**(6), 369–384 (2007)
7. Alrifai, M., Risse, T., Dolog, P., Nejdl, W.: A scalable approach for QoS-based web service selection. In: Feuerlicht, G., Lamersdorf, W. (eds.) ICSOC 2008. LNCS, vol. 5472, pp. 190–199. Springer, Heidelberg (2009)
8. Alrifai, M., Risse, T.: Combining global optimization with local selection for efficient qos-aware service composition. In: Proceedings of the 18th International Conference on World Wide Web, pp. 881–890. ACM (2009)
9. Alrifai, M., et al.: Selecting skyline services for qos-based web service composition. In: Proceedings of the 19th International Conference on World Wide Web, pp. 11–20. ACM (2010)
10. Sharkh, M.A., et al.: Resource allocation in a network-based cloud computing environment: design challenges. IEEE Commun. Mag. **51**(11), 46–52 (2013)
11. Tao, F., et al.: A parallel method for service composition optimal-selection in cloud manufacturing system. IEEE Trans. Industr. Inf. **9**(4), 2023–2033 (2013)
12. Agarwal, S., et al.: Automated data placement for geo-distributed cloud services. In: Proceedings of the 7th USENIX Conference on Networked Systems Design and Implementation, pp. 17–32 (2010)
13. Son, S., et al.: An sla-based cloud computing that facilitates resource allocation in the distributed data centers of a cloud provider. J. Supercomput. **64**(2), 606–637 (2013)
14. Xiao, J., et al.: Qos-aware service composition and adaptation in autonomic communication. IEEE J. Sel. Areas Commun. **23**(12), 2344–2360 (2005)
15. Klein, A., et al.: Towards network-aware service composition in the cloud. In: Proceedings of the 21st International Conference on World Wide Web, pp. 959–968. ACM (2012)
16. Liu, Y., et al.: Qos computation and policing in dynamic web service selection. In: Proceedings of the 13th International World Wide Web Conference on Alternate Track Papers & Posters, pp. 66–73. ACM (2004)
17. Wang, W., et al.: An improved Particle Swarm Optimization Algorithm for QoS-aware Web Service Selection in Service Oriented Communication. International Journal of Computational Intelligence Systems (1), 18–30, December 2010
18. Wang, D., et al.: A genetic-based approach to web service composition in geo-distributed cloud environment. Computers and Electrical Engineering **43**, 129–141 (2014)
19. Ma, Y., et al.: Quick convergence of genetic algorithm for QoS-driven web service selection. Computer Networks **52**(5), 1093–1104 (2008)

# Anomalous Crowd Event Analysis Using Isometric Mapping

Aravinda S. Rao, Jayavardhana Gubbi and Marimuthu Palaniswami

**Abstract** Anomalous event detection is one of the important applications in crowd monitoring. The detection of anomalous crowd events requires feature matrix to capture the spatio-temporal information to localize the events and detect the outliers. However, feature matrices often become computationally expensive with large number of features becomes critical for large-scale and real-time video analytics. In this work, we present a fast approach to detect anomalous crowd events and frames. First, to detect anomalous crowd events, the motion features are captured using the optical flow and a feature matrix of motion information is constructed and then subjected to nonlinear dimensionality reduction (NDR) using the Isometric Mapping (ISOMAP). Next, to detect anomalous crowd frames, the method uses four statistical features by dividing the frames into blocks and then calculating the statistical features for the blocks where objects were present. The main focus of this study is to understand the effect of large feature matrix size on detecting the anomalies with respect to computational time. Experiments were conducted on two datasets: (1) Performance Evaluation of Tracking and Surveillance (PETS) 2009 and (2) Melbourne Cricket Ground (MCG) 2011. Experiment results suggest that the ISOMAP NDR reduces the computation time significantly, more than ten times, to detect anomalous crowd events and frames. In addition, the experiment revealed that the ISOMAP provided an upper bound on the computational time.

A.S. Rao(✉) · M. Palaniswami
Department of Electrical and Electronic Engineering, The University of Melbourne, Parkville Campus, VIC 3010, Australia
e-mail: aravinda@student.unimelb.edu.au, palani@unimelb.edu.au

J. Gubbi
TCS Innovation Labs, Tata Consultancy Services, Bengaluru 560066, Karnataka, India
e-mail: j.gubbi@tcs.com

© Springer International Publishing Switzerland 2016
S.M. Thampi et al. (eds.), *Advances in Signal Processing and Intelligent Recognition Systems*,
Advances in Intelligent Systems and Computing 425,
DOI: 10.1007/978-3-319-28658-7_35

# 1   Introduction

Anomalous event detection is one of the important applications in crowd monitoring. Most video surveillance systems endeavor to detect such events. Video surveillance systems are also used to track targets in multi-camera video surveillance, analyze crowd behavior, provide security for people, and remotely monitor elderly people in home care. Human activity recognition places an important role in many applications, and video surveillance systems facilitate to detect such activities. Different types of human activities exist and examples include "walking," "running," "jogging," "fighting," "waving," and other. Surveillance systems are now ubiquitously installed in airports, shopping malls, public transport hubs, stadiums, concerts, etc. [24]. However, the current surveillance applications lack automated analysis of crowd behavior and detecting interested events.

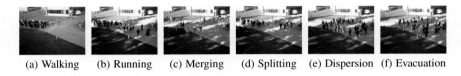

   (a) Walking    (b) Running    (c) Merging    (d) Splitting    (e) Dispersion    (f) Evacuation

**Fig. 1**   Example frames of crowd events from the PETS 2009 dataset [5].

    Automated analysis of interested "events" or "activities" using surveillance systems with intelligent video analytics can deliver a suite of tools to monitor the crowd. However, the algorithms face several challenges in object detection, tracking and detecting human activities. In the case of the human visual system, the eyes and brain coordination are so well-developed that the replication of same process has been a long-standing research in artificial intelligence systems. Many sophisticated algorithms have been proposed over the last four decades, but the algorithms fail in difficult and crowded scenarios. Some of the critical problems faced by computer vision algorithms are the loss of information from 3D to 2D during image formation at the cameras [26], nonuniform illumination, shadows, articulated motions of humans that make algorithms infeasible to track movements, occlusions that mask presence of the objects or object parts.

    The quality of the video systems has also improved from standard definition to high definition. The voluminous amount of data along with the problem of human detection and occlusions requires intelligent systems. For real-time crowd monitoring and increased presence of video analytics on cameras, it is critical that video analytics are fast and provide near real-time results. Fast video analytics on cameras saves video bandwidth by not transmitting video to a centralized server, and in addition, avoids unnecessary storage of video footage from every camera in a network. Dimensionality reduction methods provide low-dimensional representation of high-dimensional, complex data.

In this work, we present a fast approach to detection of anomalous crowd events. Two types of anomalies are being detected: (1) anomalous crowd events and (2) anomalous crowd frames. First, to detect anomalous crowd events, the motion features are captured using the optical flow [8]. A feature matrix of motion information is subjected to NDR using the ISOMAP [20]. The anomalous crowd events are then detected using the hyperspherical clustering [15]. The proposed approach detects the crowd events such as change of directions in people movements, splitting, and dispersion, etc. Next, to detect anomalous crowd frames, the method uses four statistical features (contrast, correlation, energy, and homogeneity) by dividing the frames into blocks and then using the blocks where objects were present. The main contributions of this work include: (1) a new approach to detect anomalous crowd using NDR and (2) analysis of computation time impacting the real-time crowd monitoring and video analytics.

## 2   Related Work

### 2.1   Crowd Anomaly Detection

Anomaly detection in crowd events refers to anomalous behavior of the crowd. This is a time-related event, i.e., at certain times crowd events will be normal and other times it will be abnormal (hereafter abnormal,unusual events are called as anomalous) [1, 27]. Outlier detection is one of the main approaches in anomaly detection. A comprehensive survey of anomaly detection based on vision systems is found in [14]. Mahadevan *et al.* [12] consider anomalous events as outliers and use the mixture of dynamic texture (MDT) features. The probability distribution of normal event is learned using the dynamic texture features. Temporal anomalies are detected by first using the Gaussian Mixture Model (GMM) [6] to learn the local intensities and then replacing the foreground intensities with MDT. Spatial anomalies were mapped by computing the discriminant saliency features at each video location using the MDT. The abnormality map of the video was the combined mapping results of both temporal and spatial anomalies.

Adam *et al.* [1] defined small regions in the video for monitoring the flow of moving objects. Optical flow was used to compute the motion vectors. Probability matrix of flow vectors was constructed in the defined region of normal and anomalous events. Chen *et al.* [3] utilized (Lucas-Kanade) optical flow approach to track the feature points. The orientations of the vectors were computed and stored as bins. To determine the dominant force and direction of the motion, force-field model was applied. An anomalous event was detected upon discovering a sudden appearance of force. Wang and Miao [25] used the optical flow vectors to indicate the motion points and the Kanade-Lucas-Tomasi (KLT) corners to track the features. A model was generated by using the motion patterns in different blocks. Anomalous events were detected based on the deviation from the learned normal model. Liao *et al.* [11]

used four descriptors namely crowd kinetic energy (motion intensity), histogram of motion directions, spatial distribution of motion intensity and localization between two frames for the detection of anomalous events. Liao *et al.* [11] targeted the detection of fighting events. Support Vector Machine (SVM) with Radial Basis Function (RBF) kernel was used to train the 13-dimensional feature vectors and classification of abnormal events.

Tziakos *et al.* [23] used the motion vectors to detect the abnormal events by detecting the motion vectors and then representing them on a low-dimensional manifold using the Laplacian Eigenmaps (LE) [2]. The unusual events were then classified using the supervised Mahalanobis classifier. Thida *et al.* [21] proposed to use the blocks of Histogram of Optical Flow (HOOF) feature corresponding to each frame as features to detect anomalous events. A new NDR method termed as Spatio-Temporal Laplacian Eigenmap (ST-LE), was proposed based on LE. A Novelty classifier was trained using the test samples in the low-dimensional space to classify the anomalous events.

## 2.2 Dimensionality Reduction

The dimensionality reduction methods can be broadly categorized as linear methods and nonlinear methods. Linear methods assume the existence of features on a linear subspace, whereas nonlinear methods approximate the nonlinear feature space into a linear subspace and then find low-dimensional representations. The dimensionality reduction method can be described as: given a set (data) $X = [\mathbf{x}_1, \mathbf{x}_2, \ldots, \mathbf{x}_n]$ with $\mathbf{x}_i \in \mathbb{R}^m$, find a set $Y = [\mathbf{y}_1, \mathbf{y}_2, \ldots, \mathbf{y}_n]$ with $\mathbf{y}_i \in \mathbb{R}^d$ that represents $X$ such that $d \ll m$.

Principal Component Analysis (PCA) [9] reduces the data dimensions by assuming the linear subspace. The objective function used by the PCA is to maximize the variance such that the reconstruction error from the reduced space is minimized in a least square sense. Classical Multidimensional Scaling (MDS) [22] preserves the distances between points from high-dimensional space to lower dimensions. Euclidean distance is the commonly used distance metric between points.

ISOMAP [20] finds the low-dimensional representation in three steps. At first the distance $d(i, j)$ between the neighboring points $(i, j)$ in the input space ($X \in \mathbb{R}^m$) is calculated and a weighted graph is constructed. Next, the distances $d(i, j)$ are used to calculate the geodesic distance between all the points of the input space to find the shortest path. In the last step, classical MDS is applied to weighted graph to construct a $d$-dimensional embedding from the $m$-dimensional input space $X$, where $d \ll m$. ISOMAP considers the global isometry of points between the input feature space and embedded space. ISOMAP is the nonlinear analogue of PCA.

LLE [18] is similar to the ISOMAP but constructs the neighborhood graph based on the local linear properties as opposed to the global approach considered by the ISOMAP. The same localness is maintained in the low-dimensional embedded space. The local linearity is based on the neighboring points in the feature space. Lapalacian

Eigenmaps (LE) [2] is similar to LLE, however, the weighted graph is constructed based on neighbors defined by a heat kernel. Hessian Eigenmaps (HE) [4] construct the hessian of feature space in a given neighborhood. Then a low-dimensional embedding is found based on the Hessian matrix.

# 3   Methodology

## 3.1   Preprocessing and Feature Extraction

Video frames from CCTV cameras contain high-frequency noises due to the nature of the formation of the images. In order to remove these noises, the frames are converted to grayscale. The high-frequency noise is eliminated by Gaussian applying a 2D Gaussian filter with $\sigma = 0.5$ and a block size of $5 \times 5$, which is a low-pass filter. The Gaussian filter parameters were chosen based on the method presented in [16]. The feature vector $F$ is a matrix of size $R^{m \times n}$, where $m$ indicates the feature vector length for each frame and $n$ indicates the number of frames in a given video sequence.

### 3.1.1   Anomalous Crowd Events

After removing the noise, the motion information between frames is computed using the optical flow approach [8]. Let $(x, y)$ denote a pixel in a frame $I(x, y)$. $x$ and $y$ indicate the horizontal and vertical axes, respectively. The optical flow in [8] assumes the brightness constancy in calculating the apparent motion. $E(x, y, t)$ denotes the brightness at point $I(x, y)$, where $t$ represent the time parameter. Then the optical flow vectors are given by [8]

$$E_x u + E_y v + E_t = 0 \qquad (1)$$

where $u = \frac{dx}{dt}$ and $v = \frac{dy}{dt}$ are the velocities in horizontal and vertical directions. Feature matrix $F$ is constructed by combining row vectors $u$ and $v$ into a single row vector $[u|v]$, as a row vector for each frame.

### 3.1.2   Anomalous Crowd Frames

To detect anomalous frames, the proposed method first tracks the objects and determines the centroid in each frame. Then, each frame is divided into blocks of different sizes ($8 \times 8$, $16 \times 16$, $32 \times 32$ and $48 \times 64$). Now the method compares objects' centroids with block centers and calculate the statistical features for the block where the object centroids were found. The method calculates the four statistical features for the block where the object centroids were found. Statistical features such as contrast, correlation, energy, and homogeneity [7], have been found to be useful in many crowd monitoring applications [19]. Thus the feature matrix $F$ is a sparse matrix in which the nonzero blocks indicate one of the four statical features, number of blocks is determined by the block sizes chosen, and feature matrix rows equal the number of frames in a given video.

## 3.2 Nonlinear Dimensionality Reduction (NDR)

The feature matrix $F \in R^{m \times n}$ is provide as an input to the ISOMAP. The ISOMAP first calculates the $d_F(i, j)$ distances between video frames based on the feature points, $i$ and $j$, in the input feature space. A weighted graph $G$ will be constructed based on the neighborhood $k$. During the construction of the graph, the distance between the nodes (between the feature points in this case) are based on the geodesic distance. The geodesic distance $d_M(i, j)$ is the shortest distance between two points. The weighted neighborhood graph is assumed to be isometric to lower-dimensional embedding. Next MDS is applied to the $D_G = \{d_G(i, j)\}$ to find the embedding graph whose dimensions will be $d$ such that the intrinsic geometry is preserved. The low-dimensional embedding is given by vectors $y_i$ such that the following cost function is minimized:

$$E = \left\| \tau(D_G) - \tau(D_Y) \right\|_{L_2}, \tag{2}$$

where $\tau$ is an operator that calculates the inner product between nodes and $D_Y$ is the Euclidean distance between frames in the embedded space.

## 3.3 Anomaly Detection

In this work, we use the anomaly detection scheme devised for environmental sensing applications [15]. The low-dimensional data from the ISOMAP is used to reduce the computational complexity of the data. The anomalous frames are detected based on the following parameters that are iteratively updated: (1) at first, clustering of feature points in the low-dimensional space is achieved with fixed width clustering $w$, (2) next, the clusters formed by the neighboring featuring points are clustered into a larger cluster. Based on the average inter-point distance, clusters are merged if the inter-point distance is above $\tau$ and the distance between points are greater than $\psi$ standard deviation, and (3) finally, the anomalous clusters are identified based on the $k$-nearest neighbors.

# 4 Experiment

The proposed method was implemented in MATLAB 2014 on a 64-bit Windows 7 equipped with 4 GB RAM and Intel® i7 − 2600 CPU running at 3.4 GHz. The anomaly detection algorithms were implemented in Java.

The proposed method has been tested on two datasets: (1) the PETS 2009 [5] and (2) MCG 2011 [16]. The PETS 2009 [5] dataset has three different video sequences (S1, S2, and S3) and each sequence contain different sets : L1, L2 and L3. Each set comes with different timings (such as $13-57$, $14-16$, $14-27$, $14-31$ and $14-33$). The timings refer to the hour ("hh")–minute ("mm") of data collection. There are eight different views (001, 002, . . . , 008) for each dataset time. This work uses S1.L1 (with time 13-57, View001) and S1.L2 (with time 14-06, View001) for detecting anomalous frames, and S3 (with timings 14-16, 14-31, and 14-33) for detecting anomalous crowd events, respectively. The PETS 2009 dataset has frames of size $576 \times 768$ and in color Joint Photographic Experts Group (JPEG or JPG) format. The PETS 2009 data were collected using Axis cameras. The MCG dataset was collected on four Australian Football League (AFL) matches that were held at the MCG in 2011 and Fig. 2 shows the sample frames. The data were collected using six cameras (C1-C6) at 30 fps. MCG data have a frame size of $640 \times 480$ and were in RGB color mode collected in Advanced Systems Format (asf). They were then converted to JPEG. In this work data from camera C5 were used and have highly crowded scenes, and the video length is about 20 minutes that makes it ideal to evaluate the computer vision algorithms. In addition, MCG dataset is a natural dataset (i.e., it is not a simulated environment) that provides better data to analyze crowd behavior and events.

(a)  (b)  (c)  (d)

**Fig. 2** Sample frames from MCG dataset [16].

To the best of our knowledge, crowd events such as walking, running etc. that can be clearly distinguished have been captured only by the PETS 2009 dataset. Therefore, only the PETS 2009 dataset is used detect anomalous crowd events. To detect anomalous frames, this work uses both PETS 2009 and MCG datasets. To track objects, ground truth information provided in [13] was used.

## 4.1  Results and Discussion

The feature vectors computed were of size $u + v = 3942$. The number of frames in the dataset 14-16, View-001, for the walking and running events were 222. Similarly, for 14-31 and 14-33, the number of frames were 130 and 377 respectively. The frame rate was 7 frames per second for the PETS 2009 dataset. Therefore, the size of the feature vectors for the three datasets (14-16, 14-31, 14-33) would be 222 × 3942, 130 × 3942 and 377 × 3942 respectively. In all the experiments, the parameters $k$ and $w$ were set to 3 and 1; merging threshold was set to $\tau = \frac{1}{2}w$ and $\psi = 1$, respectively.

The anomalous crowd frames resulted in video frames people where there were no people or frames which completely dissimilar compared with neighboring frames. Experiment was conducted on PETS 2009 (S1.L1, 13-57, View001 and S1.L2, 14-06, View001) and MCG (16S1, C5) datasets. The feature matrix sizes change depending on the block sizes (8 × 8, 16 × 16, 32 × 32 and 48 × 64) and number of frames. For PETS 2009 dataset, the feature matrix sizes are: 220 × 6912, 220 × 1728, 220 × 144, and 220 × 144 for 8 × 8, 16 × 16, 32 × 32 and 48 × 64 block sizes, respectively. For the MCG dataset, feature matrix sizes are: 31375 × 1200, 31375 × 300, and 31375 × 100 for 16 × 16, 32 × 32, and 48 × 64 block sizes, respectively.

Computation time is an important aspect in the real-world scenarios. Table 1 provides the details of the dataset along with feature matrix size, computational time with and without using the ISOMAP [20] for anomalous crowd event detection using hyperspherical clustering [15]. The ISOMAP calculation time also includes the time to generate low-dimensional embedding. Table 1 shows that to detect walking and running events (14-16, View001), the computational time is about 2 minutes and 40 seconds. The time for detecting anomalous events in dataset 14-31, View001, is about 1 minute and 35 seconds and in 14-33,View001 dataset, it is about 4 minutes and 36 seconds. On the other hand, the computational times are about 1.05, 1.08 and 1.5 seconds, respectively for the detection of the same anomalous events using the ISOMAP. Table 1 also shows the computational time to detect anomalous crowd frames. In the case of PETS datasets (S1.L2), we see that clustering without using the ISOMAP is almost 10 times higher for block sizes 8 × 8. In addition, we also see that the clustering computational time (without ISOMAP) reduces as the block sizes are increased. In the case of 32 × 32 the clustering computational time (without ISOMAP) approximately equals clustering with ISOMAP. In the case of 48 × 64 block size, the clustering computational time is better than using the ISOMAP. For the MCG dataset, the number of frames makes the clustering algorithm (without ISOMAP) to take longer times (nearly double). This is clearly evidenced from the experiment results provided in Table 1. For example, the contrast features of block sizes 16 × 16 require 130 seconds for clustering algorithm directly, whereas using ISOMAP, the computational time reduces to 17 seconds. Because of computational intensiveness and difficulties (memory problems) in calculating large feature matrices, the 8 × 8 have not been reported in this work.

The proposed approach detected events such as people changing the direction, people starting to run, etc (not shown due to limited space). In addition, the proposed

**Table 1** Table provides the computation time recorded to detect anomalous events. It also indicates the number of frames, frame size, and computational time required with and without using the ISOMAP. (Note: the detailed breakdown of computation time (clustering and ISOMAP)) could not be provided due to space limitations)

| Dataset | Feature (Video) | Matrix size | Block size | Computational time (in milliseconds) Clustering [15] Without ISOMAP | Clustering [15] With ISOMAP |
|---|---|---|---|---|---|
| PETS 2009 | Optical Flow (14-16) | 222 × 3942 | - | 2, 040 | 1,050 |
| | Optical Flow (14-31) | 130 × 3942 | - | 1, 350 | 1,080 |
| | Optical Flow (14-33) | 377 × 3942 | - | 4, 360 | 1,500 |
| PETS 2009 (S1.L2 14-06) | Contrast | 220 × 6912 | 8 × 8 | 25, 623 | 2, 259 |
| | | 220 × 1728 | 16 × 16 | 2, 212 | 2, 206 |
| | | 220 × 432 | 32 × 32 | 452 | 2, 682 |
| | | 220 × 144 | 48 × 64 | 121 | 2, 290 |
| | Correlation | 220 × 6912 | 8 × 8 | 23, 997 | 2, 329 |
| | | 220 × 1728 | 16 × 16 | 2, 057 | 2, 103 |
| | | 220 × 432 | 32 × 32 | 366 | 2, 701 |
| | | 220 × 144 | 48 × 64 | 113 | 2, 160 |
| | Energy | 220 × 6912 | 8 × 8 | 24, 074 | 2, 289 |
| | | 220 × 1728 | 16 × 16 | 2, 075 | 2, 166 |
| | | 220 × 432 | 32 × 32 | 390 | 4, 645 |
| | | 220 × 144 | 48 × 64 | 153 | 2, 496 |
| | Homogeneity | 220 × 6912 | 8 × 8 | 24, 123 | 2, 313 |
| | | 220 × 1728 | 16 × 16 | 2, 068 | 2, 110 |
| | | 220 × 432 | 32 × 32 | 356 | 3, 030 |
| | | 220 × 144 | 48 × 64 | 135 | 2, 129 |
| MCG 2011 | Contrast | 31375 × 1200 | 16 × 16 | 130, 000 | 16, 369 |
| | | 31375 × 300 | 32 × 32 | 13, 000 | 518 |
| | | 31375 × 100 | 48 × 64 | 2, 870 | 302 |
| | Correlation | 31375 × 1200 | 16 × 16 | 2, 934, 000 | 15, 205 |
| | | 31375 × 300 | 32 × 32 | 758, 000 | 559 |
| | | 31375 × 100 | 48 × 64 | 300, 870 | 312 |
| | Energy | 31375 × 1200 | 16 × 16 | 1, 387, 080 | 15, 354 |
| | | 31375 × 300 | 32 × 32 | 12, 499 | 507 |
| | | 31375 × 100 | 48 × 64 | 2, 525 | 311 |
| | Homogeneity | 31375 × 1200 | 16 × 16 | 1, 255, 480 | 15, 473 |
| | | 31375 × 300 | 32 × 32 | 12, 494 | 499 |
| | | 31375 × 100 | 48 × 64 | 2, 496 | 319 |

approach was also able to detect events such as splitting and dispersion. The ISOMAP computes the low-dimensional embedding based on the global geometry, which is important to maintain the connections between the frames. ISOMAP is used in this work because we consider the crowd anomalies as a global event. Another important aspect is that the clustering [15] algorithm endeavors to classify the events in a distributed manner, i.e. the algorithm [15] can be used for detecting anomalies at each camera rather than transmitting the entire video to a centralized video analytics unit. This also assists real-time and large-scale crowd monitoring [15] and the anomalous frame detection rates have been verified in [17]. The distributed hyperspherical clustering has a computational complexity of $\mathcal{O}(nN_c)$, where $n$ is the number of feature vectors and $N_c$ is the number of clusters for each camera. From Table 1, it is evident that as the $n$ increases, the computational time increases drastically. Therefore, hyperspherical clustering by itself would not be a fast solution to detect anomalies (crowd events and frames) at camera level. From Table 1, we also observe that the anomaly detection using the ISOMAP has nearly a constant time bound. The ISOMAP has a computational complexity of $\mathcal{O}(n^3)$ and therefore, serves as an upper bound for large feature matrices (large $n$). Alternatively, Self Organizing Maps (SOM) [10] could also be used for dimensionality reduction. However, SOM requires training, which is not intended in this work. This is critical and highly important for large-scale, real-time video analytics. In addition, it is obvious that increasing feature vector size and number of frames increases time complexity. Indeed, optimal feature vector size (best features) and eliminating less-informative frames are not addressed in this work.

In [17], authors had detected anomalous frames and objects using the Gray Level Co-occurrence Matrix (GLCM) and object paths respectively. It was reported that the frames where there is a relative change compared to others were detected such as scene becoming completely empty after the exist of the entire crowd. In [21], authors proposed to use HOOF to capture the crowd events and then used supervised classifier. However, in this work, we instead use the motion features and iterative clustering algorithm to detect the anomalous frames based on the crowd events. The advantage of this approach is there is no need for training. Only the neighborhood parameter $k$ and cluster width $w$ is required for clustering [15]. From the experiment, we found that for these particular datasets and features, changing of $k$ or $w$ did not have any impact on the detection of anomalous events. In [17], it was also reported that there were computational challenges because of large feature matrices. This problem of computationally demanding task was addressed in this work by the use of ISOMAP. From Table 1, it is evident that the nonlinear manifold algorithms like the ISOMAP can greatly reduce the computational cost. The ISOMAP also preserves the intrinsic geometry of the data, which is very critical in detecting the anomalous events that are temporally related.

In this work, the ISOMAP was used as it is a global dimensionality reduction approach and maintains the inter-point distances. In addition, the study used the ISOMAP parameters, such as the neighborhood parameter $k = 7$ and number of lower-dimensional representation to be equal to ten, and these parameters were determined based on the empirical knowledge. However, further work is required to determine the ISOMAP input parameters and also to examine whether these parameters could be determined automatically without manual intervention. Furthermore, more investigation into the performance of whether NDR algorithms that are global, local or a combination of both is required.

## 5 Conclusion

Anomalous event detection is one of the important applications in crowd monitoring. The detection of anomalous crowd events requires feature matrix to capture the spatio-temporal information to localize the events and detect the outliers. However, feature matrices often become computationally expensive with large number of features. In this work, a fast approach was presented to detect anomalous crowd events and frames using the ISOMAP to reduce feature space and hyperspherical clustering to detect anomalies, respectively. The main focus of the study was to understand the effect of feature matrix size on detecting the anomalies with respect to computational time. In addition, the study revealed that the ISOMAP provides a bound on the computational time and therefore, hyperspherical clustering would also provide a bound in detecting the anomalies. The proposed approach is highly relevant to crowd monitoring and for situations where deploying analytics at camera level is required.

## References

1. Adam, A., Rivlin, E., Shimshoni, I., Reinitz, D.: Robust real-time unusual event detection using multiple fixed-location monitors. IEEE Transactions on Pattern Analysis and Machine Intelligence **30**(3), 555–560 (2008). doi:10.1109/TPAMI.2007.70825
2. Belkin, M., Niyogi, P.: Laplacian eigenmaps and spectral techniques for embedding and clustering. In: NIPS, vol. 14, pp. 585–591 (2001)
3. Chen, D.Y., Huang, P.C.: Dynamic human crowd modeling and its application to anomalous events detcetion. In: 2010 IEEE International Conference on Multimedia and Expo (ICME), pp. 1582–1587. IEEE (2010). doi:10.1109/ICME.2010.5582938
4. Donoho, D.L., Grimes, C.: Hessian eigenmaps: Locally linear embedding techniques for high-dimensional data. Proceedings of the National Academy of Sciences **100**(10), 5591–5596 (2003)
5. Ferryman, J.: PETS 2009 Benchmark Data (2009). http://www.cvg.rdg.ac.uk/PETS2009/a.html
6. Figueiredo, M.A., Jain, A.K.: Unsupervised learning of finite mixture models. IEEE Transactions on Pattern Analysis and Machine Intelligence **24**(3), 381–396 (2002)
7. Haralick, R.M.: Statistical and structural approaches to texture. Proceedings of the IEEE **67**(5), 786–804 (1979). doi:10.1109/proc.1979.11328
8. Horn, B.K.P., Schunck, B.G.: Determining optical flow. Artificial Intelligence **17**, 185–203 (1981)
9. Hotelling, H.: Analysis of a complex of statistical variables into principal components. Journal of Educational Psychology **24**(6), 417–441 (1933). doi:10.1037/h0071325
10. Kohonen, T.: Self-organized formation of topologically correct feature maps. Biological Cybernetics **43**(1), 59–69 (1982)

11. Liao, H., Xiang, J., Sun, W., Feng, Q., Dai, J.: An abnormal event recognition in crowd scene. In: 2011 Sixth International Conference on Image and Graphics (ICIG), pp. 731–736. IEEE (2011). doi:10.1109/ICIG.2011.66

12. Mahadevan, V., Weixin, L., Bhalodia, V., Vasconcelos, N.: Anomaly detection in crowded scenes. In: 2010 IEEE Conference on Computer Vision and Pattern Recognition (CVPR), pp. 1975–1981. IEEE (2010). doi:10.1109/CVPR.2010.5539872

13. Milan, A.: Ground Truth - PETS 2009 (2015). http://www.milanton.de/data.html

14. Popoola, O.P., Kejun, W.: Video-based abnormal human behavior recognition–a review. IEEE Transactions on Systems, Man, and Cybernetics, Part C: Applications and Reviews $42(6)$, 865–878 (2012). doi:10.1109/tsmcc.2011.2178594

15. Rajasegarar, S., Leckie, C., Palaniswami, M.: Hyperspherical cluster based distributed anomaly detection in wireless sensor networks. Journal of Parallel and Distributed Computing $74(1)$, 1833–1847 (2014). http://dx.doi.org/10.1016/j.jpdc.2013.09.005

16. Rao, A.S., Gubbi, J., Marusic, S., Palaniswami, M.: Estimation of crowd density by clustering motion cues. The Visual Computer, 1–20 (2014). doi:10.1007/s00371-014-1032-4, http://dx.doi.org/10.1007/s00371-014-1032-4

17. Rao, A.S., Gubbi, J., Rajasegarar, S., Marusic, S., Palaniswami, M.: Detection of anomalous crowd behaviour using hyperspherical clustering. In: 2014 International Conference on Digital Image Computing: Techniques and Applications (DICTA), pp. 1–8. IEEE (2014). doi:10.1109/DICTA.2014.7008100

18. Roweis, S.T., Saul, L.K.: Nonlinear dimensionality reduction by locally linear embedding. Science $290(5500)$, 2323–2326 (2000). doi:10.1126/science.290.5500.2323

19. Srivastava, S., Ng, K.K., Delp, E.J.: Crowd flow estimation using multiple visual features for scenes with changing crowd densities. In: 2011 8th IEEE International Conference on Advanced Video and Signal-Based Surveillance (AVSS), pp. 60–65. IEEE (2011)

20. Tenenbaum, J.B., Silva, V.D., Langford, J.C.: A global geometric framework for nonlinear dimensionality reduction. Science $290(5500)$, 2319–2323 (2000). doi:10.1126/science.290.5500.2319

21. Thida, M., Eng, H.L., Dorothy, M., Remagnino, P.: Learning video manifold for segmenting crowd events and abnormality detection. In: Lecture Notes in Computer Science, vol. 6492, book section 34, pp. 439–449. Springer, Heidelberg (2011). doi:10.1007/978-3-642-19315-6_34, http://dx.doi.org/10.1007/978-3-642-19315-6_34

22. Torgerson, W.: Multidimensional scaling: I. theory and method. Psychometrika $17(4)$, 401–419 (1952). doi:10.1007/BF02288916

23. Tziakos, I., Cavallaro, A., Xu, L.Q.: Event monitoring via local motion abnormality detection in non-linear subspace. Neurocomputing $73(10–12)$, 1881–1891 (2010). http://dx.doi.org/10.1016/j.neucom.2009.10.028, http://www.sciencedirect.com/science/article/pii/S0925231210001487

24. Valera, M., Velastin, S.A.: Intelligent distributed surveillance systems: a review. In: IEE Proceedings Vision, Image and Signal Processing, vol. 152, pp. 192–204. IET (2005)

25. Wang, S., Miao, Z.: Anomaly detection in crowd scene. In: 2010 IEEE 10th International Conference on Signal Processing (ICSP), pp. 1220–1223. IEEE (2010). doi:10.1109/ICOSP.2010.5655356

26. Yilmaz, A., Javed, O., Shah, M.: Object tracking: A survey. ACM Computing Surveys (CSUR) $38(4)$, 13 (2006)

27. Zhao, B., Fei-Fei, L., Xing, E.P.: Online detection of unusual events in videos via dynamic sparse coding. In: 2011 IEEE Conference on Computer Vision and Pattern Recognition (CVPR), pp. 3313–3320. IEEE (2011). doi:10.1109/CVPR.2011.5995524

# Access Control System Which Uses Human Behavioral Profiling for Authentication

Lohit Penubaku, Jong-Hoon Kim, Sitharama S. Iyengar
and Kadbur A. Shilpa

**Abstract** "An access control device has been used for securing valuable properties and lives from sinister people" [1]. Recently, these security device technologies have been improved tremendously in providing security for people through various methods. "Nevertheless, these methods are not individually perfect to provide optimal security. These, most of the time, are not convenient for a wide range of users (i.e, the innocent users who do not pose any threat) due to access time delay and several layers of authentication. The proposed security system should exhibit capabilities that support adaptive security procedures for a diverse range of users, so most innocent users require a minimum layer of identity authentication and verification while suspicious users may require passing through some additional layers of security authentication and verification" [1]. This paper proposes "a novel smart access control (SAC) system, which can identify and categorize suspicious users from the analysis of one's activities and bio information. The SAC system observes and records users' daily behavioral activities. From the analysis of the collected data, it selectively chooses certain users for additional layers of authentication procedure and quickly isolates those individuals who might pass through scrutiny by security personnel. Due to this adaptive feature, the SAC system not only minimizes delays and provides more convenience to the users but also enhances the security measure" [1]. Moreover, a novel idea of DPIN, a concept that uses memory and analytical potency of a user to dynamic generation, and updates ones security key is proposed.

L. Penubaku(✉) · K.A. Shilpa
Reva University, Bangalore, India
e-mail: {lohitpenubaku,shilpaka}@reva.edu.in

J.-H. Kim · S.S. Iyengar
Florida International University, Miami, USA
e-mail: {kimj,iyengar}@cis.fiu.edu

© Springer International Publishing Switzerland 2016
S.M. Thampi et al. (eds.), *Advances in Signal Processing and Intelligent Recognition Systems*,
Advances in Intelligent Systems and Computing 425,
DOI: 10.1007/978-3-319-28658-7_36

419

# 1   Introduction

"The safety is the top priority of many organizations. Organizations are working on providing secure access to their resources in order to prevent any kind of loss to the organization" [1]. "Access control mechanism is used to protect both the resources and the personnel. One of the most important techniques to implement this kind of security is the door security" [1]. The essential part of any security system accurately identifies the person who enters and leaves through the door. An entrance guard can be managed with using "physical and non-physical keys. The physical access keys include situations where the user has been provided with some physical entities such as lock/key, smart cards like RFID, magnetic strips, biometric, etc" [1]. [2], [3], [4],[5]. "The non-physical keys are more like entities that cannot be touched or seen, for example, PIN, password, behavioral, etc. [6], [7].

Physical and non-physical keys have been useful in providing secure access to the resources. However, these methods are not individually perfect, to provide optimal security [8]. They are also not user-friendly and are not convenient to use for most of the users. Therefore, the proposed SAC system will support adaptability features and provide a minimum layer of authentication for the legitimate users and making it difficult for inappropriate users to breach the security system. The heart of the SAC system is to observe and record daily activities of the users and eventually use the same to create patterns for providing security adaptability. After conducting experiments and analyzing the data, the system selectively chooses suspicious users for an additional layer of security and isolates those individuals who might pass the scrutiny. Due to this adaptive feature, the SAC system not only minimizes delays, it also provides more convenience to the users and enhances the security measure at the same time" [1].

"The SAC system also uses the novel idea of dynamic PIN, which makes use of the users' memory and brain power to generate and update their security key dynamically so that the key keeps changing with time, without the need of additional devices" [1].

# 2   Related Work

There exist different kinds of access control systems. There is a distributed smart card based system for a building [4] which is Internet-centered. "Each door has an access control point, which is connected to the server and this takes care of the access. The users are given access by using the smart card, which contains the user's unique identity. This system fails to consider the situation of stolen, duplicated, forgotten, lost or impersonated cards with accuracy" [1].

"There is an access control system combining fingerprint recognition, vault scheme, and IC card technique" [1] [5]. This system does not store the user's fingerprint template in his IC card but stores the secret key locked by his fingerprint. Only the person with matching fingerprint can be able to retrieve the key successfully and, therefore, validates his identity. Even though, the security of the whole system is enhanced greatly, it does not take care of "situations where the user finds it difficult to login due to sweat in his hands or situations where there is a cut on the finger" [1].

An existing 3D face recognition system uses information like color depth data of human facial skin for detection and PCA (principal components' analysis) algorithm for recognition [9]. This method "overcomes the problem of general 2D face recognition technique, that may fail with illumination, pose, expression, makes up and age" [1]. Nevertheless, the factors such as illumination, expressions, and mechanical vibrations "affect the system significantly and the user may find difficulty in accessing the system" [1].

An authentication mechanism called ColorPIN [6] that uses an indirect input to provide security, which is completely independent of the user. Here there exists a one-to-one relationship between PIN length and the number of key presses to provide higher security. However, the system is vulnerable to intersection attacks. There is a technique to collect password based on pattern recognition and analyzing those passwords using a visualization technique [7]. The heuristics are developed based on recognized patterns. Even so, this approach was limited to a subset of pattern-based passwords.

Surveying "all the current systems and their drawbacks, a system needs to be built which will overcome the drawbacks. This can be possible by making the system intelligent and adaptable in real time. To achieve this, the system observes and record user's daily behavioral activities and uses these to create patterns for providing an adaptive feature for security. With the help of these, the SAC system assists in minimizing the delay (time taken to get access) for each user. This adaptive feature makes the SAC system intelligent, which in turn enhances the security measure" [1].

## 3 System Design

The system shown in the Fig. 1 consists of three modules: Device, Manager and Database. In Section 3.1 basic architecture of the system is introduced. Later in Section 3.2 the implementation part is discussed based on the proposed architecture.

### 3.1 System Model

The "device module is a standalone device. It basically holds all the hardware" [1] needed for the system which includes primary-level, secondary-level, user interaction level, and lock/latch hardware's. The primary-level hardware is used for the basic authentication for the system such as RFID, smart card, etc. The secondary-level hardware is used to improve the security feature of the system such as fingerprint reader, an iris scanner, etc. "The user interactive level hardware provides an interface between user and" [1] components, such as keypad, touchpad, LCD display, etc.

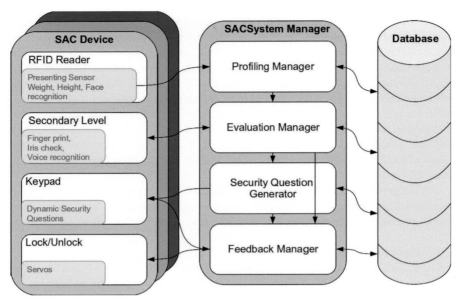

**Fig. 1** Secure Access Control System Model [1]

The manager module is considered to be the brain of SAC architecture since the data from the SAC device is processed by this module. "The SAC manager is a daemon process that runs continuously. The manager communicates" [1] between SAC device and database. This module includes sub-manager modules such as profiling, evaluation, security and feedback manager. "The profiling manager is used to handle all the data pertaining to a particular user profile. The user profile data can be a traditional key, user ID (RFID, IC card, etc.), password, pin, behavioral pattern, weight and height, skin color, tracking path and biometric information (face recognition, iris, hair color, etc.)" [1]. These data are stored in a database. The evaluation manager differentiates users with respect to their behavioral patterns. It not only evaluates data, but it validates the legitimacy of the users. "The role of the security manager is to request additional security information from the user. The feedback manager is the one which decides whether to allow or deny access for the users. The feedback manager also maintains a log table in the database. Both security and the feedback manager are controlled by the evaluation manager" [1].

The database module is used for storing user data such as profiling information, access logs which in turn will be used by the evaluation manager for assessing the legitimacy of the user. "Most of the database transactions are handled by the evaluation manager. Following all the evaluations is done, the manager updates the profiling data about the user for future evaluation" [1].

## 3.2  System Implementation

### 3.2.1  SAC Device

"This module is implemented by making use of varieties of hardware such as primary level, user interactive, and lock hardware. All the hardware components are interfaced to a controller called Robostix and this is controlled by a processor named gumstix, which is based on ARM architecture. The purpose of gumstix in the implementation is to do basic data processing and to send processed data to the SAC manager for further handling Robostix is one of the expansion modules that are designed to accompany the Gumstix in applications. Robostix is a module that will provide the required pins to read data from the sensors, by converting them from analog to digital and then forwarding them to the Gumstix" [1].

"The primary level hardware includes RFID reader, sonar, a load cell. The RFID reader comes with an antenna used for receiving ID from an RFID tag. When a user presents the card to the reader, the RFID tag gets activated because of a signal, and the tag sends its unique ID to the reader, which is used to differentiate between users and helps in creating profiling data for the respective user. The sonar is a device that uses the sound wave to calculate distance. The sonars are arranged around the RFID reader in a particular way so that the presenting pattern of the user can be recorded. The load cell is a transducer based device that when subjected to a force will cause a change in its physical form" [1].

"The only user interactive level hardware in the implementation is the keypad. The one that the system uses is classic 4X3 matrix keypad, which is not vulnerable as the system uses the concept called Dynamic PIN (DPIN). Every user has a four-digit static PIN. The DPIN concept [10] proposed by the RRL group uses this PIN to obtain DPIN. The DPIN is computed by performing an arithmetic operation on the static PIN using personal information only known to a user. Every time one does arithmetical operations it results in a different DPIN" [1].

### 3.2.2  SAC Manager

"The SAC manager is a remote server used to control the SAC device and also maintaining the system database. In the implementation, a Linux system is used as a server. The server program is completely implemented in C language. The software is split into three different parts SAC agent, SAC server and SAC DB program. The SAC server is a socket program that constantly waits for connection from the SAC device. When the server receives a connection request, it creates a client to send the information over to the SAC agent program. The SAC agent program is the actual implementation of the SAC manager. All the sub-manager functionality are implemented in this agent program. The agent program frequently communicates with the SAC DB program for evaluating and updating the database. The SAC DB can be called a database program as it is written in PostgreSQL C language. This is nothing but a C application programmer's interface to PostgreSQL. libpq is a set of library functions that allow client programs to pass queries to the PostgreSQL backend server and to receive the results of these queries" [1].

### 3.2.3 Database

The database is used to store profile data for a user. "Since the code is written in C language, PostgreSQL is used which can be queried in C language. The database has four different tables: Registration, History, Profile and Access Log. The registration table holds all the required user information like user ID, address, telephone number, static pin, etc. The profile table contains only profiling data, which is the sensor values from the sonar and the load cells. The history table has information about each user transaction such as the number of attempts made, access granted, and access denied, etc. And finally the access log table tells us about the number of connections made to the database" [1].

## 4 Operational Procedure

The operational procedure of the SAC system can be depicted "with the help of a flow diagram shown in the Fig. 2. The user needs to present his primary level ID to the SAC device. The user ID read from SAC device is sent to the SAC manager. The manager then evaluates the data received to see if the user exists in the database. If the user ID does not match to any entry in the database, it waits for a few seconds before it can acknowledge access denied. However, if the user exists, then the SAC manager evaluates the additional data received from the primary level hardware which is sent by the SAC device to the manager. The manager uses this data to evaluate profiling data previously created in the database for the user. The feedback manager is responsible for maintaining the connection between modules at all time" [1].

"If the data matches, the manager updates the database with the latest profiling data for the next evaluation for the same user which helps improve the precision of the system. If the profiling evaluation fails, then the user will be put through a secondary level of security checks" [1] by requesting the user DPIN. The DPIN will be created for all the right users, and it will be stored in the database. The data received from this level next goes through the evaluation process. "Hence, if the DPIN provided by the user matches with the one stored in the database, the user is said to clear the secondary-level, next the manager updates the profiling data and allows access for the user. If the user fails in this level, the manager checks the number of attempts on the ID and if it exceeds five times, subsequently the manager locks the user and denies access. If all the decision blocks say "yes" then the manager learns/updates' database and says access granted. The system keeps updating the database which helps in behavioral learning of the user" [1].

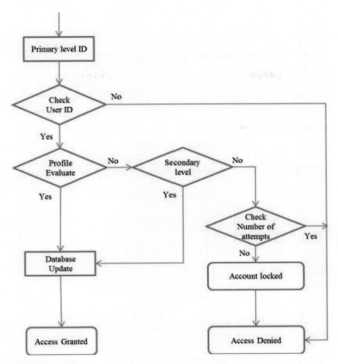

**Fig. 2** Generic flow chart [1]

## 5  System Evaluation

The Fig. 4 shows the final working prototype. "The sonars are installed in a circular fashion to collect the presenting pattern as efficiently as possible. Then there is a keypad for secondary-level authentication where the user ID matches but not the pattern, and the LCD screen for user interaction purpose. Fig. 5 is a prototype of the weight sensor which is visible on the floor" [1].

"The sonars are used to create presenting pattern for each user. In the system, RFID is used as a presenting item. The sonars are arranged around RFID reader as shown in Fig. 4. With this setup, the system could differentiate users in the way they present the RFID card" [1].

The state of the sonars shown in Fig. 6 is in its idle state. "When the user presents the RFID, the device is initiated, and the system finds a change in the values of the sonars. With this, it records the pattern as part of user profiling, below are a few expected patterns. With this proposed architecture, a drawback is found in using sonar for the system. As shown in Fig. 7 that two users may have the same pattern. To overcome this, weight is used as another parameter to differentiate between users as shown in Fig. 5" [1].

(a) Tables in Database

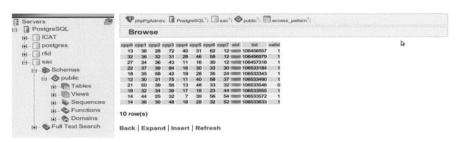

(b) Profile Table

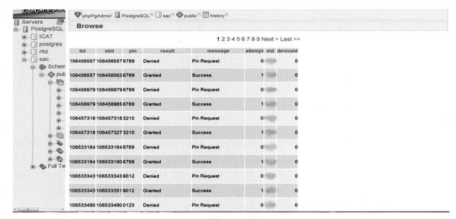

(c) History Table

**Fig. 3** Database contents [1]

"Fig. 3 shows the tables in the database. Fig. 3b is the screenshot of the presenting patterns collected, and Fig. 3c the history table which is the most important table for the system to learn the behavior of the user. Fig. 3a shows all the tables that the system makes use for evaluation. It can clearly be seen how different users have unique presenting patterns. But the system cannot see visually, hence it records the distance between the hand and the sonar to create a presenting data pattern that the system uses for evaluation. Whenever a user presents RFID, the presenting pattern

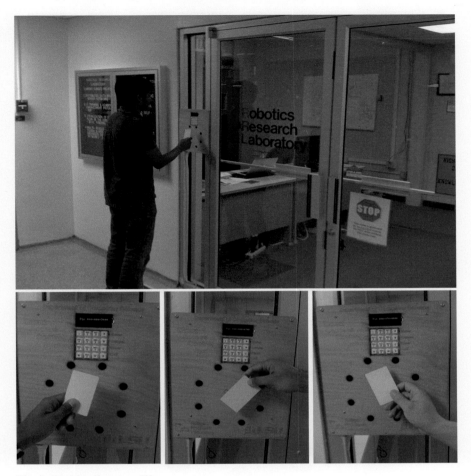

**Fig. 4** Working Prototype [1]

data is collected and stored in the profile table. Based on the final evaluation the system marks valid or not, on the pattern data stored" [1].

"Fig. 3c shows the PIN keeps changing because of the Dynamic PIN concept. Keywords like "Pin Request" means SAC manager is requesting for PIN and the respective Dynamic PIN is also stored in the table. Keywords like "Denied", "Success" and "Granted" are the decisions made by the SAC manager to keep track of the user request. Based on these decisions the SAC manager updates the profiling table to validate between the frequency of good and bad presenting patterns"[1].

"The addition of the weight sensor increased the efficiency of the system a lot. One thing that needs to be observed is the user is not actively involved in this whole process. All that the user does is present his card to the system, and the rest of the process is carried out without the user knowledge" [1].

**Fig. 5** Weight sensor prototype [1]

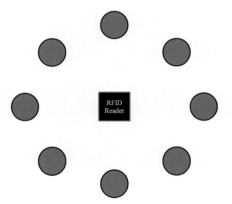

**Fig. 6** Sonars in Idle State[1]

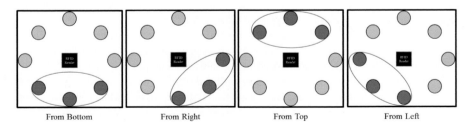

From Bottom          From Right          From Top          From Left

**Fig. 7** Possible user Patterns[1]

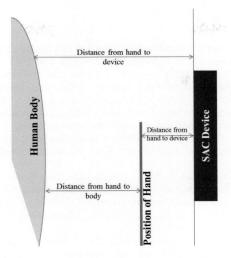

**Fig. 8** Distance Computation[1]

## 6 Conclusion

"In this paper, a novel, scalable security system that avoids the many limitations of the state-of-the-art security systems is proposed which act as a basic building block for the future generation security systems. The system is adaptive, which incorporates intelligent features and provides enhanced security. In the experimental evaluation, the SAC system showed very good performance just with fundamental primary level hardware, and it provided reliable security with minimal user interaction. A distinct feature of this system is that it implements behavioral learning so that the system can maintain the up-to-date information about the users" [1].

## 7 Future Scope

"While evaluating the data of the presenting pattern, one other parameter was observed, which can add up to the decision-making process. From the fig. 8, initially the concept considered only the distance from hand to device retrieved from the sonars. Nevertheless, with the sonars, we can also be able to retrieve the position of the body as well. Using this, we can know the distance between the hand and the body. This also is unique to a user as people have varying hand length" [1].

# References

1. Penubaku, L.: Smart access control system with behavioral profiling and dynamic pin concept, Master's thesis, Louisiana State University, Baton Rouge, June 2011
2. Roussos, G.: Networked RFID: systems, software and services. In: Computer Communications and Networks. Springer, London (2008). https://books.google.co.in/books?id=YMdguq8bAA4C
3. Astuti, W., Mohamed, S.: Intelligent voice-based door access control system using adaptive-network-based fuzzy inference systems (anfis) for building security
4. Popa, M., Popa, A., Marcu, M.: A distributed smart card based access control system. In: 8th International Symposium on Intelligent Systems and Informatics (SISY), pp. 341–346, September 2010
5. Zhang, Y., Li, Q., Zou, X., Hao, K., Niu, X.: The design of fingerprint vault based ic card access control system. In: Proceedings of the 5th WSEAS International Conference on Electronics, Hardware, Wireless and Optical Communications, EHAC 2006, pp. 172–175. World Scientific and Engineering Academy and Society (WSEAS), Stevens Point (2006). http://dl.acm.org/citation.cfm?id=1365818.1365851
6. De Luca, A., Hertzschuch, K., Hussmann, H.: Colorpin: Securing pin entry through indirect input. In: Proceedings of the SIGCHI Conference on Human Factors in Computing Systems, CHI 2010, pp. 1103–1106. ACM, New York (2010). http://doi.acm.org/10.1145/1753326.1753490
7. Schweitzer, D., Boleng, J., Hughes, C., Murphy, L.: Visualizing keyboard pattern passwords. In: 6th International Workshop on Visualization for Cyber Security, VizSec 2009, pp. 69–73, October 2009
8. Delac, K., Grgic, M.: A survey of biometric recognition methods. In: 46th International Symposium on Electronics in Marine, Proceedings Elmar 2004, pp. 184–193, June 2004
9. Qian, J., Ma, S., Hao, Z., Shen, Y.: Face detection and recognition method based on skin color and depth information. In: 2011 International Conference on Consumer Electronics, Communications and Networks (CECNet), pp. 345–348, April 2011
10. Lab, R.R.: Dynamicpin: A novel approach towards secure atm authentication

# Mathematical Morphology and Region Clustering Based Text Information Extraction from Malayalam News Videos

K. Anoop, Manjary P. Gangan and V.L. Lajish

**Abstract** Innovations in technologies like improved internet data transfer, advanced digital data compression algorithms, enhancements in web technology, etc. enabled the exponential growth in digital multimedia data. Among the massive multimedia data, news videos are of higher priority due to its rich up-to-date information and historical evidences. This data is rapidly growing in an unpredictable fashion which requires an efficient and powerful method to index and retrieve such massive data. Even though manual indexing is the most effective, it is the slowest and most expensive. Hence automatic video indexing is considered as an important research problem to be addressed uniquely.

In this work, we propose a Mathematical Morphology and Region Clustering based Text Information Extraction (TIE) from Malayalam news videos for Content Based Video Indexing and Retrieval (CBVIR). Morphological gradient acts as an edge detector, by enhancing the intensity variations for detecting the text regions. Further an agglomerative clustering is performed to select the significant text regions. The precision, recall and $F_1$-measure obtained for the proposed approach are 87.45%, 94.85% and 0.91 respectively.

## 1 Introduction

One of the most challenging tasks in the video database research is to make available effective tools for automatic video indexing and retrieval from huge video data bases. It will be more useful if the method works based on analyzing the semantic content of the video data. The literature shows several promising research outcomes are reported in the area of digital video analysis like indexing, abstraction, structural analysis, video skimming and CBVIR. Video abstraction, structural

K. Anoop · M.P. Gangan · V.L. Lajish(✉)
Department of Computer Science, University of Calicut, Malappuram, Kerala, India
e-mail: {anoopuoc,manjaryp}@gmail.com, lajish@uoc.ac.in

© Springer International Publishing Switzerland 2016
S.M. Thampi et al. (eds.), *Advances in Signal Processing and Intelligent Recognition Systems*,
Advances in Intelligent Systems and Computing 425,
DOI: 10.1007/978-3-319-28658-7_37

431

analysis and video skimming are also been incorporated with the CBVIR systems. Most of the researches in CBVIR are based on a various types of videos like sports, entertainment, and news.

The three important modalities (visual, auditory and textual) of the video data naturally lead to a semantic content and layout perspective. Among them the textual modality have a special interest as (1) they are very useful for describing the content of the video; (2) can be easily extracted and compared to other modalities to get more semantic idea about the video; (3) enables applications such as keyword-based video search and indexing and automatic video logging etc. A wide range of Text Information Extraction (TIE) methods for videos and image dataset are available in the literature and these methods are experimented with a variety of applications like document page segmentation and content analysis, address book localization, license plate localization, content based indexing, etc.

Television is one of the predominant source of delivering information to the society, in which news plays an important role. In Kerala, Malayalam is the official language and it is one among the 22 official languages of India. Total number of speakers of the language is more than 35 million spreading along the regions Kerala, Lakshadweep, and Mahe. It is designated as a Classical Language in India. DD Malayalam is the satellite channel supported by Doordarshan (India's Public Service broadcaster), which is a prominent television channel of Kerala. Literature shows that very few works are done in the area of automatic CBVIR systems on Malayalam news videos.

In this work, we propose a mathematical morphology and region clustering based text information extraction from Malayalam news videos for the purpose of content based video indexing and retrieval. Here we automatically extracted the news ticker text[1] using mathematical morphology and region clustering from the Malayalam news videos and further this can be effectively used for indexing and retrieval. The rest of the paper is organized as follows. Section 2 disuses the review on content based broadcast news video indexing. Section 3 discusses the proposed method for identifying the region of interest. Section 4 discusses the experimental result followed by the conclusion and further direction for the research.

## 2     Review on Content Based Broadcast News Video Indexing

CBVIR have a wide variety of application in present days because of the rapid growth in the video databases. Major applications are news video analysis, intelligent management of web videos, video surveillance, analysis of visual electronic commerce, sports video analysis, story based video retrieval etc.

---

[1] A news ticker text (sometimes called a "crawler" or "slide") is a primarily horizontal, text-based display either in the form of a graphic that typically resides in the lower third of the screen space on news window.

Among the massive videos, broadcast news videos have special interest because of their information rich nature. Different category of tasks, like story based segmentation, event based segmentation and analysis, etc. are performed on the news videos [1-4]. Major tasks are involved with textual, auditory and visual modalities [5-8]. Majority of textual modality based CBVIR research works in news video domain are proposed in English language [9], some of them are in Arabic [10] and Chinese [11]. Research works in the related area are also active in some Indian languages including Hindi and Telugu [12] and very few works are proposed in Malayalam language [13]. One of key issue in implementing an end-to-end CBVIR system in Indian languages is the non-availability of the good Optical Character Recognition systems.

In a recent work, M. Bertini et al. proposed a content based shot classification and extraction of high level content descriptor using caption OCR and speech recognition system for Italian news videos[14]. One of the latest work proposed by M. Daud Abdullah Asif et al. is a novel hybrid method for text detection and extraction from English news videos like CNN and BBC [9]. Harshal Gaikwad et al. [15] and Toshio Sato [16][17] proposed related works, which are based on the CNN news videos. Weiyu Zhu et al., adopted another novel approach for automatic news video categorization based on the CNN news videos [18]. Mohamed Ben Halima et al., proposed a robust approach for text extraction and recognition from Arabic news videos [11]. Few works are also reported for video indexing and retrieval in some of the Indian languages. The best evidences are the work proposed by Jawahar C. V. et al. [12] and Anoop K et al. [13]. Jawahar C.V. et al presented a video indexing method based on textual queries. In this method text is extracted from the frame and tries to find the match in words at image level using dynamic time warping (DTW). Anoop K et al. formulated an initial attempt to extract ticker text from Malayalam news videos. News video scene segmentation and key frames extraction depends on the changes in region of interest. Mathematical morphology and connected component analysis are the fundamental concepts used in this paper for TIE. The average precision and recall reported are 85.52% and 94.13% respectively.

Textual modality is an important element used for automatic semantic video indexing and retrieval. Several TIE approaches are available in the literature based on text extraction for scene image, text document, video, etc. The TIE from digital video or image deals with a of approaches say based on the adaptive thresholding and relaxation operations [19], edge information followed by connected component analysis [20], wavelet and neural network, texture analysis [22][23] and morphological approaches. Wonjun Kim et al. introduced a simplest method for text detection on edge based approach [28]. The summary of some important TIE methods are given in table1.

# 3   Text Region Identification and Extraction

We propose a method which will automatically identify the Region of Interest (ROI) and select the ticker text area from the video frames. The method is a three level process. The first level includes the text region identification. Second level is clustering to identify and extract the ticker texts. In the final stage, a normalized cross correlation (NCC) is computed to select unique tickers from a video. The working of the proposed text region identification is shown in Fig. 1. The detailed description of the important stages of the proposed method is given below.

**Table 1** Different methods of Text Information Identification and Extraction

| Category | Methodology | Publication |
|---|---|---|
| TIE from Video | Overlay | Wonjun Kim et al. [29] |
| | Multilingual | Lyu M.R et al. [30] |
| | Hybrid wavelet, Neural network | H. Li et al. [21], |
| | | Jun Ye et al. [31] |
| | Segmentation by color clustering | Hae-Kwang Kim et al. [32] |
| | Color reduction, Connected component | Jain A.K et al. [33] |
| | Connected component, Texture feature | P. Nagabhushan et al. [34] |
| TIE from scene or text Image | Horizontal spatial variance , Color segmentation | Yu Zhong et al. [35] |
| | Binairzation, Connected component, random forest classifier | Karaoglu S et al. [36] |
| | Comparison edge , Connected component, Texture | M. Swamy Das et al. [37] |
| | Discreet Cosine Transformation | Adrin et al. [38] |
| Mathematical Morphology | Morphology and histogram projection | Rodolfo P et al. [25] |
| | Morphology and SVM | R. Chandrasekaran et al. [26] |
| | Morphology | T. Pratheeba et al. and Hasan et al. [24] [27] |
| | Haar Discrete Wavelet Transform K-Mean Clustering and Morphology | Hongjiang Zhang et al. [39] |

## 3.1  Edge Assisted Text Region Identification

The system initially converts the color video frames into grayscale images and computes the corresponding morphological gradient. This morphological gradient can act as an edge detector, by enhancing the intensity variations. A square structuring element of size 5 was used for obtaining the morphological gradient. Otsu's thresholding is performed on the resultant gradient image to get a high quality binary image. The binarized image is further filtered using a Laplacian filter. Laplacian is a technique of creating a second derivative mask for highlighting the edges and other discontinuities in the binary image and hence it is efficient in identifying text regions. Laplacian of a function f(x,y) of two variable is defined as

$$\Delta^2 = \frac{\partial f}{\partial x^2} + \frac{\partial f}{\partial y^2} \tag{1}$$

The discrete Laplacian of two variables is

$$\Delta^2 f(x,y) = f(x+1,y) + f(x-1,y) + f(x,y+1)$$
$$+ f(x,y-1) - 4f(x,y) \tag{2}$$

It can be seen that the equation (2) can be implemented using the filter shown in Fig. 2.

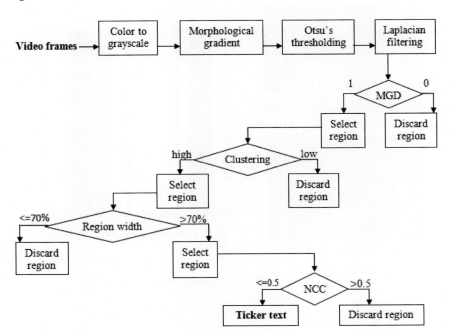

**Fig. 1** Text region identification - proposed method

## 3.2  Maximum Gradient Difference for Text Region Identification

In the next step we measured the Maximum Gradient Difference (MGD) from the filtered image. MGD is computed by subtracting the minimum gradient value from maximum gradient value. The output provides a zero value for non-text regions and non-zero value for textual regions. This helps in discarding many unwanted non-textual regions in the video frame. Fig. 3 shows the input frame and its corresponding Laplacian and MGD plot. The blue beams in the MGD plot represent position of text regions. The textual regions are further checked for ticker text, by comparing the vertical axis values of each text region beam in the frame. It is obvious that the ticker texts are always present horizontally towards the bottom of the news videos.

| 0 | 1  | 0 |
|---|----|---|
| 1 | -4 | 1 |
| 0 | 1  | 0 |

**Fig. 2** Laplacian filter or mask

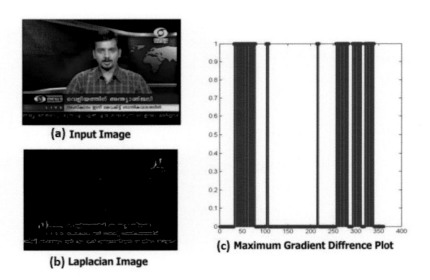

(a) Input Image

(b) Laplacian Image

(c) Maximum Gradient Diffrence Plot

**Fig. 3** (a) Input video frame, (b) corresponding Laplacian and (c) MGD plot

## 3.3 Agglomerative Clustering for Ticker Text Selection

In the second level we performed an agglomerative clustering of the text regions based on their vertical axis values. We selected three pair of text beam values from the vertical axis of MGD plot which have the highest vertical value in the frame. The three pair of values selected from each frame is clustered to discard the frame that doesn't contain any ticker text region. Agglomerative clustering is a bottom up hierarchical clustering method, which set each objects as a cluster to merge the two closest clusters and update distance matrix. The process of merging and updating the distance matrix until number of required clusters is two. The two clusters represent the frames with and without ticker texts.

Generally, a complete displayed ticker text has a horizontal length more than 75% of the total width of the frame. If the horizontal length is less than this value, we can predict this ticker is not completely displayed and can be obtained from further frames. The ticker text regions with horizontal value less than 75% can thus be discarded.

## 3.4 Candidate Ticker Text Selection Based on Normalized Cross Correlation

In third level, we perform a redundancy check for selecting the candidate ticker text by removing replication among the extracted ticker texts based on Normalized Cross Correlation (NCC). Considering the dataset, utmost three ticker text regions are present in a video frame which may repeat several times in the total video. A high NCC reports high amount of redundancy among the ticker texts. This will decrease the space required to store text information corresponding to each news video to a great extend, storing only those which are significant. The NCC between the ticker texts is computed using the following equation.

$$NCC(a,b) = \frac{1}{n}\sum_{i}^{n} \frac{(a_i-\hat{a})(b_i-\hat{b})}{\sqrt{\frac{1}{n}\sum_{i}^{n}(a_i-\hat{a})}\sqrt{\frac{1}{n}\sum_{i}^{n}(b_i-\hat{b})}} \tag{3}$$

where $\hat{a}$ and $\hat{b}$ are the average intensities of each image and, $a_i$ and $b_i$ are intensity of each pixel. Ticker region identified using this method have a single color background. The characters can be extracted using connected component analysis, shown in Fig. 4.

**Fig. 4** Ticker text extracted from a news video frame and corresponding connected components

## 4    Experimental Results

The proposed mathematical morphology and region clustering based TIE from Malayalam news videos for Content Based Video Indexing and Retrieval (CBVIR) method is experimented with a dataset of 100 one minute Malayalam broadcast news videos. The dataset is MPEG-4 compressed and collected from DD Malayalam. The frame size of the video is 360 x 480 and each video contains an average of 1500 frames. The problem reported by other approaches are the repeated selection of key frames with same ticker texts, usage of various hard thresholds, etc.[13]. The repeated selection of the key frames with same ticker text is determined in the proposed approach by using agglomerative clustering and NCC. In this section we analyze the experimental results and compare it with the standard approaches reported in the litrature.

The performance analysis of the proposed methods is carried out using precision, recall and F-score of the extracted text regions.

$$Recall = \frac{NC}{NC+NM} \tag{4}$$

$$Precision = \frac{NC}{NC+NFp} \tag{5}$$

where,$NC, NM, NFp$ are the number of correctly retrieved text regions, missed ones, and false positive respectively.

$F_1$ score or F-score is calculated as the weighted average of precision and recall. The best value of $F_1$ is at 1 and worst at 0.

$$F_1 = 2 * \frac{Precision*Recall}{Precision+Recall} \tag{6}$$

The proposed method results in an average of 15 ticker texts from a one minute video. The precision, recall and $F_1$ score obtained are 87.45%, 94.85% and 0.91 respectively. Fig. 5 shows experimental result including different stages of a video frame: input image, gray scale image, morphological gradient, Otsu's thresholded image, laplacian filtered image, connected component analysis, incomplete ticker removal and the final ticker extraction.

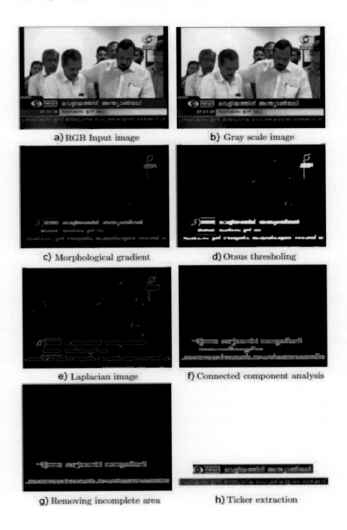

Fig. 5 (a-h). Results of proposed TIE method

Table 2 compare the results of the proposed method with other three standard approaches. Two approaches among them is based on the mathematical morphology [24][13] and the other one is an overlay based text detection and extraction method [29]. The experimented results show that the proposed method outperforms these three promising methods reported in the literature.

**Table 2** Comparison of different TIE results

| Author | Methodology | Precision (%) | Recall (%) | $F_1$- measur |
|---|---|---|---|---|
| W Kim et al. [29] | Overlay | 80.15 | 87.01 | 0.83 |
| Yassin M.Y. et al.[24] | Morphology | 78.03 | 85.05 | 0.81 |
| Anoop K et al. [13] | Morphology | 86.56 | 93.17 | 0.9 |
| Proposed method | Morphology and region clustering | 87.45 | 94.85 | 0.91 |

# 5 Conclusion and Future Direction of the Research

In the proposed approach, we identified and extracted the textual keywords from the DD Malayalam news video documents based on mathematical morphology and region clustring. The scopes open to the proposed approach is, to do away with the heuristics used explicitly for the estimation of text region and the hard thresholds used for key frame selection.

NCC is used to find the similarity of the ticker text content extracted from different frames. Thus, the frames with similar text content are eliminated. The average precision recall and F Score obtained are 89%, 95% and 0.91 respectively. The proposed method for TIE, from broadcast news videos are found robust as it gives better results with varying size, color, and fonts of text. The result is still not fully realized in the area of CBVIR in Malayalam due to the unavailability of a good Malayalam OCR engine. The further enhancements include reducing the clustering complexity in TIE, character recognition for Malayalam text indexing and retrieval, integration of textual, auditory modalities for developing a multimodal CBVIR system for Malayalam broadcast news retrieval.

# References

1. Chua, T.-S., Chaisorn, L., Hsu, W., Chang, S.-F., et al.: Story boundary detection in large broadcast news video archives: techniques, experience and trends. In: Proceedings of the 12th Annual ACM International Conference on Multimedia, MULTIMEDIA 2004, pp. 656–659 (2004)
2. Katayama, N., Satoh, S., Ide, I., Mo, H.: Topic-based inter-video structuring of a large-scale news video corpus. In: International Conference on Multimedia and Expo, vol. 3 (2003)
3. Lee, C.-H., Chaisorn, L., Chua, T.S.: The segmentation of news video into story units. In: IEEE International Conference on Multimedia and Expo, vol. 1, pp. 73–76 (2002)
4. Hauptmann, M.J., Witbrock, A.G.: Story segmentation and detection of commercials in broadcast news video. In: IEEE International Forum on, Research and Technology Advances in Digital Libraries, pp. 168–179 (1998)

5. Ye, J., Hua-yong, L., Jun-qing, Yu., Dong-ru, Z.: Content-based hierarchical analysis of news video using audio and visual information. Wuhan University Journal of Natural Sciences **6**(4), 779–783 (2001)
6. Muller, S., Eickeler, S.: Content-based video indexing of tv broadcast news using hidden markov models. In: IEEE International Conference on Acoustics, Speech, and Signal Processing, vol. 6, pp. 2997–3000 (1999)
7. Smoliar, S.W., Tan, S.Y., Zhang, H., Gong, Y.: Automatic parsing of news video. In: Proceedings of the International Conference on Multimedia Computing and Systems, pp. 45–54 (1994)
8. Huang, Q., Liu, Z., Rosenberg, A., Gibbon, D., Shahraray, B.: Automated generation of news content hierarchy by integrating audio, video, and text information. In: IEEE International Conference on Acoustics, Speech, and Signal Processing, vol. 6, pp. 3025–3028, March 1999
9. Baig, M.N., Asif, M.D.A., Tariq, U.U., Ahmad, W.: A novel hybrid method for text detection and extraction from news videos. Middle-East Journal of Scientific Research (2014). ISSN: 1990-9233, doi:10.5829
10. Ariki, Y., Teranishi, T.: Indexing and classification of tv news articles based on telop recognition. Document Analysis and Recognition **1**, 422–427 (1997)
11. Alimi, A.M., Ben Halima, M., Karray, H.: Arabic text recognition in video sequences. In: The International Conference on Informatics, Cybernetics, and Computer Applications, July 2010
12. Jawahar, C.V., Chennupati, B., Paluri, B., Jammalamadaka, N.: Video retrieval based on textual queries. In: Proceedings of the Thirteenth International Conference on Advanced Computing and Communications, Coimbatore (2005)
13. Anoop, K., Lajish, V.L.: Morphology based text detection and extraction from Malayalam news videos. In: National Conference on Indian Language Computing. CUSAT (2014)
14. Pala, P., Bertini, M., Bimbo, A.: Content based indexing and retrieval of tv news. Pattern Recognition Letters, 503–516. Elsevier (2001)
15. Kelkar, C., Khairnar, N., Gaikwad, H., Hapase, A.: News video segmentation and categorization using text extraction technique. International Journal of Engineering Research and Technology **2**(3), March 2013. ISSN: 2278–0181
16. Hughes, E.K., Smith, M.A., Sato, T., Kanade, T.: Video OCR for digital news archives. In: IEEE Workshop on Content-Based Access of Image and Video Database, Bombay, India, January 1998
17. Hughes, E.K., Smith, M.A., Satoh, S., Sato, T., Kanade, T.: Video OCR: Indexing digital news libraries by recognition of superimposed captions. Multimedia Systems. Springer (1999)
18. Liou, S.-P., Zhu, W., Toklu, C.: Automatic news video segmentation and categorization based on closed-captioned text. In: IEEE International Conference on Multimedia and Expo, pp. 829–832, August 2001
19. Akamatsu, S., Ohya, J., Shio, A.: Recognizing characters in scene images. IEEE Transactions on Pattern Analysis and Machine Intelligence **16**(2), 214–220 (1994)
20. Kankanhalli, A., Lee, C.M.: Automatic extraction of characters in complex images. Int. J. Pattern Recognition Artif. Intell. **9**(1), 67–82 (1995)
21. Kia, O., Li, H., Doerman, D.: Automatic text detection and tracking in digital video. IEEE Trans. Image Process. **9**(1), 147–156 (2000)
22. Jain, A.K., Zhong, Y., Zhang, H.: Automatic caption localization in compressed video. IEEE Trans. Pattern Anal. Mach. Intell. **22**(4), 385–392 (2000)

23. Riseman, E.M., Wu, V., Manmatha, R.: Extfinder: an automatic system to detect and recognize text in images. IEEE Trans. Pattern Anal. Mach. Intell. **21**(11), 1224–1229 (1999)
24. Karam, L.J., Hasan, Y.M.Y.: Morphological text extraction from images. IEEE Transactions on Image Processing **9**(11), 1978–1983 (2000)
25. Ren, T.I., Cavalcanti, G.D.C., dos Santos, R.P., Clemente, G.S.: Text line segmentation based on morphology and histogram projection. In: Proceedings of the 2009 10th International Conference on Document Analysis and Recognition, pp. 651–655
26. Chandrasekaran, R.M., Chandrasekaran, R.: Morphology based text extraction in images. IJCST **2**(4) (2011). ISSN: 0976–8491
27. Raja Rajeswari, S., Pratheeba, T., Kavith, V.: Morphology based text detection and extraction from complex video scene. International Journal of Engineering and Technology **2**(3) (2010).Trans. Pattern Anal. Mach. Intell. **16**(2), 214–224 (1994)
28. Kanade, T., Smith, M.A.: Video skimming for quick browsing based on audio and image characterization, Technical Report CMU-CS-95-186, Carnegie Mellon University, July 1995
29. Kim, C., Kim, W.: A new approach for overlay text detection and extraction from complex video scene. IEEE Transactions On Image Processing **18**(2), February 2009
30. Lyu, M.R., Song, J., Cai, M.: A comprehensive method for multilingual video text detection, localization, and extraction. IEEE Transactions on Circuits and Systems for Video Technology **15**(2), 243–255 (2005)
31. Hao, X., Ye, J., Huang, L.-L.: Neural network based text detection in videos using local binary patterns. In: Chinese Conference on Pattern Recognition, pp. 1–5 (2009)
32. Kim, H.-K.: Efficient automatic text location method and content-based indexing and structuring of video database. Journal of Visual Communication and Image Representation. Elsevier. doi:10.1006/jvci.1996.0029
33. Yu, B., Jain, A.K.: Automatic text location in images and video frames. In: Proceedings. Fourteenth International Conference on Pattern Recognition, vol. 2 (1998)
34. Nirmala, S., Nagabhushan, P.: Text extraction in complex color document images for enhanced readability. Intelligent Information Management **2**, 120–133 (2010)
35. Jain, A.K., Zhong, Y., Karu, K.: Locating text in complex color images. In: Proceedings of the Third International Conference on Document Analysis and Recognition, vol. 1, pp. 146–149, August 1995
36. Tremeau, A., Karaoglu, S., Fernando, B.: A novel algorithm for text detection and localization in natural scene images. In: International Conference on Digital Image Computing: Techniques and Applications (2010)
37. Govardhan, A., Swamy Das, M., Hima Bindhu, B.: Evaluation of text detection and localization methods in natural images. International Journal of Emerging Technology and Advanced Engineering **2**(6), June 2012. ISSN 2250–2459
38. Canedo-Rodriguez, A., Kim, J.H., Kelly, J., Hee Kim, J., Kim, S.H., Blanco-Fernandez, Y., Veeranmachaneni, S.K.: An efficient and accurate text localization algorithm in compressed mobile phone image domain. In: Int. Conf. on Image Processing, Computer Vision, and Pattern Recognition, July 2010
39. Smoliar, S.W., Tan, S.Y., Zhang, H., Gong, Y.: Automatic parsing of news video, multimedia computing and systems. In: Proceedings of the International Conference, pp. 45–54 (1994)

# The Impacts of ICT Support on Information Distribution, Task Assignment for Gaining Teams' Situational Awareness in Search and Rescue Operations

Vimala Nunavath, Jaziar Radianti, Tina Comes and Andreas Prinz

**Abstract** Information and Communication Technology (ICT) has changed the way we communicate and work. To study the effects of ICT for Information Distribution (ID) and Task Assignment (TA) for gaining Teams' Situational Awareness (TSA) across and within rescue teams, an indoor fire game was played with students. We used two settings (smartphone-enabled support vs. traditional walkietalkies) to analyze the impact of technology on ID and TA for gaining TSA in a simulated Search and Rescue operation. The results presented in this paper combine observations and quantitative data from a survey conducted after the game. The results indicate that the use of the ICT was good in second scenario than first scenario for ID and TA for gaining TSA. This might be explained as technology is more preferable and effective for information sharing, for gaining TSA and also for clear tasks assignment.

**Keywords** Disaster management · Smartphones · Walkie-Talkie · Indoor fire games · Coordination · Information Distribution · Task assignment · Teams' Situational Awareness · Quantitative analysis

## 1 Introduction

In emergency Search and Rescue (SAR) operations, Emergency Responders (ERs) need to rapidly share relevant information about the potential location of victims, about the situation on the ground and their capacity and progress. In addition, larger scale operations require coordination across SAR teams as well as

V. Nunavath(✉) · J. Radianti · T. Comes · A. Prinz
CIEM, University of Agder, Kristiansand, Norway
e-mail: {vimala.nunavath,jaziar.radianti,tina.comes,andreas.prinz}@uia.no

© Springer International Publishing Switzerland 2016
S.M. Thampi et al. (eds.), *Advances in Signal Processing and Intelligent Recognition Systems*,
Advances in Intelligent Systems and Computing 425,
DOI: 10.1007/978-3-319-28658-7_38

443

alignment with other emergency services such as police and medical services. SAR operations are time critical and may pose risks to the life and safety of responders. Despite the chaos characteristic for the initial phase of an emergency, decisions which are made and sent rapidly by teams, often based on incomplete or uncertain information [1].

SAR operations are often fragmented and teams on the ground carry out the work, and acquire information about the part of the environment they are working in. Particularly, in dynamic and time critical situations, it becomes difficult for the ERs to adequately decide which information might be relevant for other teams to support overall coordination. Hence, Information Distribution (ID) and Task Assignment (TA) errors may occur and hence give poor awareness of the situation [1].

In his seminal work Endsley [2], defined Situational Awareness (SA) as, "the perception of the elements in the environment within a volume of time and space, the comprehension of their meaning and the projection of their status in the near future". Sharing information that contributes to better SA – respecting also the context, perceptions, and potential interpretations – ensures that appropriate actions can be taken is therefore key to good emergency management [3]. Without SA it is difficult to make effective decisions because the perceived understanding does not correspond to the actual situation [4]. But, Teams' Situational Awareness (TSA) is defined as "The active construction of a situation partly shared and partly distributed between two or more agents, from whom one can anticipate important states in the near future [26].

Information Distribution (ID) aims at ensuring that the right people get the right information at the right time [5]. In the response to an emergency, ad-hoc communication networks arise. Information flows are dynamically evolving, data is heterogeneous and, different responders who are dispersed geographically (onsite and remotely) have to cooperate and interact to develop and maintain TSA [6].

Task allocation refers to the way the tasks are assigned, divided and coordinated within or across ERs during SAR operation based on the available information [7]. If the tasks are not assigned adequately during a SAR operation, it is difficult for the ERs to handle the emergency effectively. Moreover, the ERs will often repeat the same tasks which lead to the poor management of the emergency.

In order to achieve TSA, information distribution, and task assignment, various technologies and ICT-based tools have been developed in recent years. The widespread use of smartphones and their embedded sensing technologies have been described as a potential stepping stone towards more efficient distributed emergency response[8]. For instance, ERs can transfer the needed information from the affected area to the emergency responders or from the responders to the victims with the help of ICT. In very recent years, many number of smartphone apps, for example, Now Force [8] [9],[10],[11] , and so on  have been developed for different types of smartphones that can support firefighters and victims during a fire.  A long list of these apps can be found in [12-14].

In this paper, we address the following question: *"How does the use of Smartphone apps for communication and coordination impact on information distribution and task assignment for gaining teams' situational awareness across or within Emergency Response teams?.*" We address this question by analyzing the results of an indoor fire game, in which we simulated a search and rescue operation in a university building. We will focus especially on examining the patterns of ID and TA for gaining TSA during SAR operation.

Games played in augmented reality settings or virtual environments such as table-top exercises [15] and computer- based simulations [16] are increasingly used for emergency management training and research [17]. Particularly for emergencies, when errors result in harmful consequences, and only very mature systems can be tested in reality, games are vital to help researchers and practitioners to better understand the ID, TA, and SA patterns emerging within and across emergency response teams[18]. In this paper, we report the lessons from a search and rescue game experiment. The game was designed to enable a research team to understand the utilization of ICT for ID and TA for getting TSA.

We present questionnaire analysis to study the ICT usage in SAR operation for ID and TA for TSA. The survey results presented in this paper were analyzed after the game and have not been used for real-time error identification and adaptation. The purpose of this paper is to highlight specific aspects of response behavior. We envision that such information can be used in trainings and exercises to improve the behavior of teams and individuals and targeted design.

This paper starts with the description of the developed emergency scenario, followed by an explanation of the research methodology. We then present our findings and conclude the paper with a discussion about the findings and future research directions.

## 2 Game Design

The game was designed based on an indoor fire to investigate and study the usage of ICTs such as smartphones with installed applications for communication and for assistance and traditional walkie-talkies for ID and TA within and across rescue teams to maintain TSA during SAR. In this game, two scenarios were developed by a research team consists of 7 members. The both developed scenarios were with the noisy environment including real fire alarms, smoky corridors, and technological communication tools.

During scenario development process, three main requirements were taken into account: complexity (the scenario must be complicated enough to involve multiple teams); concreteness (the scenario must include sufficient details to allow the participants to identify the relevant actors); and realism (the scenario must be realistic)[19]. The developed scenarios were about search and rescue operation, which is as follows:

Fire accident happens in the third floor in the university building. The building consists of many students (who might be normal, disabled, and sick), books, labs and storage rooms. Several students have observed smoke, flames, and loud noise in the university building. Some of the witnesses also report fire escalation. Due to the fire, the emergency site has become chaotic and many people in the building are wounded and traumatized. The number of people within the building is un-known. But, the people who are running out of the building were giving informa-tion about the seen victims. In addition to the textual descriptions, participants will be further supplemented with a map of the floor layout in the building where the incident occurred to get an idea of the view of the floor in the first scenario. In the second scenario, the participants were given with smartphone app called Smar-tRescue consist of building a map and to know the location of the fire and vic-tims and status of the fire progression and victim [20].

## 3    Research Methodology

Empirical data for testing the research framework was collected via a game sur-vey. Three constructs were measured in this study: Information Distribution (ID) and Task Assignment (TA) for gaining Teams' Situational Awareness (TSA). The indoor fire game was designed based on the semi-structured interviews with the three different levels of Grimstad fire department officials and guidelines extracted from the two documents i.e., one is about organizing fire departments in Norway [21] and another one is about firefighters operations in smoke filled or hazardous areas [22]. Based on these documents and interview, the scenarios were developed in such way that reflects the real emergency.

   Based on the semi-structured interviews and given documents, the research team got to know that, when fire emergency occurs, only Smoke Divers (SDs) enter into the fire-filled areas to perform search and rescue operation, whereas, Smoke Leaders (SLs) led the set of SDs (who have entered into the area). Crew Manager (CM) led the set of SLs. The fire fighters organizational model is shown in Fig.1. In both sce-narios, the considered organizational model was same. But, the smartphones (with SmartRescue app) were not given to all in scenario 2 (See fig. 3).

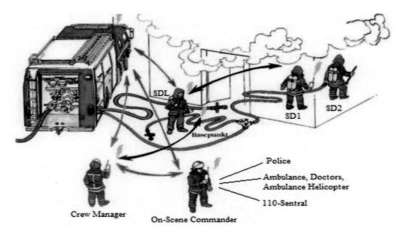

**Fig. 1** Firefighters organizational model

## 3.1 Participants

This game was performed with 23 voluntary student participants and 6 observers from Grimstad and Arendal town fire departments. In the first scenario, all participants were given with smartphones with Zello software app in it. This Zello software was used as communication, information distribution and tasks Assignment tool among students who have acted as ERs during the game to maintain the teams' situational awareness. In the second scenario, regular walkie-talkies and smartphone app called SmartRescue were given to some of the participants except medical care. In Scenario 1, 10 participants out of 23 acted as rescuers and CM, whereas, other 12 participants acted as victims. In scenario 2, the participants who acted as victims (in scenario1) acted as rescuers (in second scenario) and rescuers (in scenario1)) as victims (in scenario2).

## 3.2 ICT Tools

In this game, ERs were given with following ICT tools i.e., SmartRescue app, Zello software and traditional walkie-talkies to support and perform SAR operation.

*SmartRescue Application*: It is an Android-based smartphone app, called SmartRescue (SR) app which is developed by a group of researchers at University of Agder. This app provides the feature of sending and receiving needed data through publish-subscribe platform between ERs and victims. The data which is distributed is basically real-time data of the location and condition of the fire. The real-time data is collected through smartphone sensors and technologies which are embedded inside the smartphones such as humidity, thermometer, light sensor and barometer. This sensor data can be used to asses and predict the fire spread in the next 30 seconds. This feature works with the concept of Bayesian networks. Usually the sensor data is collected only when the users such as ERs and/or victims should install and activate smartphone app in their smartphones [8][23]. For this specific game, the sensor data was simulated.

*Zello Walkie-Talkie Application*: Zello is a free push-to-talk application for smartphones, tablets, and PCs. The reason for using this application is that, it has the feature of storing history data and we can create different levels of channels to communicate and to avoid information overload. The stored history can be used for further analysis [24].

*Traditional Walkie-Talkie*: It is a hand-held, portable, two-way portable radio receiver which is normally used by ERs for communication and coordination during emergency management [25].

## 3.3 Planned Communication for Scenario 1

The total game is conducted for 30 minutes. In this game, out of 23 participants, 10 acted as rescuers, and 1 as Medical Care Unit (MCU), and rest 12 as victims (see in fig. 2). Rescuers are divided into 3 teams and each team consists of 3 participants: one as Smoke Leaders ((SLs), and other as Smoke Divers (SDs). Crew Manager acted as a coordinator (CM). Figure 3 shows the planned communication during the game.

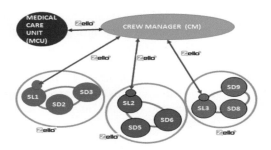

**Fig. 2** Scenario 1 communication Plan [20]

The organization of the team is as follows: Team 1consists of SL1, SD2, SD3, Team 2 consists of SL2, SD5, SD6, and Team 3consists of SL3, SD8, and SD9. All these rescuers are given with smartphones with Zello software. During the game, participants who acted as SDs are supposed to do search and rescue operation to save victims. If SDs spot any victims who are unconscious, critically wounded, then they are supposed to take the victims to either to a safe area or entrusted to MCU by reporting their health condition. When one of the teams spot victims, SDs are supposed to inform to SL and from SL to CM to gain SA and assign tasks. But, when SDs need information about the emergency, SL will provide and CM will provide to SL. In the first scenario, Zello walkie-talkie software application was used to communicate information, TA, and SA. In this app, 7 channels were created for communication (see Figure 2)

## 3.4   Planned Communication for Scenario 2

The setting of second scenario was same as first one including game time, emergency area, the structure of the ERs, their roles and goals. The only difference was the usage of ICT tools. In this scenario, the players were given with smartphones with installed application called SmartRescue. To study impact of using ICT (SmartRescue app and traditional walkie-talkie) for ID, TA, and TSA during SAR operation, we have designed the communication plan and usage of ICT tools by ERs during game, as shown in Fig. 4.

We had 3 ERs groups in each there were 2 SDs and one SL that had traditional walkie-talkie as the communicating device. During game, all 3 SLs and the medical care unit had a direct contact with the CM. SDs were in direct contact with their SLs. For distributing SR app among rescuers, we have defined three organizational models, i.e., centralized, semi-centralized, and decentralized.

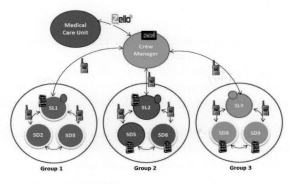

**Fig. 3** Scenario 2 communication plan [23]

Group 1 is a centralized group, as SL was provided with a smartphone with SmartRescue app and traditional walkie-talkie, whereas SDs were given with only traditional walkie-talkie. So, if the SDs need any emergency information, they have to depend on SL as the emergency information is first received and viewed by SL, and then he/she needs to pass it to their team through walkie-talkie. Group 2 is a decentralized one, as all the members of the group have smartphones with installed SmartRescue app and a traditional walkie-talkie. Group 3 is a semi-centralized group, as SDs are the one who were provided with smartphones installed with SmartRescue app, while the SL of the group acquires the information only through walkie-talkie [23].

## 3.5 Questionnaire

The players were asked the questions to rate their subjective experience for ID, TA, and for gaining TSA on a 5 point Likert scale. The questionnaire consists of 10 questions, which were divided into 3 categories (i.e., Goals): for ID, TA, and SA to better understand the ICT support for the goals during the game. Goal1(G1) to study the impacts of using ICT and walkie-talkie for information distribution; Goal 2 (G2) to investigate and compare the effects of having ICT and walkie-talkie for task Assignment; Goal3 (G3) to examine and compare the role of ICT and walkie-talkie in gaining teams' situational awareness.The questionnaire and their corresponding goals are listed in Table 1.

**Table 1** The questionnaire and the corresponding goals

| #Q | Questions | Goals |
|---|---|---|
| Q1 | Information (such as victims' status and room numbers) has been shared effectively between and among other responders. | G1 |
| Q2 | I had necessary information to think ahead about what actions to be taken. | G3 |
| Q3 | It was easy to get information from CM or SL when I needed. | G1 |

**Table 1** (*Continued*)

| | | |
|---|---|---|
| Q4 | My team received needed information to get an accurate overview of the situation at all times. | G3 |
| Q5 | I always received information about the fire development from other team members on time. | G1 |
| Q6 | The tasks were clearly assigned and I knew what I supposed to do. | G2 |
| Q7 | Each member of my team had a clear idea of the team's goals. | G2 |
| Q8 | The information I received about victim's location was clear. | G1 |
| Q9 | There were no tasks overload and none of the assigned tasks are difficult to implement | G2 |
| Q10 | I know where to get needed information from. | G1 |

## 4  Results

To achieve the above mentioned goals and evaluation purposes, a questionnaire with a set of questions was handed out to the participants (SDs, SLs, and CM) after the game for both scenarios. The questionnaire was designed to collect the responses of participants. In total, 10 players (6 SDs, 3SLs, and CM) responded to the questions for both scenarios. In both scenarios, the players were shuffled (victims acted as rescuers and vice versa) to avoid bias in the responses.

**Scenario 1 Results**
In this section, the results of the scenario 1 are presented. As can be seen in the figure, most of the responders responded contradictorily for Q1 (G1) which is relatively negative result. It might be because of players might thought using only Zello app was not sufficient as there was too much talking and sometimes irrelevant information shared. Moreover, locating the victims without SmartRescue app was difficult and time consuming.

For Q2 (G3), Only 33% of the reponders Agreed that they got necessary information to take action further. The acquired response might be given by SLs as they get information from CM who often keeps SLs updated. Other responses were mostly disagreeing as the responses might be given by SDs. The answers to this question tell that only Zello app was not sufficient enough to gain TSA. It is because of the given ICT was just to communicate within or across teams, but not for any assistance. So, it is obvious that all players did not achieve SA.

For Q3 (G1), 44% of players Agree that they received information from CM through Zello app. Based on the given answers, it can be concluded that the persons who have agreed might be SLs. As SLs have direct connection with CM (who has complete overview of the emergency situation), whereas other responses might be from SDs as they do not have any connection with CM.

For Q4 (G2), more than 50% of the players Agree that ICT (Zello app) helped them to receive needed information to get an accurate overview of the emergency situation at all times.

For Q5 (G1), more than 80% of the responders Disagreed that ICT (Zello app) helped them to receive information about fire development. From the answers, it is concluded that they do not have an overview of the fire development as it is hard to predict the fire development with human senses without any ICT support.

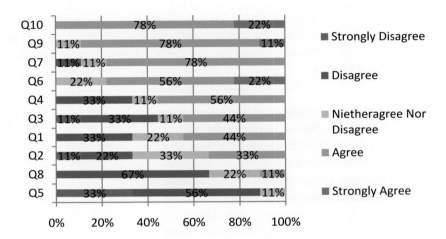

**Fig. 4** Results from Scenario 1 questionnaire answers

For Q6 (G2), more than 80% of the players Agree that tasks were clearly assigned with the help of ICT.

For Q7 (G2), 78% of the players Agree that they know teams' goals during SAR operation. It is because, the players must have understood their leaders instructions well and got proper information related to task goals.

For Q8 (G1), only 11% of the players Agree that the victims' location information was clearly distributed via ICT (Zello app), whereas, more than 80% Disagree. As per the answers to this question, the SDs might not get the exact location details from SLs and also SLs from CM. It might be because of not knowing the exact location of the victims' due to lack of ICT (SmartRescue app).

For Q9 (G2), more than 90% of the players Agree that there were no tasks overload and difficulties in tasks implementation as most of the players felt that with the help of ICT (Zello app) the tasks were distributed easily.

For Q10 (G1) in Information distribution, all of the players (100%), Agree and Strongly Agree that they know where to get information with the help of ICT (Zello app). All players have agreed because they were briefed before the game start that whom should they contact and how to use (Zello app) to get connected.

**Scenario 2 Results**

In this scenario 2 results are presnted. As you can see in the below figure, for Q1(G1), it is shown that more than 60% of the players of the game Agree, that they distributed information effectively with the help of ICTs (SmartRescue app). For this question, more than 60% of the players Agreed that ICT usage was helpful to gain the needed information and to share, whereas other players disagreed as they might found the ICT usage difficult.

For Q2 (G3), More than 50% of the players Agree that they got needed information to think ahead to make action with the help of using ICTs (SmartRescue app). Based on the answers, the one who answered Agree were must be those whose smartphones were equipped with SmartRescue app and the rest might be the one who did not get the smartphone app. (i.e., SDs in Centralized Group, SL in semi-centralized Group).

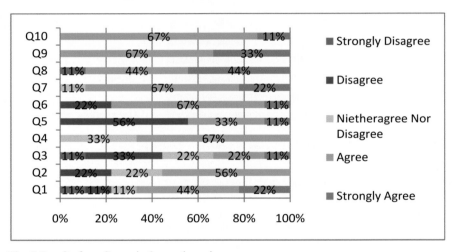

**Fig. 5** Results from Scenario 2 questionnaire answers

For Q3 (G1), more than 40% of the players disagreed and Strongly Disagree about receiving information from CM. It might be, at least one member of the centralized, decentralized and semi-centralized groups were provided with SmartRescue app and a traditional walkie-talkie for communication during the game. So, they did not need to contact CM to get information, but responders during game might have passed information about their activities to CM.

For Q4 (G3), 67% of the players Agreed that ICTs (SmartRescue app) made them get an overview of the situation, but post game discussions, players agreed that regular walkie-talkies are not that much useful as they could hear lot of unnecessary communication.

For Q5 (G1), 44% of the players Strongly Agree and Agree that ICT helped them better to get information about fire development from other team members on time. The players who have disagreed to this question might be not having

SmartRescue app with them (i.e., SDs in Group 1 and SL in Group 3), whereas, the one who has Agreed must be provided with SmartRescue app and got information from SmartRescue app.

For Q6 (G2), more than 70% of the players Agree and Strongly Agree that there were no tasks overload and difficulties in implementing the assigned tasks as the players could see easily on the ICTs (SmartRescue app) and perform the tasks. It is because ICT (SmartRescue app) provide the feature of victim's location and fire development. So, it is easy for ERs to know and perform tasks during SAR operation. Whereas, the other players who disagreed must be the one who did not have Smartphones with SmartRescue app to see instant tasks to perform as they have to wait for the instructions from their leader.

For Q7 (G2), more than 80% of the players Agree and Strongly Agree that those who use ICTs (SmartRescue app), helped them to get aware of their teams' goals.

For Q8 (G1), 88% of the players Agree and Strongly Agree that ICT (Smart-Rescue app) helped them to get to know about victims' location and distribute it with other team members. Based on the answers and observation, it is concluded that most of the players Agree that ICT was very helpful to spot the victim on the screen of the mobile to save them quickly. The other responders Disagree to this question as the one who were not provided with SmartRescue app had to wait for the information to receive from their leader (SL1 in group 1). During post game discussions, the rescuers who were not given with SmartRescue app agreed that it was difficult for them to locate exactly the victims' location without ICT help. So, they preffered to have ICT assistance to get the information and perform the tasks better.

For Q9 (G2),  All of the palyers Agree and Strongly Agree that they did not have neither tasks overload nor difficulty in performing the tasks , i.e., they had clear understanding of the tasks to perform with the help of  ICTs (SmartRescue app).

For Q10 (G1), all of the people Strongly Agree and Agree that they know where to get information from with given ICTs (SmartRescue app).

## 5      Discussion and Conclusion

In this paper, we have investigated the ICT usage for information distribution (flow) and task assignment for gaining teams' situational awareness through an emergency indoor fire game. The experimental game was designed according to workflows and processes elicited from fire fighter experts. However, the game was conducted with students, and therefore some additional instructions and training were given prior to the experiment. Intervention during the game was kept to a minimum of observations and surveys were only conducted after the game. Figure 5 and 6 shows the results of surveys, which were answered after the game for scenario 1 and scenario 2 respectively.

The results are from a quantitative analysis of a questionnaire to test ICT support on Information Distribution and Task Assignment for gaining Teams' Situational Awareness. However, the game was run by only limited number of participants to evaluate the ICT performance. Due to limited participants, we managed to collect data only from 10 participants. The results in this study were taken from limited data. Therefore, results have limited significance that needs further investigation. When we compare the results which we got about ICT usage for both scenarios: Information distribution was better in second scenario compared to first scenario as the players can see the victim's location, no.of victims in the emergency area and fire development (Goal1). So, it is easy for the players to see them on the SmartRescue application and distribute the information to the needed one.

When it comes to task Assignment (Goal 2), scenario 2 was better as the tasks were clearly graphically visible in the SmartRescue application to perform SAR operation. Whoever do not have (SDs in group1) SmartRescue app. So, the tasks were assigned by SL to SDs with the help of ICTs. For Goal 3, scenario 2 is better than scenario 1 for gaining TSA. It is because in scenario1, SLs should depend on the SDs to get information from the emergency area and CM should depend on SLs. But, when it comes to scenario2, the players can see on the screen of the mobile phone for needed information (i.e., victims' location, stastus, fire develpement and so on) in the emergency area for gaining awareness of the emergency situation. Moreover, it might be because of organizational model of Smartphone distibution and communication.

Based on the results of our study, our potential future research direction will be to run the same experiment with large number of real fire fighters to test the use of ICT-tools for communication and coordination effects on information distribution, task assignment, and teams' situational awareness across or within Emergency Response teams. The results which we get from future study will be compared with the present study to check the validity of the results.

## References

1. Netten, N., et al.: Task-adaptive information distribution for dynamic collaborative emergency response. The International Journal of Intelligent Control and Systems **11**(4), 238–247 (2006)
2. Endsley, M.R.: Measurement of situation awareness in dynamic systems. Human Factors: The Journal of the Human Factors and Ergonomics Society **37**(1), 65–84 (1995)
3. Yang, L., Prasanna, R., King, M.: On-site information systems design for emergency first responders. Journal of Information Technology Theory and Application (JITTA) **10**(1), 2 (2009)
4. Bremner, P., Bearman, C., Lawson, A.: Firefighter Decision Making at the Local Incident and Regional/State Control Levels. Human Factors Challenges in Emergency Management: Enhancing Individual and Team Performance in Fire and Emergency Services, 149 (2014)
5. Singh, P., et al.: Information sharing: a study of information attributes and their relative significance during catastrophic events. Cyber-Security and Global Information Assurance: Threat Analysis and Response Solutions. IGI Publishers (2009)

6. McEntire, D.A.: Coordinating multi-organisational responses to disaster: lessons from the March 28, 2000, Fort Worth tornado. Disaster Prevention and Management **11**(5), 369–379 (2002)
7. Gordon, D.M.: The organization of work in social insect colonies. Complexity **8**(1), 43–46 (2002)
8. SmartRescueProject.,SmartRescue project. Center for Integrated Emergency Management (CIEM) http://ciem.uia.no/project/smartrescue
9. NowForce. NowForce (2015). (cited January 2015, available from) http://www.nowforce.com/
10. Colunas, M.F.M., et al.: Droid jacket: using an android based smartphone for team monitoring. In: 7th International Wireless Communications and Mobile Computing Conference (IWCMC) (2011)
11. Kulakowski, P., Calle, E., Marzo, J.L.: Sensors-actuators cooperation in WSANs for fire-fighting applications. In: IEEE 6th International Conference on Wireless and Mobile Computing, Networking and Communications (WiMob) (2010)
12. FireRescue1. Firefighter iPhone Apps (2015). (cited January 2015, available from) http://www.firerescue1.com/firefighter-iphone-apps/
13. Bekker, W.: Apps for that-Tried and tested downloads for firefighters 2013 (2015). (cited January, available from) http://www.firefightingincanada.com/equipment/apps-for-that-14885
14. Jerrard, J.: Smartphone Apps for the Fire Service (2011). (cited January 2015, available from) http://www.firefighternation.com/article/training-0/smartphone-apps-fire-service
15. Crichton, M., Flin, R.: Training for emergency management: tactical decision games. Journal of Hazardous Materials **88**(2–3), 255–266 (2001)
16. Toups, Z.O., Kerne, A., Hamilton, W.A.: The Team Coordination Game: Zero-fidelity simulation abstracted from fire emergency response practice. ACM Transactions on Computer-Human Interaction (ToCHI) **18**(4), 23 (2011)
17. Van Ruijven, T.: Serious games as experiments for emergency management research: a review. In: Proceedings of the 8th International Conference on Information Systems for Crisis Response and Management, ISCRAM 2011, Lisbon, Portugal, 8–11 May. ISCRAM (2011)
18. Kurapati, S., et al.: Exploring shared situational awareness in supply chain disruptions. In: Proceedings of the 10th International Conference on Information Systems for Crisis Response and Management, ISCRAM 2013, Baden-Baden, Germany, 12–15 May. ISCRAM (2013)
19. Eide, A.W., et al.: Inter-organizational collaboration structures during emergency response: a case study. In: Proceedings of the 10th International ISCRAM Conference, Baden-Baden, Germany (2013)
20. Nunavath, V., et al.: Visualization of information flows and exchanged information: evidence from an indoor fire game. In: Proceedings of the 12th International Conference on Information Systems for Crisis Response and Management, Kristiansand, May 24–27. ISCRAM (2015)
21. Beredskap, D.f.s.o., Veiledning til forskrift om organisering og dimensjonering av brannvesen (2003)

22. Beredskap., D.f.s., Veiledning om røyk og kjemikaliedykking (2003)
23. Sarshar, P., Radianti, J., Gonzalez, J.J.: On the impacts of utilizing smartphones on or-ganizing rescue teams and evacuation procedures. In: Proceedings of the 11th Interna-tional Conference on Information Systems for Crisis Response and Management, ISCRAM 2015, Kristiansand, May 24–27. ISCRAM (2015)
24. Zellowalkie-talkieapp, Zello walkie-talkie software application available at: http://zello.com/app Zello walkie-talkie software application
25. Lien, Y.-N., Chi, L.-C., Huang, C.-C.: A multi-hop walkie-talkie-like emergency communication system for catastrophic natural disasters. In: Proceedings of the 2010 39th International Conference on Parallel Processing Workshops, pp. 527–532. IEEE Computer Society (2010)
26. Salas, E., et al.: Situation awareness in team performance: Implications for measure-ment and training. Human Factors: The Journal of the Human Factors and Ergonomics Society **37**(1), 123–136 (1995)

# A Hybrid Approach to Rainfall Classification and Prediction for Crop Sustainability

Prajwal Rao, Ritvik Sachdev and Tribikram Pradhan

**Abstract** Indian Agriculture is primarily dependent on rainfall distribution through-out the year. There have been several instances where crops have failed due to inade-quate rainfall. This study aims at predicting rainfall considering those factors which have been correlated against precipitation, across various crop growing regions in In-dia by using regression analysis on historical rainfall data. Additionally, we've used season-wise rainfall data to classify different states into crop suitability for growing major crops. We've divided the four seasons of rainfall as winter, pre-monsoon, mon-soon, and post-monsoon. Finally, a bipartite cover is used to determine the optimal set of states that are required to produce all the major crops in India, by selecting a specific set of crops to be grown in every state, and selecting the least number of states to achieve this. The data used in this paper is taken from the Indian Meteorological Department (IMD) and Open Government Data (OGD) Platform India published by the Government of India.

## 1 Introduction

Agriculture plays a major role in India's economy, livelihood, growth and sustenance of the country. India is one of the largest producers of agricultural products in the world. Although agriculture contributes a major part in India's GDP (gross domestic product), it has shown a slight declining trend due to various factors including in-dustrialization and increased growth of tertiary sector over the years. Nevertheless, it continues to be the primary occupation in rural India while facing numerous hurdles

P. Rao · R. Sachdev · T. Pradhan(✉)
Department of Information and Communication Technology, Manipal Institute of Technology,
Manipal University, Manipal, Karnataka, India
e-mail: {prajwalnrao,ritvik.sach}@gmail.com, tribikram.pradhan@manipal.edu

© Springer International Publishing Switzerland 2016                                    457
S.M. Thampi et al. (eds.), *Advances in Signal Processing and Intelligent Recognition Systems*,
Advances in Intelligent Systems and Computing 425,
DOI: 10.1007/978-3-319-28658-7_39

like dependency on irregular monsoon, lack of proper irrigation facilities and suitable incentives to farmers.

Agriculture in India is not uniform throughout. Due to the vast territorial expanse and changing climatic conditions throughout the year, different crops are cultivated in different seasons of the year. In India, *Kharif* and *Rabi* are the two main types of crops. *Kharif* crops are grown in the season roughly from June to September when we get monsoon rainfall and hence these crops are also called "rain-fed crops". These include Rice, Jowar, Cotton, Ragi etc. Whereas *Rabi* crops are grown in the season roughly from October to February. They are also known as "Winter crops". They include Maize, Wheat, Oil seeds etc. Apart from these two major types of crops, we have *Zaid* Crops which are grown in a season from March to May which includes mostly horticulture crops.

Indian Agriculture still depends on precipitation due to the Monsoon winds coming from the Arabian Sea and Bay of Bengal. High rainfall areas like the western ghats and southern regions of India, which receive rainfall due to Arabian Monsoon, and West Bengal Region, which receives rains from winds coming over Bay of Bengal are known for the cultivation of Kharif crops, like Rice.

The paper is organized as follows. Section 2 shows the literature survey. The methodology of Classification, Prediction, and Bipartite Cover is presented in Section 3. Section 4 shows the case studies using these methods and also discusses the results obtained. Section 5 concludes the paper and outlines areas for future research.

## 2 Literature Survey

Rainfall prediction has long been a highly researched field, with many works in this direction. The primary application being agricultural productivity, as suggested by Mahato [1] because climatic conditions like rainfall of a region can determine the nature and characteristics of the crops that can be grown. Climate change can have a negative impact on agriculture systems, as they are very sensitive to changes in temperature and precipitation, reducing both the yield of food production and quality of crops. As indicated by Kang et al. [2], impacts of climate change on crop yield are often integrated with the effects climate change has on water productivity and soil water balance. The authors point out that in the future, with increasing temperatures and fluctuations in precipitation, availability of water and crop production will decrease.

Much work has been carried out to understand the variations of rainfall during summer monsoon and its impact on the different crop growing seasons in India by Prasanna [3]. The study indicated that the total food grain yield over India during Kharif (summer) season is directly affected by variations in the precipitation during the summer monsoon between June and September. However, a similar correspondence during the Rabi (winter) season, between October-December, was not found as crops grown during this season are not only influenced by the rainfall but other factors such as water and soil moisture availability in different parts.

Therefore it becomes paramount to analyse the relationship between crop growth and rainfall patterns across different parts of India. In a study conducted by Kumar et al. [4], historic production statistics for major crops have been correlated with all-India rainfall data, and have found that production in the monsoon and post-monsoon seasons were significantly correlated to all-India summer monsoon rainfall. Additionally, work has been carried out to identify the influence of rainfall's potential predictors like Pacific and Indian Ocean sea temperatures, Darwin Sea Level pressure on crop production. Results indicated that monsoon season crops were strongly associated with three potential monsoon predictors but some spatial variations for certain crops required further study.

In order to predict rainfall, one of the popular techniques adopted is using Artificial Neural Networks (ANN). Awan et al. [5] have used Backpropogation and Learning Vector Quantization techniques for rainfall prediction in Pakistan. Training of the neural network is done using 45 years monsoon rainfall data (1960-2004), and the performance is evaluated over a test period of 5 years (2005-2009). The results show that in comparison with Multiple Linear Regression and Statistical Downscaling Models, Neural Network techniques show better performance in terms of accuracy, greater lead time and fewer required resources.

But sometimes such high accuracy in rainfall prediction requires a lot of computational resources. Nikam et al. [6] have instead suggested a data mining based approach using bayesian prediction model for rainfall prediction which is a data intensive model rather than computation one. This approach not only reduces the computation overhead, but also the processing time of very large data sets, which is done in comparatively very less time, hence claims to be more efficient. Additionally, their model proves to be a nearly accurate one in comparison with several other computationally intensive models. The model's accuracy increases with the increase in learning data, and since they've used a large dataset, the model predicts accurately. However, this approach has certain drawbacks that include the case when any predictor category is absent in the training data, the model incorrectly assumes any records having that category has zero probability. They've suggested that the accuracy of the model can be improved by making a hybrid model consisting of multiple data mining approaches, or combining data mining models with the computation based models.

## 3  Methodology

In this section we describe the method to determine the primary factors used in predicting the amount of rainfall, using correlation analysis. These factors are then used in predicting rainfall by linear and multiple regression techniques. Additionally, we describe the methods classify the data according to crop ranges and label it for suitability according to the different states across India using decision trees.

## 3.1   Correlation and Covariance

Correlation and covariance are measures to find how much two or more attributes vary in accordance to each other. Consider $X$ and $Y$, and a set of n observations $\{(x_1, y_1), ..., (x_n, y_n)\}$. The mean or expected values of $X$ and $Y$ are:

$$E(X) = \bar{X} = \frac{\sum_{i=1}^{n} x_i}{n} \quad and \quad E(Y) = \bar{Y} = \frac{\sum_{i=1}^{n} y_i}{n} \tag{1}$$

The covariance between $X$ and $Y$ is defined as

$$Cov(X, Y) = E((X - \bar{X})(Y - \bar{Y})) = \frac{\sum_{i=1}^{n}(x_i - \bar{X})(y_i - \bar{Y})}{n} \tag{2}$$

Also, $r_{X,Y}$ (Correlation Coefficient)

$$r_{X,Y} = \frac{Cov(X, Y)}{\sigma_X \sigma_Y} \tag{3}$$

where $\sigma_X \sigma_Y$ are standard deviations of $X$ and $Y$ respectively. We show that

$$Cov(X, Y) = E(X.Y) - \bar{X}\bar{Y} \tag{4}$$

If $X$ and $Y$ change together, when $X$ is greater than $\bar{X}$ (the expected value of $X$), then $Y$ will tend to be greater than $\bar{Y}$ (the expected value of $Y$), the covariance between $X$ and $Y$ is positive. Otherwise, if one of the attributes is greater than its expected value when the other attribute is lower than its expected value, then the covariance of $X$ and $Y$ is *negative*.

If $X$ and $Y$ are independent, then $E(X.Y) = E(X).E(Y)$ and the covariance is $Cov(X, Y) = E(X.Y) - \bar{X}Y = E(X).E(Y) - \bar{X}Y = 0$.

## 3.2   Prediction

Regression analysis [7] is a method used widely in statistical analysis to understand how the value of a response (dependent) variable changes when the values of the predictor (independent) variables are changed. Generally, the predictor variables' values are known, from which the response variable is predicted, and is referred to as the predicted attribute. Therefore it is widely used in prediction and forecasting related applications. A regression function is obtained after the analysis, that relates the predictor variables to the response variables.

Among the various methods used for regression analysis, two linear regression based methods have been discussed below:

### 3.2.1 Linear Regression

In Linear regression [7] we use of a single predictor variable, $x$ and a response variable, $y$. It is also referred to as *Straight Line Regression*, and the equation is,

$$y = w_0 + w_1 x \tag{5}$$

where $w_0$ and $w_1$ are coefficients of regression and the variance of $y$ is assumed to be constant.

The coefficients are solved using *method of least squares*, which minimizes the error between the actual data and the estimate by using the best-fitting straight line.

For a training set $D$, consisting of $|D|$ values of predictor variable $x$ and their associated values for response variable $y$ in the form $(x_1, y_1), (x_2, y_2), ..., (x_{|D|}, y_{|D|})$, the regression coefficients can be estimated by using the following equations:

$$w_1 = \frac{\sum_{i=1}^{|D|}(x_i - \bar{x})(y_i - \bar{y})}{\sum_{i=1}^{|D|}(x_i - \bar{x})^2} \tag{6}$$

$$w_0 = \bar{y} - w_1 \bar{x} \tag{7}$$

where $\bar{x}$ is the mean value of $x_1, x_2, ..., x_{|D|}$, and $y$ is the mean value of $y_1, y_2, ..., y_{|D|}$.

### 3.2.2 Multiple Linear Regression

Multiple linear regression [7] is an extension of linear regression, but involving more than one predictor variable. The response variable $y$ is modelled as a linear function of $n$ predictor variables or attributes, $A_1, A_2, ..., A_n$, describing a tuple, $X$. (That is, $X = (x_1, x_2, ..., x_n)$). The training data set, $D$, contains data of the form $(X_1, y_1), (X_2, y_2), ..., (X_{|D|}, y_{|D|})$, where the $X_i$ are the $n$-dimensional training tuples with associated class labels, $y_i$. An example of a multiple linear regression model based on two predictor attributes or variables, $A_1$ and $A_2$, is

$$y = w_0 + w_1 x_1 + w_2 x_2 \tag{8}$$

where $x_1$ and $x_2$ are the values of attributes $A_1$ and $A_2$, respectively, in $X$. The method of least squares shown above can be extended to solve for $w_0$, $w_1$, and $w_2$.

## 3.3 Classification

Classification is the task of assigning specific classes or categories to a group of similar items in a collection. The main goal is to accurately identify the target class for each item in the collection. Every classification begins with a data set in which class assignments are known.

---

**Algorithm 1.** Labelling season-wise rainfall values for given crop rainfall ranges

---

1: **procedure** SEASON- LABEL(*rainfall*, *Ranges*)
2: *begin*:
3:    **if** IS- IN- RANGE(*rainfall*, *Ranges*.LOWER) **then**
4:        **return** LOW
5:    **if** IS- IN- RANGE(*rainfall*, *Ranges*.AVERAGE) **then**
6:        **return** OPTIMAL
7:    **if** IS- IN- RANGE(*rainfall*, *Ranges*.HIGHER) **then**
8:        **return** ABUNDANT
9:    **if** *rainfall* < LOWER- LIMIT(*Ranges*.LOWER) **then**
10:        **return** INADEQUATE
11:    **if** *rainfall* < UPPER- LIMIT(*Ranges*.HIGHER) **then**
12:        **return** EXTREME
13: *end*

---

**Algorithm 2.** Assigning season-wise rainfall values for given crop rainfall ranges

---

1: **procedure** ASSIGN- LABEL(*States*, *Crops*, *Ranges*)
2: *begin*:
3:    *state-labels* ← {}
4:    **for** all *S* in *States* **do**
5:        **for** all *C* in *Crops* **do**
6:            *winter* ← {}
7:            *pre-monsoon* ← {}
8:            *monsoon* ← {}
9:            *post-monsoon* ← {}
10:            **for** all *year* in *S* **do**
11:                *S.winter* ← SEASON- LABEL(*C*.WINTER- RAIN, *C*.RANGE)
12:                *S.pre-monsoon* ← SEASON- LABEL(*C*.PRE- MONSOON- RAIN, *C*.RANGE)
13:                *S.monsoon* ← SEASON- LABEL(*C*.POST- MONSOON- RAIN, *C*.RANGE)
14:                *S.post-monsoon* ← SEASON- LABEL(*C*.WINTER- RAIN, *C*.RANGE)
15:            *C.winter* ← MODE(*winter*)
16:            *C.pre-monsoon* ← MODE(*pre-monsoon*)
17:            *C.monsoon* ← MODE(*monsoon*)
18:            *C.post-monsoon* ← MODE(*post-monsoon*)
19:        **return** *state-labels*
20: *end*

## 3.4  Decision Tree

Decision Tree is a useful structure for classification and prediction of large data. Its basic tree structure resembles a flowchart, having root node at the top and subsequent child nodes, each node denoting a test on an attribute, branches from these nodes represent the result of the test, and every leaf node holds a class label. In a decision tree, classification of data starts from the root node till the leaf node using top down approach.

## 3.5  Bipartite Graph and Bipartite Cover

In graph theory, a graph whose vertices can be divided into two disjoint sets $P$ and $Q$ such that every edge connects a vertex in $P$ to one in $Q$, is called a bipartite graph [8]. Vertex set $P$ and $Q$ are often denoted as partite sets. If a bipartite graph is not connected, it may have more than one bipartition. Also, a bipartite graph does not contain any odd-length cycles. It is often used in modelling relations between two different classes of objects.

Bipartite graphs are denoted by $G = (P, Q, E)$, whose partition has the parts $P$ and $Q$, with $E$ denoting the edges of the graph. If a subset of vertices in $P$, say $P_1$ connects all vertices of $Q$, then $P_1$ covers $Q$. The number of vertices in $P_1$ is said to be the size $s$ of this cover. $P_1$ becomes the minimum cover if no smaller subset of $P$ covers $Q$. Conclusively, the bipartite cover problem aims at getting minimum cover for a set of vertices in a bipartite graph.

---

**Algorithm 3.** Determining Bipartite Cover

---

1: **procedure** BIPARTITE- COVER($P$, $Q$)
2: *begin*:
3:     $s \leftarrow 0$
4:     $C \leftarrow new\_cover\_set$
5:     **for** all $i$ in $P$ **do**
6:         $p\_set[i] \leftarrow degree[i]$
7:     **for** all $i$ in $Q$ **do**
8:         $q\_set[i] \leftarrow$ FALSE
9:     **while** $p\_set[i] > 0$ *for all nodes in P* **do**
10:         $v \leftarrow$ *vertex i of P with max value in p_set*
11:         append $v$ to $C$
12:         $s \leftarrow s + 1$
13:         **for** all $w$ adjacent from $v$ **do**
14:             **if** $v$ is not reached **then**
15:                 mark $v$ as reached in $q\_set[v]$
16:                 **for** all $x$ adjacent from $v$ **do**
17:                     $p\_set[x] \leftarrow p\_set[x] - 1$
        **return** $C$
18: *end*

---

# 4 Results

In this section we show the result of applying Pearsons correlation coefficient (Section 3.1) to calculate the correlation of various predictors like Average Temperature(AT), Cloud Cover (CC), Vapour Pressure (VP), Wet Day Frequency (WDF) and Potential Evapotranspiration (PET) with Precipitation(PPT). The data used is obtained from the Indian Meteorological Department [9], for the years 1901 to 2002. We have considered the average value of all factors to get a concise result. The data used in correlation analysis is shown in Table 1.

**Table 1** Factors of precipitation used in Correlation and Regression

| Factors | Jan | Feb | Mar | Apr | May | Jun | Jul | Aug | Sep | Oct | Nov | Dec |
|---------|-----|-----|-----|-----|-----|-----|-----|-----|-----|-----|-----|-----|
| AT | 14.04 | 16.57 | 22.42 | 28.38 | 32.54 | 33.58 | 30.81 | 29.26 | 29.03 | 26.13 | 20.15 | 15.3 |
| CC | 25.74 | 21.31 | 30.79 | 24.45 | 27.34 | 43.02 | 59.19 | 58.89 | 38.21 | 16.25 | 10.08 | 18.9 |
| VP | 8.75 | 8.99 | 9.09 | 9.91 | 13.69 | 23.25 | 29.43 | 29.52 | 24.8 | 14.42 | 9.53 | 9.24 |
| WDF | 1.46 | 1.38 | 1.13 | 0.96 | 1.5 | 3.08 | 8.64 | 7.83 | 3.51 | 0.92 | 0.39 | 0.8 |
| PET | 4.45 | 5.22 | 6.41 | 7.91 | 8.62 | 8.1 | 6.64 | 5.85 | 6.3 | 6.67 | 5.81 | 4.7 |
| PPT | 14.92 | 13.11 | 8.94 | 6.74 | 12.02 | 43.97 | 177.72 | 180.75 | 93.51 | 7.75 | 4.92 | 5.09 |

Each of the factors - AT, CC, VP, WDF and PET on which PPT is dependent is correlated using the method described in Section 3.1. The data obtained after correlation analysis is shown in Table 2. It is found that the factors AT, CC, VP, and WDF are positively correlated whereas PET is negatively correlated to PPT.

**Table 2** Correlation Output

| Factors | Correlation |
|---------|-------------|
| Average Temperature (°C) | 0.469813912 |
| Cloud Cover (%) | 0.914458741 |
| Vapour Pressure (hpa) | 0.913396359 |
| Wet Day Frequency (days) | 0.976401932 |
| Potential Evapotranspiration (mm/day) | -0.02591096 |

The data used in regression analysis is shown in Table 1. The linear and multiple linear regression are carried out using the methods listed in Section 3.2.1 and Section 3.2.2. The equations obtained after applying linear regression are:
Average Temperature (AT):

$$Precipitaion = -65.51605099 + 4.545899238 * AverageTemperature \quad (9)$$

Cloud Cover (CC):

$$Precipitaion = -72.54924724 + 3.848600815 * CloudClover \qquad (10)$$

Vapour Pressure (VP):

$$Precipitaion = -67.92397235 + 7.263286477 * VapourPressure \qquad (11)$$

Wet Day Frequency (WDF):

$$Precipitaion = -14.82307806 + 23.64927015 * WetDayFrequency \qquad (12)$$

Potential Evapotranspiration (PET):

$$Precipitaion = 48.37720823 - 0.144581361 * PotentialEvapotranspiration \qquad (13)$$

The equation obtained after applying multiple linear regression are:

$$Precipitation(p) = f(A, B, C, D, E) \qquad (14)$$

$$P = 72.72602 + 10.88178 * A - 0.1534 * B - 3.29721 * C + 21.16636 * D - 46.052 * E \qquad (15)$$

where X = constant = 72.72602, A = Average Temperature, B = Cloud Cover, C = Vapour Pressure, D = Wet Day Frequency, E = Potential Evapotranspiration.

The coefficients obtained after regression analysis are shown in Table 3.

**Table 3** Results of Regression Analysis

| Key | Factor | Linear | Multiple |
|-----|--------|--------|----------|
| A | Average Temperature | 4.545899238 | 10.88178 |
| B | Cloud Cover | 3.848600815 | -0.1534 |
| C | Vapour Pressure | 7.263286477 | -3.29721 |
| D | Wet Day Frequency | 23.64927015 | 21.16636 |
| E | Potential Evapotranspiration | -0.144581361 | -46.052 |

We have taken the original rainfall value and the predicted rainfall by using our approach. The graph of actual values versus predicted values is shown in Figure 1. The horizontal axis represents the month wise mean of precipitation from 1992 to 2002. The vertical axis represents the precipitation values in millimetres(mm). It can be seen from the graph that the predicted values deviate slightly from the actual values in low precipitation months, whereas are near accurate in peak rainfall months.

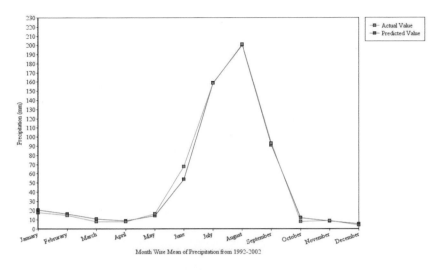

**Fig. 1** Actual Value vs. Predicted Value of Precipitation

Next, we move towards determining the suitability of crops in different states. To achieve this, we have used the concept of Decision Trees with the help of season wise rainfall data. The data set used for this has been taken from the Open Government Data Platform (OGD) India [10]. The labels for different states are chosen according to Algorithm 1. The various states are classified on the basis of rainfall ranges for each crop using Algorithm 2. The resulting table for selected states for rice crop is shown in Table 4. Similar tables can be obtained for the remaining crops.

**Table 4**  State-Wise labelled Rainfall Data for Rice Crop

| State | Winter | Pre-Monsoon | Monsoon | Post-Monsoon |
|---|---|---|---|---|
| COASTAL KARNATAKA | Inadequate | Inadequate | Extreme | Optimal |
| EAST U.P. | Inadequate | Inadequate | Inadequate | Inadequate |
| GANGETIC WEST BENGAL | Inadequate | Inadequate | Moderate | Inadequate |
| HAR. CHD & DELHI | Inadequate | Inadequate | Inadequate | Inadequate |
| MADHYA MAHARASHTRA | Inadequate | Inadequate | Inadequate | Inadequate |
| NORTH INTERIOR KAR-NATAKA | Inadequate | Inadequate | Inadequate | Inadequate |
| PUNJAB | Inadequate | Inadequate | Inadequate | Inadequate |
| SOUTH INTERIOR KAR-NATAKA | Inadequate | Inadequate | Inadequate | Inadequate |
| SUB-HIMALAYAN WEST BENGAL & SIKKIM | Inadequate | Inadequate | Optimal | Inadequate |
| WEST U.P. | Inadequate | Inadequate | Inadequate | Inadequate |

To finally determine the suitability of growing a crop in a particular state, decision trees are used with the condition attributes - winter rainfall, pre-monsoon rainfall, monsoon rainfall, and post-monsoon rainfall. Each state is classified as suitable (yes) or not-suitable (no) for growing a particular crop based on the rainfall requirements of the crop and predicted rainfall in each state. The rainfall received by each state has been categorized into inadequate, moderate, optimal, abundant and extreme. The ranges for each vary according to the crop in consideration.

Figure 2 shows the decision tree obtained for the rice crop. It is evident that rice can be grown if the monsoon is either optimal or moderate, otherwise it is dependent on post-monsoon being optimal. In all other cases the rice crop is not suitable to be grown. These rules are captured in Algorithm 4.

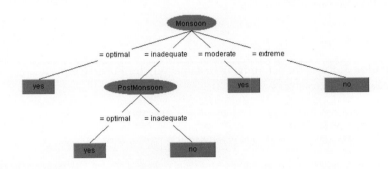

**Fig. 2**   Decision Tree for Rice Crop

---

**Algorithm 4.** Rules for determining sustainability for Rice Crop

---

1: **procedure** RICE- RULES
2: *begin*:
3:    **if** *monsoon = optimal* **then**
4:        *growrice* = YES
5:    **else if** *monsoon = inadequate* and *postmonsoon = optimal* **then**
6:        *growrice* = YES
7:    **else if** *monsoon = inadequate* and *postmonsoon = inadequate* **then**
8:        *growrice* = NO
9:    **else if** *monsoon = moderate* **then**
10:        *growrice* = YES
11:    **else if** *monsoon = extreme* **then**
12:        *growrice* = NO
13:    **else**
14:        *growrice* = NO
        **return** *growrice*
15: *end*

---

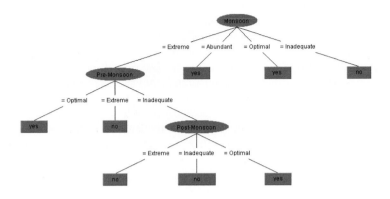

**Fig. 3** Decision Tree for Potato Crop

---

**Algorithm 5.** Rules for determining sustainability for Potato Crop

---

1: **procedure** POTATO- RULES
2: *begin*:
3:     **if** *monsoon* = *extreme* and *premonsoon* = *optimal* **then**
4:         *growpotato* = YES
5:     **else if** *monsoon* = *extreme* and *premonsoon* = *extreme* **then**
6:         *growpotato* = NO
7:     **else if** *monsoon* = *extreme* and *premonsoon* = *inadequate* and *postmonsoon* = *extreme*
    **then**
8:         *growpotato* = NO
9:     **else if** *monsoon* = *extreme* and *premonsoon* = *inadequate* and *postmonsoon* =
    *inadequate* **then**
10:        *growpotato* = NO
11:    **else if** *monsoon* = *extreme* and *premonsoon* = *inadequate* and *postmonsoon* = *optimal*
    **then**
12:        *growpotato* = YES
13:    **else if** *monsoon* = *abundant* or *monsoon* = *optimal* **then**
14:        *growpotato* = YES
15:    **else if** *monsoon* = *inadequate* **then**
16:        *growpotato* = NO
17:    **else**
18:        *growpotato* = NO
       **return** *growpotato*
19: *end*

---

From the decision tree for the potato crop, we conclude that potato can be grown if monsoon is either optimal or abundant, otherwise if monsoon is extreme it depends on pre-monsoon being optimal or post-monsoon being optimal for inadequate pre-monsoon. In all other cases the potato crop is not suitable to be grown. This is seen in the decision tree obtained in Figure 3 and all the possible set of rules has been listed in Algorithm 5.

Similarly, the decision tree and rules for nine more crops is obtained for a total of eleven crops, from which the number of crops growable in each state is determined. To determine the total number of crops growable in each state and the minimum number of states required to increase the production of these crops, bipartite graph is used. Consider, for example, the following sets:

STATES = *{West Bengal, Madhya Pradesh, Uttar Pradesh, Karnataka, Gujarat, Orissa, Tamil Nadu, Himachal Pradesh, Assam, Bihar, Haryana, Punjab, Rajasthan, Kerala, Andhra Pradesh, Maharashtra}*

and

CROPS = *{Jowar, Onion, Cotton, Rice, Rubber, Tea, Bajra, Wheat, Maize, Potato, Pulses}*

Next, we determine a subset of states that cover all the crops mentioned above. Now, let each of the states contain the following relations: *West Bengal = {Rice, Tea, Potato}, Madhya Pradesh = {Jowar, Maize, Pulses}, Uttar Pradesh = {Rice, Wheat, Potato, Pulses}, Karnataka = {Jowar, Onion, Rubber, Maize}, Gujarat = {Onion, Cotton, Bajra}, Orissa = {Rice}, Tamil Nadu = {Cotton, Rubber} Himachal Pradesh = {Tea}, Assam = {Tea}, Bihar = {Rice, Potato}, Haryana = {Wheat}, Punjab = {Rice, Wheat}, Rajasthan = {Bajra}, Kerala = {Rubber}, Andhra Pradesh = {Maize, Rice, Cotton, Jowar}, Maharashta = {Bajra, Cotton, Onion, Jowar}*

We will get the bipartite graph as follows:

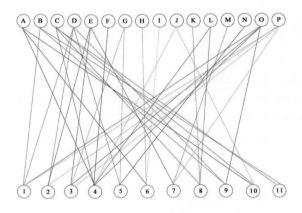

**Fig. 4** Bipartite graph showing mapping between States and Crops. KEY: A - West Bengal, B - Madhya Pradesh, C-Uttar Pradesh, D-Karnataka, E-Gujarat, F- Orissa, G-Tamil Nadu, H - Himachal Pradesh, I-Assam, J-Bihar, K-Haryana, L -Punjab, M-Rajasthan, N-Kerela, O-Andhra Pradesh, P-Maharashtra; 1-Jowar, 2-Onion, 3-Cotton, 4-Rice, 5-Rubber, 6-Tea, 7-Bajra, 8-Wheat, 9-Maize, 10-Potato, 11-Pulses.

Initially the selected set of states is empty, T = { } and all the crops are uncovered. After each iteration, we select the state which covers the most number of uncovered crops. In the first iteration, we select the state *Karnataka* with a count of 4 uncovered crops. So T = {*Karnataka*}. We mark {*Jowar, Onion, Rubber, Maize*} as covered crops.

In the second iteration, we select *West Bengal* with a maximum count of 3 un-covered crops and so T = {*Karnataka, WestBengal*}. Next, we mark {*Rice, Tea, Potato*} as covered crops.

Now from all the remaining states we remove those crops that have already been covered by {*Karntaka, WestBengal*}. On further calculation, we get the set of selected states T = {*Karntaka, WestBengal, Maharashtra, Punjab, UttarPradesh*} as the minimum set of states cover for optimal selection of states which ensures cultivation all 11 crops.

Using the bipartite graph generated, from Algorithm 3, we pick a state to see its feasibility for a particular crop. Consider for example the feasibility of cultivating rice in Karnataka. The result of the decision tree for rice crop, Figure 2, is consistent with our knowledge that Karnataka is a suitable place to grow rice crop. Further, we have made a comparison between the actual rainfall in Karnataka against predicted value from our model (Eqn. 15) for the season in which rice crop is cultivated. The graph is shown in Figure 5.

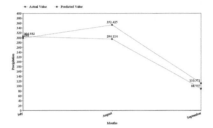 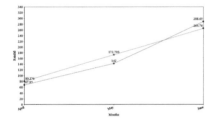

**Fig. 5** Actual vs Predicted Rainfall in Mon-soon for Karnataka

**Fig. 6** Actual vs Predicted Rainfall in Pre-Monsoon for West Bengal

Similarly we observe from the decision tree in Figure 3 that potato crop is suitable for growing in West Bengal, which is consistent with our initial assumption. The actual rainfall against predicted rainfall is also shown in Figure 6.

## 5   Conclusion

In this paper we have used a hybrid approach to classify and predict rainfall data to determine the suitability of growing a particular crop in different states across India. We have taken four seasons: Winter (January - February), Pre-Monsoon(March - May), Monsoon(June - September), Post-Monsoon(October - December). This classification helps us in choosing the best states for the productivity of a crop. Next, we have taken the correlation concept for the consideration of various parameters that are correlated with rainfall. The parameters like Average Temperature, Cloud Cover, Vapour Pressure, Potential Evapotranspiration, Wet Day Frequency are individually considered for the analysis with the decision attribute - Precipitation. After getting the positively and negatively correlated variables we have predicted the rainfall by using linear regression and multiple linear regression. Finally, we have applied the decision tree concept to classify the states for suitability of crop production with the help of season wise amount of rainfall a state receives. The application of bipartite cover has been used to find out the optimal set of states for the production of major crops. After getting the result from the above methodology, we are trying to suggest the government of India to give subsidies and exemptions to encourage the farmers of those selected states to increase the productivity of major crops.

# References

1. Mahato, A.: Climate Change and its Impact on Agriculture. International Journal of Scientific and Research Publications, IJSRP 2014, **4**(4) (2014)
2. Kang, Y., Khan, S., Ma, X.: Climate change impacts on crop yield, crop water productivity and food security - a review. In: Progress in Natural Science. Elsevier (2009)
3. Prasanna, V.: Impact of Monsoon Rainfall on the Total Foodgrain Yield over India. J. Earth Syst. Sci. **123**, 1129–1145 (2014). Indian Academy of Sciences
4. Krishna Kumar, K., Kumar, R.K., Ashrit, R.G., Deshpande, N.R., Hansen, J.W.: Climate impacts on indian agriculture. In: 24th International Journal of Climatology, pp. 1375–1393 (2004)
5. Awan, J.A., Maqbool, O.: Application of artificial neural networks for monsoon rainfall prediction. In: 6th International Conference on Emerging Technologies (ICET). IEEE Computer Society (2010)
6. Nikam, V.B., Meshram, B.B.: Modeling rainfall prediction using data mining method - a bayesian approach. In: 2013 Fifth International Conference on Computational Intelligence, Modelling and Simulation. IEEE (2013)
7. Han, Jiawei, Kamber, Micheline: Data Mining: Concepts and Techniques, 2nd edn. Morgan Kaufmann, San Francisco (2006)
8. Sahni, Sartaj: Data Structures, Algorithms and Applications in C++, 2nd edn. Universities Press, Hyderabad (2005)
9. India Water Portal. Met Data, May 2010. http://www.indiawaterportal.com/met_data (retrieved July 27, 2015)
10. Open Government Data (OGD) Platform India. Area Weighted Monthly seasonal and Annual Rainfall Data, February 2014. https://data.gov.in/catalog/area-weighted-monthly-seasonal-and-annual-rainfall-mm-36-meteorological-subdivisions#web_catalog_tabs_block_10 (retrieved July 27, 2015)

# Development and Evaluation of Automated Algorithm for Estimation of Winds from Wind Profiler Spectra

E. Ramyakrishna, T. Narayana Rao and N. Padmaja

**Abstract** National Atmospheric Research Laboratory (NARL) hosts several atmospheric measurements of which, MST Radar is being treated as a work horse, for which the wind observations up to an altitude of 100km is recorded. It is a great tool for studying the atmospheric dynamics. These radars are capable of measuring air motions over a wide range of heights with good spatial and temporal resolutions. The derivation of wind components from spectral data includes noise level estimation and then moment estimation. In the present context, Hildebrand and Sekhon method is utilized for noise level estimation for each range bin employing the physical properties of white noise. The interference due to stationary targets is reflected as the dc components which can be removed using conventional methods such as replacing the spectral power at zero frequency with the average of the power at adjacent spectral points and it is very important for distinguishing between the actual signals from interference echoes. In the light of automation of data analysis so as to obtain the wind components, it is mandatory to develop algorithms for the characterization of signal and interference echoes. In this work, an Automated Algorithm is developed for estimation of moments and is applied to Real Time Radar data from MST region near Gadanki, Tirupati.

## 1    Introduction

National Atmospheric Research Laboratory (NARL) (13.47°N, 79.18°E), has been operating a 53 MHz MST Radar [3] since 1992-93 for research like studying

E. Ramyakrishna(✉)
Space Technology, S V Unversity, Tirupati, India
e-mail: ramya.enugonda@gmail.com

T.N. Rao
National Research Atmospheric Laboratory, Gadanki, Tirupati, India
e-mail: tnrao@narl.gov.in

N. Padmaja
ECE Department, Sree Vidyanikethan Engineering College, Tirupati, India
e-mail: padmaja202@gmail.com

© Springer International Publishing Switzerland 2016
S.M. Thampi et al. (eds.), *Advances in Signal Processing and Intelligent Recognition Systems*,
Advances in Intelligent Systems and Computing 425,
DOI: 10.1007/978-3-319-28658-7_40

dynamics of lower, middle and upper atmosphere and operational applications, like supporting rocket launches from SHAR and weather forecasting.

In the proposed work, calculation of spectral moments of spectrum with composite structure is done in a slightly different way. An Automated algorithm is developed for estimation of moments and is applied to Real Time MST Radar data. This type of spectrum normally comes in the upper atmospheric region. Here the spectra shows multiple spikes and makes it difficult for identifying the peak. So, it is essential to remove the mean noise level form the spectra. The conventional peak and valley detection algorithms cannot give accurate results because of clutter and interference. To overcome this problem, the signal and interference characteristics should be studied. Multiple sets of MST Radar data recorded on varied climatic conditions is considered for analysis and the algorithm is applied to all the 6 beam directions of MST Radar data.The algorithm is developed by using MATLAB R2014b.

## 1.1   Estimation of Moments

The following are the steps that are performed for extracting moments and wind velocities.

a) Read the Doppler power spectra of MST Radar data.
b) Remove the DC using 5 point DC removal technique. It is done by re-placing the spectral power at zero frequency with the average of four adjacent points for each range bin.
c) Estimate the average noise value for each range bin by objective me-thod of Hildebrand and Sekhon [2] as given in equation 1.

$$\sigma^2 = (\Sigma f_n^2 S_n / \Sigma S_n) - (\Sigma f_n S_n / \Sigma S_n)^2,$$
$$\sigma_N^2 = F^2/12,$$
$$P = \Sigma S_n / N,$$
$$Q = \Sigma (S_n^2/N) - P^2,$$
$$R_1 = \sigma_N^2 / \sigma^2,$$
$$R_2 = P^2 / Qp,$$

$$(1)$$

Where F is the frequency spread of the spectrum, f is a spectral frequency N is the number of independent spectral densities, p is the number of lines over which a moving average is taken, and the summation is over the n spectral densities. For white noise, the ratios $R_1$ and $R_2$ should be unity. By using these conditions noise level can be estimated.

d) The calculated mean noise level is subtracted from the spectral power at each range gate.
e) Spectral Peak values are identified at each range gate and lower (m) and upper (n) bounds of the echoes are identified.

The three low order moments (zero, first and second) are computed using numerical integration [4-10]. The three low order moments represent signal strength (power), the weighted mean Doppler shift and width of the spectrum.

The expressions for the first three moments are as follows.

The $0^{th}$ moment representing the total signal power is calculated by equation (2).

$$M_0 = \sum_{i=m}^{n} P_i \tag{2}$$

The $1^{st}$ moment representing the weighted mean Doppler shift is

$$M_1 = (1/M_0) \sum_{i=m}^{n} P_i f_i \tag{3}$$

The $2^{nd}$ moment representing the variance, a measure of dispersion from the mean frequency is

$$M_2 = (1/M_0) \sum_{i=m}^{n} P_i (f_i - M_i)^2 \tag{4}$$

Where m and n are the lower and upper limits of the Doppler bin of the spectral window. $P_i$ and $f_i$ are the powers and frequencies corresponding to the Doppler bins within the spectral window.

Signal-to-noise ratio (SNR) in dB is calculated as

$$SNR = 10 \log \left( \frac{M_0}{N.L} \right) \tag{5}$$

Where N and L are the total number of Doppler bins and mean noise level respectively which on multiplication gives the total noise over the whole bandwidth.

Doppler width, which is taken to be the full width of the Doppler spectrum is calculated as:

$$Doppler Width = 2\sqrt{(M_2)} \tag{6}$$

## 1.2 Estimation of U V W components

For oblique beams, range to height conversion is done by

$$\text{Height} = \text{range} * \cos(\theta), \quad \text{range} = c * t/2 \tag{7}$$

where, $c$ is light velocity, $t$ is time shift from reference transmit pulse. The Doppler shift of each height is converted into velocity. The line-of- sight velocity $Vr$ is resolved into the components of wind vector $v = (v_x, v_y, v_z)$.

The line-of-sight component $V_d$ derived from Doppler shift $M_1$.

$$V_r = M_1 \times (\lambda/2) \tag{8}$$

$V_r$ of the velocity vector $V = (V_x, V_y, V_z)$ at a given height is expressed in terms of

$$V_r = V.i = V_x\cos\theta_x + V_y\cos\theta_y + V_z\cos\theta_z \qquad (9)$$

where, i is the unit vector along the radar beam and $\theta_x$, $\theta_y$, $\theta_z$ are the angles that the radar beam makes with the X, Y and Z axes, respectively.

$$v = \begin{pmatrix} \cos\theta_{x1} & \cos\theta_{y1} & \cos\theta_{z1} \\ \cos\theta_{x2} & \cos\theta_{y2} & \cos\theta_{z2} \\ \cos\theta_{x3} & \cos\theta_{y3} & \cos\theta_{z3} \end{pmatrix}^{-1} (v_{r1} \; v_{r2} \; v_{r3})^T \qquad (10)$$

Corresponding to east, north and vertical directions respectively. The estimate of $v$ can be obtained if the $Vr$ is obtained from three $v_{r1}$ $v_{r2}$ $v_{r3}$ orthogonal directions by matrix multiplication. Usually MST radar observations are done for more than three directions to get redundancy.

The estimate of v is determined by least-squares method, by minimizing the residual $\varepsilon_v^2 = \Sigma_{i=1}^m (v_x \cos\theta_{xi} + v_y \cos\theta_{yi} + v_z \cos\theta_{zi} - v_{di})^2$, where m is the number of beam directions. The necessary condition for $v$ is the partial derivatives of $\varepsilon_v^2$ with respect to $v$ are zero. $\partial\varepsilon_v^2 / \partial v_j = 0$, where (j=x, y, z).

$$\begin{bmatrix} V_x \\ V_y \\ V_z \end{bmatrix} = \begin{bmatrix} \sum_i \cos^2\theta_{xi} & \sum_i \cos\theta_{xi}\cos\theta_{yi} & \sum_i \cos\theta_{xi}\cos\theta_{zi} \\ \sum_i \cos\theta_{xi}\cos\theta_{yi} & \sum_i \cos^2\theta_{yi} & \sum_i \cos\theta_{yi}\cos\theta_{zi} \\ \sum_i \cos\theta_{xi}\cos\theta_{zi} & \sum_i \cos\theta_{yi}\cos\theta_{zi} & \sum_i \cos^2\theta_{zi} \end{bmatrix}^{-1} * \begin{bmatrix} V_{di}\cos\theta_{xi} \\ V_{di}\cos\theta_{yi} \\ V_{di}\cos\theta_{zi} \end{bmatrix} \qquad (11)$$

Thus, Vx, Vy, and Vz, which correspond to $u$ (zonal), v (meridional) and $w$ (vertical) components of velocity are calculated.

## 2    Signal Characteristics

Continuous weakening of echo strength, except near the tropopause and stable layers, makes it difficult to identify the echo from the background noise. In addition, the spectra at times is contaminated with the interference. Correct identification of interference and backscattered echo is the primary concern for the user or algorithm. For proper identification of interference/echo,their characteristics should be known. The Figure (1) shows the CFAD (Contoured  Frequency by Altitude Diagrams) plot between number of data points and height. It can be observed that the  maximum number of data points are more at lower heights and less at higher heights. The minimum number of data points at higher height above 10 km is less than 5 and below 10 Kms is less than 8.  Based on this , a  threshold is considered for signal detection.

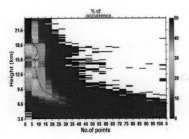

**Fig. 1** CFAD plot for Power

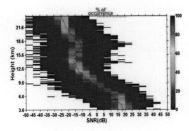

**Fig. 2**  CFAD plot for SNR

The SNR is less at higher heights. At lower heights the maximum SNR is around  15 to 20 dB and at higher heights, SNR is below -25 dB.

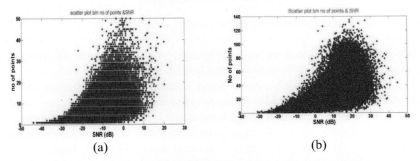

(a)                                        (b)

**Fig. 3** Scatter plot between data points and SNR (a) above and (b) below 10 km.

Although it appears from the figures (3a) and (3b) that the SNR increases with the number of data points, but there is no one-to-one relation. They do not vary linearly. Therefore, we can obtain thresholds for SNR. The threshold SNR for atmospheric echoes in case of MST radar typically varies between -12 to -15 dB depending on the noise floor. It can be possible to differentiate the actual atmospheric signal from the multiple peaks appearing in a given range bin.

# 3    Interference Characteristics

To characterize the interference in a robust way, several contaminated spectra of MST radar is used.

Figure 4 shows the typical Doppler spectra with interference band at the same Doppler point in all the range bins of the spectra [14]. From the figure, it is evident even though the band appears to be present between 12 km to 25.95 km; it is present in all the range bins, but obscured by the actual signal in the lower range bins.

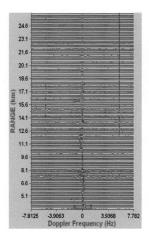

**Fig. 4**  Typical Spectral plot showing the interference bands

Both the zero band and rolling interference presents the same Doppler in all range gates. This property is used to identify them in the spectra.

The Doppler shift of such spectral peaks contains 5 or more spectral values at each altitude are examined to identify the interference. If any of those spectral peaks appear at the same Doppler in all range bins, then it is considered as interference.

After identification of interference, its characteristics viz., total power, Doppler, Doppler width, skewness and number of spectral points within the band associated with interference echoes are shown in the Figure 5 (a, b, c, d and e).

Following are the observations made from CFADS of various interference parameters.

1)  Power CFAD (Fig.5a) shows that the echo power associated with the interference primarily confines to -40 to -25 dB at all range gates.

2)  Compared to the echo power, the Doppler shift distribution is wide in (Fig .5b). The Doppler shift values vary from -4 to 4 Hz.

3)  CFAD of Doppler width (Fig.5c) shows that the maximum width possible is 0.15 Hz, with a mean and maximum possibility of 0.05 to 0.1 Hz width. It means that the width of peak associated with the interference is very narrow.

4) Skewness CFAD (Fig.5d) shows the maximum possibility of negative skewness or left elongated distribution with the mean values of about -0.5.

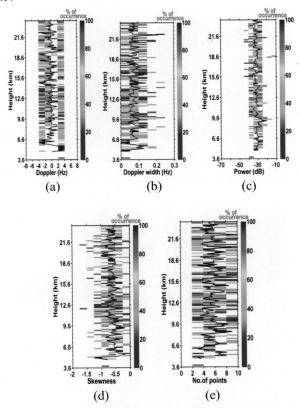

(a)                    (b)                    (c)

(d)                    (e)

**Fig. 5** CFADs for(a) Power, (b) Doppler shift, (c) Doppler width, (d) skewness and (e) number of points within the band associated with interference echoes. The overlaid solid line is the mean of the distribution.

## 4    Results and Validation

### 4.1    Estimation of Moments

The extraction of zeroth, first and second moments, Doppler width and SNR can be calculated by identifying the pure signal. After removing noise and interference from the spectrum, still it consists of some peaks in all range bins. Estimation of spectral moments for a required peak can be done by considering the east and west beam in each cycle. In the present context, an experiment specification file is so chosen that out of 24 frames with 6 beams in each scan cycle as shown in the Figure (6), four frames each in east and west direction are obtained. Since East and West beams are mirror images, the maximum peak Doppler point in east beam is same as the maximum peak Doppler point in west beams with opposite signs. By taking the median of the Doppler points in each range bin, median profile is obtained as shown in figure.

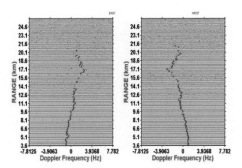

**Fig. 6** Median profile of east and west beam

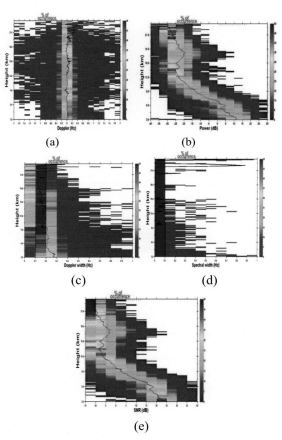

**Fig. 7** *(*a, b, c d and e*):* CFAD plots for zeroth, first and second moments spectral widths and SNR respectively.

The maximum Power is calculated from zeroth moment and Figure 7a shows the maximum power of 40 dB. The mean value of the total power ranges between -30 to 15 dB. Figure 7b shows the CFAD diagram of first moment and the maximum Doppler shift. It is observed that the Doppler shift occurs between -1 to 2 Hz and the mean Doppler shift is centralized between 0 to 1Hz.

CFAD diagram of second moment (Fig.7d) shows that the spectral width lies between 0 to 0.1 up to 15 km. In SNR CFAD (Fig.7e), the maximum SNR is about 40 dB up to 6km and decreases to -15 dB at higher range bins. The mean value of SNR is centralized between -12 dB to 23 dB.

CFAD diagram (Fig.7c) the mean value of Doppler Width ranges from 0.1 to 0.2 above 10 km and between 0.2 to 0.3, below 10 km.

## 4.2 Validation of the Results with Atmospheric Data Processor (ADP)

It is highly essential to check the performance of any new algorithm, particularly if it is an automated algorithm, by comparing with other standard techniques. In this work, the estimated winds were compared with a standard package ADP being used traditionally for atmospheric data processing in which moments and winds are estimated. [1,12,13]. The algorithm was applied on different data sets and the results were found to be satisfactory lifting the confidence levels of the algorithm. The results are shown in Figures 8 and 9.

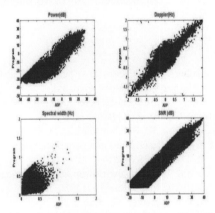

**Fig. 8** Statistical comparisons of Power, Doppler, Spectral width and SNR estimated by automated algorithm with ADP.

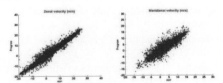

**Fig. 9** Statistical comparison of Zonal and Meridional winds estimated by automated algorithm ADP.

Fig( 8) shows the Statistical comparison of moments (Power, Doppler, Spectral width and SNR) obtained by both automated algorithm  and ADP using scatter plots. The correlation coefficient is greater than 0.8 in all the cases. Fig (9) shows the Statistical comparison of zonal velocity and meridional velocity. The correlation coefficient is 0.95 and 0.85 respectively.  At the outset, with the good correlations between the present algorithms with the standard conventional package, the proposed algorithms can be effectively utilized for the automation of the MST Radar data analysis.

## 5    Conclusion

The proposed automated algorithm was applied on multiple Radar data sets and the characteristics of echo and interference are studied as a function of height. Though one would expect the peak at only one spectral point, possibly the leakage due to FFT widens the echo. In signal characteristics, by adopting the above threshold values below and above 10 km, it can be possible to differentiate the actual atmospheric signal from the multiple peaks appearing in a given range bin. Below 10 km, SNR was found to be in the range -10 to -20 dB, while above it, the SNR ranges between -20 to -10 dB. The scattering plot between number of data points and SNR gives a good correlation of 0.965. Also, the Correlation Coefficient of 0.95 and 0.85 for Zonal and Meridional wind between the automated algorithm and expert analysis gives more confidence on the performance of the algorithm.

## References

1. Anandan, V.K.: Atmospheric Data Processor – Technical and User reference manual, National Atmospheric Research Laboratory, Gadanki
2. Hildebrand, P.H., Sekhon, R.S.: Objective determination of the noise level in Doppler spectra. J. Appl. Meterol. **13**, 808–811 (1974)
3. Rao, P.B., Jain, A.R., Kishore, P., Balamuralidhar, P., Damle, S.H., Viswanadhan, G.: Indian MST radar. I: System description and samplevector wind measurements in ST mode. Radio Sci. **30**(4), 1125–1138 (1995)
4. Woodman, R.F.: Spectral moment estimation in MST radars. Radio Sci. **20**(6), 1185–1195 (1985). doi:10.1029/RS020i006p01185
5. Anandan, V.K., Ramachandra Reddy, G., Rao, P.B.: Spectral analysis of atmospheric signal using higher orders spectral estimation technique. IEEE Trans. Geosci. Remote Sens. **39**(9), 1890–1895 (2001)
6. Anandan, V.K., Pan, C.J., Rajalakshmi, T., Ramchandra Reddy, G.: Multitaper spectral analysis of atmospheric radar signal. Ann. Geophys. **22**(11), 3995–4003 (2004)
7. Anandan, V.K., Balamuralidhar, P., Rao, P.B., Jain, A.R.: A method for adaptive moments estimation technique applied to MST radar echoes. In: Proc. Prog. Electromagn. Res. Symp., pp. 360–365 (1996)
8. Madhavi latha, G., Padmaja, N., Varadarajan, S.: Estimation of Moments for Low and High SNR Atmospheric Radar Data. i-manager's Journal on Communication Engineering and Systems (JCS), May–July 2012, ISSN Print: 2277-5102 ISSN Online: 2277-5242

9. Morse, C.S., Goodrich, R.K., Cornman, L.B.: The NIMA method for improved moment estimation from Doppler spectra. J. Atmos. Ocean. Technol. **19**(3), 274–295 (2002)
10. Padmaja, N., Varadarajan, S., Swathi, R.: Signal Processing of Radar Echoes using Wavelets & Hilbert Huang Transform. Sipij: Signal and Image Processing: An International Journal **2**(3), 101–119 (2011). ISSN: 0976 - 710X (Online); 2229 - 3922 (print)
11. Briggs, B.H.: Radar observations of atmospheric winds and turbulence: a comparison of techniques. Journal of Atmospheric and Terrestrial Physics **42**, 823–833 (1980)
12. Goodrich, R.K., Morse, C.S., Cornman, L.B., Cohn, S.A.: A horizontal wind and wind confidence algorithm for Doppler wind profilers. J. Atmos. Oceans. Technol. **19**(3), 257–273 (2002)
13. Anandan, V.K., Jagannatham, D.B.V.: An Autonomous Interference Detection and Filtering Approach Applied to Wind Profilers. IEEE Transactions on Geoscience and Remote Sensing **48**(4), April 2010

# A Real Time Patient Monitoring System for Heart Disease Prediction Using Random Forest Algorithm

S. Sreejith, S. Rahul and R.C. Jisha

**Abstract** The proposed work suggests the design of a health care system that provides various services to monitor the patients using wireless technology. It is an intelligent Remote Patient monitoring system which is integrating patient monitoring with various sensitive parameters, wireless devices and integrated mobile and IT solutions. This system mainly provides a solution for heart diseases by monitoring heart rate and blood pressure. It also acts as a decision making system which will reduce the time before treatment. Apart from the decision-making techniques, it generates and forwards alarm messages to the relevant caretakers by means of various wireless technologies. The proposed system suggests a framework for measuring the heart rate, temperature and blood pressure of the patient using a wearable gadget and the measured parameters is transmitted to the Bluetooth enabled Android smartphone. The various parameters are analyzed and processed by android application at client side. The processed output is transferred to the server side in a periodic interval. Whenever an emergency caring arises, an alert message is forwarded to the various care providers by the client side application. The use of various wireless technologies like GPS, GPRS, and Bluetooth leads us to monitor the patient remotely. The system is said to be an intelligent system because of its diagnosis capability, timely alert for medication etc. The current statistics shows that heart disease is the leading cause of death and which shows the importance of the technology to provide a solution for reducing the cardiac arrest rate. Apart from that the proposed work compares different algorithms and proposes the usage of Random Forest algorithm for heart disease prediction.

**Keywords** Health care application · Android · Wireless communication · Wearable gadget · Heart disease · Random forest

S. Sreejith · S. Rahul(✉)· R.C. Jisha
Department of Computer Science and Application, Amrita Vishwa Vidyapeetham,
Amritapuri, Coimbatore, India
{sreejiths.sasikumar,rahulsadanandan13}@gmail.com, jisha@am.amrita.edu

© Springer International Publishing Switzerland 2016
S.M. Thampi et al. (eds.), *Advances in Signal Processing and Intelligent Recognition Systems*,
Advances in Intelligent Systems and Computing 425,
DOI: 10.1007/978-3-319-28658-7_41

485

# 1    Introduction

Chronic diseases are one of the curses mankind is facing today. Heart disease, Diabetes, Stroke are some examples of chronic diseases. According to the statistics of World Health Organization [1], death due to heart disease is on top among the top 10 leading cause of death by country income group. This happens mainly because of two reasons; lack of proper medical resources in the hospital and failure of giving proper medical care to the patients at the correct time. Failure to identify the initial stages of heart disease is one of the main reasons why the medical professionals are not able to give proper medical care to the patient. Constant monitoring and early detection of chronic diseases are important to avoid the risk related to this kind of diseases [2]. This is the motivation behind developing a system that regularly monitors the health condition of patients and giving proper medical care to them at the right time. Knowing the body and the symptoms that the heart is getting worse will help people stay healthier and out of the hospital.

Mobile phones are part of people's day to day life. This system will make use of user's mobile phone as a monitoring device that monitors the health condition of patients. This work suggests the design of a health care system that provides various services to monitor the patients using wireless technology with prediction of possibility of heart disease. It mainly provides a solution for identifying heart diseases by monitoring heart rate, blood pressure and cholesterol along with various other measures.

Various health parameters of patients like Heart rate, Blood pressure and Cholesterol are transmitted to the patient's Android phone from wearable gadget using Bluetooth wireless technology. Using classification algorithm, system will predict the possibility of heart disease. If the condition becomes critical, an alarm message will be send to the patient's relatives and also to the doctors along with the current location of patient. System will show the medical history of other patients who have the same medical report of the selected patient. This is also done with the help of another classification algorithm. After analyzing the medical report, doctor will prescribe the medicine and send to the patient. System also gives a reminder to the patients about when to take medicine.

The rest of the paper is organized as follows. Section 2 gives the related work. Section 3 will explain the design and architecture of our system. Section 4 will give the implementation of each module and final section gives the conclusion. The whole methodology of our proposed system is given in Figure. 1

# 2    Related Works

The paper [2] suggests a method for developing a hardware that will transmit the readings from sensors to the mobile phone using Bluetooth wireless technology. Paper [3] presents how the health care management can be done with the help of Android mobile phone. The application will trigger an emergency alarm to the doctors and relatives in case of emergency situations. The patient will send a medical report to the doctor and after analyzing the report, doctor will send the medicine prescriptions to the patient. This application also sets a reminder to the patient to take medicine on time. The system proposed in [4], monitors the ECG (Electrocardiographic) waveforms from a wearable ECG gadget and send the measurements to doctor through patient's mobile phone. The doctor will send the medicine to the patient along with a QR code. QR code will contain the details and timings of the medicine. Utilizing the Barcode scanner, the user will read the information from the QR code and follow the medication process instructed by the doctor.

The proposed work suggests a framework for capturing various parameters like heart rate and blood pressure with the help of a wearable gadget, after analyzing these parameters system is able to trigger emergency alarm to the predefined numbers. The system compares various classification algorithms like Knn algorithm, C4.5 algorithm, Random forest and Naive Bayes and suggests the usage of the Random Forest algorithm. Also the proposed system will help the doctor to see the history of other patients' records which is similar to a medical report of a selected patient. This enables doctor to make better decision while creating medication process. These two features can be done with the help of classification algorithms. Inclusion of the classification algorithms to achieve the above two features make this work unique from other works shown above.

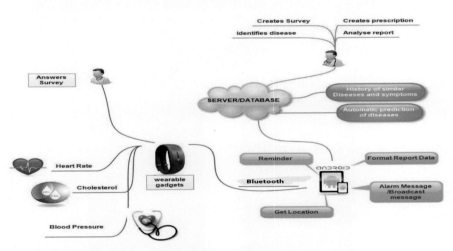

**Fig. 1** System Architecture of the proposed work

## 3    System Architecture

The system architecture consists of a wearable sensor which is basically a heart rate monitor, polar h7 Bluetooth smart heart rate sensor. Bluetooth low energy (Bluetooth smart) is the transmission technology used in polar H7 device to send data to the android app. This extends the use of Bluetooth wireless technology to devices that are powered by small, coin-cell batteries. It can transmit data up to 30 feet. We need to pair the heart rate sensor with our receiving device. Hardware part of the application will interact with the android app which supports the Bluetooth framework that allows a device to wirelessly exchange data. This is illustrated in the Fig 1. So the proposed model suggests a framework to monitor various health parameters by sensor fusion method. This work mainly used polar h7 heart rate sensor.

The system has two main parts, the client side (Patient) and the server side (Doctor). The client side is deployed on an Android Mobile phone and the server side is deployed on a server which is a computer, usually located in hospital. Usually server side will be a web interface which will be very easy for the doctor to handle huge amount of data coming from various patients. The different modules of the system is shown in Figure 2. Client side is an android app and the server side which is a web interface developed using PHP and MySQL.

Android app mainly consists of Reporting module, Alarm Service, Survey and Reminder modules. The Reporting module helps to get various health parameters like heart rate, blood pressure and cholesterol from the user. First a wearable gadget will monitor the health parameters and send to the user's Android mobile phone with the help of Bluetooth wireless technology. So the wearable gadget and Android mobile phone should be within 30 feet range in order to communicate between this two devices. The readings will be send to the doctor in the form of a medical report.

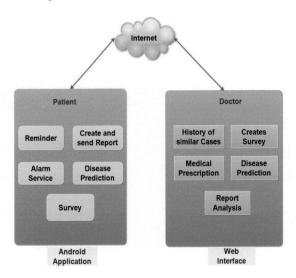

**Fig. 2** System Modules

Alarm service module will send alarm messages .When the readings of heart rate, blood pressure and cholesterol goes beyond the normal range, then an alarm message will be send to the relatives and doctor along with the current location of user. Global positioning system (GPS) technology helps to get the current location of the user. This technology is available in any Android mobile devices. The user will add the contact number of relatives and doctors when the application starts. Then during critical conditions, an alarm message will be send to the contact numbers saved by the user. The user has the option of turning off alarm service so that message will not be send to the contact numbers.

The Survey module provides a provision for the users to clear clarifications of the doctor regarding treatment. This is done in the form of question and answer survey creation. Reminder service will help the user to take the medicine at proper time by setting a reminder. From the symptoms and data send by the user to the doctor with the help of survey module, the doctor will create a prescription to the user and send it accordingly.

Server side is meant for the doctor, which contains history of similar cases. Using classification algorithm, the system will show the medical history of other patients who have similar readings just accepted from the user. This will help the doctor to identify the disease and create better medication process to the user. Medical prescription is one of the features of the proposed work. Here, the doctor will create medicine prescription to the user from the medical report sent by the user. To create better medicine prescription, doctor will analyze the report and also the history of other patients' records with similar medical report. Disease prediction module uses classification algorithm, in which system will predict the possibility of heart disease from the readings sent by the user along with various other parameters. Report analysis module is to handle huge amount of data sent by the user, the doctor will be working in a web interface usually located in hospital. Medical report send by the user will contain various information like readings of heart rate, blood pressure, cholesterol, date and time of the readings taken by the user. Creates survey feature allows the doctor to ask some additional questions to the user regarding the treatment. Sometimes doctor need some additional information while creating medication process. So the doctor will create a survey by including questions and send to the user.

# 4    Algorithm Design

## 4.1    Heart Disease Prediction

Classification algorithm is used to predict the possibility of heart disease from the health parameters taken from the user. There are various classification algorithms available for the prediction of heart disease. But it is necessary to choose the best classification algorithm which gives accurate prediction of the possibility of heart disease. Knn algorithm, C4.5 algorithm, Random forest and Naive bayes are the four classification algorithm we have chosen for the analysis. With the help of a weka data mining tool we will analyze the performance of this algorithms and choose a best one with accurate results [5]. From the University of California Irvine (UCI) machine learning repository we will take a sample dataset which consist of 303 unique records for analyzing the results of this algorithms. Type of attributes present in the dataset is shown in Table 1.

**Table 1** Dataset attributes

| Name | Type | Description |
|------|------|-------------|
| Age | Continuous | Age of the patient in years |
| Sex | Discrete | Gender<br>1=male<br>0=female |
| Cp | Discrete | Chest pain type:<br>1= typical angina<br>2=atypical angina<br>3= non angina<br>4=asymptomatic |
| Trestbps | Continuous | Resting blood pressure ( in mm Hg) |
| Chol | Continuous | Serum cholesterol in mg/dl |
| Fbs | Discrete | Fasting blood sugar>120 mg/dl:<br>1=true<br>0=false |
| Restecg | Discrete | Resting electrocardiographic results:<br>0=normal<br>1=ST-T wave abnormality<br>2=ventricular hypertrophy |
| Thalach | Continuous | Maximum heart rate achieved |
| Exang | Discrete | Exercise induced angina:<br>1=yes<br>0=no |
| Oldpeak | Continuous | ST depression induced by excercise |
| Slope | Discrete | Slope of peak exercise segment:<br>1=up sloping<br>2=flat sloping<br>3=down sloping |
| Diagnosis | Discrete | Diagnosis classes:<br>0=healthy<br>1=possible heart disease |

Out of 303 records, 138 records are with the class 1(possible heart disease) and 165 records are with class 0 (No possibility of heart disease). So we will test the performance of this algorithms on this sample dataset. In weka data mining tool, performance analysis of an algorithm is done with the help of a confusion matrix.

Confusion matrix is a specific table layout that allows visualization of the performance of various algorithms.

## 4.2   Performance Analysis- Results and Discussions

This section discusses the performance of four classification algorithms using weka data mining tool. We used tenfold cross validation testing. The result will be in the form of a confusion matrix which shows the number of correct instances and incorrect instances predicted by the algorithm. To analyze the performance we used the following measures.

- TP(True positive value):  gives the number of positive instances correctly predicted by classification model
- FN (False negative value): gives the number of positive instances wrongly predicted by classification model
- FP (False positive value): gives the number of negative instances wrongly predicted as positive instance.
- TN (True negative): gives the number of negative instances correctly predicted by classification model.
- Accuracy: It is the number of true positive and false positive predictions for the considered class. This is calculated as the sum of correct classifications divided by the total number of classifications.

Accuracy = (TP+TN)/N, where N is the total number of classifications.

- Precision: It is the measure of the accuracy predicted for specific class and calculated using the formulae.

Precision = TP/ (TP+FP)

- Recall: Also called as sensitivity, corresponds to the true positive rate of the considered class and it is calculated using the formulae.

Recall = TP/ (TP+FN)

*a)* KNN algorithm - Confusion matrix

**Table 2** KNN classifies 128 patients without heart disease out of 165 and classifies 105 patients with heart disease out of 138 respectively

| Confusion matrix | | Prediction | |
|---|---|---|---|
| | | Zero | one |
| **Model** | Zero | 128 | 37 |
| | One | 33 | 105 |

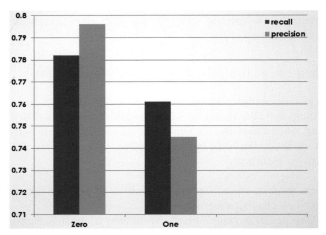

**Fig. 3** 78.2%, 76.1% of randomly retrieved document is relevant and 0.796, 0.745 probability of relevant document retrieved in a search in each class respectively.

*b)* C4.5 algorithm - Confusion matrix

**Table 3** C4.5 classifies 139 patients without heart disease out of 165 and classifies 106 patients with heart disease out of 138 respectively

| Confusion matrix | | Prediction | |
|---|---|---|---|
| | | Zero | one |
| **Model** | Zero | 139 | 26 |
| | One | 32 | 106 |

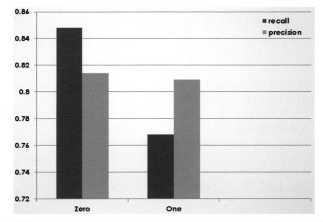

**Fig. 4** 81.4%, 80.9% of randomly retrieved document is relevant and 0.848, 0.768 probability of relevant document retrieved in a search in each class respectively.

*c)* Naive bayse algorithm

**Table 4** Naive bayse classifies 146 patients without heart disease out of 165 and classifies 109 patients with heart disease out of 138 respectively

| Confusion matrix | | Prediction | |
|---|---|---|---|
| | | zero | one |
| **Model** | zero | 146 | 19 |
| | one | 29 | 109 |

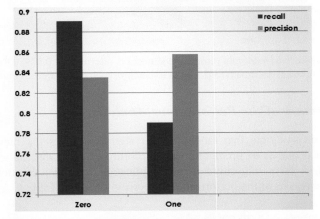

**Fig. 5** 83.5%, 85.8% of randomly retrieved document is relevant and 0.891, 0.79 probability of relevant document retrieved in a search in each class respectively.

*d)* Random forest - Confusion matrix

**Table 5** Random forest classifies 144 patients without heart disease out of 165 and classifies 112 patients with heart disease out of 138 respectively

| Confusion matrix | | Prediction | |
|---|---|---|---|
| | | Zero | one |
| **Model** | Zero | 144 | 21 |
| | One | 26 | 112 |

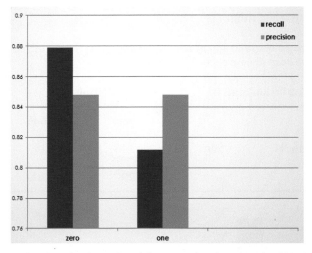

**Fig. 6** 84.8%, 84.8% of randomly retrieved document is relevant and 0.879, 0.812 probability of relevant document retrieved in a search in each class respectively.

## 4.3 Summary of the Performance Analysis

The results clearly state that random forest algorithm gives the most accurate results when it comes to predicting the possibility of heart disease. So we will choose random forest algorithm in our application for the prediction of heart disease. Final results of the performance analysis is shown in Table 6.

**Table 6** Performance Analysis results

|               | Correctly classified instances | Incorrectly classified instances |
|---------------|-------------------------------|----------------------------------|
| C4.5          | 245                           | 58                               |
| KNN           | 233                           | 70                               |
| Naive bayse   | 255                           | 48                               |
| Random forest | 256                           | 47                               |

## 4.4    *Random Forest Algorithm*

Through the bagging technique it generates a random set of training datasets for each tree [6]. In random forest, from the training data set create 200 classifiers (trees) by creating random datasets and then classify the given input in to the class which comes majority in the 200 trees outputs. For creating each tree we use gain ratio as the splitting criteria. The pseudo code of Random forest algorithm [7] is shown below.

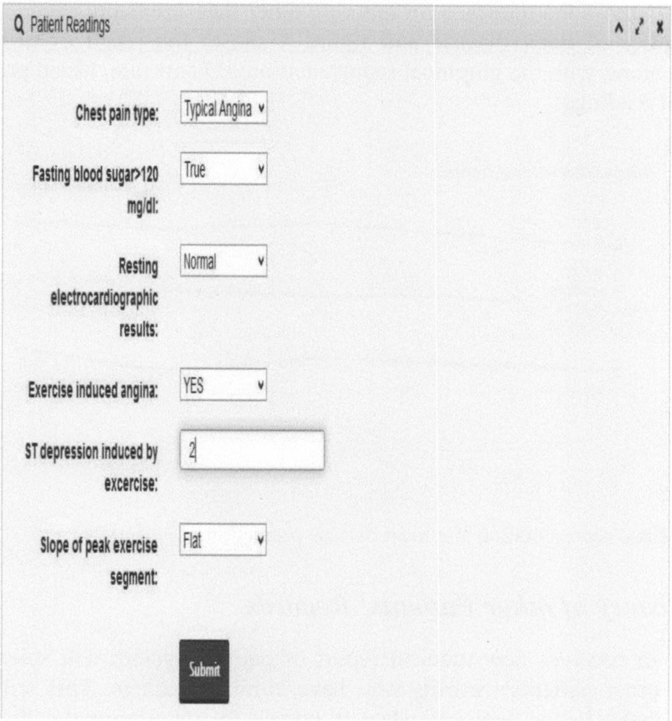

**Fig. 7** Other measurements for Heart disease prediction

To generate *c* classifiers:

**for** *i* = 1 to *c* **do**
    Randomly sample the training data *T* with   replacement to produce *Ti*
    Create a root node, *Ai* containing *Ti*
    Call GenerateTree ( *Ai* )
**end for**
**GenerateTree(A):**
**if** *A* contains instances of only one class **then**
    **return**
**else**

Randomly select x% of the possible splitting features in $N$
Select the feature $F$ with the highest information gain to split on
Create f child nodes of $A$, $A1$ ,..., $Af$ , where $F$ has $f$ possible values ( $F1$, .., $Ff$ )
**for** $i$ = 1 to $f$ **do**
Set the contents of $Ai$ to $Ti$ , where $T_i$ is all instances in $A$ that match $Fi$
Call GenerateTree( $A_i$)
   **end for**
 **end if**

Figure 7 shows the interface for entering other measurements needed to predict the possibility of heart disease and figure 8 shows the result of heart disease prediction along with the graphical representation of heart rate, blood pressure and cholesterol readings

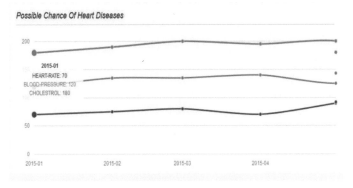

**Fig. 8** Graphical representation and heart disease prediction

## 4.5    *History of other Patients' Records*

When doctor receives new medical report of patient, system will select medical history of other patients' records who have similar readings. This will help the doctor to make better decision when it comes to identifying the diseases and giving medication instructions to the patient. Since we are dealing with categorical data, we can make use of any classification algorithm to achieve this task. We have chosen K nearest neighbor (KNN) classification algorithm to do this task. In KNN algorithm, the distance between the input data and the training data set is calculated and from the K nearest distances, selects the class which comes most frequent [8]. K nearest algorithm is shown below.

- Input: X = {(A1, B1), . . . , (An, Bn)}
- **a**= (a1, . . . , an) new input data to be classified
- FOR each labeled data (Ai , Bi ) calculate distance d(Ai , **a**)
- Sort d(Ai , **a**)in ascending order
- Select the K nearest instances to **a**: H
- Assign to the most frequent class in H

Euclidian distance is used to calculate the distance between input data and training data set

$$D(A, B) = \sqrt{\sum_{i=1}^{n} (a_i - b_i)^2}$$

The result of this algorithm is shown in Figure 9.

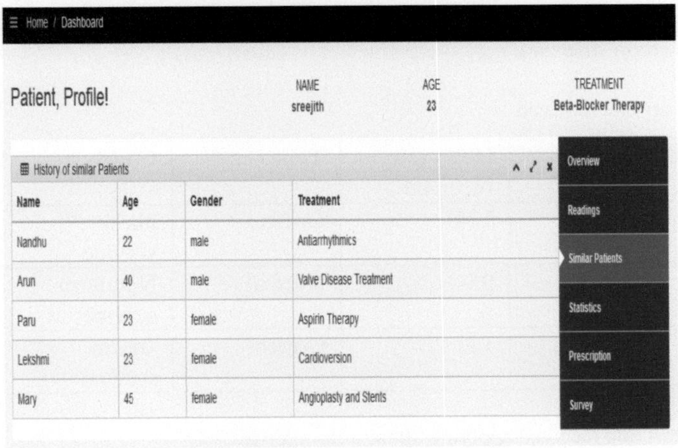

**Fig. 9** History of other patients' records

# 5 Testing and Results

The application is developed using PolarH7 sensor, Android at client side, PHP and MySQL at server side. The main objective of our proposed work is to provide a frame work for health monitoring system. We implemented heart disease prediction with emergency alarm service for doctor and caretaker, system offers other services like online feedback from doctor and patient (if any), medicine prescription and medication reminder. The system is tested in different persons with different health conditions and outputs a-re shown below.

Case 1: Critical alarm service
Table 7 shows the testing results and it shows when the alarm service is initiated.

Case 2: Heart disease prediction
The system is tested in two persons; one with heart disease and another one with normal health conditions.

a.   Person with normal health conditions

Table 8 shows the heart rate received from the person with normal health conditions. To predict heart disease, other parameters specified in Table 1 is also needed. The values of this parameters are shown in Table 9. After applying random forest algorithm, the final result will be 0 ( no heart disease).

b.   Person with heart disease

Table 10 shows the heart rate received from the person who are already diagnosed with heart disease. Table 11 shows the other readings required for predicting the heart disease. Final result after applying random forest algorithm will be 1 (Heart disease).

**Table 7** Heart rate received

| Person | Heart rate received | Type | Conclusion |
|--------|---------------------|------|------------|
| A | 124 | Critical | alarm service |
| B | 95 | Normal | No alarm service |
| C | 127 | Critical | alarm service |

**Table 8** Heart rate received applying 30 minutes interval

| Person | Heart rate received ( 30 minutes interval) |
|--------|---------------------------------------------|
| A | 95 |
| A | 97 |
| A | 93 |

**Table 9** Parameters for predicting heart disease

| Cp | 1 |
|------|------|
| Restbps | 145.0 |
| Chol | 233.0 |
| Fbs | 1.0 |
| Restecg | 2.0 |
| Thalach | 97 |
| Exang | 0.0 |
| Oldpeak | 2.3 |
| Slope | 3.0 |
| Ca | 0.0 |
| Thal | 6.0 |
| Final result | 0 |

**Table 10** Heart rate received applying 30 minutes interval

| Person | Heart rate received ( 30 minutes interval) |
|---|---|
| B | 108 |
| B | 112 |
| B | 105 |

**Table 11** Parameters for predicting heart disease

| Cp | 4 |
|---|---|
| Restbps | 160.0 |
| Chol | 286.0 |
| Fbs | 0.0 |
| Restecg | 2.0 |
| Thalach | 112 |
| Exang | 1.0 |
| Oldpeak | 1.5 |
| Slope | 2.0 |
| Ca | 3.0 |
| Thal | 3.0 |
| Final result | 1 |

# 6    Conclusion

In this paper a system has been presented that will help the people to utilize the features of healthcare management system at any time they want. Prediction of heart disease from the readings taken from the user helps them to take proper medical care as early as possible. Quality of medicine prescribed by the doctor is also improved by enabling the doctor to analyze the medical history of various patients. Here the work compares different algorithms and proposes the usage of Random Forest algorithm for heart disease prediction.

We can include various sensor fusion methods to provide an improvement over the wearable gadget. This will lead to include various health parameters.

# References

1. WHO: http://www.who.int/mediacentre
2. Saibalaji, N., Remigius, D., Vimalkumar, S.: An Ubiquitous way of Health Care Monitoring System using Android Mobile Phone. International Journal of Research in Engineering and Advanced Technology **1**(1), March 2013
3. Du, Y.,Chen, Y., Wang, D., Liu, J., Lu, Y.: An Android-Based Emergency Alarm and Healthcare Management System. In: 2011 International Symposium on IT in Medicine and Education (ITME), vol. 1

4. Hii, P.-C., Chung, W.-Y.: A Comprehensive Ubiquitous Healthcare Solution on an Android™ Mobile Device (2011). www.mdpi.com/journal/sensors, ISSN 1424-8220
5. Nahar, J., Imam, T., Tickle, K.S., Chen, Y.-P.P.: Computational intelligence for heart disease diagnosis: A medical knowledge driven approach. Expert Systems with Applications **40**(1), January 2013
6. Kulkarni, V.Y., Sinha, P.K.: Random forest classifiers: A survey and future research directions. International Journal of Advanced Computing **36**(1) (2013). ISSN: 2051-0845
7. Sirikulviriya, N., Sinthupinyo, S.: Integration of rules from a random forest. In: International Conference on Information and Electronics Engineering IPCSIT, vol. 6 (2011)
8. Sun, S., Huang, R.: An adaptive K-nearest neighbor algorithm. In: Seventh International Conference on Fuzzy Systems and Knowledge Discovery (FSKD), vol. 1 (2010)
9. JSON: http://www.json.org

# Phoneme Selection Rules for Marathi Text to Speech Synthesis with Anuswar Places

Manjare Chandraprabha Anil and S.D. Shirbahadurkar

**Abstract** This paper presents phoneme selection rules for Marathi Articulation for Anuswar places in speech synthesis. In Marathi Literature, written text contains Anuswar at different places on Marathi Alphabets. In this paper, the rules are defined for appropriate selection of phonemes for Anuswar Alphabets in the context of developing Marathi Text to Speech systems. This increases naturalness in output speech of TTS system. This paper proposes procedure for phoneme selection with the place of Anuswar on Alphabet. Decision tree algorithm is proposed to define simple rules for Articulation in Speech synthesis with high Intelligibility. The confusion matrix is defined to show performance of decision tree training and testing.

## 1 Introduction

Language is an effective media for interaction with people. Language expresses feeling, thinking, idea, emotions. Marathi grammar rules have to follow to write Marathi script and pronounce it for correct interpretation of language. Marathi belongs to Indo-Aryan language group, spoken by native people, mostly in Maharashtra, India. Presently the script used for Marathi language is known as "Balbodh" which is modified script of Devnagari Script. Text to Speech Synthesis transforms written information into spoken information for easier and efficient access of data. In a multilingual country like India, TTS helps the group of people

M.C. Anil(✉) · S.D. Shirbahadurkar
Research Scholar, Department of Electronics and Telecommunication Engineering, JSPM's
Rajarshi Shahu College of Engineering, Tathawade, Pune 411033, Maharastra, India
e-mail: cmanjare@gmail.com

M.C. Anil · S.D. Shirbahadurkar
D. Y. Patil College of Engineering, Ambi, Talegaon-Dabhade,
Pune 410506, Maharastra, India

© Springer International Publishing Switzerland 2016
S.M. Thampi et al. (eds.), *Advances in Signal Processing and Intelligent Recognition Systems*,
Advances in Intelligent Systems and Computing 425,
DOI: 10.1007/978-3-319-28658-7_42

501

who know many languages orally but not aware of the script. The main objective of this paper is to focus on one of the way to design TTS with natural speech by identifying place of Anuswar in the sentence and selection of phoneme by following proposed rules.

Further this study puts confusion matrix which helps to show performance of decision tree training and testing. To develop high quality natural speech, Concatenative speech synthesis technique can be used with basic search unit phoneme. [1]

In this paper, English translation is written for every Marathi text.

## 2    Marathi Script

Marathi script consists of 16 vowels and 36 consonants. Vowels are grouped into three groups.

a.    First group consists of 12 vowels as follows:

| a | aa(A) | i | ii(I) | u | uu(U) | e | ai | o | au | aM | aH |
|---|-------|---|-------|---|-------|---|----|---|----|----|----|
| अ | आ | इ | ई | उ | ऊ | ए | ऐ | ओ | औ | अं | अः |

These twelve vowels are used frequently. Location of "अं"( aM) affects on pronunciation of word.

b.    Second group consists of 2 vowels which are occasionally used.

ruu    lru

ऋ    लृ

c.    Third group consists of two new vowels influenced by English language.

aU    aE

ऑ    अ̃

36 consonants are divided into five groups; each containing five letters.

a.    Kanthya: These letters are Gutturals pronounced from throat called Kanthya.
b.    Taalavya: These letters are palatals  pronounced by touching the tongue to the palate.
c.    Murdhanya: These letters are cerebrals pronounced by touching the tongue to a part of upper jaw.
d.    Dantya: These letters are dentals pronounced by touching tongue to teeth.
e.    Aushthya: These letters are labials pronounced by touching the lips together.

The remaining eleven consonants are  y, r, l, v, sh, shh, s, h, L, ksh, Dnya

Out of these, four (y,r,l,v) are semi vowels. The next three (sh, shh, s) sibilants are very similar. Last two ksh, Dnya are consonants clusters. [2]

The term, "Anuswar/Shirshbindu" is described as being a voiced sound having only one 'place of articulation' - the nāsikā or nasal cavity. In other words, it is a 'pure nasal' and distinct from oral or oro-nasal sounds.

The term, "Akshar/Alphabet" is described as a text component such as a consonant or a vowel. The terms, 'text component', and 'Alphabet' may be used interchangeably. Furthermore, the term, "Akshar/Alphabet", is also called an alphasyllabary, it is a segmental writing system in which consonant–vowel sequences are written as a unit: each unit is based on a consonant letter, and vowel notation is obligatory, but secondary. The term, "Anuswar" is also called as Nasal Marker.

There are pre-defined rules for pronunciation with "Anuswar". In Marathi Literature (novel, play, story books), subtitles written for movies, subtitles written for news on Marathi television channels, written text contains Anuswaras at different places on Marathi Alphabets.   There are broadly two ways of pronunciation with Anuswar.

Clear (strong): This pronunciation is called Anuswar.
Silent (mild): This pronunciation is called Anunasik.

Now a day, people are communicating with each other through Internet, messages on mobile phone, messages on what's app. The concept of this paper will be very useful in these messaging applications. Table 1 shows two ways of Marathi text writing. In table 1, in column number two and three, two sentences are written in different way but their meaning is same. Second column shows sentences written with Marathi grammar rules. But now days, in local language speaking style, sentences are spoken by ignoring grammar rules. In Marathi Literature(novel, play, story nooks), subtitles written for movies, subtitles written for News on  Marathi Television channels written text contains Anuswara instead of Matra on Marathi Alphadets, at different places. This type of Marathi text is written in column 3. The meaning of these two sentences is exactly same only the writing style is different. In TTS system, pronunciation of text changes depending on the place of Anuswar. In TTS system, there is confusion to select correct phoneme for the word with Anuswar. Therefore some rules are defined to solve confusion problem. [3,4,5]

**Table 1** Two ways of writing Marathi Text

| Sr. No. | Sentences written following Marathi grammar rules | Sentences written in local Marathi language |
|---|---|---|
| 1 | माझे काम झाले. | माझं काम झालं. |
|   | (My work is over) | (My work is over) |
| 2 | तुझे पुस्तक मी वाचले. | तुझं पुस्तक मी वाचलं. |
|   | (I read your book.) | (I read your book.) |
| 3 | खोटे बोलू नका. | खोटं बोलू नका. |
|   | (Don't be liar.) | (Don't be liar.) |
| 4 | संथ गतीने चालत होता. | संथ गतीनं चालत होता. |
|   | (He was walking slowly.) | (He was walking slowly.) |

# 3    Rules for Pronunciation with "Anuswar"

## 3.1   *Rules for Anuswar pronunciation*

Table 2 shows some rules which are defined by Marathi Grammar for the
pronunciation of Anuswar. The pronunciation of Anuswar in Marathi depends on
the character that follows it. When Anuswar occurs, it is replaced by an
appropriate half consonant followed by it as shown in Table 2. Further we have
concentrated on some cases which are not defined in Table 2.

If Anuswar is present on other than the last Alphabet of any word, then rules
given in Table 2 are followed for pronunciation.

**Table 2** Rules for Pronunciation with Anuswar

| Pronunciation Place उच्चारस्थान | Vowels | Consonants | Half Consonants अंतस्थ वर्ण | Anunasika |
|---|---|---|---|---|
| KaNthya कंठ | a aa(A) अ,आ | **K kh g gh** क ख ग घ | h ह् | N^ ङ |
| Talavya तालु | i  ii(I) इ, ई | Ch chh j jh च छ ज झ | y, sh य, श | JN ञ |
| Muurdhanya मूर्धा | ruu ऋ | T Th D Dh ट ठ ड ढ | r, sh^ र,ष् | N ण् |
| Dantya दन्त | Lru लृ | t  th d dh त थ द ध | l, s ल, स | n न् |
| AnushThya ओष्ठ | u  uu(U) उ ,ऊ | p ph b bh प फ ब भ | | m म् |

There are two ways of pronunciation with Anuswar.
a. Clear (strong): This pronunciation is called Anuswar.
b. Light (mild): This pronunciation is called Anunasik.

If "य (y)" or "ल"(l) comes after Anuswar, it is pronounced as Anunasik "यं"(yv)
or "लं"(lv)

If "र", "व", "श", "ष", "स", "ह"  (r, v, sh, sh^, s, h)comes after Anuswar, it is
pronounced as Anunasik    "वं" (vm) [6,7]

# 4    Advanced Rules for pronunciation with "Anuswar"

If Anuswar is present on the last Alphabet of any word, it is pronounced as stop consonant.

If only one Alphabet is written with Anuswar, it is also pronounced as stop consonant. Example:

1.  हं! मी विचार करेन.   (Hmm! I will think.)  2.   होऊ दे गं! (Let it be!)

In case of extended words, last Alphabet of word and extension of Alphabet, both have Anuswar, then both the Alphabets are pronounced as stop consonants. Example: इवलंसं (small) Table 3 first row shows this examples.

In case of extended words, if word has extension, हि, स, च, पण, य, त (hi, s, ch, pan, y, t) and anuswar is present on last but one Alphabet, then Anuswar is silently pronounced. Table 3 second and third rows shows this example.

If word ends with नी, चे, ना (ni, che, na)  and Anuswar is present on last but one Alphabet, then Anuswar follows Marathi grammar rules for pronunciation shown in Table 2. E.g. विचारांनी (vicharani)

**Table 3** Different ways of Writing Anuswar in Marathi text

| Condition | Written in Native language | Written in Grammatically correct words |
|---|---|---|
| Anuswar comes on last and last but one Akshara | तिचं इवलंसं पिलू (Her small child) | तिचे इवलेसे पिलू (Her small child) |
| Anuswar comes only on last but one Akshara | आमचं चुकलंच. We mistaken.) | आमचे चुकलेच. (We mistaken.) |
| Anuswar comes only on last but one Akshara | आपलंही  कर्तव्य  आहे. (That is Our duty.) | आपलेही कर्तव्य आहे. (That is Our duty.) |

# 5    Experimental Details

In this section, we described database creation, the algorithm used to implement set of rules, phoneme selection in TTS system.

## 5.1   Database

We used CDAC Marathi text Corpus. This database has 2500 audio files. This corpus has about 15000 Marathi words. Out of these nearly 3000 different words are containing Alphabet with Anuswar at different places.

Also we have generated our own database. A database consisting of total 1500 audio files of different Marathi native speakers is prepared. The recording is done using the software Audicity sound editor. These words and sentences were recorded at a sample rate of 44100 Hz with a single (Mono) channel. The recordings were done in a silent room with minimal noise.

## 5.2   Algorithm

ID3 algorithm is used to build a decision tree. ID3 uses information gain as splitting criteria. A fundamental part of ID3 algorithm is to construct a decision tree from a dataset in which it selects attributes at each node of the tree. The growing stops when all instances belong to a single value of target value. [8,9]

1. Accept Marathi word as input text.
2. Segment each Alphabet from input word.
3. Calculate length of input word.
4. Categorize the word in following categories based on length of word
1. Multiple Alphabet Word
2. Single Alphabet Word
5. Find number of Anuswar in the word (n)
6. Locate the position of each Anuswar in the word

Loop: For i=1:n   (Loop  for number of Anuswar)

7. Categorize Anuswar based on its Position in word in following Categories
1. First
2. Intermediate
3. Last
4. Last and Last but one

8. Classify type of Anuswar between two categories
a.Silent Pronunciation
b.  Clear Pronunciation
Using following decision tree Algorithm
**If** (Location of Anuswar is **Intermediate** or **first**) **then,**
Check next Alphabet between two groups
**If** (Next Alphabet is in Group1)  **then,**
**Assign Type of Anuswar= Clear Anuswar**

**Elseif** (Next Akshar is in Group2)  **then,**
**Assign Type of Anuswar= Silent Anuswar**

**Else**
**Assign Type of Anuswar= Clear Anuswar**

**Elseif** (Location of Anuswar is Last and Length is more than one) **then**

**Assign Type of Anuswar= Silent Consonant**

**Elseif** (Location of Anuswar is Last and Length is one) **then**
**Assign Type of Anuswar= Silent Consonant**
Check the last Alphabet from two Groups
**If** (Last Alphabet belongs to Group 1) **then**
**Assign Type of Anuswar= Silent Anuswar**
**elseIf** (Last Alphabet belongs to Group 2) **then**
**Assign Type of Anuswar= Clear Anuswar;**

**Else**
**Assign Type of Anuswar= Silent Anuswar**
**End if;**
End Loop when classification of all Anuswar is finished.

## 5.3 Phoneme Selection

Rules defined in the algorithm selects appropriate phoneme for the correct interpretation of text. The main aim of this paper is to provide a speech synthesis system and method for obtaining naturalness in speech signal and to provide a text-to-speech (TTS) system which eliminates confusions in Marathi Articulation of Anuswara places in Marathi Text to Speech system.

In Table 4 and 5, attributes are defined with their values. In addition to this attributes, other features of the Marathi text plays important role in naturalness of TTS system. The work is in progress in other supportive views. This is one of the way to achieve high quality speech.

**Table 4** Attributes with their values assigned

| Attributes | Values |
|---|---|
| Length of Word | Single Alphabet, Multiple Alphabet |
| Location of Anuswar | First, intermediate, Last, last-and-first |
| Category of Anuswar | Silent Pronounced , Full Anuswar |
| Next_ Alphabet | Group1,Group2 |
| Last_ Alphabet | Group1,Group2 |

**Table 5** Alphabet Groups

| Attributes | Group1 | Group 2 |
|---|---|---|
| Next_ Alphabet | य ,ल (y, l) | र ,व ,श ,ष ,स ,ह (r, v, sh, sh^, s, h) |
| Last_ Alphabet | हि ,स ,च ,पण ,य ,त (hi, s, ch, pan, y, t) | नी ,चे ,ना (ni, che, na) |

## 6    Result and Analysis

Listening tests are conducted with five native people. The speech output of TTS system is tested and verified with confusion matrix. A confusion matrix of the phoneme selection for Anuswar is shown in Table 6. The classification accuracy of the phoneme selection for Anuswar on first and intermediate is higher than that of the other places of Anuswar. For all the cases, 91.67% of the phonemes are predicted in correctly for first and intermediate places and 89.47 of the phonemes are predicted correctly for other class. Accuracy for phoneme selection in TTS system is nearly 90%.

**Table 6** Confusion matrix showing the performance of phoneme selection

|  | Position of Anuswar in word | |
|---|---|---|
|  | Last Alphabet, Last but one in extended word | First, Intermediate |
| Anunasik (Silent pronunciation) | 89.47% | 13.16% |
| Anuswar (Clear pronunciation) | 6.67% | 91.67% |

## 7    Conclusion

In Marathi TTS system, study of Anuswar place on Alphabet is done. The decision tree learning method is used to define advanced rules for pronunciation of text with Anuswar. The main aim, proper phoneme selection is achieved. Through experiments, we removed the ambiguities in the selection of phonemes. Listening experiments have been conducted on sentences synthesized by deliberately replacing phonemes with their confused ones. This concept can be used in smart phone applications, novel reading and storytelling applications which increase naturalness in TTS system.

## References

1. Repe, M., Shirbahadurkar, S.D., Desai, S.: Prosody model for marathi language TTS synthesis with unit search and selection speech database. In: 2010 International Conference on Recent Trends on Information, Telecommunication and Computing, pp. 362–364
2. Savargaonkar, N.: Marathi Language. http://members.tripod.com/~.marathi/marathi.html
3. Godambe, T., Karkera, N., Samudravijaya, K.: Adaptation of acoustic models for improved Marathi speech recognition. In: ACOUSTICS 2013 NEWDELHI, New Delhi, India, November 10–15, 2013

4. Rama, N., Lakshmanan, M.: A Computational Algorithm based on Empirical Analysis, that Composes Sanskrit Poetry. (IJCSIS) International Journal of Computer Science and Information Security **7**(2) (2010)
5. Ramani, B., Christina, S.L., Rachel, G.A., Solomi, V.S., Nandwana, M.K., Prakash, A., Shanmugam, S.A., Krishnan, R., Kishore, S.P., Samudravijaya, K., Vijayalakshmi, P., Nagarajan, T., Murthy, H.A.: A common attribute based unified HTS framework for speech synthesis in indian languages. In: 8th ISCA Speech Synthesis Workshop, August 31–September 2, 2013, Barcelona, Spain (2013)
6. Marathi writing grammar: http://www.mymarathilearning.com
7. Samudravijaya, K.: Decision Rules for Selection of Allophones of Marathi Affricates for Speech Synthesis. 978-1-4244-3472-5/08/©2008 IEEE
8. Girish, N.J., Mishra, S.K., Chandrashekar, R., Bhowmik, P., Chandra, S., Mendiratta, S., Agrawal, M.: Developing a Sanskrit Analysis System for Machine Translation
9. Rokach, L., Maimon, O.: Top-Down Induction of Decision Trees Classifiers—A Survey. IEEE Transactions on Systems, Man, and Cybernetics—Part C: Applications and Reviews **35**(4), 476–487 (2005)

# Solving Multi Label Problems
# with Clustering and Nearest Neighbor
# by Consideration of Labels

C. P. Prathibhamol and Asha Ashok

**Abstract** In any Multi label classification problem, each instance is associated with multiple class labels. In this paper, we aim to predict the class labels of the test data accurately, using an improved multi label classification approach. This method is based on a framework that comprises an initial clustering phase followed by rule extraction using FP-Growth algorithm in label space. To predict the label of a new test data instance, this technique searches for the nearest cluster, thereby locating k-Nearest Neighbors within the corresponding cluster. The labels for the test instance are estimated by prior probabilities of the already predicted labels. Hence, by doing so, this scheme utilizes the advantages of the hybrid approach of both clustering and association rule mining.The proposed algorithm was tested on standard multi label datasets like yeast and scene. It achieved an overall accuracy of 81% when compared with scene dataset and a 68% in yeast dataset.

## 1 Introduction

Multi Label Classification (MLC) is one of the most challenging problems in the field of data mining. It fundamentally revolves around the fact that each data instance may be associated with numerous target labels. If the test instance within the dataset is coupled with only one target class, then the classification problem becomes single-label. The relevant application areas where multi label classification can be effectively applied include semantic scene classification, protein function classification, music categorization etc.

A lot of recent studies and proficient researches depicts the reality that a fine solution of this problem results in condensed human effort as well as minimal time

C.P. Prathibhamol(✉) · A. Ashok
Department of CSE, Amrita Vishwa Vidyapeetham, Amritapuri, Kollam, Kerala, India
e-mail: {prathibhamolcp,ashaashok}@am.amrita.edu

© Springer International Publishing Switzerland 2016
S.M. Thampi et al. (eds.), *Advances in Signal Processing and Intelligent Recognition Systems*,
Advances in Intelligent Systems and Computing 425,
DOI: 10.1007/978-3-319-28658-7_43

511

consumption. Two well known and traditional approaches exist for solving MLC problems. They are commonly known as Algorithm adaptation and Problem transformation method [9]. In the preliminary approach the key notion is converting the multi label problem into numerous single label problems. Whereas, in the latter approach, any existing single label classification algorithm is adapted to handle multi label problems.

In this paper, we have implemented an Algorithm adaptation approach on fundamental k-Nearest Neighbor algorithm, which is a popular classification algorithm. In this proposed method, there are mainly two stages involved: Training phase followed by a Testing phase. Initially, for the Training of the multi label dataset, we cluster the data pertaining to feature space using a well known clustering algorithm, k-means. After the execution of k-means algorithm, the normal clusters are obtained. The main advantage associated with clustering phase is that it is able to break the entire dataset into dis-joint clusters. In each of the clusters, Frequent FP-growth mining algorithm is then applied corresponding to label space. At the end of this mining, dependency rules among the labels of each cluster is generated. On the completion of training stage, the testing stage is commenced by inputting a new test data instance and checking for the appropriate cluster to which it belongs to. This is achieved by comparing the test data with the centers of all clusters using Euclidean distance. This improves the overall execution time due to the fact that the test data is then later checked only for the subsequent rules of that particular cluster only. As a result, label cardinality of the multi label dataset is significantly reduced by inferring dependency between the extracted labels. Towards the completion of the testing stage, the antecedent part of the mined rules is ultimately passed to instance based algorithm for the purpose of effective classification. An alteration done in the conventional kNN algorithm is that instead of taking into account the majority labels of k nearest neighbors, the prior probabilities of those labels are examined. If the estimated probabilities are greater than a certain threshold, those labels are predicted as that of test data instance. The rest of this paper is prepared as follows: Literature Survey and related papers are discussed in Section 2. System Architecture of our novel method is explained in Section 3. Section 4 is devoted to experiments with several standard datasets. The paper concludes with a summary and future works in Section 5.

## 2 Related Works

Considerable amount of work have been made in the field of single label classification. At the same time, efforts have also been directed to convert multi label datasets into multiple sets of single label dataset to match the existing labels.

In [1], the authors have developed two probabilistic approaches to solve multi label classification. The first approach is based on logistic regression and nearest neighbor classifiers. The second approach deals with notion of grouping related labels. The former approach is known as Method using Partial Information (MPI) and the latter approach is known as Method using Association Rules (MAR). In comparison with MAR, MPI takes large amount of time but provides accurate results.

The authors in [2] have contributed to solve MLC problems by using the frame work of Improved Conditional Dependency Networks (ICDN).This method is based on double layer based classifier chain (DCC) to make use of the label correlations in training stage and modifiers the conditional dependency networks (CDN) by initializing the entered values of the second layer with the fore casted values from the first layer during the testing stage.The experimental results from the work confirms that it reduces randomization of input for the conditional dependency networks and convergence rate is considerably improved.

Ying Yu et al. [3] proposed MLRS(Multi Label classification using Rough Sets) which contained the effect of association between the labels and the ambiguity that subsist in the mapping between the feature space and label space. A chain of experiments demonstrated the results that, for seven multi label datasets MLRS obtained better accuracy when compared with basic multi label learning algorithms such as MLkNN, BR, RAkEL etc.

Jiayang Li and Jianhua Xu [4] proposed OVODLSVM(one-versus-one decomposition strategy with double label support vector machine) is a MLC solution by using binary Support Vector Machine(SVM) to build a model of double label SVM by searching for double label instances in the margin between positive and negative instances.Hence by using a voting criteria. This method worked quite well on computational aspects and proved as an efficient solution.

In [5], the authors have conducted a research study of clustering based multi label classification (CBMLC) for the MLC problem. They have tried three clustering algorithms namely simple K-means, Expectation Maximization and Sequential Information Bottleneck algorithm. One main disadvantage of this study was that it didn't consider the dependency between labels.

In the work proposed by the authors in [6], the Apriori algorithm was used to generate all frequent item sets, compound labels with strong associations are replaced by existing single labels. In the case of classification ML-KNN was widely used. But one main limitation concerned with this methodology is that it is not suitable for weak relationship between labels.

An Improved method of Multi label Classification Performance by Label Constraints (IMCP-LC) is proposed by authors in [7]. This method mainly deals with label ranking strategy and label constraints.By using one-against-all decomposition technique, MLC problem is broken down in to numerous binary classification sub-problems. Then for each label training is done by applying binary SVM classifier.After that association rule learning method is applied to mine label constraints. In the last phase in order to correct the output from SVM classifiers, a correction model dealing with label constraints is utilized.

As discussed earlier, the main categorization of MLC problems is done in two ways including Problem Transformation (PT) and Algorithm Adaptation (AA). The authors in [8] have presented a detailed comparison between these two techniques. They have confirmed that AA based methodology is better than PT based methods. It is analyzed with respect to the results of experiments they have carried out in multi label datasets. Apart from this outcome, they have studied about the different methods of PT and AA techniques. In PT based methods, Binary relevance is suitable for fast binary classification, but a limitation is it doesn't take into account label co-relationship. But label correlation is vital as there is a possibility of label dependencies. Also, labels and their characteristics can be over lapping, so Ranking via single label is not viable. While CLR (Calibrated Label Ranking) [9] is of good quality for considering label relationship, it can't be used for unlabeled data.

Tsoumakas G and Vlahavas I [10] proposed an ensemble method for MLC known as RAKEL (Random k-Label sets). In this method each member of the ensemble is created by selecting random subset of labels. It is followed by using single label classifier for prediction of each member of the power set.

An instance based approach to multi label classification has guided to ML-KNN [11] method. This method combines the idea of the conventional kNN algorithm and Bayesian inference. Based on statistical information derived from the label sets of an unseen instance's neighboring instances, ML-KNN utilizes maximum a posteriori principle to determine the label set for the unseen instance.

In another approach as stated in [12] the authors have worked to solve the MLC problem by association rule mining. They have tried to decompose multi label datasets to extract single label rules and then combine labels with the same attributes to generate multi label rules. This experiment achieves good performance in an application to scene classification.

In [13], the authors combined multi label methods by ensemble techniques. They solved by using a combination of various multi label learners to avoid discrepancy in training sets and correlation difficulties. They have accomplished and tried to use two methods namely $EML_M$, $EML_T$ in this regard.

# 3   System Architecture

In this paper we proposed a method for algorithm adaptation. As in `Fig.1`, we have utilized simple k-means for clustering application. This is quite effective as it divides the massive dataset into disjoint clusters which considerably reduces the training time for any classification stage. By doing so we have overcome the main limitation of the work proposed by the authors in [4]. As mentioned above, in their work, they have used Apriori algorithm for mining of rules. Since FP-Growth is far better than Apriori algorithm, in terms of efficiency and resolving of label dependencies, we have utilized the former algorithm in this work. By our method, we are aiming at feature space reduction through clustering algorithm. Furthermore, label space reduction is also achieved by FP-growth algorithm.

In the testing phase as in `Fig.2`, initially for any test data instance the nearest cluster to it is identified. For this purpose we have measured Euclidean distance of the test data instance with all the other generated cluster centers. Once the test instance is known to lie in which cluster, then it is checked for the already generated label dependencies of that particular cluster.

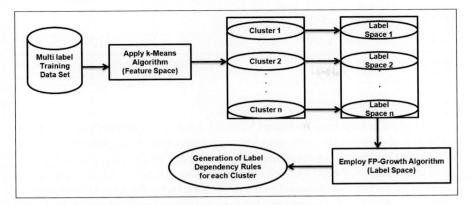

**Fig. 1** Training Phase of our proposed method. (SMLP-CNN-CL)

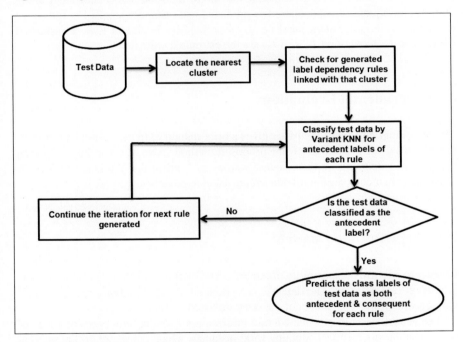

**Fig. 2** Testing Phase of our proposed method. (SMLP-CNN-CL)

Any rule of the form

$$L_1 \Rightarrow L_2 \wedge L_3 \, , \tag{1}$$

implies that if the antecedent label L1 is present, then the consequent labels or compound labels both L2 and L3 are also sure to appear along with it at the same time. A rule is considered to be strong if it satisfies a minimum support and confidence.

**Table 1** Characteristics of datasets

| Datasets | Total Labels | Total Attributes | Label Cardinality | Label Density |
|----------|--------------|------------------|-------------------|---------------|
| Yeast | 14 | 103 | 4.237 | 0.303 |
| Scene | 6 | 294 | 1.074 | 0.179 |

In the classification stage, we adopt Variant k-Nearest Neighbors (V-kNN) within the identified cluster. kNN is the algorithm by which k nearest neighbors is detected with the test instance. k is a value which is specified according to the users choice. In a conventional kNN, the test instance is assigned a class label which is the majority of k nearest neighbors class labels. The modification done to the nave kNN algorithm is that here we have considered the prior probability of antecedent labels. If the probability is greater than a particular threshold, then the antecedent label along with its consequents are considered as predicted test instances labels. For the above mentioned rule, antecedent label i.e. $L_1$ ś probability is greater than an assumed value, then $L_1$ along with $L_2$ and $L_3$ are the estimated class labels.

## 4 Experimental Evaluations

The remaining portion of the paper gives a brief insight in to the various multi label datasets used for evaluation of our proposed method. Also, the various evaluation metrics are calculated on these datasets so as to confirm the efficiency of SMLP-CNN-CL when compared with other multi label approaches.

## 4.1 Experimental Datasets

In order to confirm the feasibility of our proposed method, we have conducted experiments on real datasets. The datasets are available at http://mulan.sourceforge.net/datasets-mlc.html. The details of datasets are listed in Table 1.

The yeast dataset in essence contain information about several types of genes of one particular organism. It contains 1500 instances which consist of 103 numerical valued attributes and 14 labels.

The scene dataset contains several types of scene environmental information such as mountain, beach, sunset, fall foliage, urban and field. It comprises 1211 instances with 294 numerical valued attributes and 6 labels.Label Cardinality is more for Yeast dataset and is less for Scene. So Yeast is having more dimensionality in the label space and Scene is having less dimensionality.

## 4.2   Evaluation Metrics

In this paper, 4 measures have been selected for comparison of the proposed method with previously existing multi label classification algorithms.

In all the definitions given below, $x_i$ denotes the actual labels of the $i^{th}$ test instance. Also, $y_i$ represents the set of predicted labels for the corresponding test instance. If L is the total number of labels associated with the data set and D is the number of instances to be tested. All the evaluation measures are taken from [14]. The evaluation measures are listed as below:

### 4.2.1   Hamming Loss

This measure indicates the number of times misclassification of example-label pair occurs. It is estimated as the number of wrong labels to the total number of labels. The predicted labels are checked with respect to the original labels and are added up as 1 if they are wrong and 0 if they are correct labels associated with the dataset and D is the number of instances to be tested, the hamming loss is calculated as:

$$HL(x, y) = \frac{1}{|D|} \sum_{i=1}^{|D|} \frac{|x_i \oplus y_i|}{|L|} \tag{2}$$

### 4.2.2   Accuracy

It is measured as the number of correct labels divided by the union of predicted and true labels.

$$accuracy = \frac{1}{|D|} \sum_{i=1}^{|D|} \frac{|x_i \cap y_i|}{|x_i \cup y_i|} \tag{3}$$

### 4.2.3   Precision

This ratio estimates the number of correct matches obtained between the true and predicted label to the number of total predicted labels.

$$precision = \frac{1}{|D|} \sum_{i=1}^{|D|} \frac{|x_i \cap y_i|}{|y_i|} \tag{4}$$

**Table 2** Experimental results of yeast dataset

| Algorithms | Hamming Loss | Accuracy | Precision | Recall |
|------------|--------------|----------|-----------|--------|
| IMCP-LC | 0.190 | 0.552 | 0.678 | 0.715 |
| $EML_M$ | 0.193 | 0.500 | 0.738 | 0.553 |
| $EML_T$ | 0.197 | 0.553 | 0.682 | 0.690 |
| ML-KNN | 0.198 | 0.492 | 0.732 | 0.549 |
| C4.5 | 0.259 | 0.423 | 0.561 | 0.593 |
| Naive-Bayes | 0.301 | 0.421 | 0.610 | 0.531 |
| Binary-SVM | 0.202 | 0.530 | 0.586 | 0.633 |
| CLR | 0.210 | 0.497 | 0.674 | 0.596 |
| RAKEL | 0.244 | 0.465 | 0.601 | 0.618 |
| I-BLR | 0.199 | 0.506 | 0.712 | 0.581 |
| **SMLP-CNN-CL** | **0.111** | **0.686** | **0.921** | **0.750** |

**Table 3** Experimental results of scene dataset

| Algorithms | Hamming Loss | Accuracy | Precision | Recall |
|------------|--------------|----------|-----------|--------|
| IMCP-LC | 0.102 | 0.705 | 0.722 | 0.728 |
| $EML_M$ | 0.084 | 0.699 | 0.730 | 0.716 |
| $EML_T$ | 0.095 | 0.694 | 0.725 | 0.754 |
| ML-KNN | 0.099 | 0.629 | 0.661 | 0.655 |
| C4.5 | 0.148 | 0.576 | 0.579 | 0.588 |
| Naive-Bayes | 0.139 | 0.605 | 0.615 | 0.624 |
| Binary-SVM | 0.103 | 0.702 | 0.715 | 0.720 |
| CLR | 0.122 | 0.577 | 0.600 | 0.669 |
| RAKEL | 0.112 | 0.571 | 0.598 | 0.612 |
| I-BLR | 0.091 | 0.647 | 0.676 | 0.655 |
| **SMLP-CNN-CL** | **0.062** | **0.815** | **0.830** | **0.894** |

#### 4.2.4 Recall

This ratio estimates the number of correct matches obtained between the true and predicted label to the number of true labels.

$$recall = \frac{1}{|D|} \sum_{i=1}^{|D|} \frac{|x_i \cap y_i|}{|x_i|} \qquad (5)$$

### 4.3 Results and Discussions

We have compared the above stated measures for our proposed method with variant multi label classification algorithms. Table 2 and 3 demonstrate the facts and figures pertaining to this results. Table 4 again confirms that, when compared with many variant solutions of MLC problem, Hamming Loss is at its minimum for the proposed approach(SMLP-CNN-CL).

**Table 4** Hamming loss results of SMLP-CNN-CL and various other MLC algorithms

| Datasets | Tested algorithms | Hamming Loss |
|---|---|---|
| yeast | MPI | 0.3065 |
| scene | MPI | 0.0992 |
| yeast | MAR | 0.3335 |
| scene | MAR | 0.1219 |
| yeast | ICDN | 0.197 |
| scene | ICDN | 0.096 |
| yeast | MLRS | 0.2004 |
| scene | MLRS | 0.0927 |
| yeast | OVODLSVM | 0.181 |
| scene | OVODLSVM | 0.098 |
| yeast | **SMLP-CNN-CL** | **0.111** |
| scene | **SMLP-CNN-CL** | **0.062** |

It is evident from the results obtained that our proposed method SMLP-CNN-CL outperforms most of the existing multi label classification algorithms. In all the experiments, the parameters assumed for K-means algorithm is k=4. The minimum support threshold parameters of FP-Growth are fixed as minsup=2 and minimum confidence=75. For the efficient working of KNN algorithm, k is assumed at a value N/2. Here, N denotes the number of instances within that cluster in which test data belongs to. We have used k=N/2 as for this estimate, we got better results of classification. When k is very small, correct identification was not done as only few neighbors were considered. But as k is made very high, then variation in the results happen due to noise as numerous neighbors are being taken into account. Similarly, in the variant kNN approach, we have considered probability threshold at 0.5. This is because at this threshold, hamming loss was obtained at the minimum value. Beyond this value or at a value less than this, then the results are affected as correct identification of labels is not done. But at 0.5, we are able to attain satisfactory results. The evaluation results of C4.5, Naive Bayes and Binary-SVM algorithms on scene and yeast datasets were taken from [7] and the evaluation results of and ML-KNN, CLR, RAKEL, I-BLR were taken from [13].

As clear from Table 4, we have mainly measured Hamming Loss as the evaluation criteria for comparison of variant MLC approaches with SMLP-CNN-CL on yeast and scene datasets.

## 5 Conclusion

In this paper, we have aimed at improving multi label classification method based on SMLP-CNN-CL. As evident from the results, this proposed approach works well to give satisfactory results when compared with many other multi label classification approaches. We intend to do future work in this approach by applying other well known clustering algorithms and also apply the proposed method to various other datasets.

# References

1. Kommu, G.R., Trupthi, M., Pabboju, S.: A Novel approach for multi-label classification using probabilistic classifiers. In: IEEE International Conference on Advances in Engineering & Technology Research (ICAETR - 2014), pp. 1–8 (2014)
2. Tao, G., Guiyang, L.: Improved conditional dependency networks for multi-label classification. In: Proceedings of the Seventh IEEE International Conference on Measuring Technology and Mechatronics Automation, pp. 561–565 (2015)
3. Yu, Y., Pedrycz, W., Miao, D., et al.: Neighborhood rough sets based multi-label classification for automatic image annotation. International Journal of Approximate Reasoning **54**, 1373–1387 (2013). Elseiver
4. Li, J., Xu, J.: A fast multi-label classification algorithm based on double label support vector machine. In: IEEE International Conference on Computational Intelligence and Security (CIS 2009), vol. 2, pp. 30–35 (2009)
5. Nasierding, G., Sajjanhar, A.: Multi-label classification with clustering for image and text categorization. In: Proceedings of the Sixth IEEE International Conference on Image and Signal Processing (CISP), vol. 2, pp. 869–874 (2013)
6. Qin, F., Tang, X.-J., Cheng, Z.-K.: Application of apriori algorithm in multi-label classification. In: Proceedings of the Fifth IEEE International Conference on Computational and Information Sciences (ICCIS), pp. 717–720 (2013)
7. Chen, B., Hong, X., Duan, L., et al.: Improving multi-label classification performance by label constraints. In: Proceedings of IEEE International Joint Conference on Neural Networks (IJCNN), pp. 1–5 (2013)
8. Prajapati, P., Thakkar, A., Ganatra, A.: A Survey and Current Research Challenges in Multi-Label Classification Methods. International Journal of Soft Computing and Engineering (IJSCE) **2** (2012)
9. Fürnkranz, J., Hüllermeier, E., Mencía, E.L., et al.: Multilabel classification via calibrated label ranking. The Journal of Machine Learning **73**, 133–153 (2008). Springer
10. Tsoumakas, G., Vlahavas, I.: Random k-labelsets: an ensemble method for multilabel classification. Proceedings of IEEE Transactions on Knowledge and Data Engineering, 406–417 (2007)
11. Zhang, M.-L., Zhou, Z.-H.: ML-KNN: A lazy learning approach to multi-label learning. The Journal of Pattern Recognition Society **40**, 2038–2048 (2007). Elseiver
12. Li, B., Li, H., Wu, M., et al.: Multi-label classification based on association rules with application to scene classification. In: Proceedings of the Ninth IEEE International Conference for Young Computer Scientists (ICYCS), pp. 36–41 (2008)
13. Tahir, M.A., Kittler, J., Mikolajczyk, K., et al.: Improving Multilabel Classification Performance by Using Ensemble of Multi-label Classifiers. The Journal of Machine Learning **10**, 11–21 (2010). Springer
14. Godbole, S., Sarawagi, S.: Discriminative methods for multi-labeled classification. In: Advances in Knowledge Discovery and Data Mining, pp. 22–30. Springer (2004)

# Multi-view Robotic Time Series Data Clustering and Analysis Using Data Mining Techniques

M. Reshma, Priyanka C. Nair, Radhakrishnan Gopalapillai, Deepa Gupta and TSB Sudarshan

**Abstract** In present world robots are used in various spheres of life. In all these areas, knowledge of the environment is required to perform appropriate actions. The information about the environment is collected with the help of onboard sensors and image capturing device mounted on the mobile robot. As the information collected is of huge volume, data mining offers the possibility of discovering the hidden knowledge from this large amount of data. Clustering is an important aspect of data mining which will be explored in detail for grouping the scenario from multiple views.

## 1 Introduction

Data Mining is a powerful technology which can identify the patterns and gives insights to data. Huge amount of data is available in various fields like robotics, manufacturing etc. As time progress, since there is a rapid increase in data contained in the data store, there arises a need to provide data summarization, identify patterns from the summary, and perform necessary actions based on the findings. Data Mining can predict the pattern and future trends, to make proactive, knowledge-driven decisions.

In many fields human resource has been replaced by robots so as to obtain a precise and consistent accuracy. The robots efficiency can be made comparable to human ability by using various machine learning techniques. Robots are being deployed in manufacturing facilities for the high heat welding, continuous

M. Reshma(✉) · P.C. Nair · R. Gopalapillai · D. Gupta · T. Sudarshan
Amrita Vishwa Vidyapeetham, Bangalore, India
e-mail: {reshmamohanan5,priyanka.cnair}@gmail.com

© Springer International Publishing Switzerland 2016
S.M. Thampi et al. (eds.), *Advances in Signal Processing and Intelligent Recognition Systems*,
Advances in Intelligent Systems and Computing 425,
DOI: 10.1007/978-3-319-28658-7_44

521

handling of the heavy loads etc. In the medical field, robots are being extensively used for making precise measurements of various chemicals, surgeries and manufacturing facilities. They are also employed in areas where human intervention needs to be very minimal like handling dangerous materials.

The robot requires knowledge about its environment to perform appropriate tasks in any of the above mentioned applications. This is achieved by mining the useful information from the real time data collected by the sensors and image capturing device mounted on the robot. In this work, simple robotic indoor environment is designed and the data from the environment is captured with the help of cameras and the sensors attached to the robot. These mobile robots collects data traversing through multiple paths for obtaining different views of the environment. Since the robots move in unknown environments, prior information about the environment will not be available. Clustering can be used to identify the similarity in the unlabeled data and group the robotic environments which have similar patterns. In a real-time scenario multiple robots collect the data to accomplish a given task. When data from multiple robots are analyzed, it is important to identify whether a particular scenario seen by one robot is same or similar as a scenario seen by other robots in order to avoid redundancy in data processing. In this work a set of different environments have been created for simulating different scenarios. These environments are designed in such a way that the accuracy of techniques employed can be measured and conclusions can be arrived.

This work is organized as two sets of experiments. The objective of the initial set of experiments is to investigate the effectiveness of clustering the sensor data when the scenarios are explored in multiple robotic path. Next set of experiments focus on analysis of clustering the data obtained by combining the image and sensor data.

The rest of the paper is organized as follows. Section 2 of this paper discusses some of the earlier work done in this area. Proposed approach is given in section 3. Section 4 discusses the Experimental results. Section 5 discusses the conclusions and future work.

## 2    Related Work

The related work focuses mainly on various studies performed to understand the robotic environment and numerous time series data mining approaches. For the robot to act appropriately in an unknown environment, it needs to understand the environment first. Tim Oates et al. [2] have presented present an unsupervised learning method which enables a robot to identify and represent qualitatively different outcomes of actions. The work focuses on providing an unsupervised method based on clustering multivariate time series. An extension to the method proposed by Oates et al. is done by the authors in the work [3].This work extends the usefulness of Oates' approach in order to handle remarkably longer time series of double the dimensionality and to generate accurate classifications. Another work [1] presents the design and application of a wireless sensor network prototyping system which is able to track mobile search and rescue robots by examining the signal strength. In [6] Shivendra Mishraet al., presents much more realistic scenario where data is collected from multiple directions and more number of sensors are added in order to increase the accuracy of clustering. The clustering is performed by adopting k-medoid clustering algorithm and using Manhattan distance. The paper [7] explains fusing the data from multiple sensors and come out with the result which is heading estimation. [8] addresses the problem of object detection for the purpose of object manipulation. Various works has been studied for working with robotic image data. S. Silakari et al., [9] proposes a technique for clustering the images applying color as the feature vector by executing the block truncation coding and color moments. Swetha Sreedharan et al., [7] put forward an experimental study by applying k-medoids algorithm on the features extracted from low resolution image data collected from a set of simulated complex robotic environments. The experiments are performed offline using limited amount of data.

A survey of time series data mining can be found in the work by Esling et al. [4]. It has presented different implementation components of time series systems, namely representation techniques, similarity measures and indexing methods. An extended approach [5] of the work [2], focuses on analyzing the time series robotic data and clustering the robotic environment using DTW (Dynamic Time Warping) as a similarity measure between the time series data. In another work [11] simple robotic environments have been set up and various experiments have been carried out .Clustering is performed on time series data collected which has been obtained by applying various transformation techniques on the data collected from multiple straight paths.

In real scenarios, the robotic environment may be very complex. Therefore it is required to study complex robotic environment . The proposed work mainly focus on the clustering of a complex robotic environment from a multi view approach by collecting the data using various sensors and camera mounted on the robot. The work is organized as two stages. In the first stage, the focus is on clustering the robotic environment by using the data obtained from sensors. The second stage focuses on clustering the robotic environment by combining the robotic image data and sensor data collected from multiple direction.

## 3    Proposed Approach

Time series data is collected from a simulated robotic environment, in multiple directions and clustered to group the similar environment together.

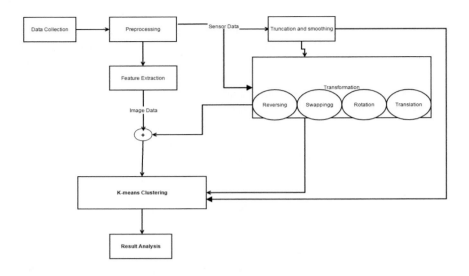

**Fig. 1** Schematic diagram of proposed approach.

## 3.1   Data Collection

A simulated robotic environment is setup [6] with a square area of dimensions
210cm x 210cm where the robot equipped with 4 IR proximity sen-
sors(IR1,IR2,IR4,IR5) and two Thermal imaging sensors(LT,RT) and moves
through multiple straight paths for data collection. Various objects of different
shape, color, temperature etc are placed in various locations on the simulated scenarios. The objects have different shapes such as spherical, rectangular, cone,
pyramid shape etc. Out of 32 objects placed in the scenario 4 of them are hot objects, 4 of them are cold objects and the rest of them are non-Thermal. The paths
are designed in such a way that the robot can navigate through the environment
along all four different directions N-S (North-South), S-N (South-North), E-W
(East-West) and W-E (West-East). Low resolution camera is mounted on the robot
to collect the images of various objects , while moving through various paths in
the environment by rotating the camera at different angles, i.e. 30°, 90° and 150°
to cover full extent of the environment. The data is captured at successive time
intervals and in different conditions, making it a time series data. Seven different
scenarios have been set up. Robot is made to explore each of these scenarios 10
times in 4 different directions. The Fig 2 shows a sample scenario. Each scenario
differs in the distribution and number of objects in it.

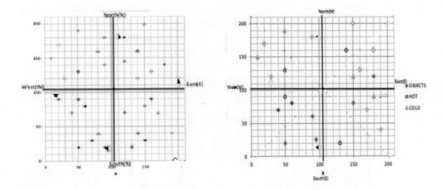

**Fig. 2** Scenario 1, Scenario 6

## 3.2    Preprocessing

In pre-processing step, the data recorded by sensors in the format of hexadecimal data is converted into a time series data. The data after preprocessing would consist of 36 attributes, which comprise of 4 distance measurements from 4 IR sensors and 16 pixel values of 4x4 images obtained from each of the two thermal imaging sensors[6]. The images collected from the robot which are of high resolution (4800x3600) pixels are converted to a low resolution images of size (48x36) pixels. Each of the images are divided into 4 tiles.Fig. 4 shows the analysis of data before smoothing and after smoothing.

## 3.3    Image Feature Extraction

The feature RGB mean is extracted from the images collected from different trials conducted where the robot is made to move along multiple straight paths through the robotic environment. RGB mean is extracted from an image which is split into 4 tiles making 12 features from an image [12]. Since the images are collected by rotating the camera in three different angles and by stopping 9 times before reaching the end of the path, altogether 324 features are generated in a trial.

## 3.4    Sensor Data Truncation and Smoothing

The Fig 3(a). shows  the data from one of the IR sensors IR1 in the trial1, trial2 and trial 3 represented using different colors. Though all these three data are collected by using the same sensor in the same scenario, there is a large variation in the data. Hence it is necessary to perform additional processing like smoothing. Smoothing identifies an appropriate parametric model to eliminate complex effects so that more precise interpolated data is obtained and is as shown in the Fig 3(b). The data collected contained data from objects that are out of the boundary of the environment and hence the data has been considered to be noisy.

To reduce the impact of this noise introduced, the data has been truncated in a way where values greater than a threshold has been reduced to the threshold value.

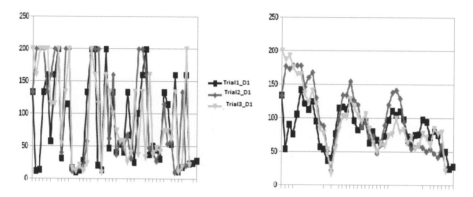

**Fig. 3** (a)Data where smoothing is not applied    **Fig 3** (b) Data where smoothing is applied

## 3.5    *Sensor Data Transformation*

As the time series data collected from same environment from different paths does not show any similarity, appropriate transformations need to be performed. Initially clustering is performed by combining the data from vertical direction i.e. N-S & S-N.N-S is taken as the reference path. To make the data collected from S-N similar to N-S reversing the reading and swapping the value of left and right sensors is performed.

Next step is to combine data from horizontal direction i.e. W-E and E-W, the same operation reversing and swapping is done. Then the data from horizontal direction is combined with vertical direction i.e. clustering is performed by combining data from 4 direction. In order to make data from horizontal path similar to vertical path some operation has to be performed as shown Fig. 4,For data collection robot moved in the intended path and collects data at regular time interval. Initial step is the calculation of position of robot as shown in formula (1).

$$X = \text{Distance traversed by robot/Total number of readings.} \qquad (1)$$

After finding the position first we need to translate y by -100 .Then we need to rotate each point (x,y) by 90 degree. After that x'(rotated x) has to be translated by +100.Then we need find the new distance using the below formula.

$$\text{Transformed distance} = x - x' \qquad (2)$$

As the data is collected from multiple directions, the same object when viewed from different paths, may have a difference in the reading. To overcome this, transformation need to be applied on the data so that these readings can be inferred to the same object. To explain in simple terms path 1(N-S) is taken as reference path, all other paths are transformed to make it overlap with the reference path as shown Fig. 5.

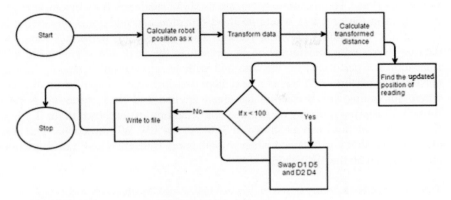

**Fig. 4** Flow diagram of transformation of horizontal direction.

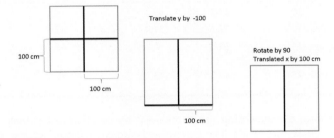

**Fig. 5** Schematic diagram of transformation.

## 3.6 k-Means Clustering

Clustering is performed using k-means[13] algorithm where the value of k is decided as 7 as there are seven scenarios in the experiment. First the sensor data alone is clustered and later it is combined with image data and then clustering is performed.

## 4 Experimental Results

The experimental setup consists of seven scenarios which represents the seven different clusters. Once the clustering is performed, the data from different scenarios need to be clustered to one of these seven different clusters. Result of the clustering algorithm is analyzed by finding how many data of same scenario are clustered together.

Table-1 shows the clustering accuracy obtained from combining the sensor data collected from two directions (N-S and S-N). The results are analyzed by varying the different attributes to see the effect of attribute selection. IR sensor data

(4 attributes) and Thermal sensor data are used (32 attribute). Initially clustering is performed in the raw data where no transformation and additional processing is applied and the result is shown in Table 1. Then clustering is performed using the data on which only transformation is applied. In the next step smoothening and truncation are performed on the data and without application of transformation, clustering is performed. Then as a next step, transformation is applied on the data where smoothing and truncation has been applied and then clustering is performed. Clustering accuracy of all these combinations are as shown in the Table 1 for NS-SN directions and as shown in Table 2. for EW-WE directions. The best clustering accuracy is obtained when smoothening, truncation and transformation all are applied on the 4IR values.

**Table 1** Clustering accuracy of sensor data combining two directions (N-S & S-N).

| Dataset | Data without transformation | Only transform | Only smoothening &truncation | Smoothening & truncation + transformation |
|---|---|---|---|---|
| 4IR attribute | 39.2 | 44.2 | 47.1 | 55 |
| 2Thermal attribute | 30.2 | 32.1 | 31.4 | 32.8 |
| 4IR + 2Thermal attribute | 44.2 | 44.2 | 35 | 35.7 |
| 4IR + 32Thermal attribute | 45 | 49.2 | 43 | 47.1 |

Table-2 shows the clustering result obtained by combining the sensor data of two direction (E-W and W-E). The experiments are carried out varying the different set of attributes. It is evident from Table-1 and Table-2 that accuracy of clustering is reduced after transformation .The reason behind drop in accuracy may be due to the reason that multiple objects may be located in same co-ordinate in some cases. Hence while the robot is moving from E-W the objects may be detected, but after transforming it to N-S these objects may fall in the same line. Hence only one object which is nearest to the robot may be detected. Other reason for accuracy drop can be due to the objects outside the boundary being detected which can be considered as noise in the data collected. It is observed that there is a 5 % improvement in accuracy when the smoothening and truncation are applied..

**Table 2** Clustering accuracy of sensor data combining two directions (W-E &E-W).

| Dataset | Without Transform | With transform | Only smoothening &truncation | Smoothening & truncation + transformation |
|---|---|---|---|---|
| 4IR attribute | 39 | 24 | 49.2 | 41 |
| 2Thermal attribute | 29.8 | 30.4 | 29.1 | 28.5 |
| 4IR + 2Thermal attribute | 45 | 37.8 | 50 | 43 |
| 4IR + 32Thermal attribute | 42 | 39 | 46.1 | 43 |

Though various techniques like smoothening, truncation and transformation are applied, the clustering accuracy has not improved significantly. Hence it is proposed to combine the features from image data together with the sensor data as the image data contains more information about the environment ( like shape and color of the objects) and the clustering accuracy has been improved as shown in the Table 3. When the image data alone is used for clustering, the accuracy obtained is 43% in NS-SN combination and 48% in EW-WE combination. When the image data is combined with the sensor data, the best accuracy is obtained using the 4 IR sensors. The thermal sensors are not contributing much in the improvement of the clustering accuracy as seen in the Table 3.

Similarly after applying various transformations it is attempted to combine the data from all the four directions and is given in the Table 4. Table 4 also shows the clustering accuracy obtained after combining the image and sensor data of all the four directions. The results are checked for various attributes. The best clustering accuracy is obtained when image data is combined with 4IR sensor data. Adding image to the sensor data has improved the accuracy of clustering by 5 to 10 %.

**Table 3** Clustering accuracy of sensor and image data combining two directions

| Dataset | N-S &S-N | | E-W & W-E | |
|---|---|---|---|---|
| | Sensor | Sensor + image | Sensor | Sensor + image |
| 4IR attribute | 55 | 58 | 41 | 51 |
| 2Thermal attribute | 32.8 | 43 | 28.5 | 31.5 |
| 4IR + 2Thermal attribute | 35.7 | 46 | 43 | 50 |
| 4IR + 32Thermal | 47.1 | 43 | 43 | 45.5 |

**Table 4** Clustering accuracy of sensor and image data combining four directions (N-S ,S-N ,E-W &W-E).

| Dataset | Accuracy | |
|---|---|---|
| | Sensor | Sensor + image |
| 4IR attribute | 30 | 41.4 |
| 2Thermal attribute | 23.9 | 31.8 |
| 4IR + 2Thermal attribute | 34 | 46 |
| 4IR + 32Thermal attribute | 34.6 | 36.2 |

# 5    Conclusion and Future Work

Initially sensor data obtained from different path are clustered separately using k-mean clustering algorithm. It is observed that the model gives better accuracy with IR proximity sensor values alone. The thermal sensors are not improving the clustering accuracy. For improving the accuracy of results smoothening and truncation is incorporated. From the result it is observed that, the model works shows best result with 4 IR attribute. Clustering is also performed by combining the sensor data with image data collected from different paths.

When data from multiple directions are combined, it is not able to produce good clustering result though various transformations are applied. This may be due to various reasons like the data is noisy and also these transformations might be insufficient for the complex robotic data. Thus it is observed that much more analysis is required to verify the actual nature of data.

It is planned to incorporate various modifications to be tried. Instead of appending the image feature with the sensor data, objects can be identified and added as the feature to the sensor data. This is expected to improve the clustering accuracy. Also revisions can be done in the way the objects are placed in the various scenarios and checked for improvement in the accuracy

The experiments if carried out in known environment might give better clustering results. So classification techniques can be applied on the data collected from the simulated robotic environment.

# References

1. Ko, A., Lau, H.Y.K.: Robot assisted emergency search and rescue system with a wireless sensor network. International Journal of Advanced Science and Technology **3**, 69–78 (2009)
2. Oates, T., Schmill, M.D., Cohen, P.R.: A method for clustering the experiences of a mobile robot that accords with human judgments. In: Proceedings of the Seventeenth National Conference on Artificial Intelligence, pp. 846–851 (2000)
3. Großmann, A., Wendt, M., Wyatt, J.: A semi-supervised method for learning the structure of robot environment interactions. In: Advances in Intelligent Data Analysis V Lecture Notes in Computer Science, vol. 2810, pp. 36–47 (2003)
4. Esling, P., Agon, C.: Time-series data mining. ACM Computing Surveys (CSUR) **45**(1), Article no. 12 (2012)
5. Radhakrishnan, G., Gupta, D., Abhishek, R., Ajith, A., Sudarshan, T.S.B.: Analysis of multimodal time series data of robotic environment. In: 2012 12th International Conference on Intelligent Systems Design and Applications (ISDA), pp. 734–739, November 27–29, 2012
6. Mishra, S., Radhakrishnan, G., Gupta, D., Sudarshan, T.S.B.: Acquisition and analysis of robotic data using machine learning techniques. In: International Conference on Computational Intelligence on Data Mining (ICCIDM 2014). Springer Publication, December 2014
7. Axenie, C., Conradt, J.: Cortically inspired sensor fusion network for mobile robot heading estimation. In: Artificial Neural Networks and Machine Learning – ICANN (2013)
8. Rigual, F., Ramisa, A., Alenya, G., Torras, C.: Object detection methods for robot grasping: experimental assessment and tuning. In: 15th Catalan Conference on Artificial Intelligence (CCIA) (2012)

9. Silakari, S., Motwani, M., Maheshwari, M.: Color Image Clustering by Block Truncation Algorithm. International Journal of Computer Science Issues, IJCSI **4**(2), 31–35
   (2009)
10. Swetha, S., Radhakrishnan, G., Gupta, D., Sudarshan, T.S.B.: Analysis of robotic
    environment using low resolution image sequence. In: International Conference on
    Contemporary Computing and Informatics, pp. 495–499 (2014)
11. Radhakrishnan, G., Gupta, D., Sindhuula, S., Shrey, K., Sudarshan, T.S.B.: Experi-
    mentation and analysis of time series data from multi-path robotic environment. In:
    2015 IEEE International Conference on Electronics, Computing and Communication
    Technologies (IEEE CONECCT 2015), Bangalore, India, July 10–11, 2015
12. Radhakrishnan, G., Meenu, M., Gupta, D., Sudarshan, T.S.B.: Clustering of robotic
    environments using image sequence data. In: ICCN-2013/ICDMW-2013/ICISP-2013
13. MacQueen, J.: Some methods for classification and analysis of multivariate observa-
    tions. In: Proc. 5th Berkeley Symp. Math. Stat. Prob, Berkely, CA, vol. 1, pp. 281–297
    (1967)

# Twofold Detection of Multilingual Documents Using Local Features

Glaxy George and M. Sreeraj

**Abstract** Twofold detection of the document images plays an important role in document image analysis. This paper presents a novel approach to detect twofold document images by extracting local features such as moment, texture and foreground pixel density. The performance of the proposed system is evaluated based on criteria of data schemes, feature and various distance metrics. Experimental results on different datasets demonstrates that proposed method is flexible enough to handle multilingual documents and provides better performance on historical, printed and handwritten documents. The performance of the proposed approach is analyzed with local features alone and better performance is observed when combined features are taken into account. Based on distance metric criteria, earth mover's distance for similarity measurement outperforms the other distance measures.

## 1 Introduction

Twofold detection of images is one of the prominent field in document image analysis to locate visually similar or identical copies of a given document in a large database. It is a challenging task for twofold document image to match due to illumination or resolution variations. In digital libraries, numerous documents being scanned and archived every day, it is essential to ensure that no duplicate exists in the database when a new document image arrives. Twofold detection deals with detecting twofold images, thereby reducing the storage cost and facilitate the indexing process.

Twofold detection of document images can be either content based or layout based of the document. The first approach extracts features from content of document image and second approach from layout of the image. In both cases, features are extracted

G. George(✉) · M. Sreeraj
FISAT, Ernakulam, India
e-mail: {glaxygeorge,sreerajtkzy}@gmail.com

© Springer International Publishing Switzerland 2016
S.M. Thampi et al. (eds.), *Advances in Signal Processing and Intelligent Recognition Systems*,
Advances in Intelligent Systems and Computing 425,
DOI: 10.1007/978-3-319-28658-7_45

and comparison is done by using any one of the distance measures. Among these existing approaches for twofold detection, content based is found to be more effective in identifying copies. In content based approach, documents are uniquely identified either by local features or global features. Local feature is extracted for each point in the input domain and global features are evaluated on sets of pixels, on a region or even on the whole document. A survey conducted by [1], figures out that local features are more distinct and has efficient discriminative power compared to global features. Hence, twofold document detection systems using local features are more efficient in identifying similar documents.

Twofold detection involves two critical issues: image representation and dissimilarity computation. Various image representation techniques have been proposed in the literature and different image representations usually entail different similarity measures. Optical Character Recognition(OCR) [3] representation is language-dependent and time-consuming. Other representations like Character shape coding [4], density distribution feature (DDF) and key block feature(KBF) [5], SIFT interest points [6] have been proposed for detecting twofold document images. The drawbacks of these approaches are incorrect similarity measure, less robust for degraded images, computational overhead respectively. Graph representation [7] have high complexity, little mathematical structure as compared to feature vectors. Hence, in this paper we are presenting a novel local feature extraction for detecting twofold documents.

The objective of this paper is to present a robust twofold document detection system that automatically identifies document copies irrespective of various illumination and resolution. To accomplish this, local features like moment, texture and foreground pixel density are extracted to represent a document image. Every object in the document is identified by using connected component analysis and features are extracted from each component. For every document, a local feature descriptor will be stored in the document database. During the testing phase, the similarity between two documents are determined by using the Earth Movers Distance [2].

The remaining portion of the paper is organized as follows. In Section 2, a brief description about different twofold detection methods and the local feature based twofold detection technique in section 3. The experimental analysis and results are presented in section 4, and finally the conclusion in section 5.

## 2   Related Works

This section, takes a look at the previous works that has been done related to twofold document image detection. Twofold detection is employed for detection of plagiarism and maintaining database integrity, information filtering and the detection of unauthorized copies. Fig. 1 shows taxonomy of twofold detection of the document images.

Twofold detection methods can be broadly classified into two categories in terms of their goals: 1)similarity based 2) generic features based methods. The first category

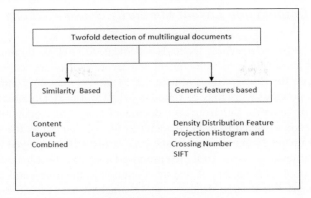

**Fig. 1** Taxonomy of twofold detection of document images.

aims at finding document images that are similar. The similarity may lie in their content, such as the documents related to the same topic, i.e. sport, finance, business etc. The second category detects document images that are similar in layout. In these two categories features of document images are extracted and comparison is performed by any one of the similarity measures.

## 2.1 Content Similarity Based Methods

In this respect, Tan et al. [8] the document image is segmented into character objects and features like Vertical Traverse Density(VTD) and and Horizontal Traverse Density(HTD) were extracted. The similarity between document images was calculated by taking the dot product of the ngram document vectors of documents. Lu et al. [9] proposed that each word in the image characterized by a primitive string making use of its Line-or-Traversal Attribute(LTA)as well as Ascender and Descender Attribute(ADA). For similarity computation between two document images, vector space model was employed.

## 2.2 Layout Similarity Based Methods

Several layout similarity based approaches were proposed in [10], including single image, image with tables and so on. An XY tree and tree-edit distance measure is employed to find similarity. Cesarini et al. [11] proposed the modified XY(MXY)tree for page classification. Hu et al. proposed a novel approach called interval encoding [12] feature to encode layout information in fixed-length vectors by exploring structural information of the image. The similarity is measured using Manhattan distance.

## *2.3 Both Content and Layout Similarity Based Methods*

Some document image matching methods is dedicated to finding duplicates, i.e. the images which are identical in both content and layout as they are scanned from the same document. But discrepancies remain due to various imaging conditions. With regard to this, previous work may be found in [4], [13]. In [4], Doermann et al. first extracted a representative line from each document image. Subsequently, the string of the shape code assigned to each character in the line served as the signature of the document image. N-grams of the shape code string were employed for indexing in an image retrieval scenario. Hull [13] proposed a set of robust local features and similarity is simply the number of features common in the query and the database document image. For a large database, the character count metric becomes less effective and also the segmentation of characters in the degraded document images is really challenging, putting the two methods above at a disadvantage.

## *2.4 Generic Features Based Methods*

The second category detects twofold document images by using more generic features extracted from the image. For example, Liu et al. [5] employed density distribution features (DDFs) and key block features (KBFs) for image representation. Meng et al. [14] combined DDF with Local Binary Pattern (LBP), Projection Histogram Features (PHFs) and Crossings Number Histogram Features (CNHFs) to represent a document image. Improvement in performance was observed when compared with that in [5] according to their experimental results. In [6], near-duplicate document images were detected based on SIFT interest points matching. A set of SIFT features were first extracted from each image. Then the number of correspondence in features between two images was obtained, based on which their similarity can be calculated. However, there are often hundreds or thousands of interest points per image, which results in excessive computational overhead.

## 3   Proposed Method

Fig. 2 shows architecture of the proposed method. Deskew and noise removal are employed for pre-processing of documents. Afterwards object segmentation or identification is performed using connected component analysis. The objects in the document can be characters, words, lines, non-text images etc. From each object in the document, local features are extracted. Local features such as moment, texture and foreground pixel density are captured. The similarity computation of document images are done by using Earth movers distance. Fig. 3 shows segmentation of document image using connected component. Each object is enclosed within a rectangle.

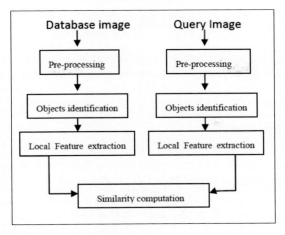

**Fig. 2** Architecture of the proposed method.

**Fig. 3** Identification of objects in an Arabic document image shown in left. Connected components are enclosed in rectangle(red color)

## 3.1 Texture Features

Texture features are efficient in identifying objects in an image. In proposed approach, Gray level co-occurrence matrix (GLCM) [15] is used for the texture analysis. GLCM of four orientations, viz., $0^o$, $45^o$, $90^o$, $135^o$ are computed for each object. Four NXN square matrices are used to represent these different orientations. Three features are calculated from each orientation, i.e. entropy, contrast as well as homogeneity. The four orientations are represented as Pj(j=1; 2; 3; 4).

$$\varepsilon_j = -\Sigma_{a=1}^{N} \Sigma_{b=1}^{N} P_j(a, b) \times log\, P_j(a, b) \tag{1}$$

$$C_j = \Sigma_{a=1}^{N} \Sigma_{b=1}^{N} (a - b)^2 \times P_j(a, b) \tag{2}$$

$$H_j = -\Sigma_{a=1}^{N} \Sigma_{b=1}^{N} \frac{P_j(a, b)}{1 + (a - b)^2} \tag{3}$$

where $\varepsilon_j$, $C_j$ and $H_j$ denotes entropy, contrast and homogeneity respectively. The average of each feature on four orientations and standard deviation is also computed.

## 3.2 Moment Features

The histogram of object $O_i$ with gray value ranging from 0 to K is computed. Based on this, the second moment for the object $O_i$, is defined as

$$M_j = \Sigma_{j=0}^{K}(j - \lambda_i)^2 \times h(j) \tag{4}$$

where $\lambda_i$ is the average gray value of object $O_i$.

## 3.3 Foreground Pixel Density Features

The foreground pixel density [7] of the object Oi is simply defined as the ratio of foreground pixels in the object with respect to the foreground pixels in the entire image:

$$F_i = \frac{number\ of\ foreground\ pixels\ in\ O_i}{number\ of\ foreground\ pixels\ in\ the\ entire\ image} \tag{5}$$

where $F_i$ is the average gray value of object $O_i$.

## 3.4 Similarity Measure

The local features of each object in the document image are captured. Afterwards average of moment, texture and foreground pixel density feature for the entire image is calculated. The signature of the document image is defined as

$$S(d_i) = \frac{(A_T + A_M + A_F)}{3} \tag{6}$$

where $A_T$, $A_M$ and $A_F$ are average of the texture, moment and foreground pixel density features respectively. The similarity of two document images is measured using the Earth Movers distance equation defined in [2]. Compared to other similarity measures such as Euclidean distance, Cosine distance, Jaccard distance, the Earth

Movers distance has significant performance in the detection of twofold document images.

## 4 Experimental Results and Validation

The proposed twofold document image detection approach was tested in an image retrieval environment. For each query issued, the relevant documents in the database was retrieved. An initial experiment of information retrieval was done with 100 document images. For each query image issued, the top ranked 20 images were retrieved along with their similarity scores.

### 4.1 Data Sets

There are no public benchmark data sets for the task of twofold document image retrieval. We constructed multilingual data sets for our study. The GW and UW2 datasets was used for performance evaluation. Experiments was also done with documents in different languages such as Malayalam, English and Arabic. Data sets covers historical, handwritten and printed documents. Images with different skew angles was also tested.

### 4.2 Performance Comparison Under Different Levels of Features

In this experiment 150 historical, 110 handwritten and 220 printed documents in different languages was tested. Document images with noise and variations in skew angle was also included. Top1 documents means similarity score, $S(d_i)$, below 15 and Top2 with score between 15 and 30. The proportion was evaluated based on ratio of retrieval rate and total documents in the database.

Twofold detection was performed under different levels of features. At first level, performance was investigated with single feature. ie, twofold detection was done with extracting moment feature, foreground pixel density and texture separately. At second level, combination of two features was investigated and gained better performance than first level of evaluation. ie, moment and texture, moment and foreground pixel density and so on. At third level, ie proposed method, all the three local features was extracted and twofold detection was computed. When three features were used, significant increase in performance was noticed. Table 1 and Table 2 shows the results obtained based on single feature and multiple features respectively.

**Table 1** Retrieval result based on single feature

|  | Texture | | Moment | | Foreground pixel density | |
|---|---|---|---|---|---|---|
|  | Top1 | Top2 | Top1 | Top2 | Top1 | Top2 |
| Number | 56 | 60 | 69 | 72 | 60 | 69 |
| Proportion | 53.3 | 57.1 | 65.7 | 68.6 | 57.1 | 65.7 |

**Table 2** Retrieval result based on combined features

|  | Moment, Texture | | Texture, Foreground pixel | | Combined | |
|---|---|---|---|---|---|---|
|  | Top1 | Top2 | Top1 | Top2 | Top1 | Top2 |
| Number | 67 | 50 | 65 | 60 | 75 | 66 |
| Proportion | 60.9 | 45.4 | 59.1 | 54.5 | 68.1 | 60.7 |

**Table 3** Average similarity based on distance measure

| Distance measure | Average Similarity |
|---|---|
| Euclidean distance | 42 |
| Cosine distance | 40 |
| Earth Movers distance | 45 |

## 4.3 Comparison on Distance Measures

The performnace evaluation on various distance measures such as Euclidean, Cosine and Earth Mover's distance are shown in Table 3. The average similarity is computed based on its performance to detect twofold document images. In this experiment,100 copies of a document image varying with resolution, skew angle and noise were tested. Out of these images,42% were retrieved as duplicate with Euclidean distance,40% with Cosine distance, and 45% with Euclidean distance.

## 4.4 Comparison on State-of-Art Methods

The performance comparison with state-of-art methods was evaluated based on the features comprised for duplicate detection. The proposed feature Texture-Moment-Foreground pixel density (TMF) was compared with four other features such as DDF [5], SIFT [6] and PHF & CNF [14]. The precision and recall was computed as follows:

**Table 4** Comparison on state-of-art features

|  | DDF | PHF | CNF | SIFT | TMF |
|---|---|---|---|---|---|
| **Historical** | | | | | |
| Precision | 57 | 55 | 62 | 69 | 72 |
| Recall | 48 | 54 | 59 | 60 | 70 |
| **Printed** | | | | | |
| Precision | 55 | 65 | 69 | 72 | 73 |
| Recall | 58 | 50 | 61 | 69 | 75 |
| **Handwritten** | | | | | |
| Precision | 47 | 52 | 58 | 60 | 69 |
| Recall | 49 | 47 | 56 | 61 | 65 |

**Table 5** Average running time in ms

| Preprocessing | Texture | Moment | Foreground pixel density | Retrieval |
|---|---|---|---|---|
| 544 | 209 | 240 | 198 | 478 |

$$precision = \frac{\{duplicate\ images\} \cap \{retrieved\ images\}}{retrieved\ images} \qquad (7)$$

$$recall = \frac{\{duplicate\ images\} \cap \{retrieved\ images\}}{duplicate\ images} \qquad (8)$$

Table 4 shows comparison with state-of-art methods and TMF. Evaluation demonstrates that SIFT and TMF features were capable for twofold printed document image detection. The precision and recall value for historical and handwritten images demonstrates the outstanding performance of the proposed feature TMF.

## 4.5 Execution Time

The average running time of proposed approach was evaluated on an Intel i5 4770 CPU @ 3.4 GHz with 16 GB RAM and implemented in Matlab. For this configuration, proposed approach consumes majority time for preprocessing and retrieval process. Average time was consumed for feature extraction. Table 5 shows average running time in ms for image preprocessing, features extraction and retrieval.

# 5 Conclusion

The detection of the twofold document image is an important issue in the field of information retrieval. This issue is addressed in this paper by exploring the local features of the document image. Initially, pre-processing operations such as deskew and noise removal are performed. The segmentation of objects in the document image is done by using connected component analysis. The local features such as texture, moment and foreground pixel density, are extracted from each connected component, which depict the signature of the document image. Subsequently by combining the three local features, a variable-length signature is generated for each image in the document database. To compute the dissimilarity between two document images, the Earth Movers Distance is employed. Since single feature alone could not lead to better result, combined features are adopted and accurate results are obtained. The effectiveness of the proposed approach is demonstrated from the extensive experiments. Further research is required to investigate the inclusion of context based twofold detection of the documents. The automatic classification of document image as structure-rich or structure-lack could also be considered to extend the proposed framework.

# References

1. Marinai, S., Miotti, B., Soda, G.: Digital libraries and document image retrieval techniques: a survey. In: Biba, M., Xhafa, F. (eds.) Studies in Computational Intelligence, vol.375, pp. 181–204. Springer (2011)
2. Rubner, Y., Tomasi, C., Guibas, L.: The earth movers distance as a metric for image retrieval. International Journal of Computer Vision **40**(2), 99–121 (2000)
3. Lopresti, D.P.: Models and algorithms for duplicate document detection. In: Proceedings of International Conference on Document Analysis and Recognition, pp. 297–300 (1999)
4. Doermann, D., Li, H.P., Kia, O.: The detection of duplicates in document image databases. Image Vis. Comput. **16**, 907–920 (1998)
5. Liu, H., Feng, S.Q., Zha, H.B., et al.: Document image retrieval based on density distribution feature and key block feature. In: Proceedings of International Conference on Document Analysis and Recognition, pp. 1040–1044 (2005)
6. Vitaladevuni, S., Choi, F., Prasad, R., et al.: Detecting near-duplicate document images using interest point matching. In: Proceedings of International Conference on Pattern Recognition, pp. 347–350 (2012)
7. Liu, L., Lu, Y., Suen, C.Y.: Retrieval of envelope images using graph matching. In: Proceedings of International Conference on Document Analysis and Recognition, pp. 99–103 (2011)
8. Tan, C.L., Huang, W., Yu, Z., et al.: Imaged document text retrieval without OCR. IEEE Trans. Pattern Anal. Mach. Intell. **24**(6), 838–844 (2002)
9. Lu, Y., Tan, C.L.: Information retrieval in document image databases. IEEE Trans. Knowl. Data Eng. **16**(11), 1398–1410 (2004)
10. Marinai, S., Marino, E., Soda, G.: Layout based document image retrieval by means of XY tree reduction. In: Proceedings of International Conference on Document Analysis and Recognition, pp. 432–436 (2005)
11. Cesarini, F., Marinai, S., Soda, G.: Retrieval by layout similarity of documents represented with MXY trees. In: Proceedings of International Workshop on Document Analysis Systems. Lecture Notes in Computer Science, vol. 2423, pp. 353–364. Springer (2002)

12. Hu, J.Y., Kashi, R., Wilfong, G.: Comparison and classification of documents based on layout similarity. Inf. Retr. **2**(2/3), 227–243 (2000)
13. Hull, J.J.: Document image matching and retrieval with multiple distortion invariant descriptors. In: Proceedings of International Workshop on Document Analysis Systems. Lecture Notes in Computer Science, pp. 379–396 (1995)
14. Meng, G.F., Zheng, N.N., Song, Y.H., et al.: Document images retrieval based on multiple features combination. In: Proceedings of International Conference on Document Analysis and Recognition, pp. 136–140 (2007)
15. Haralick, R.M., Shanmugan, K., Dinstein, I.H.: Textural features for image classification. IEEE Trans. Syst. Man Cybern.(SMC-3), 610–621 (1973)

# Part V
# Workshop on Advances in Image Processing, Computer Vision, and Pattern Recognition (IWICP-2015)

# On Paper Digital Signature (OPDS)

Sajan Ambadiyil, V. B. Vibhath and V. P. Mahadevan Pillai

**Abstract** The requirements for authenticating the identity of the sender and ensuring the content integrity in printed paper documents have been growing. Paper documents are still used in the various secure transactions. Manual verification of large amount of documents to authenticate the sender and the content is difficult. The digital signatures can be implemented on each printed paper document to fulfill these requirements. The existing technologies make use of image processing techniques for creating and verifying digital signatures and are vulnerable to physical conditions of the document. Another currently available method takes whole content of the document to achieve the requirement and requires more complex procedures for implementation. This paper presents a novel scheme for securing the authenticity and originality of certificates issued by an authority by implementing unique digital signatures created from the selected contents (words or characters) of the documents. The method assures the content integrity and sender identity authentication.

## 1 Introduction

Paper-based documents are still necessary in some situations where digital documents cannot be used effectively and efficiently. For example, documents issued by

S. Ambadiyil(✉)
Optical Image Processing and Security Products, Centre for Development of Imaging Technology,
Thiruvananthapuram 695027, Kerala, India
e-mail: ambadycdit@gmail.com

V.B. Vibhath
Electronics and Communication Department, Jawaharlal College of Engineering and Technology,
Lakkidi, Palakkad 679301, Kerala, India
e-mail: vibhathvb@gmail.com

V.P. Mahadevan Pillai
Department of Optoelectronics, University of Kerala, Kariavattom, Thiruvananthapuram 69558,
Kerala, India
e-mail: vpmpillai9@gmail.com

© Springer International Publishing Switzerland 2016
S.M. Thampi et al. (eds.), *Advances in Signal Processing and Intelligent Recognition Systems*,
Advances in Intelligent Systems and Computing 425,
DOI: 10.1007/978-3-319-28658-7_46

547

the government and legal documents, which include birth certificates, passports, driving license, must be in paper form. The development of new advanced technologies, such as scanning and printing, help the criminals to produce counterfeit documents. For checking the authenticity of paper documents, there exist several physical security features that can be added to to the paper document. Unfortunately, it does not make any sense for ensuring content integrity. Several kinds of modification can be made in variousdocuments issued by the government and several other authorities. These kind of unauthorized modifications cannot be tolerated and hence a mechanism to detect the unauthorized modifications in documents and to authorize the identity of the author of the document is required.This can be acheived by implementing digital signatures on paper documents [1]. In digital domain, the authenticity and content integrity of documents have been achieved with the help of traditional cryptographic techniques [5, 7]. But in the case of paper-based documents, the methodology adapted should entirely different and is somewhat difficult as compared to the digital domain.

An existing technology for authenticating paper documents involve the processing of the entire content in the document for creating the digital signature for that document [1]. This technology require some amount of data compression for the purpose of incorporating the digital signature into a QR code format and is thus time consuming and more complicated. The verification of the digital signature requires the optical character recognition technology to recognize the whole content in the document. These contents and QR code data are then processed to check the authenticity and content integrity. Thus, there is more chance of errors while recognizing entire characters from image of the document by using OCR. Another technology make use of image processing techniques [6] to authenticate the paper-based documents. The data are considered as an image, which is converted into its equivalent binary form. This technology requires some reference marked points during the creation of the document [6]. The points are necessary during verification. In such a method, a minor change in the whole document affects the digital signature in a large manner. Thus, the physical condition of the paper also should be considered during the verification process. The method depends on the pixels in the image taken for creating digital signature. Any change in the pixels will be reflected in the corresponding digital signature. This is the major drawback of this method.

This paper presents a method for creating and verifying digital signatures for certficates issued by the university by the effective use of optical character recognition technology to recognize the specified contents in the certificate under concern. This is achieved by specifying the more relevant text portions of the document. During the creation of digital signature at the sender, this kind of content selection is done in such a way that these selected words or characters make digital signatures for each document different from others. This unique digital signature is called OPDS (On Paper Digital Signature). The identity of the author of the certificate can also be authorized by this method. Also the above mentioned method does not require any compression [1] or reference marking operations [6]. The proposed method, allows an authorized verifier to verify the authenticity and content of the certificates from anywhere in the world by implementing a proper and secure verifier authentication mechanism. In order to ensure the originality of paper documents some additional paper-based physical security features must be added like security fibres, watermarks, holograms etc.

## 2 The Proposed Method

The proposed method aims to implement digital signatures, called on-paper digital signature or OPDS, for authenticating the certificates issued by an authorized authority. This section deals with the process of creating and verifying on-paper digital signature on paper-based certificates.

### 2.1 *OPDS Creation Procedure at Sender*

The process of creating digital signature on physical printed paper documents mainly involves three phases: OPDS creation, QR code generation and printing the generated QR code along with the original document. The detailed flow chart for creating on-paper digital signature is show in Fig. 1. The input to the creation process is a printed paper document containing a message to be secured. In order to generate a digital signature for a printed paper document, the proposed method takes only the selected characters for further processing. This selected data will be the input to the cryptographic hash function [2]. This makes the proposed method different from the existing technologies [1] to implement digital signatures on physical documents.

The first step in the OPDS creation process is the selection of the contents in the specified region of interest. In this step, the contents in the region of interest are selected in a way that they make the corresponding document unique from other documents. The sender can use these contents (characters or words) in any order, so that the verifier must know this order of the contents. In the case of an academic certificate the words that shows the name, register number, course, subject of study, grade or mark of the student, the month and year of passing are important as far as a student is concerned and the combined use of these words make it different from other students. So, in this case these words are selected for OPDS creation. Now, the selected words or characters are applied to a cryptographic hash function. This will result in a message digest. A cryptographic hash function produces a fixed length message digest from an input of variable length. The length of the message digests will depend on the type of hashing algorithm used in the process. The main feature of the hash function is its non-invertible nature [3, 4, 9]. Once a message is hashed to produce a message digest, the message cannot be recovered from the digest. This nature makes the cryptographic hash function suitable for authentication and integrity checking purposes. Here, SHA-1 (Secure Hash Algorithm-1) [11] is used to generate the message digest of 160-bit fixed length.

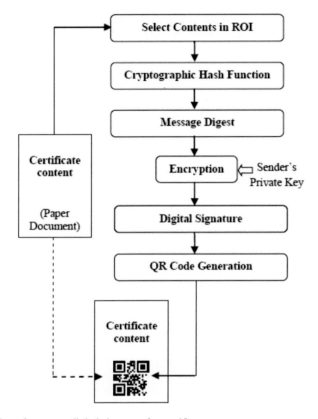

**Fig. 1** Creation of on paper digital signature for certificates

The obtained message digest is then encrypted using the private key of the sender in the third step. Public key encryption algorithms are used for this purpose [5]. They use a pair of related keys; one for the encryption and the other one for decryption. This is to authenticate the identity of the sender or owner of the document. That is, this step assures that the author or sender of the specified document is the one whose private key is used to encrypt the document. In the case of digital signatures, this step is called signing. In short, the sender signs the document by its private key, which is known only to the sender. The encryption process produces the corresponding cipher text. Here, the RSA algorithm is used to encrypt the obtained message digest. The public keys and the corresponding private keys are generated by the key generation phase of the RSA algorithm [10]. The private key and the corresponding public key for a particular certificate are stored in the database for the purpose of retrieving them during the verification process at the receiver.

The resulting cipher text is the required digital signature, i.e. OPDS, for the document under consideration. This completes the first phase of the digital signature creation. The next phase is the generation of the QR code for the created on-paper digital signature. The QR code format of digital signature is generated and is finally printed along with the paper document. Then the paper document with digital signature in the form of QR code is sent to the receiver. The receiver must know the public key of the sender for verifying the document to authenticate the identity of the sender. The verifier must also know the region of interest used by the sender to create the digital signature. The certificate format and its content may vary depending on the authority issuing the certificate. That is, the words and the layout of the certificate may different in different certficates. So the author of the certificate must communicate the region of interest to the verifier in some secure way.

## 2.2   OPDS Verification Procedure at Receiver

The document containing the QR code is received by the verifier and the verifier verifies the document for authenticating the identity of the sender and to check the content integrity. The proposed method is designed to provide online verification of a certificate containing OPDS. By this method, any authorized verifier can verify the certificate online after a user authentication process provided by the concerned authority. Thus the verifier must register their details to the concerned authority such as universities. The registration requires the mobile number, e-mail ID and other user identification details of the verifier. The authority identifies, verifies and authenticates the verifier by the use of details provided by the verifier at the time of registration. This restricts the unauthorized verifiers from verifying the certificate. When the user or verifier is authenticated then he/she is allowed to verify the certificate. The certificate issuing authority provides this kind of user verification through the website of the authority. The overall steps for verifying a particular certificate are also provided online through the same website, only after authenticating user. The flowchart for the user or verifier authentication mechanism is shown in Fig. 2.

The system provides two modes of verifier authentication mechanism: one through the registered mobile phone number and the other one through the registered E-Mail ID. The user or verifier has provision to select the mode of authentication as per the convenience of the user. The only condition is that the verifier must be a registered user either with its mobile number or with his/her E-Mail ID. When the verifier clicks the link for OPDS verification in the authority's website, he/she has to select either the phone verification or e-mail verification. On selecting either of these two modes of verification, the verifier is prompted to enter the mobile number or e-mail ID as per the previous selection of verification mode. The entered text, either the mobile number or e-mail ID, is verified with the database. If the mobile number or e-mail ID entered is not found in the database, then the verifier is not a registered user and hence not allowed to verify the certificate. Then the process terminates. Otherwise, the verifier is authenticated and the authority sends an OTP (One Time Password) to the registered mobile number in the case of phone verification or e-mail ID in

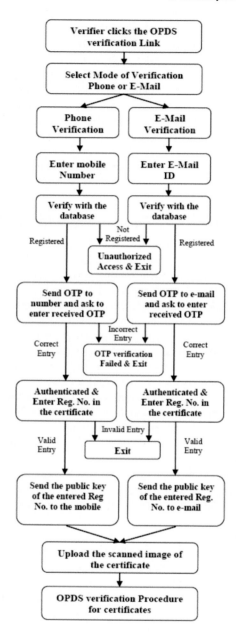

**Fig. 2** OPDS verifier authentication mechanism

the case of e-mail verification. Here, the OTP is generated as random by means of a random number generator. The OTP that has generated and sent is stored for further OTP verification.

The mechanism now asks the verifier to enter the received OTP. On entering the OTP, it is compared with the OTP that has sent to the user. OTP verification is failed in case of incorrect OTP entry by the user and hence the process terminates. When the user provides correct OTP, the OTP verification is successful and the user is now verified and found authentic. Now for verifying a particular certificate, the user or verifier must enter the register number in the certificate. On entering the register number of the certificate, the system checks for the public key corresponding to the entered register number in the database. If the register number is found in the database, then the public key corresponding to the concerned certificate is sent to the verifier either by mobile or by e-mail. Otherwise, sends a message regarding the invalid register number. The verifier must use this key to decrypt the data obtained by reading the QR code. After these steps, the verifier must upload the scanned image of the certificate. This scanned image is used for the OPDS verification. Then the verifier is prompted to the OPDS verification phase.

The Fig. 3 depicts the flow chart for the verification process. The verifier first process the message part on receiving the paper document with the QR code. The document is scanned using a scanner and converted into an image format such as JPEG. The text from images is extracted by means of optical character recognition [8]. The verifier must know the region of interest specified by the sender. The specified contents, defined by the specified region of interest, are extracted and are applied to the cryptographic hash function. This hash function must be identical to that used at the sender side for creating digital signature for that document (i.e. SHA-1). This will result in a message digest. The verifier now processes the QR code part. The QR code is decoded either by using QR code Scanner or by using mobile camera. Then the digital signature, OPDS, of that document will be obtained. The digital signature or OPDS is then decrypted to obtain the message digest by the use of RSA algorithm. The decryption is done by the use of the public key of the sender. The decryption is successful only when the public key of the person whose private key is used for the corresponding encryption, is used. That is, when the decryption is done using a public key, then it can be sure that the sender of the document is the one whose public is used for decryption.

By this way, the identity of the sender is authenticated. The content integrity can be checked by comparing the message digest obtained by decrypting the digital signature with the digest obtained by processing the message part. If both are identical, then it indicates that the contents in the document is not changed or altered. If they are not identical, the verifier has a provision to check the content integrity manually. OCR may not be accurate always. Because of the errors that occurred while recognizing the text by means of OCR, some words or characters may be recognized incorrectly. So in the verification side, this chance also has to be considered. At this situation the verifier has to manually hash the selected words or contents (by OCR) using the SHA-1 hash function and it is to be compared with the message digest obtained by reading followed by decrypting the QR code. Then the verifier can make final decision whether the document is modified or not.

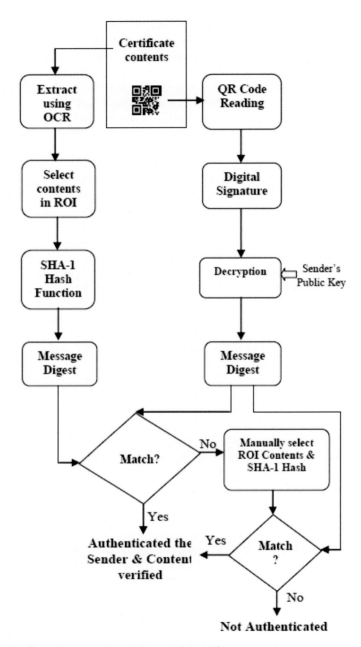

**Fig. 3** Verification of on paper digital signature for certificates

## 3  Analysis and Discussion

In order to analyze the performance of the proposed system, a number of sample documents were generated in the form of academic certificates. The generated certificates were printed on the A4 size paper together with the QR code. The main challenge in the proposed system was the accuracy of the text recognized by means of optical character recognition (OCR). In certain situations, OCR may incorrectly recognize the characters in the scanned image of the printed paper document. Thus, the performance of the proposed system is evaluated on the basis of the OCR accuracy in terms of error occurred during the text recognition from paper document. Error occurred (in percentage) while recognizing the text from the scanned image of the printed paper document is given by the ratio of the number of incorrectly recognized characters to the total number of the characters in the specified region of interest of the document.

To compare the error rate in the OCR used in the system and a most widely used OCR (here, Microsoft OneNote OCR), certificates of ten individuals were generated and were printed on the paper along with QR code. The characters in the specified region of interest are only considered for the analysis purpose. The graphical representation of the percentage of error occurred while recognizing the characters in the specified region of interest from the scanned images of ten different certificates by the use of both types of OCR is shown in Fig. 4. It can be seen from the Fig. 4 that, the OCR used in the proposed system offers more security than the Microsoft OneNote OCR.

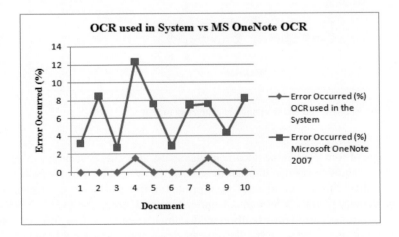

**Fig. 4** OCR used in the system v/s MS OneNote OCR

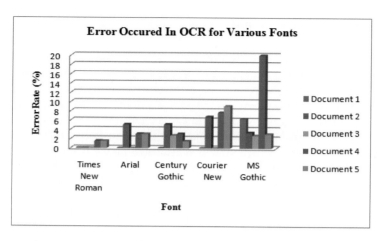

**Fig. 5** Error Occurred in OCR for various fonts

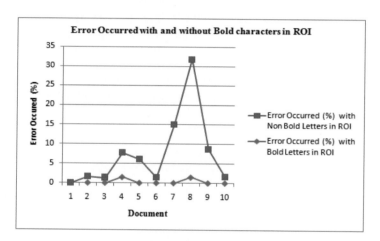

**Fig. 6** Error occurred with and without bold characters in ROI

The type of font used in the document characters also affects accuracy of optical character recognition results. In order to analyze the system for various fonts in the document, five different certificates were printed in five various fonts: Times New Roman, Century Gothic, MS Gothic, Courier New and Arial. The Fig. 5 shows the percentage error occurred for five different documents printed in five different fonts. From the plot, it can be seen that the system shows minimum error in the case of Times New Roman font.

The accuracy of the optical character recognition system also depends on whether the document contains bold characters or not. In order to find out the percentage error occurred in this case, 10 different documents were generated with and without bold characters in ROI and were printed on the paper. The graphical representation of the error occurred while recognizing the text from the scanned image for ten different documents with and without bold characters in the region of interest is depicted in Fig. 6. It can be seen from the Fig. 6 that the optical character recognition in the scanned image of the certificates with bold characters in the specified region of interest gives maximum performance than the documents with non bold characters in the region of interest.

The analysis shows that the type of OCR used in the proposed method provides good performance. The system shows minimum error by the use of Times New Roman font for the characters in the specified region and the system performance is improved by printing the characters in the specified region in bold letters.

# 4 Conclusion

The proposed method ensures the authentication and content integrity for certificates printed in paper in a simple and cost effective manner. Some words or characters that make the certificate unique are chosen and are processed further to create the on-paper digital signature. The verifier must know these words chosen at the sender side and the position of words selected to create OPDS is provided to the verifier by the signing authority during verification process. An authentication mechanism is provided to authorize the verifier and ensures that only the authorized users can verify the certificates. The verification of certificates can be done online by anyone who is authorized to do the same. Optical character recognition is applied to the scanned image of the certificate to select the specified words used at the creation process. The verification procedure compares the message digest obtained from QR code and the message digest obtained by hashing the result of OCR. Due to the fact that OCR is not accurate always, a provision for manual verification is made at the verification process. The method uses simple and less number of steps and provides easier implementation than the existing technologies. The physical condition of the document does not have much effect on the proposed method compared to the image based techniques for paper document authentication. The analysis shows that the proposed system provides minimum error rate and hence good performance. The future work will focus on incorporating the identity of the person whose name is in the certificate in the OPDS creation.

# References

1. Warasart, M., Kuacharoen, P.: Paper-based document authentication using digital signature and QR code. In: 4th International Conference on Computer Engineering and Technology (ICCET) (2012)
2. Naor, M., Yung, M.: Universal one-way hash functions and their cryptographic applications. In: Proc. 21st STOC (1995)

3. Pappu, R.: Physical one-way functions. Phd thesis, Massachusets Instutute of Tecnhology (2001)
4. Merkle, R.C.: A Certified Digital Signature. Springer-Verlag (1998)
5. Rivest, R.L., Shamir, A., Adleman, L.: A Method for Obtaining Digital Signatures and Public-Key Cryptosystems (1997)
6. Ambadiyil, S., Rajeev, V., Vakkalagadda, A., Kummamuru, P., Mahadevan Pillai, V.P.: Design and Development of on Paper Digital Signature (OPDS) using Content Image. Discovery **29**(106), 2–6 (2015)
7. Subramanya, S.R., Yi, B.K.: Digital Signatures. IEEE Potentials (2006)
8. Patel, U.: An Introduction to the process of Optical Character Recognition. International Journal of Science and Research (IJSR) **2**(5) (2013). India Online ISSN: 2319-7064
9. Stallings, W.: Cryptography and Network Security Principles and Practices. Prentice Hall (2005)
10. RSA Cryptography Standard. PKCS #1, v2.1 (2002)
11. Secure Hash Standard. FIPS 180–3 (2008)

# Scale Invariant Detection of Copy-Move Forgery Using Fractal Dimension and Singular Values

Rani Susan Oommen, M. Jayamohan and S. Sruthy

**Abstract** Digital image forgery is a nightmare in the current scenario. The authenticity of images that circulates through media and public outlets is therefore critical with a caution: "Do not believe everything you see". Copy-move forgery is the most frequently created image forgery that conceals a particular feature from the scene by replacing it with another feature of the same image. In this paper, a hybrid approach based on local fractal dimension (LFD) and singular value decomposition (SVD) to efficiently detect and localize the copy-move forged region is proposed. In order to reduce the computational complexity of the classification procedure, we propose to arrange image blocks in a B+ tree structure ordered based on the LFD values. Pair of blocks within each segment is compared using singular values, to find regions that exhibit maximum resemblance. This reduces the need for comparison to the most suspected image portions alone. The experimental results show how effectively the method identifies the duplicated region; also presents the robustness of the method to detect and localize forgery even in the presence of after-copying manipulations such as rotation, blurring and noise addition. Furthermore it also detect multiple copy-move forgery within the image.

## 1 Introduction

Images are considered as authenticated proofs or corroboratory evidence in areas like forensic studies, law enforcement, surveillance systems, journalism etc. But with the

R.S. Oommen(✉) · S. Sruthy
Sree Buddha College of Engineering for Women, Pathanamthitta, India
e-mail: ranisusan1991@gmail.com

M. Jayamohan · S. Sruthy
College of Applied Science, Adoor, India
e-mail: jmohanm@gmail.com

© Springer International Publishing Switzerland 2016
S.M. Thampi et al. (eds.), *Advances in Signal Processing and Intelligent Recognition Systems*,
Advances in Intelligent Systems and Computing 425,
DOI: 10.1007/978-3-319-28658-7_47

ease of using image editing tools even the non-professionals can alter the image contents making the problem of digital image forgery potentially a serious threat in the present scenario. Thus any individual can synthesize a fake picture and with the widely accessible Internet, the false information disseminates at a fast rate. As a consequence, facts may be distorted; public opinion may be affected, yielding a negative social impact. Therefore there is a strong demand for robust authentication methods that can discern whether an image is forged or not.

Image tampering is not a new phenomenon and can be traced back as far as the invention of photography. Many incidents of historical photographic forgeries have been reported. A widely circulated fake image was reported in 2004 where a Democratic candidate sharing a demonstration podium with a famous Hollywood actress, during American president election [5]. Another fake ad image is reported in July 2015, where Obama meeting and shaking hands with the Iranian President Hassan Rouhani, but the true fact is that Obama has never met Rouhani [20].

Three major kinds of forgery exist: image splicing, coy-move and retouching. This paper deals with the detection of copy-move image forgery. The main intention of copy-move forgery is removal of object from the scene, which is replaced by pixels from other parts of the same image. Active approaches such as digital watermarking, signature etc. can be used to detect forgery by adding some prior information at the time of image acquisition [6]. But this has limited applications. Therefore passive approaches are widely used to detect the traces of forgery as it doesn't require any prior information.

In the proposed work, a hybrid mechanism based on local fractal dimension (LFD) and singular value decomposition (SVD) feature is used to detect the particular kind of forgery. The image is classified based on the LFD values. The B+ tree organization of segments provides an effective way of processing the block information in a sorted manner with less time complexity. Then the search for similar blocks is done within each node. Singular values obtained by applying SVD to image blocks are used for block matching. Then the blocks that exhibit maximum resemblance in the singular values can be suspected as the duplicated regions. This forgery detection mechanism also considers other potential types of post-processing operations such as noise addition, blurring, rotation etc.

This paper is organized as follows: Section 2 describes few previous works reported in the copy-move forgery detection domain. Section 3 gives a brief description of the theoretical aspect of the proposed algorithm. In Section 4, the proposed detection scheme is detailed. The results and discussion is shown in Section 5. The conclusion is drawn in Section 6.

## 2  Related Work

Many prior works to copy-move forgery detection have been reported. The earliest reported copy-move detection technique was based on exhaustive search method, where the image and its circularly shifted version are overlaid looking for closely matching image segments [7]. But it was computationally expensive. Another method was based on auto-correlation, but it detects traces of forgery only if the forged region is large. Then various approaches based on block-tiling methods, key-point based methods and hybrid methods are used for the detection of forgery. J. Fridich, David Soukal and Jan Lukas suggested [7] a first block-based attempt to detect copy move forgery based on DCT coefficients. But it will not work in noisy image. Popescu and H. Farid [17] a dimensionality reduction based copy-move forgery detection by applying PCA to blocks in the image, which is robust to additive noise or lossy compression. L. Li, S. Li, H.Zhu [12] presented a method for detecting copy move forgery based on local binary patterns (LBP). The main contribution of this method is that is robust not only to the traditional signal processing operations, but also to the rotation and flipping. Key-point based approaches such as SURF, SIFT also contributes to the detection. X. Pan and S. Lyo [16] proposed a detection method based on matching image SIFT (Scale-Invariant Feature Transform) features. This method is robust to transforms such as rotation, scaling and less susceptible to noise and JPEG compression. SURF (Speeded-Up Robust Feature Extraction) based method of Copy-move Forgery in flat regions was suggested by G. Zhang, H. Wang [21]. The method can detect region duplication in non-flat region and is rotation invariant. A SIFT and Zernike moments based region duplication detection was suggested by Z. Mohamadian, A. Pouyan [14]. This hybrid approach has higher precision and is robust to Gaussian noise, JPEG compression and blurring.

# 3   Basic Theory

## 3.1   Local Fractal Dimension

Natural scenes are realistically described using fractal geometry methods. Fractal Dimension shows how densely a fractal occupies the metric space in which it lies. Thus it estimates the roughness or fragmentation of an object. Barnsley presents fractal dimension as a quantity that can measure the similarity of two fractals [1]. Different methods for estimating fractal dimension have been proposed by researchers which include box counting method, differential box counting method, blanket method, reticular cell counting method etc. Out of which the most basic and straight forward method is the box counting method. According to box-counting theorem, an image A is assumed to be divided into closed box of size $1/2^n$. Let the number of boxes which intersect A be N(A). The fractal dimension D of A can be obtained as:

$$D = \lim n \to \infty \log(N(A))/\log(2^n) \qquad (1)$$

The FD is equal to its box-counting dimension, if the object is deterministically self-similar. However, natural scenes are not the ideal deterministic fractals. The basic box counting method, since it requires the image in black and white, is not well-suited for estimating FD of gray scale images. With Differential Box Counting method, Sarkar and Choudhari [18] proposed a significant improvement over box counting which makes it better suited for estimating complexity of two dimensional gray scale images.

In Differential Box counting (DBC), an image is considered as a three-dimensional spatial surface with (x, y), denotes the pixel position on the image plane, and the third coordinate (z) denoting pixel gray level. Consider an image of size M*M. The (x, y) plane is partitioned into non-overlapping blocks of size, say, s*s. The scale of each block is r = s, where M/2s > 1 and s is an integer. On each block there is a column of boxes of size $s*s*s'$ where $s'$ is the height of each box, $G/s' = M/s$ and G is the total number of gray levels. Let the minimum and maximum gray level of the $(i,j)^{th}$ block fall in the $k^{th}$ and $l^{th}$ boxes, respectively. The boxes covering this block are counted in the number as $n_r(i, j) = l-k+1$ , where the subscript r denotes the result using the scale r. Considering contributions from all blocks, then the number of boxes $N_r$ is counted for different values of r as:

$$N_r(i, j) = \sum_{i,j} n_r(i, j) \qquad (2)$$

Then the FD can be estimated as the least squares linear fit of log ($N_r$) versus log(1/r).

Li [13] proposed an improved box-counting method which outperforms differential box counting method especially in Brownian motion images, remote sensing images and real textured images. Conci and Compass [3] also suggested an improvement over DBC method by improvising the encoding time.

In our proposed method, we use the differential-box counting method proposed by Sarkar and Choudhari [18] to calculate fractal dimension. The value of FD will be stable over a pure fractal image. But natural images are not having pure fractal nature. Hence if we estimate fractal dimensions of local areas of the image, it will be different from the FD of the entire image, and the results may vary depending on the local texture variations [11] [15]. Since the copied and pasted regions have similar texture, we initially classify the image into different texture segments based of the LFD values in the image. This confines the need of searching for similar blocks within each segment only, which reduces the overall computational complexity [18][10].

## 3.2  B+ Tree

An m-way search tree is an extended form of binary search tree where each node can have m children. A B-tree is a balanced multiway search tree in which each non-root node contains at least (m-1)/2 keys [9] [4]. In B-tree key values can be stored at any level. B+ tree is a variation of B-tree in which key information are stored at the leaf nodes. The maximum number of nodes that can be stored on a B+ tree of height h is [10]:

$$N_{max} = m^h - m^{h-1} \qquad (3)$$

The time taken for inserting and retrieving an element in a B+ tree is only O(logn). Also the keys are organized in a sorted manner in a B+tree. These two properties of B+ tree make it competent as an efficient data structure in our proposed work. The key value used is the local fractal dimension and index information of each block. The blocks are organized in a B+ tree structure ordered based on the LFD value of each block such that each segment falls in one node of the tree. Thus classification and organizing the classified group in B+ tree significantly reduce the number of block comparison in the image block similarity matching.

## 3.3 Singular Value Decomposition

Singular value decomposition is a significant topic in linear algebra that was established for rectangular real or complex matrices in the 1870s by Beltrami and Jordan. SVD is a robust and reliable matrix factorization method that provides a way of extracting algebraic and geometric invariant features from an image. SVD based image-processing techniques includes wide applications such as image de-noising, image compression, pattern analysis, image forensics, data reduction etc.

The basic mathematical theory of SVD is found in [2] [19]. Let A be an M$*$ N matrix, $A \in R^M$ of rank r, then SVD is based on the theorem from linear algebra which states that any rectangular matrix A can be decomposed into product of three matrices:

- Orthogonal matrix $U \in R^{M*M}$
- Diagonal matrix $\sum \in R^{M*N}$
- Transpose of orthogonal matrix $V^T \in R^{N*N}$, such that

$$A = U \cdot \sum \cdot V^T \tag{4}$$

where, $U \cdot U^T = 1$ and $V^T \cdot V = 1$

The matrix U is an M$*$M matrix. The columns of U are the orthonormal eigenvectors of $A \cdot A^T$. The matrix V is an N$*$N matrix. The columns of V are the orthonormal eigenvectors of $A^T \cdot A$. $\sum$ is an M$*$N diagonal matrix with singular values formed by taking square root of eigenvalues from U or V in descending order.

$$\sum = diag(\sigma_1, \sigma_2 \ldots \ldots \sigma_r) \tag{5}$$

with $\sigma_1 \geq \sigma_2 \geq \ldots \geq \sigma_r$, where $\sigma_i$- denotes singular values.

The idea of using singular values is that it allows us to represent the image in best approximation with a smaller set of values which can preserve the useful features of the original image and use less storage space of memory. This forms the basis of using SVD as a dimensionality reduction technique. Also the orthogonal matrix U and V are not unique to the matrix, but the singular values are unique for image blocks. SVD, therefore best approximate the matrix, decomposing the data into optimal estimate of signal denoted by larger singular values and noise components denoted by smaller singular values. This property of SVD takes advantage in our algorithm. Moreover, singular values represent maximum energy packing in an image that exhibits good stability even when the image suffers from minor distortion.

Singular value feature vector has following algebraic and geometric invariant properties: stability, scaling property and rotation invariance. It is also insensitive to noise. Using singular value decomposition the proposed method not only achieves dimensionality reduction of block vectors but also provides better feature that is stable, resilient to after-copying operations such as scaling, rotation and noise addition.

## 4 Proposed Scheme

Given a forged image of size M*M, initially the image is partitioned into overlapping blocks of size B*B with slide of slide-step pixels. The algorithm proceeds with the assumption that the block size is lesser than the size of the duplicated region The framework of the proposed work is shown in Fig. 1. The copy-move forged region can be detected by the following steps:

1. Pre-processing- If the input image is a color image (RGB), it is converted to gray-scale image based on the RGB to gray-scale conversion technique. The gray scale image is then tiled into overlapping blocks of fixed size.
2. Phase-1 Feature Extraction- This step extract local fractal dimension (LFD) of each block as the feature vector for the classification. LFD is estimated using the differential box-counting (DBC) method.
3. Classification- The image is classified into different texture regions based on the level-1 feature. LFD has proved to be an efficient estimate of texture.
4. B+ Tree organization- The blocks obtained in phase1 is organized in a B+-tree structure, which organizes blocks in a sorted manner. The tree is structured in such a way that each node consists of blocks in a given LFD range.
5. Phase-2 Feature Extraction- In this step, SVD is applied to each block in each node and the singular values is used as the feature vector for block comparison and matching.
6. Image Block Similarity Matching- Based on the singular value feature vectors obtained in step 5, the Euclidean distance measure is computed between each block in each node. The minimum d measure corresponds to maximum match. Let u and v corresponds to n-dimensional singular value feature vector of blocks $b_i$ and $b_j$ respectively,

$$\mathbf{u} = (u_1, u_2.....u_n)^T$$
$$\mathbf{v} = (v_1, v_2.....v_n)^T$$
$$Thend = ((u-v)^T(u-v))^{(1/2)} = (\sum\nolimits_{(i=1)}^{m}(u_i - v_i)^2) \tag{6}$$

**Fig. 1** Framework of the
proposed algorithm

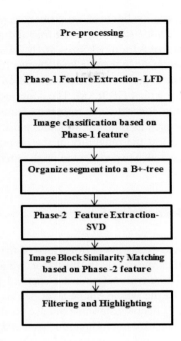

> Pre-processing
>
> Phase-1 Feature Extraction- LFD
>
> Image classification based on Phase-1 feature
>
> Organize segment into a B+-tree
>
> Phase-2 Feature Extraction- SVD
>
> Image Block Similarity Matching based on Phase -2 feature
>
> Filtering and Highlighting

If d (u, v) is greater than a threshold parameter thres, such block pairs are discarded as they non-matching. The remaining block pairs can be candidates of suspected blocks.

7. Filtering and highlighting tampered region- From suspected block pair list, consider the block pair with minimum d measure, the region is expanded and comparison is continued to its neighbourhood. Finally the result of the tamper detection is visualized by highlighting the duplicated region to a unique gray scale value.

## 5    Results and Discussion

### 5.1    Experimental Setup

We conducted the experiment on natural images of different themes. Copy move forgery was performed on the images using software like Adobe Photoshop. Considering the fact that natural images may exhibit similar regions even without forgery, the proposed algorithm is based on the assumption that duplicated region less than 1 percent(less than 16*16) of the whole image is not suspected as forged. It also assumes that the copied and the pasted segments are non-overlapping. Therefore copy-move forgery is done on the image by copying a random square of varying size and pasting to another location in the same image.

The program has been experimented with different block sizes- 16∗16, 24∗24 and 32∗32. While estimating local fractal dimension, overlapping blocks were taken at a distance of 8 for block sizes of 16, 24 and 32. This does not reduce the success of the program since there is least chance for the forged regions to be so small in size.

All the experiments are performed on a machine equipped with core i3 2.4 GHz processor with 4 GB RAM. The copy-move forgery is performed on the test images in order to conceal some important feature in the image. Various after copying operations such as blurring, rotation, noise addition etc. were also applied to the test image to test the reliability of the proposed work. Multiple copying and multiple pasting are also experimented on the image.

## 5.2   Measuring Forgery

In order to assess the performance of the copy-move detection method, we require appropriate measures. In this paper, we adopt two measures, namely precision and recall. These two measures are checked for varying block sizes 16∗16, 24∗24 and 32∗32. At the whole image level, let $T_p$ denotes the value of true positive rate of images, $F_p$ denotes the value of false positive rate of images and $F_n$ denotes the value of false negative rate. Then precision and recall can be estimated as follows:

Precision (p) is a measure for the probability that a detected forgery is truly a forgery. It denotes the exactness of the method.

$$p = T_p/(T_p + F_p) \tag{7}$$

Thus, $p \propto (1/F_p)$ i.e. High precision indicates low false positive rate.

Recall (r) is a measure for the probability that the forged image is detected. It denotes the completeness of the method.

$$p = T_p/(T_p + F_n) \tag{8}$$

Thus, $r \propto (1/F_n)$ i.e. High precision indicates low false negative rate.

## 5.3   Test for Detection Rate and Visual Results

### 5.3.1   Normal Copy-Move Forgery

Fig. 2 shows the example of applying normal copy-move forgery. Fig. 2a denotes the original image, 2b denotes the forged image and 2c denotes the result of detection. Table 1 shows the first basic test of the measures involving original and forged image without any additional modifications.

**Table 1** Performance measure on normal copy-move forgery

| Measure | 16∗16 | 24∗24 | 32∗32 |
| --- | --- | --- | --- |
| p | 0.9975 | 0.9934 | 0.9912 |
| r | 0.9811 | 0.9279 | 0.9222 |

1.2 a     1.2 b     1.2 c

**Fig. 2** Detection results with normal Copy-Move operation

### 5.3.2 With Modification Applied

Fig. 3 shows the visual detection result when rotation is applied before pasting is applied. Fig. 3a denotes the original image, 3b denotes the forged image and 3c denotes the result of detection. Table 2 shows the test of the measures involving original and forged image when rotation is applied before pasting the copied segment.

a     b     c

**Fig. 3** Detection results with rotation operation

**Table 2** Performance measure when rotation applied

| Measure | 16∗16 | 24∗24 | 32∗32 |
| --- | --- | --- | --- |
| p | 0.9834 | 0.9728 | 0.9557 |
| r | 0.9411 | 0.9220 | 0.9014 |

### 5.3.3 With Multiple Copy-Move Forgery

Fig. 4 shows the visual detection result when multiple copy-move operation is performed on the image. Fig. 4a denotes the original image, 4b denotes the forged image and 4c the result of detection. Table 3 shows the precision and recall estimates when multiple copy-move operation is applied.

a                              b                              c

**Fig. 4** Detection results with multiple copy-move operation

**Table 3** Performance measure when multiple copy-move applied

| Measure | 16* 16 | 24* 24 | 32* 32 |
|---------|--------|--------|--------|
| p | 0.9957 | 0.9791 | 0.9553 |
| r | 0.9505 | 0.9562 | 0.8835 |

Table 4 depicts the comparative figures of precision and recall estimates of the proposed method with the SVD-DWT method proposed by Liu& Feng [8]. The measures have been computed for normal copy-move forgery.

**Table 4** Comparison of the method under normal copy-move operation

| Measure | DWT-SVD | LFD-SVD |
|---------|---------|---------|
| p | 0.9955 | 0.9975 |
| r | 0.9655 | 0.9811 |

## 6 Conclusion

Detecting forged portions within a natural scene with texture complexity is not an easy task. We have attempted to implement a new hybrid method using the strength of fractal dimension along with singular value decomposition. Experimental results show that the method is effective in images even after post-copying manipulations.

The only challenge with the method is the high computation time required for estimating fractal dimension, which we have successfully reduced to a great extend minimizing the comparison steps by making use of B+ tree arrangement of image blocks sorted in the order of local fractal dimension.

Results show that the experiments with minimum block size gives better accuracy in identifying forgery, but with the cost of high computation time.

The main contribution of this work is that it carries out copy-move detection in image bocks with same texture instead of the whole image, which greatly narrows comparison of the similar blocks, thereby reducing the computational complexity. Also, using singular values for block comparison exploits its properties such as rotation invariance, stability etc. Experimental results show the effectiveness of the proposed work, also its robustness against after-copying operations, and detection of multiple copy-move forgery.

# References

1. Barnsley, M.: Fractals Everywhere, 2nd edn. Academic Press, Cambridge (1993)
2. Cao, L.: Singular value decomposition applied to digital image processing. Division of Computing Studies, Arizona State University Polytechnic Campus, Mesa, Arizona State University polytechnic Campus (2006)
3. Conci, A., Campos, C.F.J.: An efficient box-counting fractal dimension approach for experimental image variation characterization. In: Proceedings of IWSIP (1996)
4. Samanta, D.: Classic Data Structures. Prentice Hall of India, New Delhi (2000)
5. Fonda, K., Fakery, P.: The Washington PostWriting. http://www.washingtonpost.com/wp-dyn/articles (cited February 28, 2004-1)
6. Fridrich, J.: Methods for tamper detection in digital images. In: Multimedia and Security, Workshop at ACM Multimedia, vol. 99 (1999)
7. Fridrich, A.J., Soukal, B.D., Lukas, A.J.: Detection of copy-move forgery in digital images. In: Proceedings of Digital Forensic Research Workshop (2003)
8. Liu, F., Feng, H.: An efficient algorithm for image copy-move forgery detection based on DWT and SVD. International Journal of Security and Its Applications **8**(5), 377–390 (2014)
9. Horowitz, E., Sahni, S.: Fundamentals of data structures. Pitman (1983)
10. Jayamohan, M., Revathy, K.: Domain classification using B+trees in fractal image compression. In: 2012 National Conference on Computing and Communication Systems (NCCCS). IEEE (1989)
11. Keller, J.M., Chen, S., Crownover, R.M.: Texture description and segmentation through fractal geometry. Computer Vision, Graphics, and Image Processing **45**(2), 150–166 (1989)
12. Li, L., et al.: An efficient scheme for detecting copy-move forged images by local binary patterns. Journal of Information Hiding and Multimedia Signal Processing **4**(1), 46–56 (2013)
13. Li, J., Du, Q., Sun, C.: An improved box-counting method for image fractal dimension estimation. Pattern Recognition **42**(11), 2460–2469 (2009)
14. Mohamadian, Z., Pouyan, A.A.: Detection of duplication forgery in digital images in uniform and non-uniform regions. In: 2013 UKSim 15th International Conference on Computer Modelling and Simulation (UKSim). IEEE (2013)
15. Novianto, S., Suzuki, Y., Maeda, J.: Near optimum estimation of local fractal dimension for image segmentation. Pattern Recognition Letters **24**(1), 365–374 (2003)
16. Pan, X., Lyu, S.: Detecting image region duplication using SIFT features. In: 2010 IEEE International Conference on Acoustics Speech and Signal Processing (ICASSP). IEEE (2010)
17. Popescu, A.C., Farid, H.: Exposing digital forgeries by detecting duplicated region. Technical Report, TR2004-515, Dartmouth College, Computer Science, August 2004

18. Sarkar, N., Chaudhuri, B.B.: An efficient differential box-counting approach to compute fractal dimension of image. IEEE Transactions on Systems, Man and Cybernetics **24**(1), 115–120 (1994)
19. Sadek, R.A.: SVD Based image processing applications: State of the art, contributions and research challenges. arXiv preprint arXiv: 1211.7102 (2012)
20. Kertscher, T.: Barack Obama met with Iran president, says PAC backing Wisconsin Sen. Ron Johnson for re-election (2015). http://www.politifact.com/wisconsin/statements (cited July 24, 2015)
21. Zhang, G., Wang, H.: SURF-based Detection of Copy-Move Forgery in Flat Region. International Journal of Advancements in Computing Technology **4**(17), 8 (2012)

# Adaptive Nonlocal Filtering for Brain MRI Restoration

V.B. SuryaPrasath and P. Kalavathi

**Abstract** Brain magnetic resonance images (MRI) plays a crucial role in neuro-science and medical diagnosis. Denoising brain MRI images is an important pre-processing step required in many of the automatic computed aided-diagnosis systems in neuroscience. Recently, nonlocal means (NLM) and variants of these filters, which are widely used in Gaussian noise removal from digital image processing, have been adapted to handle Rician noise which occur in MRI. One of the crucial ingredient for the successful image filtering with NLM is the patch similarity. In this work we consider the use of fuzzy Gaussian mixture model (FGMM) for determining the patch similarity in NLM instead of the usual Euclidean distance. Experimental results with different noise levels on synthetic and brain MRI images are given to highlight the advantage of the proposed approach. Comparison with other image filtering methods our scheme obtains better results in terms of peak signal to noise ratio and structure preservation.

## 1 Introduction

Brain magnetic resonance image (MRI) has been widely used in neuroscience and medical diagnosis. Automatic processing and analysis of the MRI images require various image processing techniques. One of the basic image analysis tasks is noise

V.B. Surya Prasath(✉)
Computational Imaging and Visualization Analysis(CiVA) Lab,
Department of Computer Science, University of Missouri, Columbia, MO 65211, USA
e-mail: prasaths@missouri.edu
https://goo.gl/ZaLRb8

P. Kalavathi
Department of Computer Science and Applications, Gandhigram Rural Institute – Deemed
University, Gandhigram, Dindigul 624302, Tamil Nadu, India
e-mail: pkalavathi.gri@gmail.com

© Springer International Publishing Switzerland 2016                                                       571
S.M. Thampi et al. (eds.), *Advances in Signal Processing and Intelligent Recognition Systems*,
Advances in Intelligent Systems and Computing 425,
DOI: 10.1007/978-3-319-28658-7_48

removal from obtained MR images which is the basis for further image segmentation and analysis [1, 2, 3]. Better image filtering without introducing artifacts is a must for neuro-informatics data processing workflow and devising image specific filters are an active research area. Image processing literature is rich in image filtering with global smoothing techniques such as anisotropic diffusion [4, 5, 6, 7], total variation [8, 9, 10], and a wide variety of neighborhood based approaches [11].

Recently, nonlocal methods are gaining popularity due to their better performance and adaptability to various noise models. One of the important nonlocal filter is the nonlocal means (NLM) proposed by Buades et al [12]. The main premise of NLM filter is to use instead of point (pixel) similarities utilize patch (neighborhood around the pixel to be denoised) since natural image contain self similarity of different features. NLM filter has been adapted to various other image domains and noise models [13]. There exists some works in denoising brain MRI images which are corrupted by Rician noise [14, 15]. These depend on modifying the weights computation using localized mean, variance, and improved blockwise implementation. In particular, classical Euclidean distance was used to simplify the local window search and parameters adjusting weights are manually tuned. In this work, we show that the fuzzy GMM can be used effectively for determining the parameters of NLM for image restoration. This, in turn, helps us obtain improved denoised results in brain MRI images as indicated by experimental results. Compared with previous NLM filter for brain MRI [15], as well as classical anisotropic diffusion [6], total variation [9] devising filters indicate that we obtain better results in terms of higher peak signal to noise ratio and structural similarity with ground truth.

The rest of the paper is organized as follows. Section 2 introduces the fuzzy GMM driven nonlocal means filter adapted for brain MRI restoration. Section 3 sets up the experimental results on brain MR images along with comparison of related denoising filters. Finally, Section 4 concludes the paper.

## 2   Fuzzy GMM Driven Nonlocal Means

### 2.1   Nonlocal Means (NLM) Filter

Nonlocal means (NLM) is one of the well-known and state of the art denoising filter in image processing. It is based on patch based similarity and utilizes Euclidean distances to compare patches for denoising a pixel at the center. We briefly recall the fundamental principles of NLM filter and refer to [12] for more details. Let $u_0 : \Omega \rightarrow [0, 255]$ be a given noisy (gray scale) image, and $\Omega \subset \mathbb{R}^2$ the image domain of size $m \times n$. The NLM filter is given by the following,

$$u(i, j) = \frac{\sum_{k,l} W_{i,j}(k, l)u_0(k, l)}{\sum_{k,l} W_{k,l}}, \quad \forall (i, j) \in \Omega. \tag{1}$$

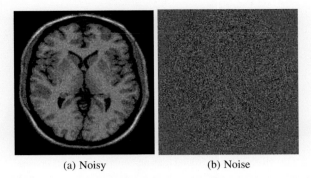

(a) Noisy                                    (b) Noise

**Fig. 1** Synthetic data used for validation with Rician noise. Example of the Brainweb Database. (a) T1-w image corrupted with a Rician noise at 9%, and (b) the amount of noise added. Noise image is amplified for visualization purpose.

Here the weights are computed using the Euclidean distance,

$$W_{i,j}(k,l) = \exp \frac{-\left\| u_0(P_{i,j} - P_{k,l}) \right\|_{\sigma}^2}{\rho^2} \tag{2}$$

Note that the $\|\cdot\|_{\sigma}^2$ is Gaussian weighted Euclidean distance where $\sigma$ is the standard deviation. The $P_{i,j}$ is a patch of size $r \times r$, centered at pixel $(i, j) \in \Omega$. Thus, we see that denoising happens using the similarity of patches on the patch space $\mathcal{P}$ and the the metric on similarity of patches involved is crucial. We further note that NLM filter (1) works well in removing Gaussian noise but for other noise models it needs to be adapted appropriately. Adaptation of NLM filter for MRI images was undertaken by Coupé et al [15] who made modifications to the original NLM filter by changing the Gaussian weighted Euclidean distance with classical Euclidean distance and used a neighborhood independent formulation. The weights they considered are given by,

$$W_{i,j}(k,l) = \exp \frac{-\left\| u_0(P_{i,j} - P_{k,l}) \right\|^2}{2\beta\rho^2 \left| P_{i,j} \right|} \tag{3}$$

where $\left| P_{i,j} \right|$ is the size of the neighborhood, and $\beta > 0$ is an adjusting parameter which needs to be manually tuned.

## 2.2  Fuzzy GMM for Patch Similarity

One of the main ingredient in NLM filter is local patch similarity and various adaptions can be considered for comparing disparate patches. We utilize here the fuzzy Gaussian mixture model (FGMM) [16] for determining similarity of local patches with NLM filter for Rician noise removal. Note that the traditional NLM [12] or

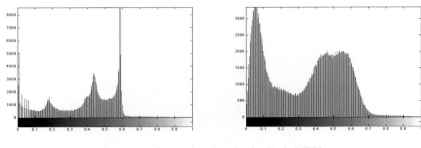

(a) Histograms of ground truth and noisy Brain MRI images

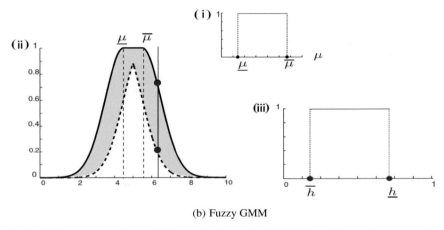

(b) Fuzzy GMM

**Fig. 2** (a) Histograms of brain MRI given in Figure 1(a-b). (b) Type-2 fuzzy Gaussian mixture model involves Gaussian primary membership functions (i). (ii) Mean boundaries (iii) Primary membership function as an interval. Image adapted from [16].

the adapted NLM [15] are based on Euclidean distances which are used to compare patches, see Eqns. (2-3). We use type-2 FGMM to model the parameter estimation via expectation-maximization (EM) [17]. The final similarity of patches is computed using earth movers distance of the computed FGMM parameters. Traditional GMM is prone to noise and in contrast FGMM obtains better results under a footprint of uncertainty and extreme outliers [16]. Here we briefly recall the FGMM and refer to [16] for detailed analysis of the effect of fuzzification and comparison with traditional GMM. FGMM uses $M$ mixture components of multivariate Gaussians and for a $d$ dimensional observation vector $\mathbf{x} = (x_1, \ldots, x_d)$ is given as[1],

$$p(\mathbf{x}|\Theta) = \sum_{i=1}^{M} \alpha_i \, \mathcal{N}(\mathbf{x}|\mu_i, \Sigma_i) \qquad (4)$$

---

[1] Without loss of generality, we assume diagonal covariance matrix $\Sigma$. For a non-diagonal symmetric matrix $\Sigma$ there exists orthogonal matrix $\mathcal{O}$ and diagonal matrix $\Lambda$ such that $\Sigma^{-1} = \mathcal{O}\Lambda\mathcal{O}$.

where $\Theta = (\alpha_1, \ldots, \alpha_M, \mu_1, \ldots, \mu_M, \Sigma_1, \ldots, \Sigma_M)$, $\sum_{i=1}^{M} \alpha_i = 1$, and

$$\mathcal{N}(\mathbf{x}|\mu_i, \Sigma_i) = \frac{1}{\sqrt{(2\pi)^d |\Sigma_i|}} \exp\left[-\frac{1}{2}(\mathbf{x} - \mu_i)' \Sigma_i^{-1}(\mathbf{x} - \mu_i)\right] \qquad (5)$$

with mean $\mu_i \in [\underline{\mu}_i, \overline{\mu}_i]$, $\forall i$. Each of the function (5) is the Gaussian primary membership function $h$ with uncertain mean. Figure 2(a) shows the histogram of brain MRI image given in Figure 1(b) which is used as the observation vector in FGMM. The EM algorithm is started with initial number of Gaussian components of 20 and decreased if $\alpha_m$ is less than 0.01, we refer to [16] for more regarding FGMM and its implementation details. Figure 2(b) provides an overview of type-2 FGMM along with its primary membership functions as intervals.

## 3 Experimental Results

### 3.1 Setup

The experiments were conducted in MATLAB and we used the NETLAB toolbox for FGMM part [18]. We used the BrainWeb simulated brain database[2] for the experiment with synthetic image (see Figure 1(a)). We followed the implementation provided by [15] along with the default parameters provided for adapted NLM filter (only the EM parameters were tuned in our proposed method), for other compared schemes their parameters were set according to peak PSNR values. Experiments are conducted with MATLAB on a Mac Laptop with 2.3GHz Intel Core i7, 8GB RAM. It took on average less than 2.5 minutes (without multithreading) for an MRI data of size $181 \times 217 \times 181$ voxels.

### 3.2 Real Data

First in Figures 3-4 we show two example restorations on real T1 weighted brain MRI images filtered by our proposed scheme. Both the input images contain unknown amount of Rician noise with first Figure 3(a) containing less density of noise than the second one in Figure 4(a). The close-up of restoration results of both these noisy brain MRI images, Figure 3(b)-4(b), indicate good preservation of the cerebellum contours, and corpus callosum structure respectively. We see that our restoration results provide structure preservation without noise and we believe this will further help automatic segmentation and quantitative analysis of such structures.

---

[2] http://www.bic.mni.mcgill.ca/brainweb/

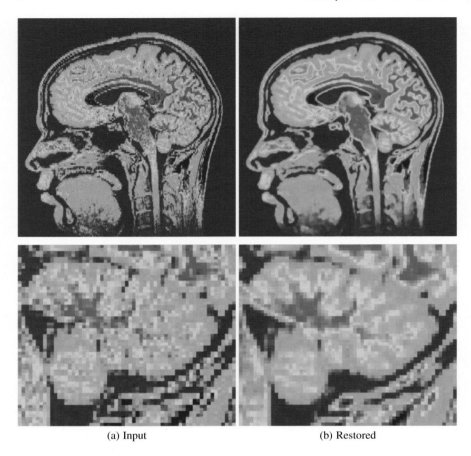

(a) Input                                                (b) Restored

**Fig. 3** Real T1 weighted MRI image restored with proposed FGMM-NLM filter show the advantage of our denoising on specific brain tissue structure (cerebellum). (a) Input image. (b) Proposed filter result. Top row: Full images. Bottom row: Close-up of cerebellum. Given in scaled color (jet map) visualization for differentiating tissue structures clearly.

## 3.3 Comparisons

We compare the FGMM-NLM with BNLM (optimized version) of [15], anisotropic diffusion [6], and total variation [9]. For this purpose, we consider the synthetically generated normal brain MRI T1-w image from the BrainWeb database, see Figure 1. To test different filters we use a Rician noise corrupted image (Figure 1(b)) at 9%. This means the Gaussian noise in the complex domain is used with $\mathcal{N}(0, \nu \frac{9}{100})$, where $\nu$ is the values of the brightest tissue in the image (150 for T1 weighted image used here).

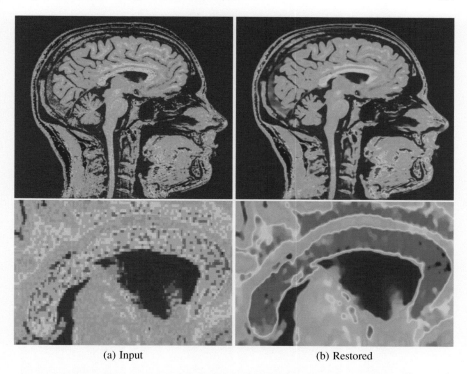

(a) Input                                          (b) Restored

**Fig. 4** Real T1 weighted MRI image restored with proposed FGMM-NLM filter show the advantage of our denoising on specific brain tissue structure (corpus callosum). (a) Input image. (b) Proposed filter result. Top row: Full images. Bottom row: Close-up of corpus callosum. Given in scaled color (jet map) visualization for differentiating tissue structures clearly.

We visually compare the effect of different filters on a Rician corrupted (9% noise level, see Figure 1(b) for a visualization of the amount of noise). Figure 5 provides the close-up of denoised images, their corresponding residue, and method noise as images. Figure 5(a) shows a close-up region of the restored results. As can be seen by comparing Figure 5(b-c) overall our method preserves salient structures and removes noise effectively. In particular the putamen, globus pallidus regions (top left and right), and ventricles are kept intact in our method (Figure 5(d)) in contrast to staircasing observed in anisotropic diffusion (Figure 5(a-b)), or smoothed appearance in block optimized NLM (Figure 5(c)). The amount of noise removed given in Figure 5(e) indicate that our method do not remove salient structures as in block optimized NLM method of [15] which are mistakenly removed as noise in anisotropic diffusion and total variation filters in particular. Figure 5 shows the difference between ground truth image and the filtered images from various schemes. Our scheme shows no structural differences whereas the other methods show distinct tissue characteristics which indicates the loss of brain tissues when the filters where applied.

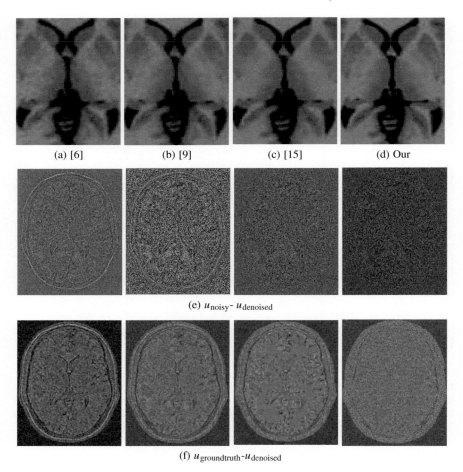

(a) [6]          (b) [9]          (c) [15]          (d) Our

(e) $u_{\text{noisy}}$- $u_{\text{denoised}}$

(f) $u_{\text{groundtruth}}$-$u_{\text{denoised}}$

**Fig. 5** Comparison with Anisotropic Diffusion [6], Total Variation [9], and Block Optimized NL-means [15] and our denoising filters on BrainWeb T1-weighted image at 9% of Rician noise given in Figure 1(b). Left to right: (a-d)Restoration results with [6], [9], [15], and our method, close-up of the denoised results. (e) Residue, the amount of noise removed. (f) Difference between ground-truth and denoised images. Better viewed online and zoomed in.

We further added different Rician noise levels (to BrainWeb T1weighted images) to compare final restoration accuracy, and check the robustness of different methods [6, 9, 15] with our FGMM-NLM. To quantitatively compare the different denoising filters we utilize the peak signal to noise ratio (PSNR, dB) error metric which is given by the following formula,

$$\text{PSNR}(u) = 20 * \log_{10}\left(\frac{u_{max}}{\sqrt{\text{MSE}}}\right) dB \qquad (6)$$

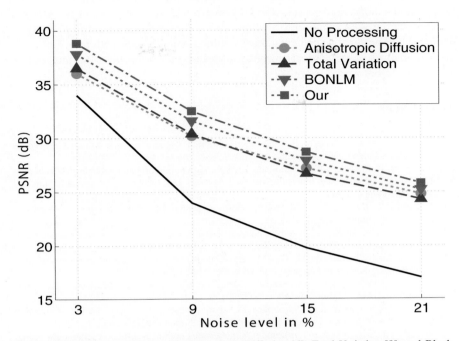

**Fig. 6** PSNR (dB) comparison with Anisotropic Diffusion [6], Total Variation [9], and Block Optimized NL-means [15] (denoted as BONLM), and our denoising filters on BrainWeb with different Rician noise levels.

where $\text{MSE} = (mn)^{-1} \sum \sum (u - u_O)^2$, with $u_O$ is the original (noise free) image, $m \times n$ denotes the image size, $u_{max}$ denotes the maximum value, for example in 8-bit images $u_{max} = 255$. A difference of $0.5\,dB$ can be identified visually. Higher PSNR value indicates optimum denoising capability. Figure 6 provides a comparison of different methods for the BrainWeb brain MRI image for a range of synthetically added noise levels (compare with Figure 15 in [15]). Our FGMM-NLM performs substantially better than [6, 9] and obtained improved PSNR results over [15]. Moreover, the performance is consistent across different noise levels as indicated by the increased PSNR values and visual comparisons support our assertion that FGMM-NLM can preserve important brain structures and remove noise robustly than other methods.

## 4 Conclusion

We considered a nonlocal means filter driven by fuzzy Guassian mixture model parameter estimation. By using type-2 fuzzy GMM our method removes noise without compromising salient brain tissue structures. Comparison with anisotropic diffusion, total variation and block optimized nonlocal means filters our approach improves the peak signal to noise ratio across different Rician noise levels. We believe our filtering method can be used as a preprocessing step for further brain MRI image analysis tasks such as automatic segmentation. Extending our approach with principal component analysis based nonlocal means for handling multispectral MRI images is a promising future work.

# References

1. Somasundram, K., Kalavathi, P.: Brain segmentation in magnetic resonance human head scans using multi-seeded region growing. Imaging Science Journal **62**, 273–284 (2014)
2. Kalavathi, P., Prasath, V.B.S.: Automatic segmentation of cerebral hemispheres in MR human head scans. International Journal of Imaging Systems and Technology - Neuroimaging and Brain Mapping (2015)
3. Kalavathi, P., Prasath, V.B.S.: Methods on skull stripping of MRI head scan images - A review. Journal of Digital Imaging (2015)
4. Prasath, V.B.S., Singh, A.: A hybrid convex variational model for image restoration. Appl. Math. Comput. **215**, 3655–3664 (2010)
5. Prasath, V.B.S., Vorotnikov, D.: On a system of adaptive coupled PDEs for image restoration. Journal of Mathematical Imaging and Vision **48**, 35–52 (2014)
6. Perona, P., Malik, J.: Scale-space and edge detection using anisotropic diffusion. IEEE Trans. Pattern Anal. Mach. Intell. **12**, 629–639 (1990)
7. Prasath, V.B.S., Urbano, J.M., Vorotnikov, D.: Analysis of adaptive forward-backward diffusion flows with applications in image processing. Inverse Problems **31**(105008), 30 (2015)
8. Prasath, V.B.S., Singh, A.: Well-posed inhomogeneous nonlinear diffusion scheme for digital image denoising. Journal of Applied Mathematics (2010)
9. Rudin, L., Osher, S., Fatemi, E.: Nonlinear total variation based noise removal algorithms. Physica D **60**, 259–268 (1992)
10. Prasath, V.B.S., Vorotnikov, D., Pelapur, R., Jose, S., Seetharaman, G., Palaniappan, K.: Multiscale Tikhonov-total variation image restoration using spatially varying edge coherence exponent. IEEE Transactions on Image Processing **24**, 5220–5235 (2015)
11. Gonzalez, R.C., Wood, R.: Digital image processing. Addison-Wesley, Reading (1993)
12. Buades, A., Coll, B., Morel, J.M.: A review of image denoising methods, with a new one. Multiscale Modeling and Simulation **4**, 490–530 (2005)
13. Prasath, V.B.S., Delhibabu, R.: Color image restoration with fuzzy Gaussian mixture model driven nonlocal filter. In: AIST, pp. 131–139 (2015)
14. Manjón, J.V., Carbonell-Caballero, J., Lull, J.J., Garciá-Martí, G., Martí-Bonmatí, L., Robles, M.: MRI denoising using non-local means. Med. Image Anal. **12**, 514–523 (2008)
15. Coupé, P., Yger, P., Prima, S., Hellier, P., Kervrann, C., Barillot, C.: An optimized blockwise nonlocal means denoising filter for 3-D magnetic resonance images. IEEE Transactions on Medical Imaging **27**, 425–441 (2008)
16. Zeng, J., Liu, Z.Q.: Type-2 fuzzy graphical models for pattern recognition. Studies in Computational Intelligence, vol. 666. Springer-Verlag, Berlin (2015)
17. Bilmes, J.A.: Gentle Tutorial of the EM algorithm and its application to parameter estimation for Gaussian mixture and hidden Markov models; Technical Report TR-97-021; International Computer Science Institute, Berkeley, CA, USA (1998)
18. Nabney, I.: NETLAB: Algorithms for pattern recognitions. Springer, London (2002)

# Automatic Leaf Vein Feature Extraction for First Degree Veins

S. Sibi Chakkaravarthy, G. Sajeevan, E. Kamalanaban
and K.A. Varun Kumar

**Abstract** Leaf vein is one of the most important and complex feature of the leaf used in automatic plant identification system for automatic classification and identification of plant species. Leaves of different species have different characteristic features which help in classification of specific plant species. These features help the botanists in identifying the key species of the plants from its leaf images more accurately. Vein feature is one of the most important complex features of leaf in plant species. In this paper we proposed a new feature extraction model, to extract the vein features from the leaf images. The proposed system using Hough lines stems the extraction of vein feature from the leaf images by plotting the lines over the first degree veins. Angle of lines from the primary vein to the secondary vein is considered as the input parameter for processing the extracted vein features. The centroid vein angle is considered to be the primary feature. The vein feature was given as the input to the neural network for efficient classification and the results were tested with 15 species of plants taken from "leafilia" data sets.

**Keywords** Leaf vein · Venation · Leaf recognition system · Leaf identification system · Leaf classification · Leafilia

## 1   Introduction

Plant is one of the most important long lasting species of the planet earth with various medicinal features. Hence identifying such plant species from huge number is still a challenging task. However automatic plant identification system

S.S. Chakkaravarthy(✉) · E. Kamalanaban · K.A.V. Kumar
Department of Computer Science and Engineering, VelTech Rangarajan
Dr.Sagunthala R&D Institute of Science and Technology, Chennai, India
e-mail: {sb.sibi,kamalanaban2009,varun.kumar300}@gmail.com

G. Sajeevan
Department of Computer Science and Engineering, SS & DM Group,
Centre for Development of Advanced Computing (C-DAC), Pune, India
e-mail: sajeevan@cdac.in

© Springer International Publishing Switzerland 2016
S.M. Thampi et al. (eds.), *Advances in Signal Processing and Intelligent Recognition Systems*,
Advances in Intelligent Systems and Computing 425,
DOI: 10.1007/978-3-319-28658-7_49

581

without human intervention is desirable and the optimality should reflect by automatically recognizing plants from leaf images. Plant identification from its leaf images is one of the most challenging tasks for the botanists due to the large number of plant species. In last few years, various methodologies have been proposed by various researchers in plant characterisation and leaf recognition. Various features and feature descriptors are considered to identify the exact plant species. In this paper, a novel methodology is proposed to extract a vital leaf feature called "leaf-vein".

The paper was organized as follows. The detailed scope of the plant recognition system is clearly defined in the section "Leaf recognition system". The detailed view of leaf venation is explained in the section "Vein structure". And a detailed methodology to extract the vein structure defined was the cutting edge of the paper. The literature review included in the section "leaf vein extraction" denotes the scope of the vein structure and vein feature consideration for plant recognition.

## 1.1 Leaf Recognition System – Related Work

Leaf recognition system is used to identify the plant species and in botanical aspect, for plant characterisation. Traditional plant recognition system focus only on contour and shape based methodologies. In recent research, various researchers/research forums have gained a lot of access in the context of obtaining leaf venation and vein like object features (Edwin Correa et al., 2014). ShayanHati et.al proposed a semi-automatic plant recognition system, in which more than 10 botanical features of the leaf was considered to classify the plants from its leaf images (Amanda Ash et al.)

In plant identification, leaf features plays vital role in identifying the plant species and plant classification. The research methodologies in such systems include shape, Texture, Colour and features. Out of those parameters, feature based plant classification is an optimal method in identifying plant species. Colour based and texture based methodologies works seasonally, while feature based methodologies are used often throughout the span. Our further investigation continued with feature based methods, some of the efficient features obtained from the leaf are width ratio, length, centroid, apex angle, base angle, moment deviation ratio, aspect ratio, circularity, vein structure, margin, blade etc. However leaf with same shapes and complicated contours makes the system more complex in identifying the species. These systems are traditional which focus fully on contour based leaf recognition or shape based leaf recognition. In most of the cases, it is hard to recognize two leaves from unique species which are identical in nature. The main feature of the leaf is vein and vein pattern, it is highly unique feature related to the plant structure. Since vein feature extraction is one of the challenging tasks in identifying the plant species, in recent years very little research is carried out with the vein features. Hence an optimal methodology to recognize the vein feature is desirable.

Vein structure is grouped into order of three degrees, the first degree states the primary veins which are visible to human eye, second degree denotes the secondary structure of veins and third degree states the tertiary structure of the vein. Considering the order, the vein structure can be classified up to $2^{nd}$ degree visibility to human eye. In this paper we proposed a novel methodology to recognize the first degree vein structures as feature for plant classification and characterisation. Traditionally vein feature can be extracted using various edge operators like Perwitt, Sobel , Laplacian, Gaussian and Canny etc. Using these operators the information about the variation of grey level or intensity of pixels (RGB level) can be obtained. In general, vein feature is somewhat differ from edges. In vein, some of the pixels oriented to leaf images belong to edge level. Here pixel orientation may be subjected to variation in grey level and proportional to vein pixel width. Since the vein pixel width is key factor to differentiate the primary vein from its accessory vein, hence the edge information is not sufficient to obtain the full orientation of the vein structure in the leaf.

## 1.2  Leaf Vein Extraction

Many image processing techniques(Wang Xue et.al) are involved to extract the structure/ feature from the plants; it is one of the hardest jobs to extract those features using traditional techniques. Some of the features are evaluated based on geometrical feature and morphological feature. Leaf vein, leaf blade, apex angle & base angle, margin etc. are some complicated features of the leaf which helps in identifying the plants more accurately. Here in this paper, one of the complicated feature "leaf veins" is considered. In this area of research, a limited number of authors have proposed their valuable work on particular complexity (leaf vein). Hong Fu et.al proposed a two stage approach for leaf vein extraction, here the intensity histogram is applied to extract the tough region of pixels from the vein structure of the leafs, then the pixel is given as the input to the neural model and trained. Pixel with second order positive derivatives and second order negative derivates are considered to extract the vein structure. Then the histogram structure is applied and preliminary segmentation by means of thresholding is used to extract the vein structure. Leaf vein extraction is one of the key factors for identifying the plant organs, Yun Feng Li et.al proposed a new scheme in leaf vein extraction using snake technique with Cellular neural network. Then Yun Feng Li et.al applied a prior knowledge over similar characteristics of the leaf with implicit parameters like contour, corners etc which helps in extracting leaf venations and outlines of the leaf. Combined thresholding and neural network approach is one of the most robust techniques used to extract the leaf vein patterns. Hu et.al proposed a scheme with thresholding based on intensity value and histogram equalisation which helps in figuring out the leaf vein structure. Compared with the previous methodology Hu et.al proposed scheme is more robust in vein extraction but fails in extracting feature level component in vein structure. Yan Li et.al used FastICA algorithm for vein order mapping and independent component analysis for leaf vein extraction.

$$W(k) = \frac{W^*(k)}{\sqrt{W^*(k)^T \, Cw^*(k)}} \text{ final formulation for leaf venation by Yan Li et.al,}$$

Where $W^*(k)$ is the complex conjugate of $W(k)$

Various high pass filters with contour emphasizing are used to extract the leaf vein structures. Some of the optimal filters used to extract the vein structures are Sobel, Canny and Perwitt operators. FFT with high band pass was used to extract the second and third degree vein structures. Xiaodong et.al proposed a new scheme based on hue & intensity with mathematical morphology to extract leaf veins. In this method colour difference between leaf vein and mesopyll is estimated with the outcome of contrasting colour of leaf and concolourous leaf. Contrasting colour leaf needs only hue information to extract the vein structures whereas concolourous leaf needs hue and intensity to extract the vein structures. This method is limited only to high quality leaf images; the methodology is fully based on colour and green instances present in the leaf images.

## 2    Vein Extraction using Hough Lines

In this paper we proposed a robust technique (stated in Figure 1) to extract the vein feature from the leaf images; Hough line transform is used to extract the first degree (category) vein features. Here it is used to find the imperfect instances in the leaf objects within its shape by voting at explicit parameter space. Then the Hough transform transforms those parameter spaces with local maxima function to get accumulator space, which explicitly finds the coordinates of the straight line present in the leaf images. General line equation, $MX+B$ is applied to generate lines for the coordinates acquired by accumulator space variants. The angle of the lines generated from the coordinates is considered as the feature for the first degree vein structure of the leaf. The average sum of the angles are taken as the feature input for the leaf recognition system (In this paper we considered "leafilia (Shayan et al., 2013) – a plant recognition system").

### 2.1   *Algorithm*

```
Function
leafveinextraction(feature[0..n],img,img_hough_line)
  Input image:img
  Output image:img_hough_line
  Greyimage = rgb2gray (img);
  Varfeature [0..n] y
  H=Highpassfilter(Greyimage,'highpass')
  P=getPolarCoordinates(H);
  P1=localgradient(p);
  applyhoughTransform(p1);
  for i from 1 to H
```
$$y[i] = \left(-\frac{\cos\theta}{\sin\theta}\right)x + \left(\frac{r}{\sin\theta}\right)$$

```
for each line[r_θ,θ] passing through (x_0,y_0)
draw line r_θ=   x_0.cosθ + y_0.sinθ
f=calc (avg (angle (r_θ)))
Feature (f);
End
```

## 2.2   *Experimental Results and Discussions*

Leaf images with white background are collected and processed with the proposed method. A total of 15 species of plants with more than 400 leaf samples are taken into consideration. The good quality images with the resolution of 200 – 300 dpi are taken as the input. Figure 2 shows the grey scale conversion of the image using standard formulae 0.2989*R + 0.5870*G + 0.1140*B. Figure 3 shows the leaf vein extraction and Figure 3(h) shows the feature extraction and its coordinate angles from the leaf images. Finally the average of sum of the angles is considered as the input feature for plant recognition. Hough transform for the input leaf image and the point of line generation for the first degree vein structure were shown in the Figure 4(a) where the most of the pixel coordinate belongs to primary vein pixel (first degree). Detailed results are denoted in the Figure 4 and Figure 5.

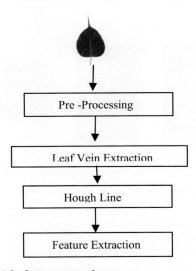

**Fig. 1** Flow Chart for Leaf Vein feature extraction

The proposed method yields better result than the traditional methods in literature and a detailed comparison has been included in the section "Result comparison". The method represented here outperforms with better prediction than the traditional edge detection operator such as Sobel, Prewitt and Laplacian with the conventional strips of lines over the vein structure. The proposed methodology has high compact over the missing vein structure (secondary and tertiary vein structure), on the other side the results obtained from the traditional edge detectors which are discussed above yields only the edge pixel information about the vein which is inadequate to automate the system and it may leads to false positive edge point detection in vein pixel such as missing of vein edge points due to edge discontinuity.

## 3 Performance Analysis

Initially four gradient values are considered from the high pass filter and taken as the input to detect the edge pixels. Here our hypothetical assumption is vein pixel is slightly brighter than the background pixel where the background pixel is dark region, hence the colour orientation of the vein pixel will results in better intensity. This gradient level of FFT will yield more robustness.

### 3.1 Feature Extraction

In the proposed methodology, feature consideration from the vein structure is more complicated task. Here we employed local gradient function in Hough lines to extract vein features by drawing the straight lines over the first degree vein structure using x, y coordinates.

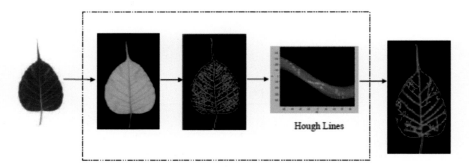

**Fig. 2** Working Model of the proposed methodology

The coordinates is calculated by measuring the intensity of the brighter pixel (vein pixel (i,j)) by comparing the darker region, hence the intensity measure is at the instance of the vein pixel and darker region will be the background (probably all black pixels). The local gradient intensity was defined as follows

$$\text{Local gradient Max function} = \frac{\max[\,0, \text{VeinPixell}(i,j) - \text{Intensity }(i,j)]}{\text{VeinPixell}(i,j)}, \quad C(i,j) \geq 0$$

$$\text{Local gradient Min function} = \frac{\min[\,0, \text{VeinPixel2}(i,j) - \text{Intensity }(i,j)]}{\text{Intensity}(i,j)}, \quad C(i,j) \leq 0$$

Here i,j defines the x,y coordinates of the pixel, $C(i,j)$ is used to determine whether the intensity of the pixel is contrast or dark. Then we calculate the average intensity of the brighter pixel in the background with $C(i,j) \geq 0$.

The average intensity of the brighter pixel in the foreground is also computed. The estimation of the brighter pixel results in the vein pixels. The difference between the brighter pixel intensity and average intensity results in the contrast of the brighter pixels (Vein pixel). Thus the orientation of the brighter pixel results in determining the vein structure along with its pixel coordinates. $\max(0, \text{VeinPixell}(i,j))$ is used to calculate the primary vein structure by evaluating the first order vein pixel, thus the coordinates of the all brighter pixel are evaluated and the angle to the sub vein pixel coordinates are estimated by calculating the local gradient min function, structure of each instances for maximum sub vein coordinates are also recorded.

# 4 Implementation Results

Table 1.Computing time for Edge detector and proposed method (Hough Lines)

| Method | Time in ms |
|---|---|
| Sobel | 0.10 |
| Prewitt | 0.30 |
| Canny | 0.20 |
| Bottom –Hat | 1.37 |
| Top –Hat | 1.45 |
| Roberts | 1.24 |
| Proposed – Hough Line | 0.20 |

**Fig. 3** Leaf Vein structures

a)   Vein structures of **Ficus benghalensis**
b)   Hough lines using $r_\theta = \quad x_0 \cdot \cos\theta + y_0 \cdot \sin\theta$
c)   Angle of coordinates generated in the vein structure
d) – g) Edges extracted by the Sobel, Prewitt, Canny, and Zero Cross Operator
h)   Feature Extraction using proposed methodology

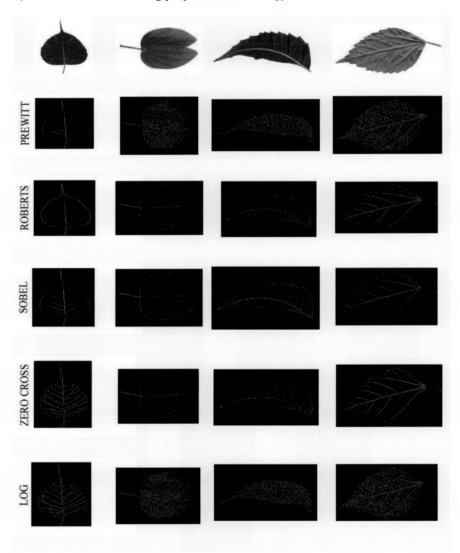

**Fig. 4** Results of Leaf Vein Extraction using various operators

**Fig. 5** Hough Transformed leaf images with its feature pattern

    a)   Hough transformed Leaf Image
    b)   Coordinate angles for Feature

## 5    Vein Coordinates angle

    a)   **Ficus benghalensis** - Leaf image
    b)   Coordinate angle
    c)   Vein structure
    d)   1°Vein extraction

Figure 6 clearly denotes the co-ordinate angles of the primary first degree vein. Here the first degree vein was extracted as the input feature and lines generated on the vein structure are considered to be the feature. The angle of the vertical instance i.e., mid vein structure to the actual vein structure which denotes the coordinate angles of the vein structure. Then the deviation of the angle is measured.

**Fig. 6** Example of Vein Coordinates angle

The main feature to be considered is taken from the centroid angle of the vein from the mid vein to the primary vein which is centroid to it range. The angle is taken as the input feature, when the angle deviation in the centroid of mid vein accomplishes higher variation, then the angle nearer to the base is considered and taken as the input angle. The angle of the deviation from the mid vein to the primary vein is calculated to get the angle value and finally the sum of the angles is considered to be the feature parameter. The angle of deviation varies from one species to another species. The two major angles mainly acquiring from the mid vein to all the primary vein are acute angle and Obtuse angles. The detailed vein deviation ratio and vein angle range is denoted clearly in the figure 6.

**Table 1** Plant species and its scientific classification

| S.No | Scientific Name | Average of Sum of Interior angles |
|:---:|:---:|:---:|
| 1 | *Ficus benghalensis,Ficus religiosa* | 45° |
| 2 | Ocimum tenuiflorum | 15° |
| 3 | *Azadirachta indica* | 10° |
| 4 | *Elaecarpus ganitrus* | 55° |
| 5 | *Neolamarckia cadamba* | 17° |
| 6 | *Hibiscus rosa-sinensis* | 25° |
| 7 | *Spruce* | 0 |
| 8 | *Bauhinia racemosa* | 55° |
| 9 | *Sophora Microphylla* | - |
| 10 | *Murraya koenigii* | 11° |
| 11 | *Jasminum* | 16° |
| 12 | *Trigonella foenum-graecum* | 18° |
| 13 | *Ixora coccinea* | 13° |
| 14 | *Citrus × limon* | 12° |

# 6 Conclusion

Here we conclude the paper by proposing an optimal methodology to extract primary first degree vein structure from the leaf images. The proposed methodology is robust and applicable to any plant recognition systems. By comparison stated in the section "result comparison", our proposed scheme performs with better results. Figure 4 states the comparison results of the proposed method with the other existing methods using edge operators. The main idea is to extract the leaf vein feature and the method performs better results for first degree vein structure.

The proposed methodology was implemented using JAVA. The methodology was tested with popular data sets like "Flavia data set, leafilia data set and normal plant indexed datasets". According to the observation the performance was compared with various edge detection operators, it yields better results than those operators. The Computational time taken to compute leaf vein structure of a single leaf image is as same as canny edge detector i.e., both computes in 0.20ms, the proposed methodology using Hough lines yields better accuracy than edge operators and feature consideration of the Hough line methodology is a high extent to accuracy level. Experimental result demonstrates the proposed methodology yields 95% accuracy for good resolution and visible leaf images. The proposed method works fine for all the vein structured leaves. The demonstrated results of the variation of angles from one species to another species are showed in the figure 6. The future enhancement is to automatically extract second degree and third degree complex vein structure from the leaf images.

# References

1. Ash, A., Ellis, B., Hickey, L.J., Johnson, K., Wilf, P., Wing, S.: Manual of Leaf Architecture. Publication by Smithsonian Institution
2. Correa, E., Green, W., Jaramillo, C., Jaen, M.C.R., Salvador, C., Siabatto, D., Wright, J.: Protocol for leaf image acquisition and analysis, version 2, May 13, 2014. www.ctfs.si.edu/data/documents/LeafScan_WAGreen_draft.pdf
3. Ehsanirad, A.: Plant Classification Based on Leaf Recognition. International Journal of Computer Science and Information Security **8**(4), 78–81 (2010)
4. Fu, H., Chi, Z.: A two-stage approach for leaf vein extraction. In: IEEE Int. Conf. Neural Networks & Signal Processing, China, December 14–17 (2003)
5. Fu, H., Chi, Z.: Combined Thresholding and neural network approach for vein structure extraction from leaf images. IEEE Proc.-Vis. Image Signal Process **153**(6), December 2006
6. Huai, Y., Li, J., Wang, L., Yang, G.: Plant leaf modelling and rendering based-on GPU. In: IEEE Int. Conf. ICISE 2009 (2009)
7. Yahiaoui, I., Mzoughi, O., Boujemaa, N.: Leaf shape descriptor for tree species identification. In: IEEE Int. Conf. ICME (2012)
8. Hossain, J., Amin, M.A.: Leaf shape identification based plant biometrics. In: ICCIT (2010)
9. Chaki, J., Parekh, R.: Plant leaf recognition using shape based features and neural network classifiers. In: IEEE Int. Conf. I JACSA 2011 (2011)
10. Valliammal, N., Geethalakshmi, S.N.: Hybrid image segmentation algorithm for leaf recognition and characterization. In: IEEE International Conference on Process Automation, Control and Computing (PACC) (2011)
11. Pham, N.-H., Le, T.-L., Grard, P., Nguyen, V.-N.: Computer aided plant identification system. In: IEEE International Conference on Computing, Management and Telecommunications (ComManTel) (2013)
12. Mzoughi, O., Yahiaoui, I., Boujemaa, N.: Petiole shape detection for advanced leaf identification. In: IEEE ICIP 2012 (2012)
13. Mishra, P.K., Maurya, S.K., Singh, R.K., Misral, A.K.: A semi automatic plant identification based on digital leaf and flower images. In: IEEE Int. Conf. ICAESM 2012 (2012)

14. Revathi, P., Hemalatha, M.: Classification of cotton leaf spot diseases using image processing edge detection techniques. In: IEEE International Conference on Emerging Trends in Science, Engineering and Technology

15. Nguyen, Q.-K., Le, T.-L., Pham, H.: Leaf based plant identification system for android using surf features in combination with bag of words model and supervised learning. In: IEEE Int. Conf. ATC 2013 (2013)

16. Janani, R., Gopal, A.: Identification of selected Medicinal plant leaves using Image features and ANN. In: IEEE Int. Conf., ICAES 2013 (2013)

17. Hati, S., Sajeevan, G.: Plant Recognition from Leaf Image through Artificial Neural Network. International Journal of Computer Applications (0975 – 8887) **62**(17), January 2013. Data set: Leafilia

18. Prasad, S., Kumar, P., Tripathi, R.C.: Plant leaf species identification using curvelet transform. In: IEEE Int. Conf. ICCCT 2011 (2011)

19. Wu, S.G., Bao, F.S., Xu, E.Y., Wang, Y.-X., Chang, Y.-F., Xiang, Q.-L.: A leaf recognition algorithm for plant classification using probabilistic neural network. In: IEEE International Symposium on Signal Processing and Information Technology (2007)

20. Pan, S., Kudo, M., Toyama, J.: Edge detection of tobacco leaf images based on fuzzy mathematical morphology. In: Int. Conf. ICISE (2009)

21. Wang, X., Ma, T.-M.: The application of edge detection algorithm based on rough sets in the detection of soybean target leaf spot. In: IEEE Int. Conf., ICCASM 2010 (2010)

22. Wang, L.: Identification based on color and texture of the soybean leaf nitrogen diagnostic model. In: IEEE Proceedings of the 29th Chinese Control Conference (2010)

23. Zheng, X., Wang, X.: Fast leaf vein extraction using Hue and intensity information. In: IEEE International Conference on Information Engineering and Computer Science, ICIECS 2009 (2009)

24. Zheng, X., Wang, X.: Leaf vein extraction using a combined operation of mathematical morphology. In: IEEE 2nd International Conference on Information Engineering and Computer Science (ICIECS) (2010)

25. Li, Y., Zhu, Q., Cao, Y., Wang, C.: A leaf vein extraction method based on snakes technique. In: Chongqing University, Chongqing, 400044, P.R. China. IEEE (2005)

26. Li, Y., Chi, Z.: Leaf vein extraction using independent component analysis. In: IEEE International Conference on Systems, Man, and Cybernetics, October 8-11, 2006, Taipei, Taiwan (2006)

27. Yahiaoui, I., Herve, N., Boujemaa, N.: Shape based image retrieval in botanical collections. In: Lectures Notes in Computer Science. Springer (2006)

28. Zulkifli, Z.: Plant Leaf Identification Using Moment Invariants & General Regression Neural Network. Universiti Teknologi Malaysia (2009)

29. Sun, Z., Lu, S., Guo, X., Tian, Y.: Leaf vein and contour extraction from point cloud data. In: Int. Conf. on Virtual Reality and Visualization (2011)

30. Wang, Z., Chi, Z., Feng, D.: Shape based leaf image retrieval. IEEE Proceedings - Vision, Image and Signal Processing

31. Web source: Leafsnap Leafgui

32. Dataset: Leafilia, A semi-automatic plant recognition system developed by CDAC, Pune, India

# Singular Value Decomposition Based Image Steganography Using Integer Wavelet Transform

Siddharth Singh, Rajiv Singh and Tanveer J. Siddiqui

**Abstract** Transform domain Steganography techniques embed secret message in significant areas of cover image. These techniques are generally more robust against common image processing operations. In this paper, we propose an image Steganography method using singular value decomposition (SVD) and integer wavelet transform (IWT). SVD and IWT strengthen the performance of image Steganography and improve the perceptual quality of Stego images. Results have been taken over standard image data sets and compared with discrete cosine transform (DCT) and redundant discrete wavelet transform (RDWT) based image Steganography methods using peak signal to noise ratio (PSNR) correlation coefficients (CC) metrics. Experimental results show that the proposed SVD and IWT based method provides more robustness against image processing and geometric attacks, such as JPEG compression, low-pass filtering, median filtering, and addition of noise, scaling, rotation, and histogram equalization.

**Keywords** Image steganography · Data hiding · Singular value decomposition · Integer wavelet transform · Chaotic sequence

S. Singh(✉)
Department of Electronics and Communication Engineering,
National Institute of Technology, Delhi, India
e-mail: siddharthjnp@gmail.com

R. Singh
Department of Computer Science, Banasthali University, Banasthali, India
e-mail: jkrajivsingh@gmail.com

T.J. Siddiqui
Department of Electronics and Communication, University of Allahabad, Allahabad, India
e-mail: siddiqui.tanveer@gmail.com

© Springer International Publishing Switzerland 2016
S.M. Thampi et al. (eds.), *Advances in Signal Processing and Intelligent Recognition Systems*,
Advances in Intelligent Systems and Computing 425,
DOI: 10.1007/978-3-319-28658-7_50

# 1    Introduction

Recent advancement in the field of multimedia technology and sharing, copying and distribution of digital data through internet posed several challenges on data security. Digital Steganography [1-4] is the possible way handle these challenges and provides a secure way of data communication. The solution involves hiding secret information, called copyright mark, in a cover file so that the very existence of the secret message is not detectable. The hidden information is later extracted and used to ensure rightful ownership and authentication of the digital content. The cover file can be image, audio or video; the most commonly being the image files because; (i) digital images are being used quite frequently on the internet, (ii) the size of image is large, and (iii) digital images usually contain redundant bits. So, we can hide secret data easily in a digital image without being suspected by human visual system. The commonly used image file formats which are used for Steganography are graphic interchange format (GIF), joint photographic expert group (JPEG) and bitmap format (BMP) [10]. Steganography [5, 6] has got a wide range of applications in digital media which has been a source of motivation for this work.

The key issues in the design of image Steganography scheme are robustness, imperceptibility and security. Robustness refers to the capability of hidden data to survive both intentional manipulation which aim to destroy the hidden information and unintentional manipulation, which do not aim to remove the hidden data. The data hidden in host file has to be undetectable. This requires that embedding has to be done in a manner so that no visible distortion occurs in the Stego image. Otherwise, one can guess that the host image has been modified.

In order to achieve these goals of Steganography, several approaches have been proposed in both spatial and transform domain. The spatial domain techniques are computationally simple and less robust against intentional or unintentional attacks [7] whereas transform domain techniques require more computations but are more robust against common image processing attacks such as JPEG compression, low-pass filtering, addition of noise etc.

In this paper we limit our scope to transform domain and propose integer wavelet transform (IWT) based image Steganography method which combines singular value decomposition (SVD). IWT provides faster computation of wavelet coefficients whereas SVD enables to have better perceptual quality of Stego images with greater robustness against common image processing attacks. Thus, by combining SVD and IWT, we have improved performance of image Steganography. To validate the performance of the proposed method, we compare it with discrete cosine transform (DCT) and redundant discrete transform (RDWT) based image Steganography methods using peak signal to noise ratio (PSNR) and correlation coefficient (CC) metrics.

The rest of paper is organized as follows. Section 2 introduces background concepts of IWT, SVD and chaotic sequence. Section 3 describes the proposed method. Results are discussed in Section 4. Finally, the conclusions have been made in Section 5.

# 2     Preliminaries

## 2.1   Integer Wavelet Transfrom (IWT)

The use of traditional wavelet transform for data hiding results in high computational complexity and rounding error. Integer wavelet transform (IWT) [18, 19] uses integer coefficients, thereby avoids rounding error, and leads to perfect reconstruction. IWT is obtained with simple modification of discrete wavelet transform (DWT) implemented using lifting scheme proposed by Sweldens [20] which involves three main steps, namely: splitting, predication, and update. Lifting scheme uses simple filtering operations to modify odd and even sample sequences. The modifications are described by the following expressions [21]:

$$d[n] = d_0[n] - \left\lfloor \frac{s_0[n+1] + s_0[n]}{2} \right\rfloor \tag{1}$$

$$s[n] = s_0[n] + \left\lfloor \frac{d[n-1] + d[n] + 2}{4} \right\rfloor \tag{2}$$

where $s[n], d[n]$ are even and odd samples respectively. After the IWT, the even samples become low frequency coefficients and the odd samples become high frequency coefficients. The mathematical structure of lifting is based on correlation between data, the sample sequence.

## 2.2   Singular Value Decomposition (SVD)

An image $X$ can be considered a $M \times N$ matrix with non-negative scalar values. Like other matrices it can be decomposed using singular value decomposition (SVD) [14, 15] into two orthogonal matrices $U$ and $V$ and a diagonal matrix $S$ of singular values of $X$. This decomposition is mathematically described by the following expression [16]:

$$X = U * S * V^T \tag{3}$$

The two major advantages of using SVD in image steganography are: first, a small variation of singular values of image does not affect the visual perceptual quality and second, it provides robustness against common signal and image processing attacks [17].

## 2.3   Chaotic Sequence

The most significant feature of a chaotic system is its sensitivity to initial conditions. Due to this sensitivity, a chaotic system exhibits random behavior, even though it is deterministic. Therefore, a chaotic map can be used to generate a large number of non-periodic, noise-like yet deterministic and reproducible sequences [11]. Chaos theory can be used in data hiding to generate pseudo random sequence for spreading data. Various non-linear dynamic systems are used to generate the chaotic sequence, e.g., logistic map, henon map and tent map [12, 13]. In this chapter, we use logistic map to generate random sequences. The recurrence relation for the logistic map is [11] expressed as:

$$x_{k+1} = r \times x_k (1 - x_k), 0 < x < 1 \qquad (4)$$

where $0 < r \le 4$ is bifurcation parameter.

## 3    The Proposed Method

The proposed method consists of two parts: embedding and recovery. For embedding, cover image is decomposed using IWT and SVD is applied on its low frequency (LL) component. Then logo image is scrambled using chaotic sequence and embedded into SVD matrix of LL component. Inverse IWT and SVD provide resultant Stego image. Whereas for recovery of logo image, Stego image is first decomposed by IWT and SVD sequentially and inverse scrambling is performed to obtain original logo image. The steps of embedding and recovery processes are described in following sections.

### 3.1    *Embedding Process*

1. Read Cover image $X$ of size $N \times N$ and Logo image (L) of size $M \times M$.
2. Perform 1-level IWT on $X$ to decompose it into $X_{LL}, X_{LH}, X_{HL}, X_{HH}$ and then SVD is applied on its LL subband, $X_{LL}$

$$X_{LL} = U_{X_{LL}} * S_{X_{LL}} * V_{X_{LL}} \qquad (5)$$

3. Logo image ($L_o$) is scrambled using chaotic sequence and then embedded into singular matrix $S_{X_{LL}}$ of transformed coefficients of LL subband of $X$ follows:

$$S'_{X_{LL}} = S_{X_{LL}} + k * L_o \qquad (6)$$

where $k$ is the gain factor used to specify the strength of embedding data and is set to 4 empirically. Please note that the size of $L_s$ is less than $X_{LL}$. We deal with this by padding extra bits to it.

4. Inverse SVD and inverse integer wavelet transform is applied on $S'_{X_{LL}}$ and then get stego image $X_S$.

## 3.2    *Recovery Process*

1. Read stego image $Y_s$
2. Perform 1-level IWT on $Y_s$ and then SVD is applied on its LL sub-band, $Y_{X_{LL}}$

$$Y_{S_{LL}} = U_{Y_{S_{LL}}} * S_{Y_{S_{LL}}} * V_{Y_{S_{LL}}} \qquad (7)$$

3. Obtained singular values of scrambled logo from following expression:

$$S_s^r = \frac{S_{Y_{S_{LL}}} - S_{X_{LL}}}{k} \qquad (8)$$

where $S_{Y_{S_{LL}}}$, $S_{X_{LL}}$ are singular values of stego and cover image respectively.
4. Perform inverse scrambling to get original logo image.

## 4    Results and Discussions

To test the performance of the proposed techniques, we conducted a series of experiments with standard cover images Lena, Pepper of size 512 × 512 and logo image (JKAU) of size 64 × 64. For experiments, we have used one level decomposition level of IWT, since higher level decomposition may decrease the information content of the wavelet subbands and hiding capacity as well.

Performance evaluation of the proposed method has been done using peak signal to noise ratio (PSNR) and correlation coefficient (CC). PSNR is used in this work to calculate the stego image quality. For gray level image, PSNR is given by

$$PSNR = 10 \log_{10} \left( \frac{255^2}{MSE} \right) \qquad (9)$$

Mean Square Error (*MSE*) between the original image $I$ of size M × N and the stego image $I_s$ is calculated as follows:

$$MSE = \frac{1}{MN} \left[ \sum_{i=1}^{M} \sum_{j=1}^{N} (I(i,j) - I_s(i,j))^2 \right] \qquad (10)$$

The correlation coefficient (*CC*) is used to measure the similarity between original logo image ( $L_o$ ) and the recovered logo image ( $L_r$ ) of size $r \times s$ . It is defined as follows:

$$CC = \frac{\sum\limits_{i=1}^{r}\sum\limits_{j=1}^{s}(L_o(i,j)-\bar{L}_o)(L_r(i,j)-\bar{L}_r)}{\sqrt{\sum\limits_{i=1}^{r}\sum\limits_{j=1}^{s}(L_o(i,j)-\bar{L}_o)^2}\sqrt{\sum\limits_{i=1}^{r}\sum\limits_{j=1}^{s}(L_r(i,j)-\bar{L}_r)^2}} \qquad (11)$$

where $\bar{L}_o$, $\bar{L}_r$ are mean of original and recovered logo image. Higher value of CC means original and recovered image are highly correlated which is desirable.

Imperceptibility is one of the important parameters for evaluation of image Steganography technique. This can be shown in terms of PSNR value. High PSNR value implies better imperceptibility of the Steganography method. In addition to this, the globally accepted PSNR value for good picture quality is greater than 40 dB [2-3].

In order to show the effectiveness of the proposed method, we have shown Stego images (Lena and Pepper) with logo images in Fig. 1 and their corresponding PSNR values are given in Table 1. From Table 1, it can be easily observed that PSNR values of Stego Lena and Pepper images are greater than 52. Therefore, the embedding does not cause perceptual degradation in Stego Lena and Pepper images and can be easily verified from Fig. 1. Thus, imperceptibility is maintained which is one of the key requirements of any embedding scheme.

Further, the proposed method has been tested for robustness against JPEG compression, gaussian low-pass filtering, median filtering, addition of gaussian noise, cropping, rotation, resizing and histogram equalization attacks. Again, experiments have been performed over Lena and Pepper images. The Stego Lena and corresponding recovered logo images are shown in Fig. 2. From Fig. 2, it can be seen that recovered logo images and their corresponding Stego images are not much affected by image processing attacks.

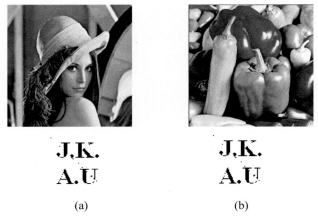

(a)                                           (b)

**Fig. 1** Stego and recovered logos images: (a) Lena with logo JKAU, PSNR= 52.40 dB, (b) Pepper with logo JKAU , PSNR = 52.41 dB.

However, this visual analysis is not sufficient. For quantitative analysis of the proposed method CC values of Lena and Pepper images against different attacks are given in Table 2 and compared with different transform domain image steganography techniques like DCT based algorithm [8], RDWT based algorithm [9]. It is also evident from the Table 2 that the CC values obtained from proposed method is better than DCT based algorithm [8] and RDWT based algorithm [9] for JPEG compression, gaussian low-pass filtering, median filtering, addition of gaussian noise, cropping, rotation, resizing and histogram equalization attacks. Thus, on the basis of the CC values shown in Table 2, the proposed method outperforms DCT and RDWT based image Steganography methods. The dashed values in the Table 2 indicate that these values are not available for particular attacks.

**Table 1** PSNR and CC values of Stego images for the proposed method.

| Performance Metric | Lena Image | Pepper Image |
|:---:|:---:|:---:|
| PSNR | 52.40 | 52.41 |
| CC | 0.9826 | 0.9887 |

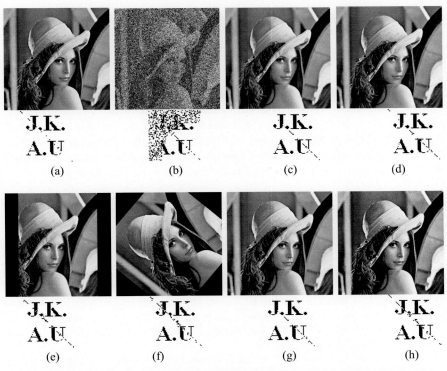

**Fig. 2** Stego and recovered logo images after after (a) JPEG compression (Q-50), (b) Gaussian noise addition (0.5), (c) Gaussian Filter [3×3], (d)Median Filter[3×3], (e) Cropping (25% middle portion), (f) Rotation (50⁰), (g) Resizing(1/2), (h) Histogram equalization.

**Table 2** Correlation Coefficients of recovered logo from proposed method, DCT based data hiding algorithm [8], RDWT based data hiding algorithm [9].

| Attacks | The proposed method | | DCT method [8] | | RDWT method [9] | |
|---|---|---|---|---|---|---|
| | **Lena** | **Pepper** | **Lena** | **Pepper** | **Lena** | **Pepper** |
| **JPEG compression (Q-50)** | 0.9752 | 0.9850 | 0.5021 | 0.5528 | 0.9140 | 0.9468 |
| **Gaussian Noise (var =0.5 )** | 0.5510 | 0.5089 | 0.3002 | 0.2907 | 0.9540 | 0.9861 |
| **Gaussian Low Pass Filter[3×3]** | 0.9670 | 0.9658 | 0.8397 | 0.8037 | 0.9371 | 0.9810 |
| **Median Filter [3×3]** | 0.9610 | 0.9633 | 0.6701 | 0.6298 | 0.9527 | 0.9835 |
| **Cropping(25% both side)** | 0.9434 | 0.9403 | 0.7443 | 0.7320 | 0.9540 | 0.9861 |
| **Rotation(50°)** | 0.9230 | 0.9303 | --- | --- | 0.9571 | 0.9886 |
| **Resizing(1/2)** | 0.9657 | 0.9693 | 0.6032 | 0.5919 | 0.9501 | 0.9861 |
| **Histogram Equalization** | 0.9387 | 0.9392 | -- | -- | 0.9475 | 0.9797 |

# 5    Conclusions

In this paper, we have proposed SVD and IWT based image Steganography method which uses chaotic sequence for scrambling logo. IWT with SVD increases the robustness against geometrical as well as image processing attacks. Moreover, the use of IWT makes our scheme computationally competitive. The experimental results show that the IWT and SVD based technique is more imperceptible and achieves higher security and robustness against Gaussian and median filtering, addition of noise, cropping, resizing, rotation, and histogram equalization attacks compared to DCT and RDWT based techniques for data hiding.

# References

1. Fridrich, J.: Steganography in digital media: principles, algorithms, and applications. Cambridge University Press (2010)
2. Atawneh, S., Almomani, A., Sumari, P.: Steganography in digital images: common approaches and tools. IETE Technical Review 30(4), 344–358 (2013)
3. Cheddad, A., Condell, J., Curran, K., Kevitt, M.P.: Digital image Steganography: survey and analysis of current methods. Signal Processing 90(3), 727–752 (2010)
4. Cox, J., Miller, M.L., Bloom, J.A., Fridrich, J., Kalker, T.: Digital watermarking and Steganography. Elsevier (2008)
5. Katzenbeisser, S., Petitcolas, F.A.P.: Information hiding techniques for Steganography and digital watermarking. Artech House Inc., Norwood (2000)
6. Johnson, N.F., Jajodia, S.: Exploring Steganography: seeing the unseen. IEEE Computer 31(2), 26–34 (1998)

7. Westfield, A., Pfitzmann, A.: Attacks on steganographic systems. In: Lecture Notes in Computer Science, vol. 1768, pp. 61–75. Springer (2000)
8. Singh, S., Siddiqui, T.J.: A security enhanced robust steganography algorithm for data hiding. International Journal of Computer Science Issues **9**(3), 131–139 (2012)
9. Singh, S., Siddiqui, T.J.: Robust image data hiding technique for copyright protection. International Journal of Information Security and Privacy **7**(2), 44–56 (2013)
10. Singh, S., Siddiqui, T.J.: Transform domain techniques for image steganography. Information Security in Diverse Computing Environments, 245–259. IGI global (2014)
11. Devaney, R.: An Introduction to chaotic dynamical systems, 2nd edn. Addison-Wiesley (1989)
12. Sun, X.H., Dai, Y.W., Wang, Z.Q.: The generation of chaotic sequence and its application to image encryption. Journal of Nanjing Normal University (Engineering and Technology) **4**(1), 56–78 (2004)
13. Jessa, M.: The period of sequences generated by tent- like maps. IEEE Transactions on Circuits and Systems I: Fundamental Theory and Applications **49**(1), 84–88 (2002)
14. Kakarala, R., Ogunbona, P.O.: Signal analysis using a multiresolution form of the singular value decomposition. IEEE Transactions on Image Processing **10**(5), 724–735 (2001)
15. Liu, R., Tan, T.: An SVD-based watermarking scheme for protecting rightful ownership. IEEE Transactions on Multimedia **4**(1), 121–128 (2002)
16. Chang, C.C., Tsai, P., Lin, C.C.: SVD-based digital image watermarking scheme. Pattern Recognition Letters **26**(10), 1577–1586 (2005)
17. Wang, M.S., Chen, W.C.: A hybrid DWT-SVD copyright scheme based on K-mean clustering and visual cryptography. Computer Standard and Interfaces **31**(4), 750–762 (2009)
18. Calderbank, A.R., Daubechies, I., Sweldens, W., Yeo, B.L.: Wavelet transforms that map integers to integers. Applied Computation Harmonic Analysis **5**(3), 332–369 (1998)
19. Arsalan, M., Malik, S.A., Khan, A.: Intelligent reversible watermarking in integer wavelet domain for medical images. Journal of System and Software **85**(4), 883–894 (2012)
20. Sweldens, W.: The lifting scheme: a construction of second generation wavelets. SIAM Journal Mathematical Analysis **29**(2), 511–546 (1998)
21. Adams, M.D., Kossentini, F.: Reversible integer-to-integer wavelet transforms for image compression: performance evaluation and analysis. IEEE Transactions on Image Processing **9**(6), 1010–1024 (2000)

# Unsupervised Learning Based Video Surveillance System Established with Networked Cameras

R. Venkatesan, P. Dinesh Anton Raja and A. Balaji Ganesh

**Abstract** The paper presents an autonomous video surveillance system for tracking moving objects in a networked camera environment. The system is validated to identify an authenticated users' vehicle which is provided with unique sticker as well as vehicle registration number. The multi-camera tracking is implemented on the basis of decentralized hand-over procedure between adjacent cameras. The object of interest in the source image is learnt as single tracking instance and eventually shared among other cameras in the network, autonomously. Thus, the moving objects are continuously tracked without the advent of central supervision and it can be scaled up higher for monitoring of vehicle traffic and other remote surveillance applications.

**Keywords** Networked cameras · Optical character recognition · Unsupervised learning · Hand over · Video surveillance

## 1 Introduction

Video surveillance is an active area in computer vision which plays an important role in many applications, including industrial automation, robotics, autonomous vehicle navigation and human machine interfaces [1, 2].

It is categorized into three, namely manual, semi-autonomous and fully-autonomous based on the level of human intervention in tracking the object of interest [3]. In fully-autonomous system, computer vision algorithms supported with hardware can replace the presence of human operators [4]. Also, it guarantees wider area surveillance through distributed networks and significantly reduces the hard monotonous workload by directing attention to a specific portion of the data

R. Venkatesan · P.D.A. Raja · A.B. Ganesh(✉)
Electronic System Design Laboratory, Velammal Engineering College, Chennai, India
e-mail: {venky88an,pdaraja}@gmail.com, abganesh@velammal.edu.in

© Springer International Publishing Switzerland 2016
S.M. Thampi et al. (eds.), *Advances in Signal Processing and Intelligent Recognition Systems*,
Advances in Intelligent Systems and Computing 425,
DOI: 10.1007/978-3-319-28658-7_51

603

[5, 6]. In a multi-camera environment, the tracker or tracking agent follows the object over the camera network by migrating from one camera to the next that results restricted occlusion [7, 8]. Nevertheless the features, several issues, such as short initialization time, internal state of the tracker and robustness are to be considered to implement an autonomous system in a distributed multi-camera environment especially for traffic surveillance [9, 10]. Embedded smart cameras that combine video sensing, analyzing, processing and communicating over the network are found in many reports [11, 12, 13]. The involvement of Internet of Things (IoT) further extends the possibility of miniaturized hardware components over the wider networks [14]. The adaptation of Cloud based video tracking with additional scalability, inter-operability and security is demonstrated by the researchers [15, 16]. Recently, the combination of cloud-IoT paradigm suggested as ubiquitous access to the recorded video and enabled complex surveillance analyses in the Cloud. The object tracking methods are divided into four groups, namely region-based tracking; active-contour-based tracking; feature-based tracking and model-based tracking [17, 18, 19]. For vehicle classification in traffic surveillance both optical character recognition (OCR) and feature-based tracking are applied predominantly. The applications of such system include, gathering of traffic flow statistics, finding stolen cars, curbing unlawful traffic activities, controlling access to authorized vehicles and fixing up parking charges [20, 21].

Each of these methods has its own merits as well as demerits [1, 2, 3, 4]. In feature-based tracking, the features of image, such as shape, colour, size, orientation, position, coordinates, etc., are extracted as feature-vectors. When compared to other tracking mechanisms, this method focuses only on the featured vectors thus minimizes the computational load; however it suffers from similar attributes in extracted information. In this study the combination of OCR and feature-based tracking is implemented in a multi-camera distributed network for traffic surveillance. The object of interest classification is based on the vehicle's registration number as well as unique sticker provided to identify the authenticated users. The tracking agent has been configured and object of interest is circulated among the computing systems in the network. Thus, the surveillance system offers decentralized continuous tracking mechanism to follow the object of interest within the network autonomously. The vehicle tracking system is successfully implemented and the results, including delay time are presented. The sticker is designed with numeral character and logo with unique colour pattern. A database is generated with the details of authenticated users' information, such as name, vehicle registration number, sticker identity number, driving licence number and address of the user. The feature-based tracking is implemented using colour thresholding, Fourier descriptors, shape descriptors and local binary pattern (LBP) with binary particle classification.

## 2   Experimental Setup

Fig.1 describes the multi-camera distributed surveillance system to classify an authenticated user's vehicle in a network. The network is comprised of both embedded imaging systems and computers, which are interconnected with multiple cameras. The tracking algorithms are embedded into all the computing systems in the network only when a camera finds a vehicle without sticker identity.

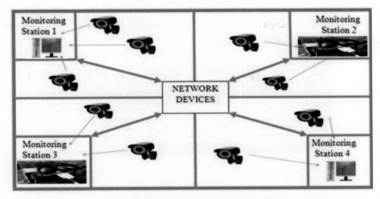

**Fig. 1** Block Diagram representation of an automated traffic surveillance system in a multi-camera distributed network

The algorithm of real time vehicle surveillance has been implemented and tested in a wider campus network. The state machine diagram is developed using Unified Mark-up Language (UML) to describe the functional overview of video surveillance algorithm which is shown in Fig. 2.

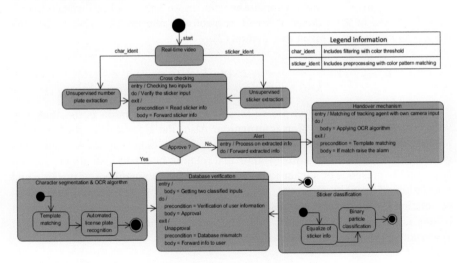

**Fig. 2** State machine representation of traffic surveillance

The source frame from each camera does have the information of both sticker id and registration number which are eventually to be extracted. Fig. 3 shows the license plate information along with prepared sticker id. The sticker id extraction is based on the textural parameters and shape descriptors where as character detection is performed using OCR technique. The OCR is proven mechanism which computes correlation coefficients regardless of scaling, translation and position of license plate over complex background. The results are then compared with the stored authenticated user database. The system is programmed to notify autonomously to other systems in case of mismatching of user information and also absence of valid sticker id.

**Fig. 3** Vehicle license plate with sticker id

The extraction of sticker id process starts from finding of Gaussian distribution and moment of invariants [22, 23] in the source frame. The probability of observing the current pixel value at time $t$ is modeled as a mixture of $k$ Gaussian distributions which is given in Equation (1).

$$p(X_t) = \sum_{i=1}^{k} w_{i,t} \times \phi(X_t, \mu_{i,t}, C_{i,j}) \tag{1}$$

Where $w_{i,t}$, $\mu_{i,t}$, $C_{i,j}$ are the estimate weight, mean value and covariance matrix of $i_{th}$ Gaussian in the mixture at time t, respectively and the Gaussian probability density function that is given in Equation (2).

$$\phi(X_t, \mu_{i,t}, C_{i,j}) = \frac{1}{(2\pi)^{n/2} c^{1/2}} \exp(-1/2(X_t - \mu_t)^T c^{-1}(X_t - \mu_t)) \tag{2}$$

The color and shape descriptors [23, 24] provide centre point in the sticker id with corresponding orientation information. Also, the algorithm is configured to identify the critical factors, such as symmetry, feature detail, positional and background information. Both shift-invariant and rotation invariant matching is applied to detect the presence of sticker id in the captured frame regardless of position and translation. The shift invariant matching is based on the normalized cross-correlation [24]. Consider, an image f (x, y) of size M×N and a sub-image w(x, y) of size K x L where K ≤ M and L ≤ N. The correlation between w(x, y) and f(x, y) at a point (i, j) is given in the Equation (3).

$$C(i, j) = \sum_{x=0}^{L-1} \sum_{y=0}^{K-1} w(x, y) f(x+i, y+j) \tag{3}$$

Where i = 0, 1…M − 1, j = 0, 1… N − 1 and the summation are taken over the region in the image where w and f overlaps. The character extraction from license plate involves color thresholding and shape descriptors and eventually

OCR module is applied. In color thresholding, the license plate is extracted from background image by enforcing threshold interval for each color planes. The computed binary output image exactly localizes the license plate information regardless of scaling, translation and position. The shape descriptors involve clustering, entropy and moments of invariants which are independent of position, rotation or scale. Binary particle classification identifies an anonymous binary model by comparing set of significant textural features with classes of known samples.

Analysis of Textural features has fulfilled by Local Binary Pattern (LBP) operator [25]. Invariance of illumination changes, image rotation and textural scale changes is the main approach of LBP operator. Local image contrast and local texture pattern (LBP code) are the features provided by LBP operator. Local contrast measure is obtained by subtracting the average of gray levels below the center pixel from the average of gray level above (or equal to) the center pixel. Local texture pattern is determined by thresholding a neighborhood by the gray scale value of its center pixel in the form binary code (LBP code). Fig. 4 shows the working of local binary pattern operator. Consider a pixel (center pixel) as a threshold with eight neighbors. Thresholding of neighborhood pixels has been done by the center pixel value 3. After thresholding, the binary code has been identified. Finally the LBP code of decimal value has been resolved from the binary value. On the other hand the local image contrast is attained and shown in the Fig. 4.

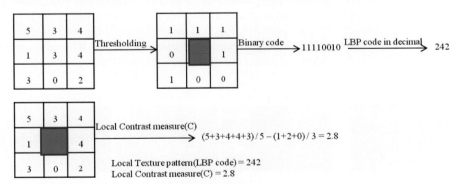

**Fig. 4** Working of local binary pattern (LBP) operator

There are two phases are involved in particle classification, namely training and identification. Training phase trains the machine learning algorithm to understand the types of samples to be classified during the identification phase. Training phase also provides the feasibility to add new samples with existing classes and also supports creation of new classes. Thus, it scales up the features in existing application. The identification phase involves retrieval of region of interest by employing gray scale co-occurrence matrix (GLCM) in order to get the spatial dependence of gray scale values at different angles i.e. 0°,45°,90° and 135°. Each matrix is run through probability density functions to calculate different textural parameters [24, 26], such as contrast, homogeneity, entropy, energy and momentum. Eventually, for the classification some of the sorting algorithms, including minimum mean distance, nearest neighbour and

K-nearest neighbour are used. The resultant feature vectors are fed into support vector machine (SVM) which is considered as a supervised learning technique. In this study, SVM is used to analyze the feature vectors and to recognize the patterns that are eventually utilized in binary classification and regression analysis. Optical character recognition (OCR) is the process by which the system learns the type of character and/or pattern in the region of interest of source image [27, 28]. A character set file is generated during the training phase in OCR. Also, new samples are updated dynamically with new font information.

## 3 Results and Discussions

In this study, an unsupervised distributed camera network is established to track any abnormality event triggered by any network terminal. The detected anomaly is forwarded as tracking agent which is transferred to all other terminals in the network simultaneously. The results are found to be robust irrespective of vehicle deformations, varying illumination and background. The tracking instance acts autonomously and follows the target over the camera network with minimum delay time. The computer networks is formed using the terminals which has Intel i3 core processor (speed up to 2.60 GHz) and 4GB RAM integrated with camera i.e. Basler SCA1390-17GC. The cameras have, frame rate 17 fps and resolution of 1392x1040 pixels. The developed algorithms also tested using NI EVS1464RT which made as a standalone embedded digital platform. Fig. 5 shows the captured frame, localization of specific information and segmentation and shape extraction. The measurements are taken at different time period and environment illumination conditions.

**Fig. 5** Real time autonomous sticker number extraction and recognition. (a) Original image, (b) sticker number localization, (c) segmentation and shape extraction

The experiments are performed at various illumination conditions as well as focusing angle at $30^0$ both left and right directions which are shown in Fig. 6.

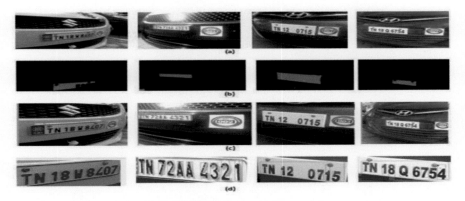

**Fig. 6** Real time autonomous license plate extraction and recognition inclined at 30° (a) Original image, (b) Color thresholding at 30° left/right, (c) Localization of plate at 30° left/right, (d) OCR over number plate at 30° left/right

Fig. 7 shows real time camera video sequence is obtained and unsupervised extraction of vehicle license plate number and unique sticker identity number is done. Optical character recognition permits to read the number from color thresholded number plate and extracted sticker identity is given to the LBP operator and the features obtained from LBP are given to the binary particle classifier which has been used for extraction of précised staff's/student's name from the described owner's details (such as STAFF NAME, LICENSE PLATE NUMBER, STICKER IDENTITY NUMBER, STAFF ID ENTITY NUMBER, DESIGNATION, STUDEN IDENTITY NUMBER).

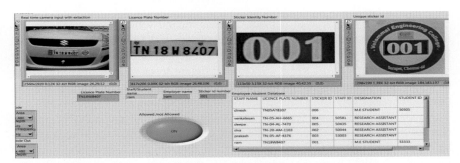

**Fig. 7** Tracking and recognition of concerned vehicle and its identity measure

Fig. 8 shows the performance of the developed algorithm while the vehicle enters without unique sticker identity number and decision taken where corresponing vehicle allowed or not allowed.

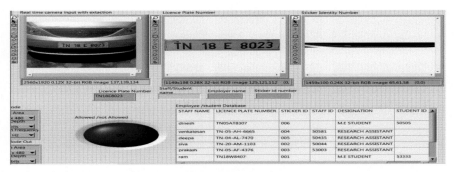

**Fig. 8** Unauthorised vehicle entry and algorithm performance

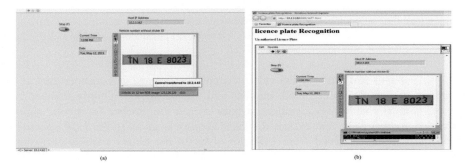

(a)                                              (b)

**Fig. 9** Handover mechanism attained after finding unauthorized vehicle number (a) Source station template transferred to next target station, (b) Template received at the target station transferred from source station using dedicated URL

Fig. 9 (a) shows extracted number of unauthorized vehicle recognition and its information transfer from source to next target station. Extraction of unauthorized license plate number will be passed from source station to the specified target station for identifying the same vehicle in different location over the broad range of area within the network. Transformation of template in a camera network has been done effectively from source station to target station with in the camera network. On the other hand the user can easily identify the unauthorized vehicle number from the remote location. The main convincing technology is each standalone camera in the network can provide their controls and information to a web page through URL. From the dedicated URL, the remote user can control the particular camera and monitor its activity as shown in Fig. 9(b).

**Table 1** Recognition rate for the developed algorithm applied at different illumination condition and varying angles

| Type of license plate | Camera position | Light Condition | Number of Images | Number of success templates acquired at target session | Number of success | | Percentage % |
|---|---|---|---|---|---|---|---|
| | | | | | Vehicle number plate | Sticker id number | |
| White License Plate | Straight | 10-12am | 25 | 25 | 24 | 24 | 94 |
| | 30° from left/right | 10-12am | 25 | 25 | 23 | 23 | 92 |
| | Straight | 1-5 pm | 25 | 25 | 21 | 23 | 88 |
| | Straight | Night | 11 | 11 | 9 | 9 | 82 |

The obtained results are shown in Table 1to be capable of finding the character regardless of font style and size also capturing angle and illumination. It should be notified that the implemented algorithm shown decreased performance over factors such as bends, broken and blurred characters in license plate. The average execution time to execute a frame is calculated as 0.5 Seconds, where as for the localization of character and sticker id requires 0.1 Seconds and for the mapping, the algorithm requires 0.3 Seconds.

## 4    Case Study

The developed algorithm has been successfully tested using the IVY LAB surveillance video dataset [29, 30]. The dataset contains the human facial information which is meant for video surveillance and tracking of unauthenticated users in the campus environment. The results are found similar and shown in Fig. 10 and 11.

a                    b                    c

**Fig. 10** Results of tracking an intended person in consecutive frames from a- c

**Fig. 11** Tracking of intended person in consecutive frames from a-c

# 5 Conclusion

In this study, a real time automated vehicle surveillance system is demonstrated in a campus network. The networked cameras are successfully traded off and exchanged the abnormal findings such as mismatched sticker id and unauthenticated vehicle successfully among others, without manual operator intervention. The data-logging is also performed in the database. The results are obtained by varying the capturing angles, light illumination and style and size of characters in license plate.    The algorithm is found to be robust and successfully integrated also as embedded system platform. The algorithm is found to be very useful in any campus environments, such as factory, educational institution, military establishments and commercial organizations.

**Acknowledgements** The authors wish to thank Department of Science and Technology for awarding a project under Cognitive Science Initiative Programme (DST File No.: SR/CSI/09/2011) to Velammal Engineering College, Chennai through which the work has been implemented and also thanks IVY LAB surveillance video dataset used for algorithm verification.

# References

1. Joshi, K.A., Thakore, D.G.: A survey on moving object detection and tracking in video surveillance system. International Journal of Soft Computing and Engineering **2**(3), 44–48 (2012)
2. Kim, I.S., Choi, H.S., Yi, K.M., Choi, J.Y., Kong, S.G.: Intelligent visual surveillance—a survey. International Journal of Control, Automation and Systems **8**(5), 926–939 (2010)
3. Patel, C., Shah, D., Patel, A.: Automatic number plate recognition system: a survey. International Journal of Computer Application **69**(9), 21–33 (2013)
4. Vinod, M., Sravanthi, T., Reddy, B.: An adaptive algorithm for object tracking and counting. International Journal of Engineering and Innovative Technology **2**(4), 64–69 (2012)
5. Quaritsh, M., Kreuzthaler, M.: Autonomous multicamera tracking on embedded smart camera. EURASIP Journal on embedded system (2006)
6. Kim, J.S., Yeom, D.H., Joo, Y.H.: Intelligent unmanned anti-theft system using network camera. International Journal of Control, Automation and System **8**(5), 967–974 (2010)

7. Appiah, K., Hunter, A., Owens, J.: Autonomous real time surveillance system with distributed IP cameras. In: 3rd ACM/IEEE International Conference on Distributed Smart Cameras, Como, Italy (2009)

8. Clarot, P., Ermis, E.B., Jodoin, P.M., Saligrama, V.: Unsupervised camera network structure estimation based on activity. In: 3rd ACM/ IEEE Conference, Como, Italy (2009)

9. Leistner, C., Sterzacher, A.: Visual online learning in distributed camera networks. In: Second ACM/IEEE International Conference on Distributed Smart Cameras, Stanford, CA, pp. 1–10 (2008)

10. Chung, K.W.: Adaptive learning for target tracking and true linking discovering across multiple non-overlapping cameras. IEEE Transaction on Multimedia 13(4), 625–638 (2011)

11. Lin, C.H., Wolf, M., Kout Soukos, X.: System and software architecture of distributed smart cameras. ACM Transaction on Embedded Computing Systems 9(4) (2010)

12. Shirmo Hammadi, B., Taylor, C.J.: Distributed target tracking using self-localizing smart camera network. In: Proceedings of the Fourth ACM/IEEE International Conference on Distributed Smart Cameras, New York, USA, pp. 17–24 (2010)

13. Kulkarni, M., Wadekar, P., Dagale, H.: Real time object tracking system using distributed smart cameras. In: International Conference on Distributed Smart Cameras, Atlanta, USA (2010)

14. Geetha, B., Gokul, K.: Cloud based anti-vehicle theft by using number plate recognition. International Journal of Engineering Research and General Science 2(2), 147–151 (2014)

15. Rao, P., Saluia, P., Sharma, N.: Cloud computing for internet of things and sensing based applications. In: Sixth International Conference Sensing Technology (ICST), pp. 374–380 (2012)

16. Zaslavsky, A., Perera, C., Georgakopoulos, D.: Sensing as a service and big data (2013). http://arxiv.org/abs/1301.0159

17. Pu, B., Zhou, F., Bai, X.: Particle filter based on color feature with contour information adaptively integrated for object tracking. In: Fourth International Symposium on Computational Intelligent and Design, Hangzhou, pp. 359–364 (2011)

18. Salhi, A., Jammoussi, A.Y.: Object tracking system using camshift, meanshift and kalman filter. World Academy of Science Engineering and Technology 64, 32–38 (2012)

19. Divya, K.N., Danti, A.: Recognition of vehicle plate number and retrieval of vehicle owner's registration details. In: International Journal of Innovation Research in Technology and Science, pp. 61–66 (2012)

20. Meshram, P., Indurkar, M., Raj, R., Chitare, N.: Automated license plate recognition using regular expression. International Journal of Engineering Research and Application, pp. 18–22 (2014)

21. Sharma, G., Sood, S., Gaba, G.S., Gupta, N.: Image recognition system using geometric matching and contour detection. Proceedings with International Journal of Computer Applications 51(17), 48–53 (2012)

22. Hu, W., Zhou, X., Li, W., Luo, W., Zhang, X., Maybank, S.: Active contour based visual tracking by integrating colors, shapes and motions. IEEE Transaction on Image Processing 22(5), 1778–1792 (2013)

23. Petrosyan, A.: Vision system for disabled people using pattern matching algorithm. In: Proceedings of the Seventh International Conference on Computer Science and Information Technologies, pp. 343–346 (2009)

24. Zhu, S., Yuille, A.: Region competition unifying snakes, region growing and bayes for multiband image segmentation. IEEE Trans. Pattern Analysis Mech. Intelligent **18**(9), 416–423 (1996). Cambridge, MA
25. Wei, L., Luo, D.: A biologically inspired computational approach to model top-down and bottom-up visual attention. Optik-International Journal for Light and Electron Optics **126**(5), 522–529 (2015)
26. Fajas, F., Farhan, Y., Remya, P.R., Ambadiyil, S.: A neural network based character recognition system for Indian standard high security number plates. International Journal of Image Processing and Visual Communication **11**(1), 32–39 (2012)
27. Rasheed, S., Naeem, A., Ishaq, O.: Automated number plate recognition using hough lines and template matching. In: Proceedings of the World of Congress on Engineering and Computer Science, San Francisco, USA, vol. 1 (2010)
28. Babu, C.N.K., Nallaperumal, K.: An efficient geometric feature based license plate localization and recognition. International Journal of Imaging Science and Engineering **2**(2), 189–194 (2008)
29. Sodemann, A.A., Ross, M.P., Borghetti, B.J.: A review of anomaly detection in automated surveillance. IEEE Transactions **42**, 1257–1272 (2012)
30. Available: http://ivylab.kaist.ac.kr/demo/vs/dataset.html

# Automatic Left Ventricle Segmentation in Cardiac MRI Images Using a Membership Clustering and Heuristic Region-Based Pixel Classification Approach

Vinayak Ray and Ayush Goyal

**Abstract** This study demonstrates an automated fast technique for left ventricle boundary detection in cardiac MRI for assessment of cardiac disease in heart patients. In this work, fully automatic left ventricle extraction is achieved with a membership clustering and heuristic region-based pixel classification approach. This novel heuristic region-based and membership-clustering technique obviates human intervention, which is necessary in active contours and level sets deformable techniques. Automatic extraction of the left ventricle is performed on every frame of the MRI data in the end-diastole to end-systole to end-diastole cardiac cycle in an average of 0.7 seconds per slice. Manual tracing of the left ventricle wall in the multiple slice MRI images by radiologists was employed for validation. Ejection Fraction (EF), End Diastolic Volume (EDV), and End Systolic Volume (ESV) were the clinical parameters estimated using the left ventricle area and volume measured from the automatic and manual segmentation.

## 1 Introduction

Heart dysfunction is one of the most prominent diseases worldwide. Cardiac imaging allows cardiologists to examine the heart of a patient to assess cardiac function. Assessment of cardiac function is commonly obtained with estimation of the left ventricle ejection fraction, a measure indicating the efficiency of the left ventricle's function of pumping blood out from the heart into the body through the aorta. Calculation of the left ventricle ejection fraction [1-2] requires measuring the volume of the left ventricle at end systolic and end diastolic phases. Left ventricle volume and area measurements can be obtained with segmentation of the left ventricle. Several

V. Ray · A. Goyal(✉)
Amity University Uttar Pradesh, Sector 125, Noida, U.P., India
e-mail: ray.vinayak@gmail.com, agoyal1@amity.edu

© Springer International Publishing Switzerland 2016
S.M. Thampi et al. (eds.), *Advances in Signal Processing and Intelligent Recognition Systems*,
Advances in Intelligent Systems and Computing 425,
DOI: 10.1007/978-3-319-28658-7_52

615

modalities such as ultrasound, X-ray, single photon emission computed tomography (SPECT), computed tomography (CT), and MRI exist for imaging the heart. However, MRI is the preferred modality [3-5] since the other modalities require nephrotoxic contrast injection and cause exposure to radiation. The high resolution of MRI images has also been found useful for precise left ventricle area and volume measurements. Multi-frame cardiac MRI across the full end-diastole to end-systole to end-diastole cardiac cycle is routinely utilized to capture the contraction and expansion of the left ventricle. From the segmentation of the left ventricle in all the frames across the full cardiac cycle, the area and volume can be calculated, from which clinical parameters such as stroke volume, end-diastolic volume, end-systolic volume, and ejection fraction can be estimated. Cardiologists in practice trace the left ventricles themselves, a highly tedious and human error prone procedure. Often there may be discrepancies and variability in the segmentations of one cardiologist versus another due to human error in outlining the left ventricle in the presence of papillary muscles, trabecular muscles, and other such intricate components at the boundary of the left ventricular cavity. Hence there is a need for automatic segmentation of the left ventricle to save the precious time of cardiologists and remove human error.

Several methods have been presented in literature for automatic segmentation of the left ventricle. Deformable methods such as active contours and level sets are very commonly published for left ventricle segmentation [6-8]. However, these methods have the disadvantage of requiring manual initialization of the contour. They are also time-consuming as the contour evolves towards the boundary at each iteration. Region growing methods have the same problem of requiring a manual seed point initialization from which the region expands to include neighboring pixels of similar intensity, texture, or gray level. Edge detection methods are not robust enough for left ventricle segmentation because the boundary of the left ventricle is not clearly defined at many points. Hence, we present a pixel classification approach that is based on membership clustering and heuristic region-based segmentation. This combined algorithm is fast because it does not require iterations and it is fully automatic as it does not require any manual initialization. This method first classifies each pixel as belonging to the foreground or background based on which cluster it has highest membership to. The clustered image is thresholded automatically since the membership clustering divides pixels into foreground and background. Connected regions in the clustered image are detected and labeled and the region which is closest to the image centre and most circular is detected as the left ventricle. The circularity is measured as the inverse of eccentricity and the distance to the image centre is used to measure how close the region is to the image centre. These heuristics works since the left ventricle is relatively the most circular region in the cardiac MRI images as compared to the right ventricle, myocardium, and other structures in the heart. Also, given that the left ventricle is the object or region of interest, MRI images generally have the ventricles in the centre of the image. Post the segmentation across the full cardiac cycle, the area and volume of the left ventricle was calculated at the end-diastolic and end-systolic image slices. These parameters were used to calculate stroke volume and ejection fraction. These clinical parameters were calculated using the manual segmentation performed by expert radiologists and automatic segmentation performed by the method presented in this paper for comparison and validation.

## 2 Methodology

The left ventricle in the cardiac MRI images was extracted using the procedure shown in Fig. 1. The subsections below give the details of each step.

**Fig. 1** Diagram of the proposed procedure for rapid completely automatic left ventricle segmentation from cardiac MRI. Pixels are classified using membership clustering. Heuristic connected component analysis is used to label the connected regions. Noise structures are removed through median filtering. The left ventricle region is defined by selecting the connected region which is most circular and closest to the image centre. Tracing of the boundary of the left ventricle region can then be overlaid on the original input MRI image.

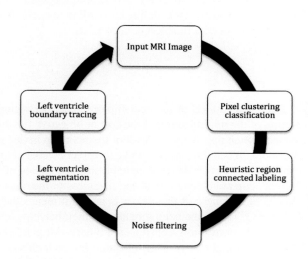

### 2.1 Automatic Segmentation Algorithm

Pixels in the image were classified using membership or fuzzy c-means clustering. Two clusters were formed - background and foreground. This membership clustering is a method for thresholding the foreground from the background. Post the clustering step, all the foreground pixels were divided into labeled connected regions. Region-labeling separates the ventricular connected regions from the foreground. Noise structures are removed from the connected regions using median filtering. The left ventricle connected region is heuristically differentiated from the other connected regions based on the two assumptions that the left ventricle connected region will be centermost in the image and will be most circular in shape. So the distance from the image centre and eccentricity are the properties of the connected labeled regions that are used as the heuristics to define the left ventricle region. Once the left ventricle region is defined, the list of pixels at the boundary can be used for overlaying the segmented left ventricle boundary on the original MRI image.

## 2.2 Pixel Clustering

Fuzzy c-means or membership clustering groups similar pixels into clusters. Similarity measures include distance, connectivity, and pixel intensity. Whereas in hard clustering, a pixel belongs to only one cluster, in fuzzy c-means or membership clustering, a pixel can belong to more than one cluster. The belongingness or membership to a particular cluster is greater if the membership level is greater for that cluster. Membership clustering or fuzzy c-means groups $n$ pixels into $c$ belongingness clusters using a similarity measure. Each cluster has a cluster centre defined for it. A partition matrix consists of levels of belongingness or membership $w$, i.e. membership of pixel $x$ to cluster $k$, defined according to eq. 1. Minimization is performed on the objective or cost function defined in eq. 1:

$$w_k(x) = \frac{1}{\sum_j \left( \frac{d(center_k, x)}{d(center_j, x)} \right)^{\frac{2}{(m-1)}}} \tag{1}$$

Each pixel's level of membership or cluster belongingness is updated at each iteration of the minimization of the cost or objective function in eq. 1. The iteration is stopped when the change in level of membership in one iteration is below a specific threshold. When the iteration ends, the pixel is labeled to the cluster it has the highest belongingness or level of membership to. Values of belongingness or levels of membership are denoted by $w$ and fuzziness of the cluster is denoted by $m$ in eq. 1. When the fuzziness of the cluster is larger, it means the levels of membership will be lesser. If $m = 1$, the belongingness or membership can take on only two values, 0 or 1 i.e. the partitions or clusters are binary. All pixels have a set of coefficients of levels of belongingness or membership for each cluster. The centre of the clusters, denoted as $c$ are defined as the weighted average of all the pixels in the cluster:

$$c_k = \frac{\sum_x w_k(x)^m x}{\sum_x w_k(x)^m} \tag{2}$$

The membership level $w$ of a pixel $x$ to a cluster $k$ is inversely proportional to the distance between the pixel and the cluster centre. The cluster fuzziness $m$ determines the weight for each cluster centre.

The steps for the fuzzy c-means clustering or membership clustering are enumerated below:

1. The number of clusters is defined, in the case of segmentation, two.
2. The centres of each cluster are determined.
3. The partition matrix consisting of membership levels is updated on each iteration.
4. The iterations continue until the change in membership is less than a certain threshold.
5. At the end of the iterations, each pixel is labeled to belong to the cluster for which it has least distance to the cluster centre of and largest membership level to.

# 3  Results

Fig. 2 shows the procedure of the left ventricle automatic extraction proposed in this study using fuzzy c-means membership clustering and connected region labeling pixel classification. This method obviates iterations or user seed point initialization and is thus fast and easily applicable for segmenting the left ventricle over several MRI image frames rapidly.

**Fig. 2**  The procedure of the fast completely automatic left ventricle segmentation presented in this paper is shown here stepwise. After the MRI image is loaded in, membership clustering on the input image divides the pixels into foreground or background. Connected component analysis on the clustered image labels each connected region. Median filtering removes noisy structures and regions. Next is heuristics-based selection of the left ventricle region with the least eccentricity (inverse of circularity) and least distance of connected region centre from image centre (closest to image centre). Finally boundary tracing of the left ventricle region allows visualizing the boundary on the original input MRI image.

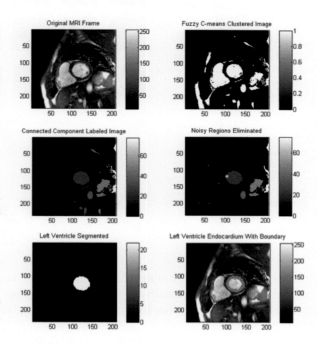

Fig. 3 illustrates that this method completely automatically loads in and performs left ventricle segmentation in all slices in the full cardiac cycle from end-diastole to end-systole to end-diastole in the cardiac MRI images.

Fig. 4 graphs a plot of the left ventricle volume over all the frames in the full cardiac cycle from end-diastole to end-systole to end-diastole. The volume was calculated from the area, which is the number of pixels in the left ventricle region. The left ventricle volume is computed using the following formula:

$$V = A p_h p_w p_t \qquad (3)$$

The left ventricular volume $V$ ($mm^3$) is equal to the number of pixels $A$ in the left ventricle region multiplied by the pixel dimensions, height $p_h$ ($mm$), width $p_w$ ($mm$), and thickness $p_t$ ($mm$). The plot in Fig. 4 shows the systolic and diastolic phases of the

**Fig. 3** The left ventricle is segmented completely automatically ( shown as red circles) throughout all frames of the complete cardiac cycle from end-diastole to end-systole to end-diastole. The procedure is fully automatic and loads in and processes each image frame without any user selection or manual intervention. The proposed method in this paper works successfully in segmenting the left ventricle in all frames of the complete cardiac cycle from end-diastole to end-systole to end-diastole.

**Fig. 4** Plot of the LV volume calculated from automatic and manual segmentations. LV volume (cubic millimeters) is the LV area in pixels multiplied by pixel dimensions (millimeters squared) multiplied by slice thickness (millimeters).

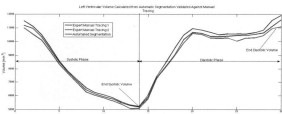

contraction and expansion of the left ventricle. The left ventricle volume calculation using the automatic segmentation presented in this paper is validated with the left ventricle volume calculations using the manual tracing-based segmentations by two experts.

Table 1 tabulates the clinical parameters calculated with the automatic and manual tracing-based segmentation of the left ventricle. The end-systolic volume (ESV) and end-diastolic volume (EDV) can be measured at the end-systole and end-diastole points from the left ventricular volume plot in Fig. 4. Using these parameters, we obtain the stroke volume (SV) as follows:

$$SV = EDV - ESV \tag{4}$$

The ejection fraction (EF) is calculated as follows:

$$EF = \frac{SV}{EDV} 100 \tag{5}$$

For validation of the left ventricle extraction and segmentation method presented in this paper, manual tracing-based segmentation by two experts was used. As shown

**Table 1** Calculation of clinical parameters end diastolic volume, end systolic volume, stroke volume, ejection fraction, etc. using automatic and manual segmentation.

| Method | ED Area (mm$^2$) | ES Area (mm$^2$) | ED Volume (mm$^3$) | ES Volume (mm$^3$) | Stroke Volume (mm$^3$) | Ejection Fraction |
|---|---|---|---|---|---|---|
| Manual 1 | 1924.41 | 869.90 | 11546.50 | 5219.43 | 6327.06 | 0.55 |
| Manual 2 | 1824.80 | 847.33 | 10948.85 | 5083.96 | 5864.88 | 0.54 |
| Automatic | 1851.37 | 863.27 | 11108.23 | 5179.59 | 5928.64 | 0.53 |

in Figs. 5-8, the correlation coefficients between automatic and manual tracing-based segmentations were higher (0.945 on average) than the correlation coefficients between the two manual tracing-based segmentations (0.855 on average) for all four slices tested. The correlation between automatic versus manual segmentation is higher than the correlation between the two manual segmentations. The error margin between the automatically segmented left ventricle boundary and volume and the manually traced boundary and volume is less than the error margin between the two manually traced boundaries and computed volume. Hence, the automatic method is sufficiently accurate if considering the manual tracing by experts as the gold standard for validation.

The validation of the automatic left ventricle extraction method presented in this paper has been done in three ways. Firstly, comparison was performed of the automatic left ventricle segmentation to manual tracing-based left ventricle segmentation by two experts. The results of the comparison are shown in Figs. 5-8 and they show

**Fig. 5** Slice 1 test of automatic compared to manual segmentation. Correlation coefficients between the automatic (red) and two manual segmentations (blue and yellow) were 0.987 and 0.945, respectively, which were higher than the correlation coefficient 0.932 between the two manual segmentations.

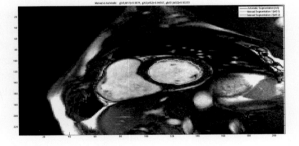

**Fig. 6** Slice 2 test of automatic compared to manual segmentation. Correlation coefficients between the automatic (red) and two manual segmentations (blue and yellow) were 0.932 and 0.951, respectively, which were higher than the correlation coefficient 0.785 between the two manual segmentations.

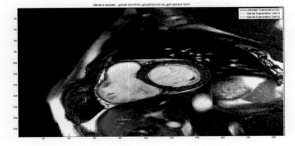

**Fig. 7** Slice 3 test of automatic compared to manual segmentation. Correlation coefficients between the automatic (red) and two manual segmentations (blue and yellow) were 0.844 and 0.994, respectively, which were higher than the correlation coefficient 0.805 between the two manual segmentations.

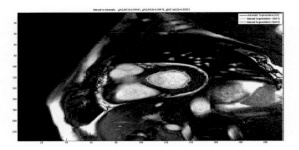

**Fig. 8** Slice 4 test of automatic compared to manual segmentation. Correlation coefficients between the automatic (red) and two manual segmentations (blue and yellow) were 0.984 and 0.916, respectively, which were higher than the correlation coefficient 0.897 between the two manual segmentations.

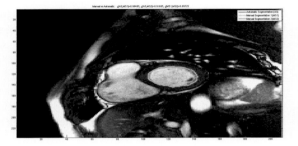

that the correlation between the automatic and manual segmentation is higher than the correlation between manual segmentations by two experts. Since, the individual errors between the automatic and two manual segmentations were less than the error between the two manual segmentations, the method is sufficiently accurate for left ventricle segmentation. Secondly, clinical parameters were calculated using the automatic left ventricle segmentation and the manual tracing-based segmentations. The clinical parameters calculated with the automatic segmentation were close to and well within the range of the values obtained with the manual tracing-based segmentations. Finally, the left ventricular volume calculated from the automatic segmentation was plotted and it falls well between the left ventricular volumes obtained from the manual tracing-based expert segmentations.

## 4   Discussion

The novel fuzzy c-means membership clustering and connected heuristic region-labeling based method in this paper is fast, completely automatic, and user-intervention-free. The pixel clustering and connected region labeling extract the left ventricle rapidly in 0.7 seconds (on average) for one slice, which results in extraction of the left ventricle in all 30 frames of a full cardiac cycle in under 21 seconds. Therefore this method has the advantage of speed and automation, since it is fast, user-intervention-free, and iteration-free. The technique presented in this paper for left ventricle extraction has a much less computational cost and better performance in terms of speed than level sets based or active snakes based deformable model methods.

# References

1. Bellenger, N.G., Burgess, M.I., Ray, S.G., Lahiri, A., Coats, A.J.S., Cleland, J.G.F., Pennell, D.J.: Comparison of left ventricular ejection fraction and volumes in heart failure by echocardiography, radionuclide ventriculography and cardiovascular magnetic resonance. Are they interchangeable? European Heart Journal **21**(16), 1387–1396 (2000)
2. Germano, G., Kiat, H., Kavanagh, P.B., Moriel, M., Mazzanti, M., Su, H.T., Train, K.F.V., Berman, D.S.: Automatic quantification of ejection fraction from gated myocardial perfusion SPECT. Journal of Nuclear Medicine **36**(11), 2138 (1995)
3. Kaus, M.R., von Berg, J., Weese, J., Niessen, W., Pekar, V.: Automated segmentation of the left ventricle in cardiac MRI. Medical Image Analysis **8**(3), 245–254 (2004)
4. Lynch, M., Ghita, O., Whelan, P.F.: Automatic segmentation of the left ventricle cavity and myocardium in MRI data. Computers in Biology and Medicine **36**(4), 389–407 (2006)
5. Lynch, M., Ghita, O., Whelan, P.F.: Segmentation of the left ventricle of the heart in 3-D+ t MRI data using an optimized nonrigid temporal model. IEEE Transactions on Medical Imaging **27**(2), 195–203 (2008)
6. Paragios, N.: A level set approach for shape-driven segmentation and tracking of the left ventricle. IEEE Transactions on Medical Imaging **22**(6), 773–776 (2003)
7. Lynch, M., Ghita, O., Whelan, P.F.: Left-ventricle myocardium segmentation using a coupled level-set with a priori knowledge. Computerized Medical Imaging and Graphics **30**(4), 255–262 (2006)
8. Xu, C., Pham, D.L., Prince, J.L.: Image segmentation using deformable models. Handbook of Medical Imaging **2**, 129–174 (2000)

# Mixed Noise Removal Using Hybrid Fourth Order Mean Curvature Motion

**V.B. Surya Prasath and P. Kalavathi**

**Abstract** Image restoration is one of the fundamental problems in digital image processing. Although there exists a wide variety of methods for removing additive Gaussian noise, relatively few works tackle the problem of removing mixed noise type from images. In this work we utilize a new hybrid partial differential equation (PDE) model for mixed noise corrupted images. By using a combination mean curvature motion (MMC) and fourth order diffusion (FOD) PDE we study a hybrid method to deal with mixture of Gaussian and impulse noises. The MMC-FOD hybrid model is implemented using an efficient essentially non-dissipative (ENoD) scheme for the MMC first to eliminate the impulse noise with a no dissipation. The FOD component is implemented using explicit finite differences scheme. Experimental results indicate that our scheme obtains optimal denoising with mixed noise scenarios and also outperforms related schemes in terms of signal to noise ratio improvement and structural similarity.

## 1 Introduction

Gaussian noise removal from digital images by partial differential equations (PDEs) is a well studied area. Edge preservation along with selective smoothing has made PDE based image restoration an attractive option in mathematical image processing.

V.B. Surya Prasath(✉)
Computational Imaging and Visualization Analysis (CiVA) Lab,
Department of Computer Science, University of Missouri, Columbia, MO 65211, USA
e-mail: prasaths@missouri.edu
http://goo.gl/F6ZmVE

P. Kalavathi
Department of Computer Science and Applications, Gandhigram Rural Institute – Deemed
University, Gandhigram, Dindigul 624 302, Tamil Nadu, India
e-mail: pkalavathi.gri@gmail.com

© Springer International Publishing Switzerland 2016
S.M. Thampi et al. (eds.), *Advances in Signal Processing and Intelligent Recognition Systems*,
Advances in Intelligent Systems and Computing 425,
DOI: 10.1007/978-3-319-28658-7_53

625

Among a wide range of PDE based models applied to image denoising we mention the total variation TV model studied by Rudin et al. [1] and related adaptive models [2–4], Perona and Malik anisotropic diffusion [5] and its variants [6–10], mean curvature motion (MCM) [11, 12], and fourth order diffusion (FOD) [13], see [14] for a review.

Although these diffusion based schemes perform better in general for images which are corrupted by white additive Gaussian noise they perform poorly on other noise (multiplicative) scenarios without necessary modifications. Few works have considered mixed noise case in a PDE framework, as it is challenging than handling pure multiplicative noise alone. One of the interesting model is proposed by Kim [15] where a hybrid scheme is studied by combining the TV along MMC for impulse and Gaussian noise removal. Unfortunately, the model is derived from the total variation diffusion and hence it inherits the well-known staircasing artifact associated with traditional TV based restorations [16].

In this work we consider a similar hybrid combination akin to [15] where we replace the total variation with a general fourth order diffusion [13] component which improves results. The model is utilized for removing mixed impulse and Gaussian noise with efficient numerical implementation done by using a combination of essentially non-dissipative (ENoD) and explicit finite differences [17]. In the first step we use ENoD based iterative implementation of MMC and after a few iterations we utilize the general fourth order diffusion until convergence. Experimental results are given in mixed noise corrupted images and comparison with different models indicate we obtain better performance and edge preservation.

The rest of the paper is organized as follows. Section 2 gives the proposed hybrid denoising model. Section 3 provides experimental support on various images corrupted by mixed impulse and Gaussian noises. Finally, Section 4 concludes the paper.

## 2 Hybrid PDE for Mixed Noise

We first recall the mean curvature motion (MCM) [11] with general diffusion coefficients [18]. It is given by a second order diffusion model and is proven provide better speckle-free restoration results. For a given image $u : \Omega \subset \mathbb{R}^2 \to R$ with $\Omega$ the image domain, $u_0$ the noisy input image, the MCM is written as the PDE,

$$\frac{\partial u(x, t)}{\partial t} = |\nabla u(x, t)| \; \nabla \cdot (g(|\nabla u(x, t)|)\nabla u(x, t)) - (u(x, t) - u_0(x)), \quad (1)$$

where $x = (x_1, x_2)$ pixel co-ordinates, $\nabla = (\partial_{x_1}, \partial_{x_2})$ is the gradient operator, and $g(\cdot)$ is the diffusion coefficient which is a decreasing function. For example, the anisotropic diffusion [5] based coefficient can be chosen $g(s) = (1 + s^2/K)^{-1}$ with $K > 0$, and the MCM provides a scale space of solutions $\{u(x, t)\}_{t=0}^{\infty}$. Despite its success in image denoising it can create staircasing artifacts and fourth order

diffusion (FOD) models [13] can overcome this by utilizing higher order derivatives
to drive the diffusion flow. The FOD PDE is written as,

$$\frac{\partial u(x,t)}{\partial t} = -\Delta \cdot (g(|\Delta u(x,t)|)\Delta u(x,t)) - (u(x,t) - u_0(x)), \qquad (2)$$

where $\Delta = (\partial_{x_1 x_1}, \partial_{x_2 x_2})$ is the Laplace operator. Moreover, both the MCM and FOD
works fairly well under additive Gaussian noise though these can not perform well
for removing multiplicative noise.

To achieve both staircasing free restorations and to handle mixed noises that
maybe present in digital images we propose a generalized MCM and FOD model.
We consider the hybrid mean curvature motion fourth order diffusion, i.e. combining
MCM (1) and FOD (2), PDE along initial and boundary conditions,

$$\begin{cases} \dfrac{\partial u(x,t)}{\partial t} = -|\nabla u|^\alpha \, \Delta \cdot \left( \dfrac{\Delta u(x,t)}{\|\Delta u(x,t)\|^{1+\omega}} \right) - \beta(u(x,t) - u_0(x)), & x \in \Omega, \ \alpha, \beta, \omega \geq 0 \\ u(x,0) = u_0(x), & x \in \Omega, \\ \dfrac{\partial u}{\partial n} = 0, & x \in \partial\Omega. \end{cases} \qquad (3)$$

Note that the propose model in (3) becomes the traditional fourth order diffusion
model when $\alpha = 0$ and $\omega = 1$; fourth order MCM for $\alpha = 1$, $\omega = 1$. Following,
Kim [15], we call our proposed fourth order model (3) $\alpha\beta\omega4$ (ABO4)-*model*. The
hybrid algorithm we propose is uses the following two steps:

1. Begin with $\alpha = 1$, $\beta = 0$, and $\omega = 0$ and solve (3) using ENoD discretization
   scheme, and let the solution obtained be $u^M$.
2. After 1 or 2 iterations,

   a. assign $u_0 = u^M$
   b. reset $\alpha = 1 + \omega$, $\omega > 0$, and $\beta > 0$
   c. Run this Fourth order MCM upto convergence.

We remark that as in [15] we used ENoD for edge stopping to the MCM to eliminate
impulse noise first. The second step then involves solving a fourth order generalized
MCM which then removes the Gaussian noise effectively without artifacts associated
with TV diffusion model done in [15]. Thus, we obtained improved restoration as
will be vindicated by numerical examples in the next section. For more details on
the ENoD we refer to [15] and [17, 19] for convergence results of explicit finite
differences and wellposedness of the PDE (3) will be reported elsewhere.

## 3   Experimental Results

All the parameters were tuned to obtain best possible result in terms of best signal to
noise ration (SNR) and the stopping time of the diffusion flows is based on tolerance of
$|u(x, t-1) - u(x,t)| \leq 10^{-12}$. Figure 1 show an example result using our scheme
for the *Elaine* gray scale image with size $256 \times 256$. The input image, Figure 1(a) is

(a) Original                                    (b) Noisy

(c) Step1                                       (d) Step2

**Fig. 1** Our proposed two stage mixed noise removal scheme on *Elaine* image. (a) Noise-free original image for comparison (b) Noisy image (Gaussian variance 400, impulse strength 20%). Close up of the image after the first step ENoD based fourth order MCM result (c) and the final result with our ABO4 scheme (3), SNR=16.74 *dB* (d). Better viewed online and zoomed in.

corrupted with Gaussian noise of variance 400 followed by a random-valued impulse noise of probability 20%. As can be seen in Figure 1(c), the first step1 of removing the impulse noise using ENoD based MCM is effective and does a decent job in removing the impulse pixels. The final step2 based result using the generalized FOD smoothing removes Gaussian noise and preserves edges as can be seen in the close-up Figure 1.

We next show in Figure 2 the restoration results obtained with Perona-Malik diffusion (PMD) [5], total variation diffusion (TVD) [1], and Kim's ABO [15] for the *Peppers* gray scale image of size 256 × 256. In comparison our ABO4 performs

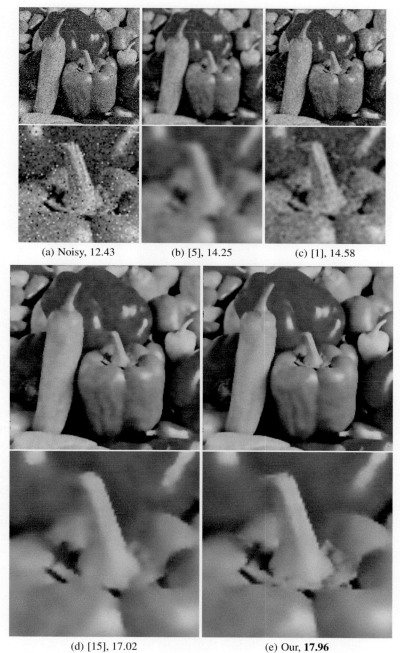

(a) Noisy, 12.43  (b) [5], 14.25  (c) [1], 14.58

(d) [15], 17.02     (e) Our, **17.96**

**Fig. 2** Comparison of mixed noise removal on the *Peppers* image for various schemes. SNR (in *dB*) values are given for each result. (a) Noisy image (Gaussian variance 200, impulse strength 20%) (b) PMD [5] (c) TVD [1] (d) ABO [15] (e) Our ABO4 (3). Top row: full images, Bottom row: Close-up. Better viewed online and zoomed in.

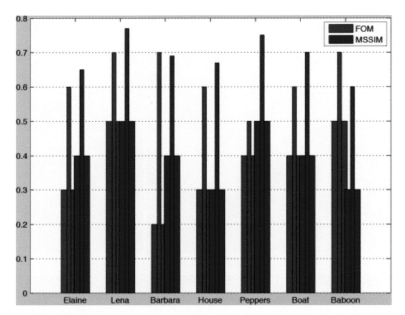

**Fig. 3** FOM/MSSIM improvement over the ABO model of Kim [15] for different images corrupted by Gaussian noise of variance 200 and impulse noise of probability 20%. The thin bars (length) at the top indicate the amount of improvement in FOM and MSSIM values.

well without artifacts, over-smoothing and obtains better SNR values as indicated in each sub-figure.

Figure 3 shows the improved obtained between the ABO model of [15] and our proposed ABO4 using Pratt's figure of merit (FOM) [20] and mean structural similarity (MSSIM) metric [21] for different test images. The FOM is a measure of edge preservation whereas MSSIM measures the overall measure of structural similarity between the original and the restored images. Values closer to one for FOM and MSSIM indicates superior performance and as can be seen in Figure 3 our model improves the original formulation for all the images. Extensions to handle multichannel images [22–25] and simultaneously deblurring images [26] define our future works in this direction. Test images, further visualizations, and extra results are available online [27].

## 4  Conclusion

In this paper we proposed a new hybrid PDE model which can remove the mixture of Gaussian and impulse noises from digital images. By utilizing a combination mean curvature motion and fourth order diffusion an edge preserving image smoothing method is obtained. Experimental results indicate the proposed model works well for a variety of images without artifacts associated with other models and gets better signal to noise ratio and structural similarity with noise-free images for the final denoised images.

# References

1. Rudin, L., Osher, S., Fatemi, E.: Nonlinear total variation based noise removal algorithms. Physica D **60**(1–4), 259–268 (1992)
2. Prasath, V.B.S., Singh, A.: A hybrid convex variational model for image restoration. Applied Mathematics and Computation **215**(10), 3655–3664 (2010)
3. Prasath, V.B.S.: A well-posed multiscale regularization scheme for digital image denoising. International Journal of Applied Mathematics and Computer Science **21**(4), 769–777 (2011)
4. Prasath, V.B.S., Vorotnikov, D., Pelapur, R., Jose, S., Seetharaman, G., Palaniappan, K.: Multiscale Tikhonov-total variation image restoration using spatially varying edge coherence exponent. IEEE Transactions on Image Processing **24**(12), 5220–5235 (2015)
5. Perona, P., Malik, J.: Scale-space and edge detection using anisotropic diffusion. IEEE Transactions on Pattern Analysis and Machine Intelligence **12**(7), 629–639 (1990)
6. Prasath, V.B.S., Singh, A.: Controlled inverse diffusion models for image restoration and enhancement. In: First International Conference on Emerging Trends in Engineering and Technology (ICETET), Nagpur, India, pp. 90–94 (2008)
7. Prasath, V.B.S., Singh, A.: Multispectral image denoising by well-posed anisotropic diffusion scheme with channel coupling. International Journal of Remote Sensing **31**(8), 2091–2099 (2010)
8. Prasath, V.B.S., Singh, A.: Well-posed inhomogeneous nonlinear diffusion scheme for digital image denoising. Journal of Applied Mathematics **2010** (2010). 14pp, Article ID 763847
9. Prasath, V.B.S., Vorotnikov, D.: Weighted and well-balanced anisotropic diffusion scheme for image denoising and restoration. Nonlinear Analysis: Real World Applications (2013)
10. Prasath, V.B.S., Urbano, J.M., Vorotnikov, D.: Analysis of adaptive forward-backward diffusion flows with applications in image processing. Inverse Problems **31**(105008) (2015). 30pp
11. El-Fallah, A.I., Ford, G.E.: On mean curvature diffusion in nonlinear image filtering. Pattern Recognition Letters **19**(5–6), 433–437 (1998)
12. Ruuth, S.J.: Efficient algorithms for diffusion-generated motion by mean curvature. Journal of Computational Physics **144**(2), 603–625 (1998)
13. You, Y.L., Kaveh, M.: Fourth-order partial differential equation for noise removal. IEEE Transactions on Image Processing **9**, 1723–1730 (2000)
14. Aubert, G., Kornprobst, P.: Mathematical problems in image processing: Partial differential equation and calculus of variations. Springer-Verlag, New York (2006)
15. Kim, S.: PDE-based image restoration: a hybrid model and color image denoising. IEEE Transactions on Image Processing **15**(5), 1163–1170 (2006)
16. Prasath, V.B.S., Vorotnikov, D.: On a system of adaptive coupled pdes for image restoration. Journal of Mathematical Imaging and Vision **48**(1), 35–52 (2014)
17. Prasath, V.B.S., Moreno, J.C.: Feature preserving anisotropic diffusion for image restoration. In: Fourth National Conference on Computer Vision, Pattern Recognition, Image Processing and Graphics (NCVPRIPG 2013), India, December 2013
18. Prasath, V.B.S., Singh, A.: An adaptive anisotropic diffusion scheme for image restoration and selective smoothing. International Journal of Image and Graphics **12**(1) (2012). 18pp
19. Prasath, V.B.S., Moreno, J.C.: On convergent finite difference schemes for variational - PDE based image processing. Technical report, ArXiv (2013)
20. Gonzelez, R.C., Woods, R.E.: Digital Image Processing, 2nd edn. Pearson Inc., Boston (2002)
21. Wang, Z., Bovik, A.C., Sheikh, H.R., Simoncelli, E.P.: Image quality assessment: from error visibility to structural similarity. IEEE Transactions on Image Processing **13**(4), 600–612 (2004)

22. Prasath, V.B.S., Singh, A.: Multichannel image restoration using combined channel information and robust M-estimator approach. International Journal Tomography and Statistics **12**(F10), 9–22 (2010)
23. Prasath, V.B.S.: Weighted Laplacian differences based multispectral anisotropic diffusion. In: IEEE International Geoscience and Remote Sensing Symposium (IGARSS), Vancouver, BC, Canada, pp. 4042–4045, July 2011
24. Pope, K., Acton, S.T.: Modified mean curvature motion for multispectral anisotropic diffusion. In: IEEE Southwest Symposium on Image Analysis and Interpretation, Tucson, AZ, USA, pp. 154–159, April 1998
25. Moreno, J.C., Prasath, V.B.S., Neves, J.C.: Color image processing by vectorial total variation with gradient channels coupling. Inverse Problems and Imaging (2015)
26. Prasath, V.B.S., Singh, A.: Ringing artifact reduction in blind image deblurring and denoising problems by regularization methods. In: Seventh International Conference in Advances in Pattern Recognition (ICAPR), Kolkata, India, pp. 333–336 (2009)
27. October 2015. http://sites.google.com/site/suryaiit/research/aniso/abo4

# An Implicit Segmentation Approach for Telugu Text Recognition Based on Hidden Markov Models

D. Koteswara Rao and Atul Negi

**Abstract** Telugu text is composed of aksharas (characters). The presence of split and connected aksharas in Telugu document images causes segmentation difficulties and the performance of the Telugu OCR systems is affected. Our novel approach to solve this problem is using an implicit segmentation for recognizing words. The implicit segmentation approach does not need prior segmentation of the words into aksharas before they are recognized. Since the Hidden Markov models (HMM) are successfully applied for phoneme recognition with no prior segmentation of the speech into phonemes in the automatic speech recognition applications. In this paper, we report on the use of continuous density Hidden Markov Models for representing the shape of aksharas to build Telugu text recognition system. The sliding window method is used for computing simple statistical features and 450 akshara HMMs are trained. We use word bigram language model as contextual information. The word recognition relies on akshara models and contextual information of words. The word recognition involves finding the maximum likelihood sequence of akshara models that matches against the feature vector sequence. Our system recognizes words with split and connected aksharas. The performance of the system is encouraging.

## 1 Introduction

An optical character recognition (OCR) system does conversion of textual data images to machine readable form. Telugu is a widely spoken language in India and

D.K. Rao(✉)
Department of Information Technology, Mahatma Gandhi Institute of Technology,
Hyderabad, India
e-mail: dkrao@mgit.ac.in

D.K. Rao · A. Negi
School of Computer and Information Sciences, University of Hyderabad, Hyderabad, India
e-mail: atulcs@uohyd.ernet.in

© Springer International Publishing Switzerland 2016
S.M. Thampi et al. (eds.), *Advances in Signal Processing and Intelligent Recognition Systems*,
Advances in Intelligent Systems and Computing 425,
DOI: 10.1007/978-3-319-28658-7_54

633

mainly spoken in Andhra Pradesh and Telangana states of India. Most of the previous research efforts to build Telugu OCR systems [4, 8, 11] involve segmentation of document images into connected components before sending them to the classifier for recognition.

Telugu text is composed of aksharas (characters). An akshara is a unique writing unit formed from four groups of glyphs correspond to vowels, vowel modifiers, consonants and consonant modifiers called as achchus, hallus, maatras, and vaththus respectively [11]. Telugu document images may contain split and connected aksharas due to several reasons. The reasons for split aksharas are binarization, degradation, and the usage of metal shapes for different glyphs in printing. The causes for connected aksharas are the presence of noise, nature of the Telugu script that allows the usage of vowel and consonant modifiers. When the Telugu document images contain split and connected aksharas, extraction of connected components becomes troublesome.

We use words in our speech. We usually read words rather than characters in our real life. Like man reads words, we need machines with the ability of recognizing words. In our research work, we deal with how to address the split and connected akshara problems using a word recognition method [4]. The word recognition methods can be basically divided into analytical and holistic methods. The analytical methods can be further categorized into explicit segmentation and implicit segmentation methods. The explicit segmentation separates the word image at akshara boundaries, whereas implicit segmentation uses the sliding window to divide the word image into frames of fixed width. The frames are grouped into aksharas by recognition.

The HMM tool successfully makes speech recognition possible with no prior segmentation of speech into either phonemes or words. The major benefits with HMM are support for automatic training, no need of segmentation at the character and word levels, and training as well as recognition methods can be language-independent [2]. Since implicit segmentation methods do not require prior segmentation of Telugu words into aksharas before they are recognized. In this paper, we describe Telugu text recognition using an implicit segmentation method based on Hidden Markov Models. We use word bigram language model as contextual information. The word recognition relies on akshara models and contextual information of words. Our approach works, though the word images are composed of split and connected aksharas.

## 1.1 Related Work

Many research efforts in recognizing split and connected characters focus on how to avoid segmentation at the character and word levels. The HMMs do not require prior segmentation of data into words and characters. Many researchers use the HMM tool, which supports segmentation through recognition, for the recognition of split and connected characters.

The recognition of words with connected and degraded characters using the Hidden Markov Model and level building search reported in the paper [3]. The states of character HMMs are represented by sub-character segments. For each segment, they compute a feature vector representing stroke and arc features. The level building search consists of matching a sequence of character models to an observation sequence in a maximum likelihood manner.

The recognition of merged characters through avoiding segmentation of text images into words and characters presented in [5]. They extract features from the sequence of vertical scan lines of characters using FFT. A set of left-right discrete HMMs are trained for representing English character shapes. Because of the segmentation difficulties of cursive Arabic script, discrete HMMs used for recognizing Arabic text in [7]. The simple statistical features such as intensity, intensity of horizontal derivative and intensity of vertical derivative are computed using sliding window method.

The Hindi documents recognition using HMMs described in [10]. Here each character shape is represented as continuous density HMM with state emission probabilities determined by the Gaussian mixture. The system involves training of character models on script-independent statistical features extracted from the line images using sliding window. The text recognition is performed through searching the sequence of characters that is most likely generated by the input feature-vector sequence.

The recent work demonstrated in [13] for the recognition of scale and orientation invariant words in graphical documents using HMMs. They compute local gradient histogram features from frames in a path of sliding window. In another recent work presented in [9] for Telugu character recognition. They report that the major degradation of the system performance, which usually occur because of binarization and improper character segmentation, is due to broken characters. The feedback from the distance measure used by the classifier and orthographic properties of Telugu script are exploited to address the problems of broken characters and improper character segmentation respectively.

This paper is organized as follows. In Sect. 2, we give the challenges of Telugu script. The topology of akshara Hidden Markov Models are discussed in Sect. 3. We describe the proposed Telugu text recognition system in Sect. 4. The experimental results are reported in Sect. 5. Finally, we give the conclusion in Sect. 6.

## 2 Challenges

The Telugu OCR systems are facing challenges due to different properties of the Telugu script such as like-shaped aksharas, hundreds of commonly used aksharas, and shape of the aksharas changes with the use of vowel and consonant modifiers.

Telugu language has its own beautiful script, which is by nature a phonetic one. Every spoken sound unit has corresponding written unit called as akshara (character). Telugu text is composed of aksharas and the beauty of the script is coming from its good curved aksharas. There is less inter-akshara shape difference among aksharas. The basic akshara set of Telugu consists of 52 letters with 16 vowels and 36 consonants. The aksharas can be in one of the forms such as V, C, CV, CVC,

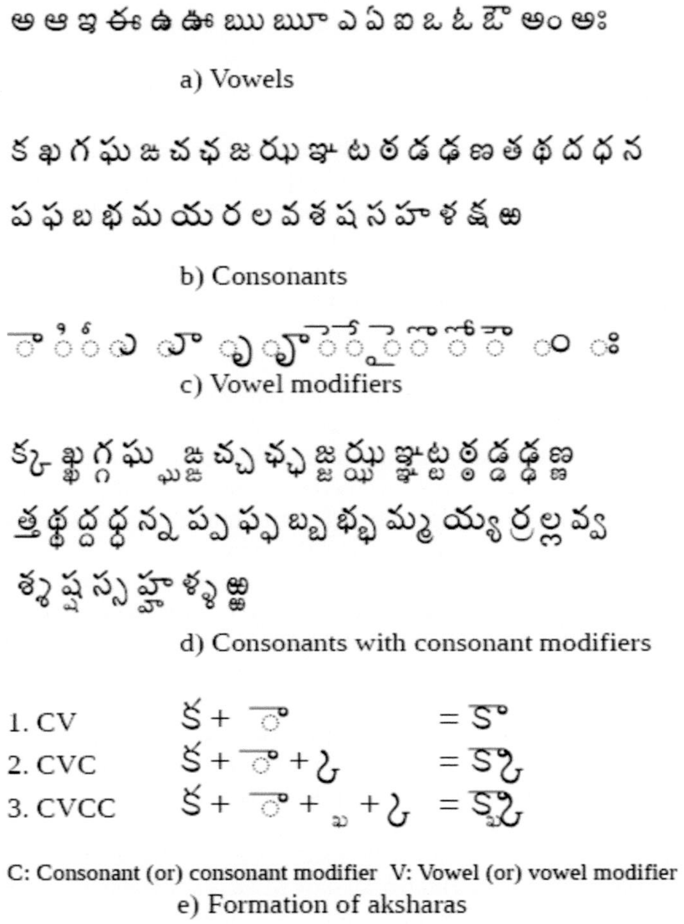

అ ఆ ఇ ఈ ఉ ఊ ఋ ౠ ఎ ఏ ఐ ఒ ఓ ఔ అం అః

a) Vowels

క ఖ గ ఘ ఙ చ ఛ జ ఝ ఞ ట ఠ డ ఢ ణ త థ ద ధ న

ప ఫ బ భ మ య ర ల వ శ ష స హ ళ క్ష ఱ

b) Consonants

c) Vowel modifiers

d) Consonants with consonant modifiers

| | | | |
|---|---|---|---|
| 1. CV | క + ో | | = కో |
| 2. CVC | క + ో + ్ | | = కో్ |
| 3. CVCC | క + ో + ్ + ్ | | = కో్్ |

C: Consonant (or) consonant modifier   V: Vowel (or) vowel modifier

e) Formation of aksharas

**Fig. 1**   The Telugu basic akshara set and akshara formation

CVCC etc., where C is a consonant or its modifier and V is a vowel or its modifier. The basic akshara set and formation of aksharas are given in Fig. 1. The commonly used aksharas are approximately between 5000 and 10000 [11].

   The vowel modifiers are usually placed over the consonants, whereas the consonant modifiers are placed at below-left, below, or below-right sides of the consonants. The use of vowel and consonant modifiers forms various akshara shapes that may result in connected aksharas. The connectedness of aksharas also depends on the type of the font. In Fig. 2, we show example word images with split, connected aksharas and their vertical projection profiles. The vertical projection profile is the sum of black pixels along every column that can be used to describe whether aksharas are split or connected.

**Fig. 2** Examples of word images with connected, split aksharas and their vertical projection profiles: a) The two aksharas are connected b) Among three aksharas, the third akshara has a split

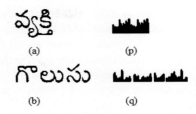

(a)

(p)

(b)

(q)

## 3 Akshara Hidden Markov Models

We use the Hidden Markov models (HMM) to represent the shape of Telugu aksharas. For each akshara, we train a continuous density HMM. The words of the system are described as sequences of akshara HMMs. We also compute word bigram language model to exploit contextual information for recognizing words. The Hidden Markov model tool kit (HTK) is used for building our system [16].

An HMM is a doubly stochastic process [12] that can be defined by a number of states $N$, a vector of initial probability of states $\pi$, a state transition matrix $N \times N$ and an observation probability distribution function. One stochastic process describes the transition between states, whereas another determines the probability of states that generate observations. The HMMs can be basically divided into continuous density HMMs and discrete HMMs depending on whether the observation probability distribution is continuous or discrete. In our system, the shape of an akshara is represented by a continuous density HMM. The state of an akshara HMM represents a frame of the akshara shape. The emission probabilities of states are determined by the Gaussian mixture and the parameters of the state are mean and variance vectors. For state $j$, the probability of generating observation $o_x$ at position $x$ along writing order is given by the Eq. 1.

$$b_j(o_x) = \frac{1}{\sqrt{(2\pi)^n|\sum|}}e^{-\frac{1}{2}(o_x-\mu)^T\sum^{-1}(o_x-\mu)} \tag{1}$$

In Fig. 3, we show the topology of akshara HMM. In our system, we use 8-state left-right akshara HMMs where states 1 and 8 are empty states. The empty states enable the connectivity between models for describing the words in the system. We define all akshara models with the same number of states for training with ease [1].

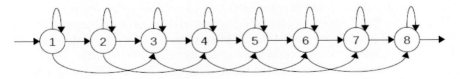

**Fig. 3** The topology of akshara HMM

The initial akshara models are defined using one Gaussian per state and refined by increasing the number of Gaussians per state.

A word bigram is a sequence of two words and the word bigram language model is used to determine each word in the sequence given its one of the predecessor. For each word sequence: $xy$, the word bigram probability $p(x, y)$ is computed based on the number times word $y$ follows word $x$ and the number of times word $x$ appears.

## 4 Proposed System

The purpose of our system is to do Telugu text recognition using implicit segmentation thereby overcoming segmentation problems that cause due to the presence of split and connected aksharas in the Telugu text. The use of implicit segmentation involves decomposing the word images into frames of fixed width and grouping the feature vectors computed from frames into aksharas by recognition.

In our system, we compute simple statistical features such as intensity and derivative of intensity similar to [7] using sliding window method on training set of word images and use continuous density HMMs [10] to model akshara shapes. We also compute word bigram language model on word label files of training word images to incorporate contextual information in performing recognition. The word recognition relies on akshara models and word bigram language model. The Telugu printed text recognition system consists of four major stages: preprocessing, feature extraction, akshara HMMs training and word recognition. The former two stages are common for the later two stages. The description of the stages is given in the following subsections.

### 4.1 Preprocessing

The preprocessing stage involves performing tasks such as noise removal, thresholding, line segmentation and word segmentation to prepare data for the next stage. The input to the preprocessing stage is document images and the output of this stage is word images.

Since noise pixels do not give any meaning to the actual image and may typically arise in document images due to image acquisition process, aging or photo copying of documents. The salt and pepper noise is the most common form of noise. First, noise from input document images is removed by applying median filter with $3 \times 3$ window size. The filtering process replaces each pixel value by the median of the neighborhood pixels and then the thresholding method of Otsu is used to convert Gray scale images into their binary equivalents. For each binary image, frequency of black pixels along x-axis called Horizontal Projection Profile (HPP) is computed in order to segment lines. Once the lines are segmented, frequency of black pixels along y-axis called Vertical Projection Profile (VPP) is computed for each line image to extract word images. Thus the word images are extracted and given to the feature extraction stage.

## 4.2 Feature Extraction

In the feature extraction stage of the system, different types of statistical and language independent features are computed similar to [7, 10]. We intend to compute features using sliding window method [13] rather than extracting structural features such stokes, arcs, etc.

Since the statistical feature computation is easy compared to the structural feature extraction. The statistical feature computation does not fail due to the noise in document images, whereas the structural feature extraction involves structural analysis of patterns and may give incorrect features due to the noise. We compute the statistical features by applying sliding window method as shown in Fig. 4. In our experiments, the width of the narrow sliding window and horizontal overlapping are 4-pixels and 1-pixel respectively. The line height threshold is 80-pixels and the sliding window is vertically divided into 20-cells. During feature extraction, a sliding window is moved in left-to-right writing order along each input word image and three statistical features: intensity, derivative of intensity along x-axis and derivative of intensity along y-axis are computed using Sobel operator from each cell. Then the three sub-vectors correspond to three features are combined into 60-dimensional feature vector. The feature vector sequences computed on training word images are

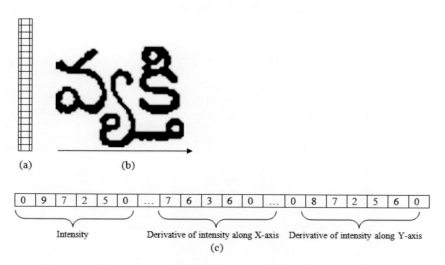

**Fig. 4** Sliding window method: a) Moving window b) A word image with an arrow that indicates reading order c) The format of feature vector

the input to akshara HMMs training stage, while feature vector sequences computed on testing word images are the input to word recognition stage.

## 4.3 Akshara HMMs Training

The akshara HMMs training stage consists of estimating mean and variance parameters of the Gaussian components to represent the states of akshara models. We use Baum-Welch algorithm based tool of HTK called HERest to estimate the parameters.

The resources and tasks involved in the training of akshara HMMs are shown in Fig. 5(a). The input to training stage are the resources such as prototype akshara HMMs, word images, word label files that are mapped into akshara level label files. The resources are depicted using rounded rectangles, whereas the rectangles are used to represent the tasks such as model initialization, parameter estimation and model refinement. During training, an HMM is needed for each unique akshara that appears in the training word images. Hence the prototype akshara HMMs are defined with the typical structure. In model initialization, the mean parameter of all states of prototype models is set to the vector of zeros, whereas the variance parameter of every Gaussian component is set to the global variance. The HCompV tool of HTK is used to estimate the global variance from the feature vector sequences computed on training word samples. This implies uniform segmentation of each feature vector sequence during the first cycle of parameter estimation [16]. Thus the akshara models are initialized for all the aksharas that appear in the training set.

The next task is to estimate the parameters of the states of akshara HMMs. The parameter estimation requires the training word images and akshara level label files

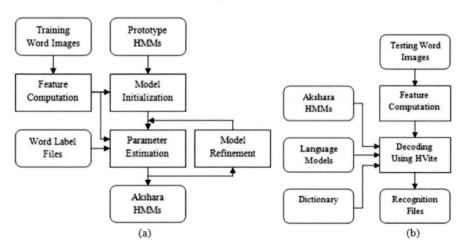

(a)                                                                                             (b)

**Fig. 5** Telugu text recognition system: a) Training of akshara HMMs b) Implicit segmentation approach for word recognition

where each label file consists of akshara labels. It involves making a single composite HMM by joining all akshara HMMs and processing the training samples to collect statistics that are used to simultaneously estimate the parameters of akshara models [16]. The model refinement involves manipulating akshara models followed by re-estimation of the parameters.

## 4.4  Word Recognition

Each word in the system is described by an akshara HMM sequence. The word recognition is based on the akshara models and word bigram language model. The word recognition stage consists of using a Viterbi based decoder to match feature vector sequences computed on the test set of word images against akshara HMM sequences.

The standard Viterbi algorithm can be used to match a feature vector sequence against a set of akshara HMMs to find the maximum likelihood akshara HMM [6], whereas HVite tool of HTK allows to match a feature vector sequence against a set of akshara HMM sequences [15]. In Fig. 5(b), we show how to recognize word images using implicit segmentation approach. We use rounded rectangles to depict the resources, whereas rectangles to denote the tasks involved for performing word recognition. The implicit segmentation approach does not depend on prior segmentation of the words into aksharas before they are recognized. The decoder requires feature vector sequences, trained akshara HMMs, language models and dictionary. The dictionary consists of words where each word is described as a sequence of akshara HMMs. The word bigram language model is used to specify the word network. The word network consists of a set of nodes connected by arcs. The nodes denote the words of the system and the arcs represent the transitions between words. In our word network, the words are placed in parallel with a loop-back that allows recognition of any word sequence.

The decoder finds the path through the network, which has the highest log probability, using Token passing algorithm [15]. A token represents a partial path from frame 0 through frame $t$. Initially a token is placed in every possible start node at frame 0. For each frame step, the tokens are propagated and the log probability is incremented by transition and emission probabilities when they reach an emitting HMM state. At the end of each step, all tokens are discarded except the best one. The best token represents the maximum likelihood sequence of akshara HMMs matching against the feature vector sequence. During recognition, the feature vector sequences are computed on all test set of word images. Then the feature vector sequences are given to the decoder. For each feature vector sequence, the decoder matches against a set of akshara HMM sequences using HVite and finds the maximum likelihood sequence of akshara HMMs.

# 5   Experimental Results

Since the commonly used Telugu aksharas are in the order of hundreds. The training of akshara HMMs to represent different akshara shapes using Hidden Markov Model approach demands large training data. We use document images of Telugu corpus of a consortium project to prepare training set.

The training set includes word images segmented from 41 real document images scanned from a popular Telugu book. It also includes word samples, which contain split and connected aksharas, segmented from document images of different books. The training set does not include very poor quality word images, English words printed in Telugu script, and word images that are in bold and italic. We use 7090 training word images including about 400 word samples with split and connected aksharas. The number of unique words in the training set are 4981. These word images are used to train 450 akshara HMMs correspond to different akshara shapes found in the training set. The word bigram language model is computed from word label files of the training set.

We use HResults tool of HTK to perform result analysis for evaluating both akshara and word recognition performances of Telugu text recognition system. First, the word recognition files produced by the decoder are converted into akshara level recognition files using HLEd tool. Then the converted files are given to the HResults tool where each file consists of akshara labels. For each akshara level recognition file, result analysis involves matching against the actual akshara level label file using dynamic programming and counting substitution, deletion and insertion errors. We use a standard metric called percentage correct for measuring the performance [16]. Let $N$ be the total number of labels in all akshara level recognition files, $D$ be the number of deletion errors and $S$ be the number of substitution errors. The percentage correct is given by the Eq. 2.

$$Percentage\ correct = \frac{N - D - S}{N} \times 100\% \tag{2}$$

The test set consists of two collections of word images. One collection has 364 normal words and another has 67 words containing split and connected aksharas. The Telugu word and akshara recognition results are given in Table. 1. We also report the recognition results of 67 words with split and connected aksharas using

**Fig. 6**   a) Sample word recognition results b) Examples of recognition failures

**Table 1** The Telugu word and akshara recognition results

| Systems | Type of words | Count | Words %Correct | Aksharas %Correct |
|---|---|---|---|---|
| Our system | Normal words | 364 | 23.90 | 55.06 |
| | Words with split and connected aksharas | 67 | 19.40 | 60.20 |
| Tesseract OCR | Words with split and connected aksharas | 67 | 56.06 | 79.70 |

popular Tesseract OCR system [14]. The much difference between percent correct of aksharas and words in our system is mainly due to word insertion errors. The sample word recognition results are shown in Fig. 6(a) and the text equivalents are given below the images for the four examples of correctly recognized words. In Fig. 6(b), we show the examples of recognition failures. The text equivalent of first word image has an error. The text of second word image contains a word insertion error and the third word is wrongly recognized.

## 6 Conclusion

The proposed system offers Telugu printed text recognition despite the presence of split and connected aksharas in the Telugu document images. The system does not depend on prior segmentation of the words into aksharas before recognition and gives word recognition files that are converted into Unicode based text.

We report in this paper a novel Telugu printed text recognition system using an implicit segmentation method based on Hidden Markov models (HMM). Our system mainly based on continuous density HMMs to represent the shape of aksharas and word bigram language model. We compute the word bigram language model by excluding the words that consist of numbers, punctuation symbols and special characters. The system is used to recognize Telugu words with split and connected aksharas. The performance of the system is encouraging and the much difference between percent correct of aksharas and words in our system is mainly due to word insertion errors. Our future work includes experiments on improving the performance of the system by increasing training data, refining models and using more powerful features. We also intend to enhance our system for recognizing line images by exploiting word trigram language model similar to [10].

**Acknowledgements** The authors thank the Ministry for Communications and Information Technology (MCIT), New Delhi, Government of India, for providing financial support under the grant No. 14(6)/2006-HCC(TDIL).

# References

1. Aas, K., Eikvil, L., Andersen, T.: Text recognition from grey level images using hidden markov models. In: Hlaváč, V., Šára, R. (eds.) Computer Analysis of Images and Patterns. Lecture Notes in Computer Science, vol. 970, pp. 503–508. Springer, Heidelberg (1995)
2. Bazzi, I., Schwartz, R., Makhoul, J.: An omnifont open-vocabulary ocr system for english and arabic. IEEE Transactions on Pattern Analysis and Machine Intelligence **21**(6), 495–504 (1999)
3. Bose, C., Kuo, S.S.: Connected and degraded text recognition using hidden markov model. In: Proceedings of the 11th IAPR International Conference on Pattern Recognition, 1992, Conference B: Pattern Recognition Methodology and Systems, vol. II, pp. 116–119 (1992)
4. Dutta, S., Sankaran, N., Sankar, K., Jawahar, C.: Robust recognition of degraded documents using character n-grams. In: 2012 10th IAPR International Workshop on Document Analysis Systems (DAS), pp. 130–134 (2012)
5. Elms, A.: A connected character recogniser using level building of hmms. In: Proceedings of the 12th IAPR International Conference on Pattern Recognition, 1994, Conference B: Computer Vision amp; Image Processing, vol. 2, pp. 439–441 (1994)
6. Elms, A., Procter, S., Illingworth, J.: The advantage of using an hmm-based approach for faxed word recognition. International Journal on Document Analysis and Recognition **1**(1), 18–36 (1998)
7. Khorsheed, M.S.: Offline recognition of omnifont arabic text using the hmm toolkit (htk). Pattern Recogn. Lett. **28**(12), 1563–1571 (2007)
8. Krishnan, P., Sankaran, N., Singh, A., Jawahar, C.: Towards a robust OCR system for indic scripts. In: 2014 11th IAPR International Workshop on Document Analysis Systems (DAS), pp. 141–145 (2014)
9. Kumar, P.P., Bhagvati, C., Agarwal, A.: On performance analysis of end-to-end OCR systems of indic scripts. In: Proceeding of the Workshop on Document Analysis and Recognition, pp. 132–138. ACM, New York (2012)
10. Natarajan, P., MacRostie, E., Decerbo, M.: The BBN byblos hindi OCR system. In: Govindaraju, V., Setlur, S.R. (eds.) Guide to OCR for Indic Scripts, Advances in Pattern Recognition, pp. 173–180. Springer, London (2010)
11. Negi, A., Bhagvati, C., Krishna, B.: An OCR system for telugu. In: ICDAR, pp. 1110–1114. IEEE Computer Society (2001)
12. Rabiner, L.: A tutorial on hidden markov models and selected applications in speech recognition. Proceedings of the IEEE **77**(2), 257–286 (1989)
13. Roy, P., Roy, S., Pal, U.: Multi-oriented text recognition in graphical documents using hmm. In: 2014 11th IAPR International Workshop on Document Analysis Systems (DAS), pp. 136–140 (2014)
14. Tesseract: http://code.google.com/p/tesseract-ocr/
15. Young, S.: The HTK Hidden Markov Model Toolkit: Design and Philosophy (1993)
16. Young, S., Evermann, G., Gales, M., Hain, T., Kershaw, D., Liu, X.A., Moore, G., Odell, J., Ollason, D., Povey, D., Valtchev, V., Woodland, P.: The HTK Book (for HTK Version 3.4). Cambridge University Engineering Department (2006)

# Detection of Copy-Move Forgery in Images Using Segmentation and SURF

V.T. Manu and B.M. Mehtre

**Abstract** In this era of multimedia and information explosion, due to the cheap availability of software and hardware, everyone can capture, edit, and publish images, without much difficulty. Image editing done with malicious intentions known as image tampering, may affect individuals, society, economy and so on. Copy-move forgery is one of the most common and easiest image tampering method which involves copying a patch of an image and pasting it within the same image. The purpose of this may be to conceal some objects in the image or conceal the artifacts of image editing. In this paper, we propose a new method to detect copy-move tampering in images, without prior image information, using an over complete segmentation and keypoint detection. It is evident from the experimental results obtained by testing it on standard datasets, that the proposed method is tolerant to postprocessing operations like blurring, JPEG compression, noise addition and so on. Also, our method is effective in detecting copy-move forgery where copied portions are subjected to various geometric transformations, like translation, rotation and scaling.

**Keywords** Copy-move forgery · Image segmentation · SURF

## 1 Introduction

Digital images finds a plethora of space on the internet from being photographs on social networking sites to critical financial documents. With most people using their

V.T. Manu(✉) · B.M. Mehtre
Institute for Development and Research in Banking Technology (IDRBT),
Hyderabad 500057, India
e-mail: {vtmanu,bmmehtre}@idrbt.ac.in

V.T. Manu
School of Computer Science and Information Sciences(SCIS), University of Hyderabad,
Hyderabad 500046, India

© Springer International Publishing Switzerland 2016
S.M. Thampi et al. (eds.), *Advances in Signal Processing and Intelligent Recognition Systems*,
Advances in Intelligent Systems and Computing 425,
DOI: 10.1007/978-3-319-28658-7_55

smartphones and tablets than their desktop computers or laptops for their computing needs; capturing, posting and circulating images on social media platforms and photo sharing sites have increased tremendously. Along with this, a considerable number of powerful image manipulation programs and apps are in the market both free and open source and otherwise; simplifying the effort of a potential perpetrator to tamper images for the good or the bad.

There are plenty of ways that an image can be forged or tampered, of which the most easiest for the manipulator and hardest for the detector is the copy-move forgery. Copy-move forgery involves copying a particular area of the image and pasting it on another area on the same image. As this happens within the same image it may be likely that most of the image parameters remains the same, say for example the illumination condition and the texture, making the task challenging for a detector. The intention of such an attack may be to hide a particular detail in the image while exploiting the aforementioned observation to its maximum benefit.

A classical example of an copy-move forgery and its detection is presented in Fig. 1, related to a controversial photo of Iran's so-called successful missile test that ran in newspapers and websites worldwide during July 2008. The photo shows four missiles launching into the sky, with the third missile appearing to be the copy-move of the second missile. This was done to hide a launcher on the ground with an unfired missile. Image tampering detection methods are classified into two based upon the detector's cognizance of the prior image information- active and blind. The former being informed and the latter being uninformed about the image information. Examples of active methods are steganography and watermarking, wherein, certain information required for proving the genuineness of the image are intentionally embedded into it. But this methods has a few limitations, like a dedicated hardware which may be expensive. Also, a forensic investigator may not expect to get an image which has been undergone an active method. Thus in most of the realistic situations, active methods cannot be depended upon.

Blind or passive methods uses the information inherently present in a given image to find traces of tampering.

(a)                                            (b)

**Fig. 1** Iran's missile launch test. (a) The doctored image from the website of Sepah News, the media arm of the Iranian Revolutionary Guards. (b) The original version released on the online service of the Iranian daily, Jamejam Today.

In this paper, we introduce a new blind method to detect copy-move forgery in an image by first segmenting the image into different regions and later looking for matches by finding interesting points within them.

## 2 Related Work

In copy-move forgery detection algorithms, we search for visually similar regions in a given image. So the forgery detection problem is a matching problem [5].

The seminal works in image tampering detection started with [4] and [5]. There are two basic approaches to search for matching of such regions (Fig. 2). They are:

- **Key/interest point detection:** Identify and select regions of high entropy.
- **Block tiling:** Tile the image into overlapping blocks. The shapes of the blocks can be any geometrical structure, like a circle, square or a rectangle.

Feature vectors are computed for every key-point in the former and for every overlapping blocks in the latter. Though there are large number of works on copy-move

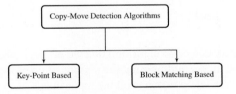

**Fig. 2** Common workflow of copy-move detection algorithms

forgery detection in general and specifically based on keypoint based approaches, we were interested on the ones which used SURF(Speeded-Up Robust Features) as the keypoint descriptor. Bo et al. [8] proposed a work which detect copy-move regions by matching the SURF descriptors vectors using distance measure and calculating the ratio of the nearest neighbor and the second-nearest neighbor and grouping them if they are greater than a predefined threshold. The limitation of this work is that it is not able to automatically locate tampered regions and its boundary.

In [9], the authors have extracted the SURF keypoints and represented using a data structure called k-d tree. k-d tree helps to efficiently search for nearest neighbors which represent areas of duplication. But it fails to localize small copy-moved areas. They performed the testing on some images from the dataset by [3] but have not presented the details of the results in terms of precision and recall. So it is not possible for us to make a comparison with their performance.

# 3   Background of the Proposed Copy-Move Detection Algorithm

In our work, we adopt a keypoint based method to detect copy-move forgery using SURF [2] which is an improvised or rather time-efficient version of Scale Invariant Feature Transform (SIFT) [6]. Both does scale space analysis, for which SIFT approximates Laplacian of Gaussian (LoG) with Difference-of-Gaussian (DoG) whereas SURF approximates LoG with Box Filters. Convolution with box filter can be computed parally for different scales with the help of integral images and also the use of Haar wavelets makes SURF faster than SIFT making it a better candidate for real-time applications, but compromising accuracy than SIFT. It is a well known fact that SURF is capable of giving better keypoints than SIFT when subjected to image quality degradation.

Segmentation of images is done using the algorithm based on the concept of superpixels using a method called simple linear iterative clustering (SLIC) [1] which performs local clustering of pixels in the 5-D space defined by the L, a, b values of the CIELAB color space the $x$, $y$ pixel coordinates.

# 4   Proposed Method

The copy-move forgery detection algorithm that we propose takes an image as input and outputs the same image highlighting the regions of copy-moved areas.

We employ two parallel approaches on the input image as mentioned below:

1. Key point generation using SURF [2] ("Key Points" block in Fig. 3).
2. Segmentation to generate the region map of the image for which we use the simple linear iterative clustering (SLIC) [1]. The resulting region label matrix ("Region Labels" block in Fig. 3) is stored and used in later stage of our method.

As copy-move forgery involves copying a region of the image and pasting it at some other location within the same image, the SURF keypoints at these locations may have almost same features. So, in order to find if a particular keypoint occurs elsewhere in the image, we compute the correlation among the SURF features. The steps till here has been represented as a flowchart in Fig. 3.

We make an assumption that the copy-pasted regions are separated by a distance which is proportional to the dimensions of the input image. For example, in an image whose dimensions are more than $400 \times 400$ sq. pixel units, the copy moved region may be approximately 100 to 120 pixel units apart. Based on this assumption, we compute the Euclidean distance between the keypoint locations and those keypoints having value above a particular threshold are considered for later analysis.

These keypoints are analyzed for replication by computing the keypoint feature correlation among each other and those keypoints having value greater than a particular threshold are assigned a particular label. This is nothing but a clustering operation

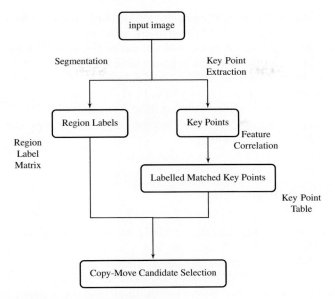

**Fig. 3** Schematic Diagram of Proposed Method.

based on feature correlation. The resulting clusters are stored as a table which we call as the keypoint table, which contains each SURF keypoint and its corresponding label. For example, if $i^{th}$ keypoint matches with $j^{th}$ and $k^{th}$ keypoints, then $i$, $j$ and $k$ has the same label. The block named "Labelled Matched Key Points" in Fig. 3 summarizes the steps discussed in this paragraph.

Now, we have to find the regions to which the corresponding keypoint belongs, for which we lookup the region label matrix which is obtained as the result of the segmentation algorithm as discussed earlier in this section.

The block named "Copy-Move Candidate Selection" in Fig. 3 is been discussed below. We travese through the keypoints in the keypoint table in a particular fashion, so we have to keep track of the keypoints visited. For that, keep a vector $V$ of size $1 \times M$, where $M$ is the length of keypoint table. Each of the elements of this vector are initialised to zero. To indicate whether a keypoint is a candidate for copy-move forgery, keep a vector $L$ having the same dimension of $V$, initialised to zero. For each keypoint $i$ in the keypoint table, if $V(i) = 0$, find the region label from the region label matrix and make $V(i) = 1$. For each keypoint $j$ in the keypoint table not having the same cluster label of $i$, which comes in the same region of $i$ or closer to the region boundary of $i$ are taken. Find its cluster label and check whether there are keypoints with the same cluster label of $j$ in the same region or region boundary of keypoints having same cluster label of $i$.

The keypoints thus obtained are saved for further processing. $j^{th}$ keypoint is pushed into a stack, $keep_j$. Similarlly, keypoints satisfying the above mentioned property having cluster label $j$ are kept in a stack, $keep_{jj}$ and that of $i$ are kept in a stack, $keep_{ii}$. Perform this for all the neighboring keypoints of $i$. Once the entire set of neighbors of $i$ are processed, $i$ has to be updated. So we pop an element from $keep_j$, if the stack is not empty and update $i$ with the popped element. If $keep_j$ is empty, update $i$ with the first keypoint $k$ from keypoint table where $V(k) = 0$. If $keep_{jj}$ and $keep_{ii}$ are non-empty then $L(i)$ and $L(j)$ are assigned the value 1. Until $keep_{jj}$ is not empty, pop out an element to $jj$ and make $V(jj) = 1$ and $L(jj) = 1$. Similarlly, perform this operations for $keep_{ii}$. Perform these operations till $V(i) = 1, \forall i$.

The keypoints having $L(i) = 1$ are candidate keypoints of copy-move forgery, hence display them in the input image.

## 5   Results and Discussions

The proposed method is implemented in MATLAB R2013a(8.1.0.604), performed on Intel Core i5-3230M CPU @ 2.60GHz × 4. The threshold of correlation which is discussed in Sect.4 is experimentally taken as 0.975.

As Fig. 4(a) is an example of real life copy-move forgery (discussed in Sect.1), it is given as input to our proposed system and the result obtained is shown in Fig. 4(b) where the region within the green circles indicate forgery. Also, a classic image (Fig. 5(a) & (b)) which appeared in the seminal work [5] depicting a canopy copy-moved to conceal a truck, is tested and the result obtained is presented in Fig. 5(c). We tested our algorithm on the dataset used in [10] and the CoMoFoD dataset [7]. In the former, the copy-moved areas were subjected various affine transformations

(a)                                              (b)

**Fig. 4**  Iran's missile launch test. (a) Doctored image (b) Result obtained using proposed method.

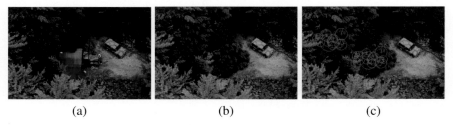

(a)                                 (b)                                 (c)

**Fig. 5**  "Jeep" image. (a) Original image with a truck visible (b) Copy-move forged image where truck is concealed using a portion of foliage (c) Result obtained using proposed method.

like translation, rotation and scaling. In the latter, the forged images were subjected to postprocessing methods like JPEG compression, noise adding, image blurring, brightness change, color reduction and contrast adjustments, after the copy-moved areas were subjected to different geometric transformations which include translation, rotation, scaling, distortion and combination of two or more transformation mentioned before.

We carried out the testing on the dataset mentioned in [10] fully and the details are summarized in Table 1. To test the performance of our algorithm on images in which postprocessing is applied, we used the CoMofoD dataset. Therefore, we limited the testing only on translation copy-move forged images with various postprocessing techniques mentioned before.

The result obtained using the proposed method on an image from the dataset used in [10], named "im9.bmp"(Fig. 6(a)) is presented in Fig. 6. The image in Fig. 6(b) is a forged image made from Fig. 6(a) by copy-move forgery by translation. Fig. 6(e) is a forged image created from Fig. 6(a) by copying, rotating by 270° and pasting. Similarly, the Fig. 6(e) made from Fig. 6(a) by copying, scaling by a factor 2 and pasting. The white regions in Fig. 6(c, f & i) represents the regions of forgery corresponding to the forged images in Fig. 6(b, e & h), provided by the developers of the dataset for reference . The green circles in Fig. 6(d, g & j) represents the regions of forgery corresponding to the forged images in Fig. 6(b, e & h) .

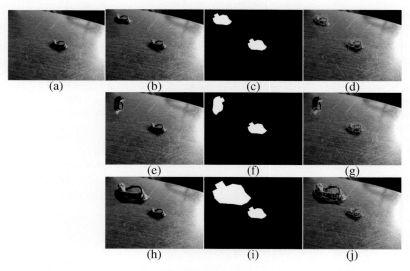

**Fig. 6** Copy-move forgery detection on an image from the the dataset used in [10]. (a) Original image (b) Forged image (Translation) (e) Forged image (Rotation by 270°) (h) Forged image (Scaling by a factor 2) (d,g,j)Result obtained using proposed method (regions of copy-move forgery shown in green circles ) (c,f,i) Region of copy-move (provided)

The result obtained using the proposed method on the image from the CoMoFoD dataset, named "035_F.png"(Fig. 7(b)) which is a forged version of "035_O.png" (Fig. 7(a)) is presented in Fig. 7. Fig. 7(c) is the copy-move mask provided by the developers of the dataset. Fig. 7(d)-(i) represents the postprocessed versions of "035_F.png": Brightness Change, Contrast adjustments, Color reduction, Image blurring, JPEG compression and Noise adding.

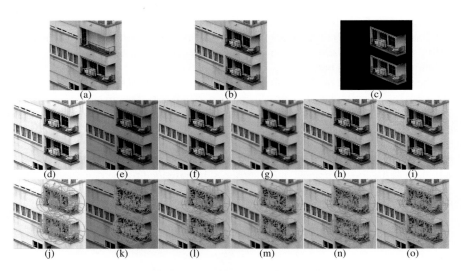

**Fig. 7** Copy-move forgery detection on an image from the CoMoFoD dataset [7]. (a) Original image (b) Copy-moved image (c) Regions of copy-move (d) Brightness Change, (lower bound, upper bound)= (0.01, 0.8) (e) Contrast adjustments, (lower bound, upper bound)= (0.01, 0.8) (f) Color reduction, intensity levels per each color channel = 128 (g) Image blurring, averaging filter $= 5 \times 5$ (h) JPEG compression, factor = 100 (i) Noise adding, $\mu = 0$, $\sigma^2 = 0.009$ (j-o) Copy-move detected output images of d to i respectively, with keypoints in the forged areas represented with green circles.

To evaluate the performance of the proposed scheme, we calculated the true positive (TP) which refers that a forged image is detected as fake and true negative (TN) which refers to that the authentic image is detected as original as shown below.

$$\text{Precision} = \frac{TP}{TP + FP} \tag{1}$$

$$\text{Recall} = \frac{TP}{TP + FN} \tag{2}$$

$$\text{Accuracy} = \frac{TP + TN}{TP + FN + FP + TN} \tag{3}$$

The values of True Positive(TP) Rate, True Negative (TN) Rate and Accuracy computed by testing our method on the dataset in [10] is given in Table 1. The precision, recall and accuracy of our proposed method on this dataset are 99.4%, 91.64% and 91.56% respectively.

The test results on the CoMoFoD are summarized in the Table 1. Since the dataset is very complex, 30 out of 40 translated copy-move forged images are detected. As given in the table, CoMoFoD uses compression factors ranging from 20 to 100 in steps of 10, noise addition with $\mu = 0$ and $\sigma^2 = (0.009, 0.005, 0.0005)$, image blurring using average filters, brightness change with specific lower bound and upper bounds, color reduction with particular intensity levels per color channel and contrast adjustment with specific lower bound and upper bound. Among the 30 detected images, our method is robust to compression, contrast adjustment, brightness change and color reduction. The algorithm is slightly sentsitive to blur and noise.

**Table 1** Performance Evaluation of the Proposed Method under various Transformation/Postprocessing categories

| Dataset | Transformation/ Postprocessing Category | Results | | |
|---------|------------------------------------------|---------|---|---|
| | | Tamper Parameters | TRUE Positives | FALSE Negatives |
| | Translation | – | 46 | 4 |
| [10] | Rotation | [-5°, 5°] | 212 | 8 |
| | | [-25°, 25°] | 150 | 10 |
| | | [0°, 360°] | 211 | 9 |
| | Scaling | [0.25, 2] | 131 | 29 |
| | | [0.75, 1.25] | 139 | 21 |
| [7] | JPEG Compression | [20, 100] | 270 | 90 |
| | Noise Addition | (0.009, 0.005, 0.0005) | 80 | 40 |
| | Image Blur | [3×3, 5×5, 7×7] | 84 | 36 |
| | Brightness Change | [(0.01, 0.95), (0.01, 0.9), (0.01, 0.8)] | 97 | 23 |
| | Color Reduction | [32, 64, 128] | 95 | 25 |
| | Contrast Adjustments | [(0.01, 0.95), (0.01, 0.9), (0.01, 0.8)] | 98 | 22 |

The performance of the proposed method is not compared with that in [10] and [7], as the former used a different approach for performance evaluation based on pixel level classification and the latter doesn't provide complete information regarding their performance on postprocessed tampered images. Also, the performance of the former is better than ours, but they have not tested with postprocessing. SURF based works [8] and [9], that we discussed in 2 have not provided detection accuracies, hence it also cannot be compared.

# 6 Conclusions

Copy-move forgery is the most common image tampering technique, which is done by copying a part of the image and pasting it within the same image to conceal a part of the image. We have presented examples of such forgeries in real world scenarios and developed an efficient algorithm that can detect them even if some postprocessing is done to counter the detection. We used a combination of segmentation and SURF keypoints for detection of copy-move forgeries in images by clustering the keypoints. We tested our method on two datasets- the datset used in [10] and CoMoFoD. The purpose of using the former was to test the performance on the basis of precision, recall and accuracy and the latter to verify its tolerance towards postprocessing operations after performing a copy-move forgery. The result on the former is found to be good as evident from the experimental results. But the results on the latter is not as much as the former considering the complexity of the images in it. Our future plan is to test and improve the accuracy of the algorithm on complex copy-move forged images.

# References

1. Achanta, R., Shaji, A., Smith, K., Lucchi, A., Fua, P., Süsstrunk, S.: SLIC superpixels. Tech. rep. (2010)
2. Bay, H., Tuytelaars, T., Van Gool, L.: SURF: speeded up robust features. In: Computer Vision–ECCV 2006, pp. 404–417. Springer (2006)
3. Christlein, V., Riess, C., Jordan, J., Riess, C., Angelopoulou, E.: An evaluation of popular copy-move forgery detection approaches. IEEE Transactions on Information Forensics and Security 7(6), 1841–1854 (2012)
4. Farid, H.: Detecting digital forgeries using bispectral analysis (1999)
5. Fridrich, A.J., Soukal, B.D., Lukáš, A.J.: Detection of copy-move forgery in digital images. In: Proceedings of Digital Forensic Research Workshop (2003)
6. Lowe, D.G.: Distinctive image features from scale-invariant keypoints. International Journal of Computer Vision 60(2), 91–110 (2004)
7. Tralic, D., Zupancic, I., Grgic, S., Grgic, M.: CoMoFoD—new database for copy-move forgery detection. In: 2013 55th International Symposium on ELMAR, pp. 49–54. IEEE (2013)
8. Bo, X., Junwen, W., Guangjie, L., Yuewei, D.: Image copy-move forgery detection based on SURF. In: 2010 International Conference on Multimedia Information Networking and Security (MINES), pp. 889–892. IEEE (2010)
9. Shivakumar, B., Baboo, L.D.S.S.: Detection of region duplication forgery in digital images using SURF. IJCSI International Journal of Computer Science Issues 8(4) (2011)
10. Ardizzone, E., Bruno, A., Mazzola, G.: Copy-Move forgery detection by matching triangles of keypoints. IEEE Transactions on Information Forensics and Security 10(10), 2084–2094 (2015)

# Automatic Pattern Recognition for Detection of Disease from Blood Drop Stain Obtained with Microfluidic Device

Basant S. Sikarwar, Mukesh Roy, Priya Ranjan and Ayush Goyal

**Abstract** This paper investigates automatic detection of disease from the pattern of dried micro-drop blood stains from a patient's blood sample. This method has the advantage of being substantially cost-effective, painless and less-invasive, quite effective for disease detection in newborns and the aged. Disease has an effect on the physical properties of blood, which in turn affect the deposition patterns of dried blood micro-droplets. For example, low platelet count will result in thinning of blood, i.e. a change in viscosity, one of the physical properties of blood. Hence, the blood micro-drop stain patterns can be used for diagnosing diseases. This paper presents automatic analysis of the dried micro-drop blood stain patterns using computer vision and pattern recognition algorithms. The patterns of micro-drop blood stains of normal non-diseased individuals are clearly distinguishable from the patterns of micro-drop blood stains of diseased individuals. As a case study, the micro-drop blood stains of patients infected with tuberculosis have been compared to the micro-drop blood stains of normal non-diseased individuals. The paper delves into the underlying physics behind how the deposition pattern of the dried micro-drop blood stain is formed. What has been observed is a thick ring like pattern in the dried micro-drop blood stains of non-diseased individuals and thin contour like lines in the dried micro-drop blood

B.S. Sikarwar(✉) · M. Roy
Department of MAE, ASET, Amity University, Noida 201313, U.P., India
e-mail: bssikarwar@amity.edu, mkroy89@gmail.com

P. Ranjan
Department of EEE, ASET, Amity University, Noida 201313, U.P., India
e-mail: pranjan@amity.edu

A. Goyal
Department of ECE, ASET, Amity University, Noida 201313, U.P., India
e-mail: agoyal1@amity.edu

© Springer International Publishing Switzerland 2016
S.M. Thampi et al. (eds.), *Advances in Signal Processing and Intelligent Recognition Systems*,
Advances in Intelligent Systems and Computing 425,
DOI: 10.1007/978-3-319-28658-7_56

stains of patients with tuberculosis infection. The ring like pattern is due to capillary flow, an outward flow carrying suspended particles to the edge. Concentric rings (caused by inward Marangoni flow) and central deposits are some of the other patterns that were observed in the dried micro-drop blood stain patterns of normal non-diseased individuals.

**Keywords** Pattern recognition · Tuberculosis · Dried blood microdrop stain pattern · Microfluidic device · Physical properties of blood

# 1  Introduction

The pathological application of dehydrated blood stain specimen on blotter paper for the quantification of chemical constituents has emerged as a novel minimally invasive option for the screening of disease from blood samples, which can be used for affordable diagnosis of patients' disease. Analysis of dried blood stains has been known to doctors [1, 2], and medical researchers are using this technique for screening of various diseases cautiously after experimental studies and clinical trials on this method [3-13]. Once validation is proved with experimental trials and clinical tests, this disease screening method using dried blood stain pattern analysis can provide a cost-effective affordable healthcare tool for rapid, and/or instant pathological screening of disease from a patient's blood sample. Scientifically, any chemical constituent in either the plasma or serum component of blood will also be present and can be detected and quantified from a dehydrated micro-droplet of that same blood sample, subject to the conditions that the chemical constituents to be analyzed in the dehydrated blood stain will have stability to dehydration. Analysis of whether a chemical constituent is stable in storage is also necessary if this method is to be used for extensive epidemiological screening of a disease in a population. Once the stability analysis and experimental validation and clinical trials establish this method, it can foster development of a novel disease screening or detection tool based on a microfluidic device that collects blood droplet from patient's blood sample and measures the physical properties of blood and detects patterns in the dehydrated blood stain to recognize which disease the patient has.

Previously published work demonstrating that the collection of a blood micro-droplet with filter paper is now as precise as microcapillary tubes and microfluidic pipettes [3] makes this research feasible to detect disease from the physical properties of blood micro-droplets of very precise volume obtained from a patient's blood sample. The precise collection of a blood micro-droplet and staining on filter paper and its processing is now a regimented procedure [2]. During the drying process of blood samples on filter paper, cellular components get ruptured, further blood spots are reconstituted and after this it is subsequently released in to the solution. To overcome this problem additional extraction method may be required for certain chemical constituents. There are two other issues which may come with the analysis of dried blood drops – elution efficiency of chemical constituents being measured and preciseness of micro-droplet volume of collected blood sample. Notwithstanding these drawbacks, dehydrated blood stain method has many benefits, one of which is the almost pain-free and minimally invasive drawing of blood with a small prick in a finger capillary as compared to the highly invasive and painful collection of blood through a needle puncturing a vein. Another advantage of the dried blood stain method is that there is no need of doing standard centrifugation and separation techniques during sampling, by virtue of which it can be performed by any technician with lack of training in centrifugation and separation techniques. As per experimental study, it is observed that at room temperature most analytes after drying are stable up to one week. The dried blood stain method also has low infectious hazard because all those viruses which may be active in the liquid state of blood and its constituents become non-infective due to disruption of their capsid upon dehydration. Although intuitively one may consider dehydrated blood sample obtained via finger capillary pricking has different properties than blood sample suctioned from a needle puncturing a vein via a syringe, several published papers report high positive correlation coefficients between levels of insulin, hs-CRP (C reactive protein), and triglyceride measured from dried blood stains and blood serum [4]. Given that measurement of chemical constituents from dried blood stain is as valid as their measurement of serum, the physical properties of dried blood stains will be highly correlated to those of serum. Based on this assumption, this paper builds upon prior research on dried blood stains to detect disease from blood stains.

Dehydrated blood stain disease detection is a highly cost-effective and rapid method for large-scale epidemiological screening in rural areas using an affordable healthcare diagnosis device which will ease the burden on the relatively low-budgets allocated for disease centers and clinics in developing countries. In this research, we are proposing a low cost method for disease detection from blood stains which will reduce the need for collection and transportation of liquid blood serum.

Blood is a complex non-Newtonian, biological fluid, half of which is plasma comprising of RBC (red blood cells) and WBC (white blood cells), particles in the size of micrometers [14-15]. It has shear-thinning property, i.e. viscosity decreases upon increased shear strain. Particle concentration in the blood can be measured by hematocrit, computed as red blood cell volume of plasma over net volume of blood [16-17]. Measuring the physical properties of blood has been the subject of research worldwide for more than two decades. This research has a direct contribution to biomedical science because it has been found that blood physical properties play an important role in many human and cattle diseases [18-21].

Biomedical research in published literature [19-21] has demonstrated that the patterns of dried microdrops of blood and other such fluids were distinct for normal versus diseased patients. This kind of research [18] focuses on the drying process of a blood drop and the pattern formation subsequent to the drying for distinguishing patients suffering from a particular disease from non-diseased individuals. This work has found the dried blood stain is completely of a different pattern for diseased patients in comparison to the normal healthy individuals (see Fig. 1). The above-mentioned researchers reported that the physical properties of blood differ for healthy and ill patients.

Against this background, this paper presents analysis of the drying stains of various blood samples and correlates the disease with the pattern of the drying blood micro-drop stain.

## 2    Materials and Methodology

Biomedical researchers reported that different patterns of blood stains were observed in non-diseased and diseased individuals. Fluid dynamics research reported that the pattern of stains relates to the physical properties of blood. Hence, the drying process of blood is largely influenced by the physical properties of blood. Physical properties depend on percentage of live red blood cell presence whole volume of blood that was varying samples of blood. Our research combined these scientific communities' research that leads to correlate the diseases tuberculosis with the pattern of drying stains of blood. For such promising technique, simple, robust and low-cost micro-fluidic device was designed and fabricated to capture pattern of drying stains of various sample of blood.   Theory and experimental details of drying process are given below:

### 2.1    Dried Blood Microdrop Phenomenon

In drying blood drops, the drop surface evaporative flux causes a radially outward flow in the drop, which transports blood cells. When a pinned contact line of the drop is formed, the blood that evaporates from the contact line region must be replenished by liquid from the interior of the drop. A capillary mechanism drives the flow: in order to maintain the drop's spherical-cap shape dictated by surface tension, a compensating flow is required to refill the liquid that evaporates from the contact line so pinned. Contact line pinning occurs by pre-existing roughness of the substrate. On top of this, suspended particles that arrive at the line of contact due to effect of blood stain reinforce pinning and thereby generate a self-pinning mechanism.

The amount of flow inside an evaporating drop can be calculated from mass conservation once the evaporative flux as:

$$J = -D\nabla C \tag{1}$$

The diffusion coefficient of vapor in the surrounding gas is D. This coefficient is different for different blood sample (healthy and ill peoples). To find the evaporative flux from the vapor concentration field, $\partial c / \partial t$ in equation (2) is solved subject to boundary conditions,

$$\frac{\partial c}{\partial t} = D \cdot \nabla^2 c \tag{2}$$

To illustrate drying process of blood drop, we consider a drop that is axisymmetric with a contact line that is pinned. This pinned line of contact can be defined using the cylindrical coordinate system (r; z), as shown in Figure 1. Inside this drop, we define an infinitesimally small control volume of width $dr$, located at a distance $r$ from the drop center. Law of conservation of mass necessitates that the rate of change in the amount of liquid inside this control volume equals the net inflow of liquid into the control volume, minus the amount of liquid hat has evaporated from the drop surface. This can be expressed as:

$$\frac{\partial h}{\partial t} = -\frac{1}{r}\frac{\partial}{\partial r}Q - \frac{1}{\rho_b}J \qquad (3)$$

where the local drop height is given as $h$(r; t), time is given as $t$, volume flow is given as $Q$ and blood density is given as $\rho$ . Hence, the decrease in drop height of a volume element equals the evaporative flux through the free surface of the element plus the net outflow to the neighboring elements.

The authors [22-25] present the details of solution of these equations (1-3) elsewhere. They found that during evaporation the fluid moves outward from the drop centre to the three-phase receding contact line. A collides deposited at the edge of the drop. The maximum concentration of fluid was observed close to three phase contact line.

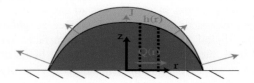

**Fig. 1** Schematic representation of an axisymmetric drop that is evaporating, defined with cylindrical coordinate system (r; z). Flow Q(r) inside the drop is driven by the flux J evaporating from the surface of the drop with height profile *h(r)*. The dashed lines indicate the control volume of width dr.

## 2.2    *Dried Blood Microdrop Experimental Setup*

In this work's experimental setup, a sessile blood drop evaporates spontaneously inside the environmental chamber under control conditions (see Fig. 2 a). Specimen of blood was collected from patients with tuberculosis infection and normal healthy individuals. They were collected in 10 ml sterile tubes containing 1% sodium heparin as anticoagulant. The tubes were stored in a refrigerator at 5°C. The samples used in this manuscript were taken from normal healthy, anemic, and tuberculosis infected donors.

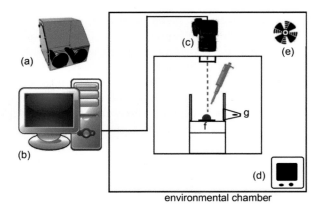

**Fig. 2** Schematic diagram of equipment used for the capture of the pattern of stains of a sessile blood microdrop on an aluminum substrate - (a) actual view of environmental chamber (b) computer, c) EOS 70 HD digital camera, (d) refrigeration unit for controlling humidity and temperature inside the chamber, (e) fan to recirculate air inside the chamber, (f) sanded aluminum substrate, and (g) two cold light sources.

A schematic of the setup used in this experimental study is shown in Fig. 2. The drops for every blood sample were gently placed on the aluminum surface in a controlled condition inside the environmental chamber at RH=50% and T=32°C. The aluminum surfaces were sanded. The sandpaper used had a grain size of 600 (CAMI standard), which results in alternations to the surface smaller than the resolution of the 3D microscope. The average surface roughness was 1 µm, as characterized by the digital 3-D microscope. The surfaces were cleaned with ethanol and wiped with micro fabricated cloth before 2 hours of commencement of experiment. Surface tension, wettability, hematocrit, viscosity, and other such blood physical properties were measured.

The mass $m(t)$ was calculated with the standard Mettler Toledo XS 105 balance having a 0.001 mg resolution at 10 Hz in the dehydration procedure. Evaporation rate $J(t)$ was computed with the derivative of the equation of mass loss (eq. 2). In parallel, a macro-lens attached Canon EOS 70D camera was used to capture images of the drying blood drop. Images of an area of 22.3 mm × 14.9 mm were captured at 1 fps (frames per second) with a pixel resolution of 5184 × 3456 pixels, adequate for imaging the drying, gelling, and cracking of the sessile drop.

At the start of the dehydration of the drop, the fluid is colloidal and homogeneous. Hence, a thin outer boundary can form upon drying on the aluminum substrate plate. Simultaneously, Marangoni convection causes colloidal particles such as RBCs to collect at the receding triple line of the sessile drop. This causes deposition of solids, referred to as the corona. What remains is serum that lacks any colloidal particles and creates patterns that are small in size and that stick to the substrate plate subsequent to drying.

Images were captured and stored in ROM memory. They were further analyzed as presented in this paper with the automatic pattern recognition algorithm using image processing functions. The image acquisition captured the droplet in resolution fine enough for recognition of patterns in the dried blood microdrop stain. A cold cathode back light was employed (with a luminous intensity of 900 candela) as the light source on the blood drop to eliminate reflection. Light density also affected the evaporation and deposition of the drying blood drop.

*Protocol of Carrying Out Experiments:*

*Surface Preparation – Aluminum*

1. Have the aluminum tiles and cut as 70x70 (thickness is 2 mm) mm squares surfaces. Polish the surface of each tile with a 600 grit sandpaper for approximately 5-10 minutes (the machining time depends on the surface quality of the tile, so carefully check the uniformity of the superficial machining.
2. After each tile, remove the residues on the sandpaper by gently tapping the polisher on a wooden surface. Change the sandpaper every 10 tiles, approximately.
3. Place the polished substrates on the work table with a cover on top to minimize foreign particles on the finished surfaces.
4. Approximately two hours before the experiment, wash the aluminum tile with isopropanol, then deionized water, and dry them with compressed air.
5. Check with the level that the target platform is horizontal.
6. Calibrate the video-image system by recording several images with an appropriate graduate scale, for the conversion of pixels to mm.
7. Save the images mentioning the objective type, camera and magnification in the file name.
8. Set the volume micro-pipit.
9. Check that the micro-pipit is perpendicular to the deposit surface. The alignment should be checked by first ensuring that the tip of micro-pipette is perpendicular to the holder bar by viewing the tip of pipette from multiple angles, then with a spirit level, check that the bar itself is horizontal.
10. Prepare a spreadsheet (Excel) with tables preformatted for all the data that will be collected during the steps: " Blood Preparation" and "Blood Property Measurements" and "Impact Experiments".

*Blood Preparation*

The "Blood Properties Measurements" and "Experiment" should be conducted simultaneously to measure physical properties of blood. The plastic test tubes (10 ml) containing blood sample should be stored in the fridge when not used. The temperature inside the fridge should be kept around 3.5°C (between 3 and 4 on the knob inside the lab refrigerator) in order for the blood to be maintained within 2 to 8°C recommendation. The blood should not be used more than a week after the bleeding date. Put the plastic tubes containing the blood on the rocker approximately 60 minutes before starting experiment. Before starting experiment, measure temperature with a traceable thermometer to verify blood temperature is the same as room temperature (+/-0.5C).

*Blood Property Measurement Procedure*

On starting the experiment, measure (and record) the temperature of the blood, the actual room temperature, and the relative humidity in the room, close to experimental setup. Also record the time at which these measurements are taken.Measure the blood hematocrit with the HemataSTAT® II (~10 minutes). Measure the blood viscosity (~1 hour) (Error <5%): syringe pump micro-capillary viscometer is used to measure the viscosity of blood at seven values of shear rate (10,100, 500, 1000, 5000 and 10000). Measure the equilibrium surface tension of blood (Error$_\pm$–2%).

## 2.3    Drying Blood Microdrop Pattern Formation

The formation of patterns was investigated from drying of various samples of blood on aluminum substrate. In the blood microdrop sample of the healthy normal patient blood, strong radial flow was observed carrying live blood cells towards the line of wetting that is pinned, where flux of evaporation was the greatest caused by geometry that was wedge-shaped. A ring of nano-particles is formed over the buildup (pinning of the contact line) and wetting (receding of the wetting line) stages. It was illustrated that substrate heterogeneities, both physical and chemical, influenced the pattern of the blood microdrop stain. Additionally, drop convection (Marangoni) caused by gradients of surface tension influence the flux and final blood microdrop stain deposition. This leads to deposition at the droplet's centerpoint instead of the boundary. The physical phenomena of deposition bumps in the center and ring-like formations occur in the drying blood microdrops of normal healthy patients.

## 2.4    Dried Blood Microdrop Samples

Blood drop drying sample collection of normal, tuberculosis infected, and anemic patients were carried out. Sample (a) is of 27 year old man having no disease and sample (b) is of 29 years old man having tuberculosis disease. It was found that the blood sample (a) and blood sample (b) have different stain patterns, as shown in Fig. 3 below.

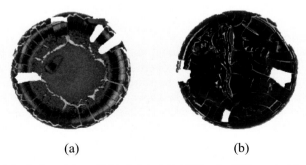

(a)                                                          (b)

**Fig. 3** Completely drying stains of blood: (a) stain of blood of 27 years old man having no disease and (b) stain of blood of 29 years old man having tuberculosis disease.

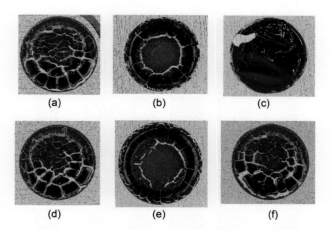

**Fig. 4** Patterns of various stains, all dried at same environment conditions (RH 55%, room temperature 32 °C). Blood stain patterns of 45 year old anemic man (a), 45 year old normal woman (b), 45 year old man with TB (c), 55 year old anemic woman (d), 55 year old normal man (e), and 45 year old anemic man (f).

# 3    Results

Figure 5 illustrates the automatic image processing of the blood stains of patients with TB, anemic patients, and normal individuals without any disease. The patients with TB have a different blood microdrop stain pattern than the anemic patients, that in turn are different than the normal non-diseased individuals. The blood microdrop stain pattern of the normal non-diseased individuals has a thicker periphery and thicker cracks, whereas the pattern of the blood microdrop stain of the diseased individuals has a uniform thickness throughout the field. The blood microdrop stain pattern of the normal non-diseased individuals has a non-uniform pattern, where the center is thinner and the periphery or edge is thicker, whereas the blood microdrop stain pattern of the diseased TB patients has a uniform pattern.

   The first column of Fig. 5 shows the original blood microdrop stain, the second column shows the grayscale version, the third column shows the edges detected, the fourth column shows the contour lines, the fifth column shows the labeled contour lines, and the sixth and final column shows the absolute image difference between the blood microdrop stain images of the diseased patient and normal non-diseased patient.

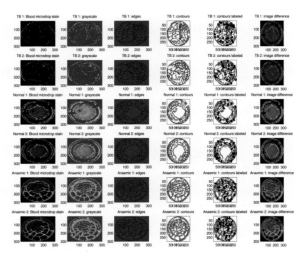

**Fig. 5** Stain patterns of microdrop blood samples of patients with TB and normal non-diseased individuals. Column 1 contains the images of the blood microdrop stain. Column 2 shows the grayscale version of the blood microdrop stain. Column 3 draws the edges / lines detected in the blood microdrop stain. Column 4 plots the contours of the blood microdrop stain. Column 5 plots the contours labeled. Column 6 shows the difference of the diseased blood stain with the average of the normal blood stain.

Table 1 tabulates the mean square error (MSE), root mean square error (RMSE), and chi-squared ($\chi^2$) distance between the images of the blood microdrop stains of the normal and diseased patients as a dried blood microdrop stain image-based measure of disease.

**Table 1** Image difference between the dried blood microdrop stains of patients and dried blood microdrop stain (average) of normal non-diseased computed as mean square error (MSE), root mean square error (RMSE), and chi squared distance ($\chi^2$), computed as average difference of all pixels.

| Patient | Mean Square Error (MSE) | Root Mean Square Error (RMSE) | Chi Squared Distance ($\chi^2$) |
|---|---|---|---|
| TB Patient 1 | 150.3398 | 12.2613 | 347.5144 |
| TB Patient 2 | 145.3065 | 12.0543 | 334.0356 |
| Normal Patient 1 | 36.6869 | 6.0570 | 197.5191 |
| Normal Patient 2 | 56.4104 | 7.5107 | 205.1641 |
| Anemic Patient 1 | 88.0961 | 9.3860 | 213.4854 |
| Anemic Patient 2 | 93.1196 | 9.6499 | 221.8208 |

The results in Table 1 demonstrate that the image difference / distance computed as mean square error (MSE), root mean square error (RMSE), or chi squared ($\chi^2$) distance between the images of dried blood microdrop stain of a patient and normal healthy case can be used to distinguish a diseased patient from normal non-diseased individual. This image difference/distance between the patterns of dried blood microdrop stain of patient versus normal also can be used as a disease severity measure, since the distance or difference is smallest for normal patients, bit more for anemic patients, but largest for TB patients.

# 4    Conclusions

The research in this work corroborates the biomedical research that reports that different patterns stains of blood are observed for normal healthy versus diseased patients. Fluid dynamics research reported that the pattern of stains highly depends on physical properties of fluid. Our research combined these scientific communities' research that leads to correlate the diseases such tuberculosis, with physical properties of blood. For such promising technique, simple, robust and low-cost micro-fluidic devices, a micro-capillary viscometer and a pendant drop tensiometer will be designed and fabricated to measure the viscosity, surface tension and wettability of blood.  In this paper, we represented the drying pattern of blood stains as an alternative method to diagnose pathological information from biological fluids, in this case the blood of diseased and non-diseased individuals. The blood samples were untreated, collected with a highly precise microfluidic pipette, and the micro-droplets were allowed to deposit onto filter paper. Mechanical properties of blood such as viscoelasticity, viscosity, wettability, and surface tension, measured in the experimental setup, influenced the drying of the blood micro-droplets after deposition, drying, and crack formation. The pattern of the dried blood micro-droplets and the crack formation or fracturation therein was found to be different for infected versus normal healthy cases.

The difference in images between the normal and diseased dried blood microdrop stains can clearly distinguish a normal non-diseased individual from a diseased patient. The difference between the patient and average normal blood stain was largest for the blood stains of TB patients, then anemic patients, and smallest for non-diseased normal individuals. Hence, the image difference in blood stain of a patient with blood stain of average normal individuals can be explored as a measure of disease obtained from the dried blood microdrop stain of patients. This is a fairly less-invasive method and will be explored by the authors with a larger sample dataset of patient and normal non-diseased dried blood microdrop stains. Additionally, we will explore the difference in the patterns in dried blood microdrop stains of patients suffering from different infections such as dengue, malaria, TB, etc. to analyze if the dried blood microdrop stain pattern can be used to identify disease type from the blood micro-droplet of a patient.

The author's future recommendation is to create an affordable healthcare diagnostic tool for patients suffering from infections, in particular tuberculosis, which is a global endemic. This diagnostic tool can aid DOTS centers and TB clinics in the rapid detection of tuberculosis and for obtaining a severity measure of the disease for correct diagnosis, timely drug administration, effective treatment, and periodic diagnosis of the effectiveness of the medication. Such a TB detection diagnostic tool can be an automatic microscope based blood, sputum, and urine stain screening device. This device or set of devices can comprise of detection of TB bacteria from the sputum smear stains [26] and pattern recognition of dried blood microdrop stains for diagnosis of TB from a droplet of the patient's blood sample, as presented in this paper.

**Acknowledgements** The authors would like to acknowledge Prof. Dr. Sarman Singh, Head of the All India Institute of Medical Sciences (AIIMS, New Delhi) Clinical Microbiology Department, Dr. Vandana Garg, Physician of Internal Medicine, Max Hospital, Noida, U.P., and the pathological laboratory of J.S. Hospital, Noida, U.P. for providing data and insight into tuberculosis infection.

# References

1. Guthrie, R., Susi, A.: A simple phenylalanine method for detecting phenylketonuria in large populations of newborn infants. Pediatrics **32**(3), 338–343 (1963)
2. Lakshmy, R.: Analysis of the use of dried blood spot measurements in disease screening. Journal of Diabetes Science and Technology **2**(2), 242–243 (2008)
3. Mei, J.V., Alexander, J.R., Adam, B.W., Hannon, W.H.: Use of filter paper for the collection and analysis of human whole blood specimens. The Journal of Nutrition **131**(5), 1631S–1636S (2001)
4. Kapur, S., Kapur, S., Zava, D.: Cardiometabolic risk factors assessed by a finger stick dried blood spot method. Journal of Diabetes Science and Technology **2**(2), 236–241 (2008)
5. Jinks, D.C., Minter, M., Tarver, D.A., Vanderford, M., Hejtmancik, J.F., McCabe, E.R.: Molecular genetic diagnosis of sickle cell disease using dried blood specimens on blotters used for newborn screening. Human Genetics **81**(4), 363–366 (1989)
6. Chamoles, N.A., Niizawa, G., Blanco, M., Gaggioli, D., Casentini, C.: Glycogen storage disease type II: enzymatic screening in dried blood spots on filter paper. Clinica Chimica Acta **347**(1), 97–102 (2004)
7. Zytkovicz, T.H., Fitzgerald, E.F., Marsden, D., Larson, C.A., Shih, V.E., Johnson, D.M., Strauss, A.W., Comeau, A.M., Eaton, R.B., Grady, G.F.: Tandem mass spectrometric analysis for amino, organic, and fatty acid disorders in newborn dried blood spots a two-year summary from the new england newborn screening program. Clinical Chemistry **47**(11), 1945–1955 (2001)
8. Crossle, J., Elliot, R.B., Smith, P.: Dried-blood spot screening for cystic fibrosis in the newborn. The Lancet **313**(8114), 472–474 (1979)
9. Chace, D.H., Kalas, T.A., Naylor, E.W.: Use of tandem mass spectrometry for multianalyte screening of dried blood specimens from newborns. Clinical Chemistry **49**(11), 1797–1817 (2003)
10. Alvarez-Muñoz, M.T., Zaragoza-Rodríguez, S., Rojas-Montes, O., Palacios-Saucedo, G., Vázquez-Rosales, G., Gómez-Delgado, A., Torres, J., Muñoz, O.: High correlation of human immunodeficiency virus type-1 viral load measured in dried-blood spot samples and in plasma under different storage conditions. Archives of Medical Research **36**(4), 382–386 (2005)

11. Gelb, M.H., Turecek, F., Scott, C.R., Chamoles, N.A.: Direct multiplex assay of enzymes in dried blood spots by tandem mass spectrometry for the newborn screening of lysosomal storage disorders. Journal of Inherited Metabolic Disease **29**(2–3), 397–404 (2006)

12. Li, Y., Scott, C.R., Chamoles, N.A., Ghavami, A., Pinto, B.M., Turecek, F., Gelb, M.H.: Direct multiplex assay of lysosomal enzymes in dried blood spots for newborn screening. Clinical Chemistry **50**(10), 1785–1796 (2004)

13. Zhang, X.K., Elbin, C.S., Chuang, W.L., Cooper, S.K., Marashio, C.A., Beauregard, C., Keutzer, J.M.: Multiplex enzyme assay screening of dried blood spots for lysosomal storage disorders by using tandem mass spectrometry. Clinical Chemistry **54**(10), 1725–1728 (2008)

14. Brutin, D., Sobac, B., Nicloux, C.: Influence of substrate nature on the evaporation of a sessile drop of blood. AME, Journal of Heat Transfer **134**, 061101:1–061101:8 (2012)

15. MacDonell, H.L.: Bloodstain Patterns, 2nd edn. Laboratory of Forensic Sciences, Corning (2005)

16. Yakhno, T.A.: Drying drop technology as a possible tool for detection leukemia and tuberculosis in cattle. Journal Biomedical Science and Engineering **8**, 1–23 (2015)

17. Deegan, R.D.: Pattern formation in drying drops. Physical Review E **61**, 475–485 (2000)

18. Brutin, D., Sobac, B., Loquet, B., Sampol, J.: Pattern formation in drying drops of blood. Journal of Fluid Mechanics **667**, 85–95 (2011)

19. Savina, L.: Crystalline Structures of Serum of Healthy and Ill Patients, p. 96. Krasnodar, Soviet Kuban (1999)

20. Shabalin, V.N., Shatokhina, S.N.: Morphology of Biological Fluids, p. 304. Khrisostom, Moscow (2001)

21. Rapis, E.: Changing the physical non equilibrium phase of complex plasma proteins film in patients with carcinoma. Technical Physics **47**, 510–512 (2002)

22. Sikarwar, B.S., Shrama, S.K., Shukla, R.K., Ranjan, P.: Parametric study of sessile drop evaporation at atmospheric condition. In: Accepted in Proceedings of the 17th ISME Conference, ISME-17, IIT Delhi, India, October 3–4, 2015

23. Ghosh, A., Sikarwar, B.S., Attinger, D.: Microfluidic measurements of physical properties of blood for forensic studies. In: ASME 12th International Conference on Nanochannels, Microchannels, and Minichannels, Chicago, Illinois, USA, August 3–7, 2014. Paper number: FEDSM2014-22181

24. Sikarwar, B.S., Ghosh, A., Attinger, D.: Simple, low cost microfluidic measurements of physical properties of blood for forensic studies. In: ASME 12th International Conference on Nanochannels, Microchannels, and Minichannels, Chicago, Illinois, USA, August 3–7, 2014. paper number: FEDSM2014-22182

25. Attinger, D., Sikarwar, B.S., Chirstophe, F.: On the importance of drop impact studies in forensic applications. In: ASME 2014, 4th Joint US-European Fluids Engineering Division Summer Meeting, Chicago, Illinois, USA (August 3–7, 2014) paper number: FEDSM2014-22097

26. Goyal, A., Roy, M., Gupta, P., Dutta, M.K., Singh, S., Garg, V.: Automatic detection of mycobacterium tuberculosis in stained sputum and urine smear images. Archives of Clinical Microbiology (in press, 2015)

# Part VI
# Workshop on Signal Processing for Wireless and Multimedia Communications (SPWMC'15)

# STAMBA: Security Testing for Android Mobile Banking Apps

**Sriramulu Bojjagani and V.N. Sastry**

**Abstract** Mobile banking activity plays a major role for M-Commerce (Mobile-Commerce) applications in our daily life. With the increasing usage on mobile phones, vulnerabilities against these devices raised exponentially. The privacy and security of confidential financial data is one of the major issues in mobile devices. Android is the most popular operating system, not only to users but also for companies and vendors or (developers in android) of all kinds. Of course, because of this reason, it's also become quite popular to malicious adversaries. For this, mobile security and risk assessment specialists and security engineers are in high demand. In this paper, we propose STAMBA (Security Testing for Android Mobile Banking Apps) and demonstrate tools at different levels. These supported tools are used to find threats at a mobile application code level, communication or network level, and at a device level. We give a detailed discussion about vulnerabilities that help design for further app development and a detailed automated security testing for mobile banking applications.

## 1 Introduction

Android mobile application and operating system security has been clearly explained in [1, 8, 9], but some improvements are are to be made to the implementation of android security because versions of android operating system was started with the Cupcake 1.5, now KitKat 4.4, expected in future is KeyLimePie 5.0 [7]. Recent developments in android mobile operating system have been tested and demonstrated by drozer framework [14] and wire shark packet analyzer [5]. Android work based on

S. Bojjagani(✉) · V.N. Sastry
Centre for Mobile Banking (CMB), Institute for Development and Research in Banking Technology (IDRBT), Hyderabad, India
e-mail: sriramulubojjagani@gmail.com, vnsastry@idrbt.ac.in

S. Bojjagani
School of Computer and Information Sciences, University of Hyderabad, Hyderabad, India

© Springer International Publishing Switzerland 2016      671
S.M. Thampi et al. (eds.), *Advances in Signal Processing and Intelligent Recognition Systems*,
Advances in Intelligent Systems and Computing 425,
DOI: 10.1007/978-3-319-28658-7_57

Linux kernel operating system and its mobile applications are written in an eclipse of Java language run with built-in Application Programming Interfaces (API'S) [34]. Android uses a security framework that consists of application sandboxing, secure inter-application communication, cryptographic API's, application signing [8]. But these countermeasures for security against vulnerability may not be effective. Even though we have existing malware detection mechanisms, they are failed to eliminate the android mobile threats totally. These android malware threats change with a timeline, some common examples of malware threats found in android devices in 2014 are Torec, DroidPack, DriveGenie, OldBoot [7]. Android intents and permissions framework for security mechanism provides a guard between software and hardware resources. Intents in android is a communication model for launching the activities and services, but we should take care about the applications exported and services exported [4, 21]. Intent spoofing attack is the most common found in android, it leads to the broadcast receivers, exported applications and exported services [20], we examine the mobile applications not only at android application level, but also we examine the android applications at network level and device level. Many security threats have been found including confidential information shared unauthorized parties because of poor SSL (Secure Socket Layer) encryption, and insufficient transport layer protection, improper session handling these threats are well described in OWASP (or Open Web Application Security Project) [16]. And other unexpected behavior.

The rest of the paper is organized as follows. Section 2 discusses the related work. Section 3 considers threat scenario and vulnerability analysis for mobile banking apps. Section 4 describes the proposed testing strategy. Finally, Section 5 concludes the paper.

## 2   Related Work

Mobile banking applications based on Android, iOS, Windows platforms, have been tested by others in the last few years. Chakraborti et al. [3] proposed a security framework for mobile apps for any enterprise. This paper provides the only literature review on possible threats and vulnerabilities. Marforio et al. [18] described security indicators for detecting threats against phishing attacks in mobile platforms and possible countermeasures. This proposed framework describes application phishing, web phishing but doesn't deal with vulnerabilities in the code or app level, and communication level. Kathuria et al. [15] deals with the challenges in android application development. This proposed app mainly focuses on the user-centric level, without giving details about mobile security testing. Felt et al. [11] clearly describes android permissions and applied some automated testing techniques to android version 2.2 for determining the maximum permissions that are needed for an application and compares those permissions with actually required permissions. Likewise, he examined 940 Android apps using the tool of Stowaway and detected that 1/3rd of them are over privileged. He et al. [12] analyzes forty-seven: Android, iOS mobile apps in survey regarding SD cards, logging, Bluetooth, content provider, usage of cloud services, The internet.

Related work closest to our mobile banking applications from [6, 10, 13, 17]. Hu et al. [13] propose a set of six criteria for identifying and evaluating killer apps for mobile payment, banking, emerging mobile commerce applications and services. This paper doesn't achieve certain challenges of privacy and security concerns in mobile banking, and incompatible of mobile communication. Fahl et al. [10] examine various most popular free apps and investigation of the current state of SSL/TLS (Secure Socket Layer/Transport Layer Security) for android, he used a tool MalloDroid that detect potential threats against MITM (Man-In-The-Middle) attacks. Lee at el. [17] describes the complete literature review on the investigation of features and security in mobile banking. This framework of design helps and recommends for security and privacy issues involved in current mobile banking services. However, this framework doesn't analyze the real time scenarios possible in mobile banking services. Delac et al. [6] develop an attacker-centric model for different mobile platforms such as Android and iOS. The designed threat model addresses 3 key features of mobile device security, 1. Goals of attacker's, 2. Attack vectors, 3. Mobile malware.

Apart from other studies, we focus on examining several mobile banking Android applications by static code analysis and dynamic analysis using the tools of ApkAnalyser [22], Mercury [23] or (Drozer) [24] for static analysis, Wireshark [5, 27], Burp Suite [28] for dynamic analysis and found 356 exploitable vulnerabilities. Our testing approach moves further than previous related work: because others have tested with a dynamic testing strategy with one tool, but we focus on code or app level, communication level, and device-level testing. This makes a novel study for security in mobile banking applications and it is helpful for code developers for further enhancing the security measurements.

# 3 Threat Scenario and Vulnerability Analysis for Mobile Banking Apps

Before testing the mobile banking app's, initially we define the context. Mobile banking apps are more securable, user-friendly, and immediate mobile payment system.

## 3.1 Threat Scenario

A complete modeling of threat and vulnerability analysis is beyond the scope of our effort, but we suggest some points that bring out a neat framework for secure mobile banking. We concerned on apps for code level and network level, device level, because they are most important for the banker side and at customer's point of view.

- **Untrusted party learning of bank data:** Unauthorized persons gain the bank information belonging to an individual customer. They access not only secure data but also monitor the network.
- **Tampering with bank data:** An adversary tamper or alters the bank data, by performs replay and man-in-the-middle attacks in the communication media or network.
- **Customer chooses wrong bank app** Here the end user or customer chooses a wrong app, that app is installed in his/her mobile phone and giving valid credentials to the untrusted bank. Then the adversary plays all attacks on to the original bank and end user. This type of threat is called as a phishing attack.

The above three types of threats represent the violation of the security features such as data integrity, confidentiality, authorization, authentication, and non-repudiation.

## 3.2  Identifying the Attack Surface and Analysis of Vulnerabilities

Figure 1 shows a typical threat scenario for mobile banking apps, initially we install mobile banking apps in a mobile device (smartphone). The smartphone stores the bank apps data internally in a file system, database backups.

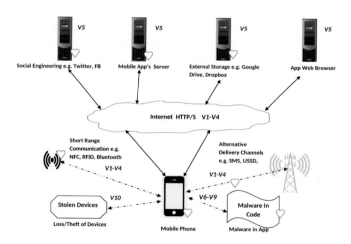

**Fig. 1**  A Typical Threat Scenario for mobile banking app in context. The heart represents fine-grained banking data

In Figure 1 shows several types of individual servers that are connected to the network. Some apps connect to the social engineering server such as Twitter, Facebook (FB) etc. Many fake mobile banking apps are connected directly to the payment gateway to a dedicated app web server for allowing the transactions of upload valid credentials to the bank server, backup, and data synchronization. The app's connect to external storage services e.g.Google Drive, Dropbox. Finally, app web browser is used to search the rate and amount payable for apps. This framework of threat scenario helps to identify the attack surface for various threats and suggests where attacker look for loopholes existing in the network. The mobile risks or vulnerabilities (denoted as V1-V10) are:

- **V1. Masquerade:** This attack is caused because of not establishing a secure connection between client and server, and improper SSL/TLS i.e., poor transport layer protection. Because of this reason the adversary performs session hijacking, and stealing session ID's.
- **V2. Man-In-The-Middle:** He is able to tap the secure data, manipulate, integrate with own data on a network and send to victims.
- **V3. Replay:** Attacker sends subsequent retransmission or delayed messages in the network.
- **V4. Traffic Analysis, Wi-Fi Sniffing:** The attackers always monitor the network traffic and observe packets are in encrypted format or not, sometimes they capture the packets.
- **V5. Browser Exploits:** It is also called as SSL BEAST (Browser Exploit Against SSL/TLS), he will be able to leverage weaknesses implemented in cipher block chaining to exploit the SSL protocol. Based on this reason the attacker easily read the encryption form data in plain text.
- **V6. SQL Injection:** The attacker identifies weak tables in the database possibly by SQLite can be subject to inject in web applications.
- **V7. Lack of binary protection:** Typically, an attacker will analyze and apply some reverse engineering tools for getting the original source code, then modify that code and perform hidden functionality.
- **V8. Broken Cryptography:** The code contains weak cryptography algorithms, easily breakable passwords, and poor key management process leads adversary easily steal the algorithms and theft secure data.
- **V9. Data Leakage:** Agents easily exploit the vulnerabilities by malware exist in the mobile of legitimate apps, or physical access of the device by an adversary from the victim's mobile device.
- **V10. Insecure Data Storage:** An Attacker that has attained a theft of mobile device, malware or another re-modified app acting on the attacker's behalf that executes on the smartphone.

Other possible attacks such as DoS (Denial of Service), phishing, pharming, protocol attacks, etc.) are currently not included.

# 4 Testing Strategy

Our testing mechanisms is divided into 4 parts: i. Static analysis ii. Dynamic analysis iii. Web app server security, iv. Device forensic. Apart from the four mechanisms of testing, we consider three levels of security testing 1. App Level, 2. Communication Level, and 3. Device Level. Table 1 shows how our testing three levels addresses four mechanisms. The possible vulnerabilities from V1-V10 represented in the previous section are shown in a table. For each vulnerability, the countermeasures or precautions to be taken by the developer, end-user, and banker is shown in Table 1.

- **Application Level or App Level:** This level identifies the vulnerabilities in code, for android applications all the files are .apk only, so we put this .apk file on ADB (Android Debugging Bridge) and possible vulnerabilities identified at attack surface. For this level, we consider both static and dynamic analysis testing mechanism.
- **Network or Communication Level:** An attacker captures or alters the packets in the network. So in this level we used dynamic analysis.
- **Device Level:** Sometimes the devices e.g. mobile device, SD card, personal log files stolen by an adversary, compromised devices, and physical threats comes under this category.

**Table 1** Strategy for Security Testing of Mbank Apps

| Testing Levels | Testing Mechanism | Supported Tools | Vulnerabilities | Countermeasures |
|---|---|---|---|---|
| App Level | Static and Dynamic | Drozer ApkAnalyser virustotal | V6-V9 | Check Uniform Resource Locator (URL) Caching for both HTTP request and response, Application backgrounding, Use strong cryptography algorithms with appropriate key lengths. |
| Communication Level | Dynamic Analysis Web app server security | Drozer WireShark Burp Suite TCPDUMP | V1-V5 | Apply SSL/TLS for transport channels. Use digital certificates signed by a trusted Certificate Authority (CA) provider. Do not send confidential data over alternate delivery channels (e.g, SMS, MMS, or notifications). |
| Device Level | Device Forensic | Sleuth Kit | V10 | Never store personal credentials on the SD card, For databases concern SQLcipher for SQLite data encryption mechanism. |

## 4.1 Static Analysis

Static analysis does not involve opening a file, or running the code or reverse engineering it is based on data contained in the APK (Android application package file). Static analysis involves identifying and querying cryptographic hash values, such as MD5 (Message Digest), metadata, strings, extract the apps permissions. The tools virustotal [30], androguard [25] is used for this analysis.

- **Antivirus Scans and Aliases:** Antivirus scans and aliases help to analyze a threat, identify date and time, comments, votes, and a list of aliases. Aliases acknowledge that stamp are other common related names attributed to the same code. Antivirus scan results, samples, blogs, and so on. In our test of various apps, the results are shown in Figure 3 and Figure 4 respectively. In Figure 4 reveals one Trojan horse detected out of 54 engines.

- **Broadcast Receivers:** Attackers often hold needed data about an apps attack surface and offer attackers to perform many wrong things, from arbitrary code execution to proliferating information. Broadcast receivers respond for both software and hardware-level events. They get notifications for these events through intents. The Drozer [24] tool identifies these receivers at the attack surface.

- **Activities and Services** These are the application components, and services that facilitate user interaction. It is useful for identifying which application, services can be released without permissions during an application security assessment because any of them provide access to confidential data or cause an app to crash if launched in the wrong context.

- **Content Providers (V6, V8, V9) :** In any database architecture, content providers have the ability to perform malicious operations into their SQLite databases or any file stores. They identify weak URI (Uniform Resource Identifier) in the database and perform all Structure Query Language (SQL) operations. For this content providers, we use the same tool drozer. The test app results on the attack surface of activities, broadcast receivers, content providers, services exported is shown in the Figure 2.

- **Certificate Information (V1-V5):** All apps must be signed because android uses x.509 certification otherwise apps will not install. And tester verifies the given certificates valid by the CA (Certification Authority), time period, and serial no. To verify these certificate procedures we use [29].

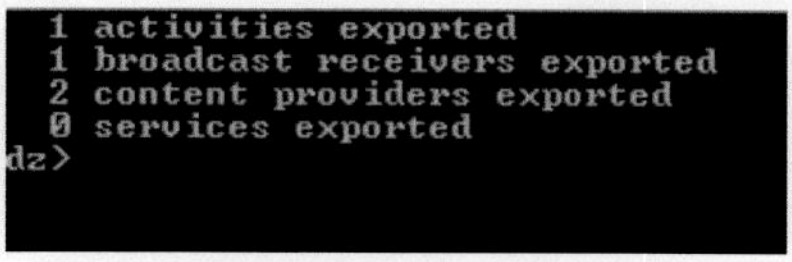

```
1 activities exported
1 broadcast receivers exported
2 content providers exported
0 services exported
dz>
```

**Fig. 2** An attack surface contains vulnerabilities in the app

## 4.2 Dynamic Analysis

Dynamic or behavioral analysis is mostly manual testing. In this analysis we create a dummy account for testing the credentials of user Account no, MPIN (Mobile PIN), user ID, Password, OTP (One Time Password) all these are invalid. From this account details, we can perform transactions of transferring an amount from one account to another account.

- **Insufficient Transport Layer Protection (V1-V5, V9) :** When designing any application, the information is exchanged in a client and server environment. To transmit data, it must traverse the device carrier or network devices (intermediate

(a) Scan Results

(b) Detailed Results

**Fig. 3** Virus total scan without virus results and detailed tesults

(a) Scan Results

(b) Detailed Results

**Fig. 4** Virus total scan with virus results and detailed results

routers or nodes, proxy's, cell towers etc.). Threat agents might exploit vulnerabilities such as monitor Wi-Fi (Wi-Fi Sniffing), compromised network, capture
sensitive data or malware in mobile etc. These threats are because of insufficient
transport layer protection. So we test this by Wireshark [27], tPacketCapture [31],
Burp Suite [28]. The results are shown in Figure 5, and Figure 6 respectively. These
two figures shows that the highlighted with yellow color data is not in encrypted
format, all the sensitive information is represented as a plain text.

• **Secure backup directories, files, and logging:** Here we test backup directories,
  files or log information contained encrypted data or not using the tool of adb [26].

**Fig. 5** TCP Stream in unencrypted format: Amount, Mobile no, Account no

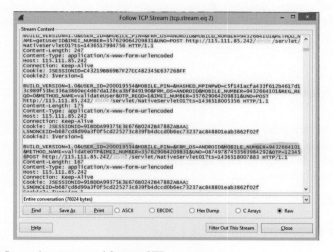

**Fig. 6** TCP Stream in unencrypted format: OTP

## 4.3 Web App Server Connection

In this testing mechanism apps are directly connected to dedicated web server to backup files or synchronize bank data. For this analysis again we used a dummy account of customer on the web site.

- **Web server connection (V1-V4):** Normal HTTP connection allows an adversary easily read the customer sensitive data; by history of URLs (Uniform Resource Locater) check the connection with secure transport of (HTTPS:)
- **Web server authentication and authorization (V1-V2):** Further we investigate the effectiveness of the server, e.g. an attacker understands how the authentication procedure is vulnerable, they create bypass or fake authentication by submitting transaction service requests to the mobile application backend server and bypass any direct interaction with the mobile application. This submission process is typically done via mobile malware within the device or botnets owned by the attacker. For this analysis we use again, Wireshark [27], and tPacketCapture [31].

## 4.4 Device Forensic

Using this testing mechanism, suggests a complete investigation of file structures of the android operating system, services, data storage, and external devices [32]. For complete forensic analysis, no tool extracts all the possible information when the device is locked [33]. We conduct forensic analysis on mobile devices using the sleuth kit [2].

## 5 Case Study

Case study section deals with an overview of our testing for android mobile banking apps. For testing the bank apps we need a .apk files. And these apps which are directly available from the Google's Play Store.The mobile devices for testing on two mobile platforms, android and J2ME.

## 5.1 Test Bed and Other Tools Used

**For Android Environment**

- Mobile Phone Used: HTC Desire X ( 1.2 GHz Dual Core, 768 MB RAM )
- Samsung Galaxy wave- II (1GHz Dual Core, 1 GB RAM)
- Android version "KitKat"
- Eclipse ADT IDE integrated with Android SDK
- Drozer Agent [24], VirusTotal, ApkAnalyser, androguard
- Wireshark, Burp Suite, TCPDUMP, tpacketcapture, Sleuth Kit

We connect a mobile device with the laptop running Ubuntu operating system version 12.04. Install above static tools on mobile device and for dynamic testing, tools installed on the laptop.

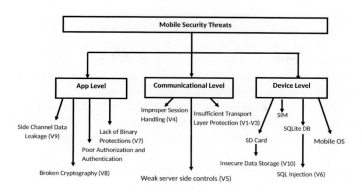

**Fig. 7**  Coverage vulnerabilities with OWASP

## 5.2   App Preparation

We tested several financial institute apps like banking and other apps. For static analysis testing, verify the vulnerabilities at attack surface but for dynamic analysis test banks create dummy accounts with some amount. These dummy account credentials are same for the original account, in the sense if customer account no is 9 digits, then dummy account no is also nine digits. Other valid information for doing transactions such as OTP (One Time Password), login id, password, and an amount in payer's account.

## 5.3   Discussion

The verification and validation of our testing strategy are demonstrated by identifying several cryptographic parameters, and privacy issues. Our testing strategy is compared with OWASP (Open Web Application Security Project) [19] and it covers entirely or partly 6 of 8 static and 4 of the 7 dynamic categories. This cross compare testing mechanism is shown in Figure 7.

## 6   Conclusions and Future Work

Our security testing strategy for mobile banking applications gives clear evidence of cryptography features such as authentication, data integrity, confidentiality, non-repudiation. We test and analyze many apps and found 356 exploitable vulnerabilities. This paper is helpful for those who are in the android app development and android malware threat analysis. The contribution of our work suggests, the app developer should concern the cryptography features, then implement the app in a secure environment. Our work shows that automated security testing for mobile banking applications is expensive, because the large amount of manual work is needed. Some tools for automated security testing cover only a minimum number of vulnerabilities, and every year the new versions of device OS launch into the market. So, many updations are needed and fully automated testing is desirable.

As future work, we will focus on testing mobile banking apps in the environment of iOS, Windows. Enhancing the security and safety parameters for mobile banking apps will be needed to concern the fully automated spectrum of issues.

# References

1. Blasco, J.: Introduction to android malware analysis (2012)
2. Carrier, B.: The sleuth kit (TSK) (2010). http://www.sleuthkit.org/sleuthkit/
3. Chakraborti, S., Acharjya, D., Sanyal, S.: Application security framework for mobile app development in enterprise setup (2015). arXiv preprint arXiv:1503.05992
4. Chin, E., Felt, A.P., Greenwood, K., Wagner, D.: Analyzing inter-application communication in android. In: Proceedings of the 9th International Conference on Mobile Systems, Applications, and Services, pp. 239–252. ACM (2011)
5. Combs, G.: Wireshark: Go deep (2009). homepage for wireshark
6. Delac, G., Silic, M., Krolo, J.: Emerging security threats for mobile platforms. In: MIPRO, 2011 Proceedings of the 34th International Convention, pp. 1468–1473. IEEE (2011)
7. Dunham, K., Hartman, S., Quintans, M., Morales, J.A., Strazzere, T.: Android Malware and Analysis. CRC Press (2014)
8. Enck, W., Octeau, D., McDaniel, P., Chaudhuri, S.: A study of android application security. In: USENIX Security Symposium, vol. 2, p. 2 (2011)
9. Enck, W., Ongtang, M., McDaniel, P.: Understanding android security. IEEE Security & Privacy **1**, 50–57 (2009)
10. Fahl, S., Harbach, M., Muders, T., Baumgärtner, L., Freisleben, B., Smith, M.: Why eve and mallory love android: an analysis of android ssl (in) security. In: Proceedings of the 2012 ACM Conference on Computer and Communications Security, pp. 50–61. ACM (2012)
11. Felt, A.P., Chin, E., Hanna, S., Song, D., Wagner, D.: Android permissions demystified. In: Proceedings of the 18th ACM Conference on Computer and Communications Security, pp. 627–638. ACM (2011)
12. He, D.: Security threats to android apps. Ph.D. thesis, Masters thesis, University of Illinois at Urbana-Champaign (2014)
13. Hu, X., Li, W., Hu, Q.: Are mobile payment and banking the killer apps for mobile commerce? In: Proceedings of the 41st Annual Hawaii International Conference on System Sciences, pp. 84–84. IEEE (2008)
14. Hunt, R.: Security testing in android networks-a practical case study. In: 2013 19th IEEE International Conference on Networks (ICON), pp. 1–6. IEEE (2013)
15. Kathuria, A., Gupta, A.: Challenges in android application development: A case study (2015)
16. King, J.: Android application security with owasp mobile top 10 2014. Ph.D. thesis, Masters thesis, Luleå University of Technology (2014)
17. Lee, H., Zhang, Y., Chen, K.L.: An investigation of features and security in mobile banking strategy. Journal of International Technology and Information Management **22**(4), 2 (2013)
18. Marforio, C., Masti, R.J., Soriente, C., Kostiainen, K., Capkun, S.: Personalized security indicators to detect application phishing attacks in mobile platforms (2015). arXiv preprint arXiv:1502.06824

19. Mobile Security Testing Guide: https://www.owasp.org/index.php/OWASP_Mobile_ Security_Project#tab=M-Security_Testing//
20. Nauman, M., Khan, S., Zhang, X.: Apex: extending android permission model and enforcement with user-defined runtime constraints. In: Proceedings of the 5th ACM Symposium on Information, Computer and Communications Security, pp. 328–332. ACM (2010)
21. Ongtang, M., McLaughlin, S., Enck, W., McDaniel, P.: Semantically rich application-centric security in android. Security and Communication Networks **5**(6), 658–673 (2012)
22. https://github.com/sonyxperiadev/ApkAnalyser
23. https://www.labs.mwrinfosecurity.com/tools/2012/03/16/mercury
24. https://www.mwrinfosecurity.com/products/drozer/
25. https://code.google.com/p/androguard//
26. https://developer.android.com/tools/help/adb.html
27. https://www.wireshark.org (accessed February 20, 2015)
28. https://portswigger.net/burp/ (accessed February 20, 2015)
29. https://www.opnessl.org/ (accessed March 11, 2015)
30. https://www.virustotal.com/ (accessed May 10, 2015)
31. https://play.google.com/store/apps/details?id=jp.co.taosoftware.android.packetcapture (accessed May 10, 2015)
32. Walnycky, D., Baggili, I., Marrington, A., Moore, J., Breitinger, F.: Network and device forensic analysis of android social-messaging applications. Digital Investigation **14**, S77–S84 (2015)
33. Wang, Y., Alshboul, Y.: Mobile security testing approaches and challenges. In: 2015 First Conference on Mobile and Secure Services (MOBISECSERV), pp. 1–5. IEEE (2015)
34. Wei, X., Gomez, L., Neamtiu, I., Faloutsos, M.: Permission evolution in the android ecosystem. In: Proceedings of the 28th Annual Computer Security Applications Conference, pp. 31–40. ACM (2012)

# An Efficient and Secure RSA Based Certificateless Signature Scheme for Wireless Sensor Networks

Jitendra Singh, Vimal Kumar and Rakesh Kumar

**Abstract** Certificateless signature cryptography is an efficient approach studied widely because it eliminates the need of certificate authority in public key infrastructure. There are number of security algorithms based on PKC (Public Key Cryptography), ID based certificate less schemes using bilinear maps assumptions. But implementation costs of these schemes are very high. On the other hand RSA is a scheme widely applied in real-life scenarios. So in this paper we proposed a RSA based certificate less signature scheme especially for wireless sensor networks. We have shown that this scheme is secure in random oracles model. A random oracles model is a way to proof the security of cryptographic algorithms which use hash functions and exponential operations in algorithms. Security of this scheme is closely related to discrete logarithm problem. Our scheme provide security against type I attack, type II attack, it also ensures integrity, non-repudiation and authentication.

**Keywords** Certificateless signature · Cryptography · Discrete logarithm problem random oracles model · RSA · Type I · Type II attack · Wireless sensor networks

## 1 Introduction

Express progress in the field of technology made it feasible to foster wireless sensor networks technology [1]. Wireless Sensor Networks (WSN) are comprised of large number of tiny sensor nodes with constrained resources in terms of processing power, energy and storage. WSN can be used in various applications mainly environmental monitoring, medical, military, and agriculture [1].

J. Singh(✉) · V. Kumar · R. Kumar
Department of Computer Science,
Madan Mohan Malviya University of Technology, Gorakhpur 273010, India
e-mail: {jitendra6890,vimalmnnit16,rkiitr}@gmail.com

© Springer International Publishing Switzerland 2016
S.M. Thampi et al. (eds.), *Advances in Signal Processing and Intelligent Recognition Systems*,
Advances in Intelligent Systems and Computing 425,
DOI: 10.1007/978-3-319-28658-7_58

685

Since, devices used in sensor networks are not tamper resistant, so adversary can gain its physical access easily. Hence, the main objective is to protect the data from unauthorized access, which can be done by using some security mechanisms [6, 9]. The technology faces lots of security problems as it has a wireless mode of communication and access to such sensor devices is quite easy [32]. There are two approaches to restrict the unauthorized access in to the network: symmetric cryptography and asymmetric cryptography.

In traditional Public Key Infrastructure (PKI), the user selects a public key but it needs to be validated by a trusted third party known as Certificate Authority (CA) [3]. The CA provides a digital certificate to tag the public key with the user's identity. PKI has a problem of high computation and storage. To avoid this, Shamir introduced the concept of Identity-based Infrastructure. Many researchers are using bilinear pairing assumption to construct a certificate-less signature [7]. But the problem with these bilinear pairing techniques is that implementation of pairing is way harder than RSA based implementations. As we know that RSA technique has been applied in different atmospheres for decades. Jianhong Zang et al. [6] proposed a RSA based certificate-less scheme under Strong RSA and Discrete Logarithm problem. Drawback of this scheme was that it was not secure under Type I attack if we provide enough power to the attacker.

In cryptography, proofs of security schemes are often relative to the computational hardness of some well-known mathematical problems which are hard to solve. So for an efficient functionality of scheme to work well, it is necessary to use an idealized model. Random Oracle Model is a well-known computational model for giving proof about the security of a cryptography scheme. This model was formalized by Bellare and Rogaway [21] in 1993. In Random Oracle Model a given hash function is replaced by publically accessed random oracle which is also known as theoretical Black Box to analyze the security scheme. This random oracle gives response (True) to each unique query from a fixed output domain randomly.

The organization of the rest of the paper is as follows: section 2 of the paper contains the related work and our contribution for this scheme. Section 3 contains proposed RSA based scheme followed by Security analysis in random oracles, complexity and performance analysis in section 4. Last section 5 of paper contains conclusion future scope followed by references for the paper.

## 2 Related Work

To provide security in wireless infrastructure-less medium researchers are using bilinear pairing techniques, public key cryptography, id based cryptography and other techniques [5, 6, 7, 9, 11]. Researchers presented lots of schemes for security using these techniques.

While studying the resource constrained wireless sensor networks, some [6] have focused specially on attacks and vulnerabilities in wireless sensor networks. Wood and Stankovic [6] discussed a lots of attacks based on denial of service for WSNs, and indicated some possible solutions for these attacks. Karlof and Wagner [7] focused on attacks in routing layer, and showed how some present WSN protocols were vulnerable to these attacks. They provide valuable guidelines for proposal of our scheme. Since, the pairing operation is the most expensive operation, so researchers needed to find an alternative solution. In 2009, Wang et al. [20] proposed a scenario where pairing is not needed to be computed at sign phase, it precomputes and publishes as the system parameters. But, this is not the solution for the removal of pairing operation. In 2011, He et al. [18] developed an efficient short CLS scheme without pairing. After that some schemes were proposed based on analysis of Elliptic Curve Discrete Logarithm Problem [11, 21].

Jinhong Zang et al. [5] proposed an efficient RSA based Certificate-less scheme on 2011 which used Strong RSA assumption and discrete logarithm problem to prove the security of scheme. Their scheme consists of 7 polynomial time algorithms. For the public and private key generation of the nodes, they used a secret value randomly selected by the node which is only known to that node only to remove the key escrow problem in ID based crypto-systems. According to Zang et al. their scheme was secure under Type I and Type II attacks but if we provide enough power to attacker, their scheme is not secure under type I attack. Other than this their scheme also has the problem of Signer's public key which was identified by Chin-Chen et al. [8].

To overcome the insecurity form Type I attack in [5], Gaurav Sahrma et al. [9] presented a new certificate-less scheme. In this scheme they modified Zang scheme by modifying $R_1 = [(H_0(ID))^e]^{r_1}$ to $R_1 = x_{ID}^e[(H_0(ID))^e]^{r_1}$ and corresponding value of $u_1$. By doing this they were able to secure their scheme against Type I attack by this modification increased their signing phase computation cost. Sharma et al. also used strong RSA assumption in random oracle model to prove the security of their scheme against Type I and Type II attacks. Researchers provided different variations on developing and improving the performances of certificate-less signature scheme in [12, 14, 16, 17, 18 and 29].

## 2.1   Our Contribution

A key challenge during the design of this was how to propose a scheme which can provide maximum security with less overhead and to minimize computational complexity. So in this paper we are taking one assumption that BS is working as KGC and verifier which is fixed and contains unlimited power source which is a possible scenario in case of WSNs. As we know that RSA based schemes are easy to implement as compared to pairing schemes, so we are modifying Zang et al. [5] RSA based scheme with less computation and making our proposed scheme secure against Type I attack also. We also reduced no. of phases in our proposed scheme. Security complexity of our scheme is closely related to definition given below.

***Definition (discrete logarithm problem (DLP)):*** *"Suppose $n = pq$ is an RSA modular number which satisfy $p = 2p' + 1$, $q = 2q' + 1$ and $g \in Z_n^*$ where $g$ represents as generator of order $p'q'$, given the elements $g$, $y$, $n$, its goal is to compute exponent $x$ such that $y = g^x \bmod n$."*

# 3    Proposed Scheme

In this section, we describe the proposed RSA based certificate-less signature scheme for wireless sensor network.

## 3.1   Network Model

In this scenario we will use a homogenous wireless network as shown in Fig. 1. Some assumptions will be as follows:

1)      All sensor nodes will be homogenous in nature.
2)      Base Station will perform as KGC (Key Generation Center) and verifier because every communication will be between sensor nodes and base station.
3)      Base Station will have unlimited energy.

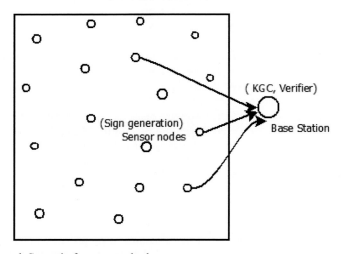

**Fig. 1** Network Scenario for proposed scheme

## 3.2   Proposed Certificate-less Scheme

- **Setup Phase:** Given a security parameter $1^k$ as input, a RSA group $(n, p, q, e, d)$ is generated, where $p'$ and $q'$ are two large prime numbers which satisfy $p = 2p' + 1$, $q = 2q' + 1$ and $n = pq$ which is a RSA modular number and e is the public key for Key Generation Center (KGC) which is $e < \varphi(n)$ and satisfies $\gcd(e, \varphi(n)) = 1$ and $ed = 1 \bmod \varphi(n)$, where $\varphi(n)$ is the Euler's phi function. We will be using two cryptographic hash functions H and $H_0$ which satisfy $H_0 : \{0,1\}^* \to Z_n^*$ and $H : Z_n^4 \times \{0,1\}^* \to \{0,1\}^s$ where $s$ is a security parameter. The master secret key is $d$ and the public parameters of system is $Param = \{n, e, H, H_0\}$.

- **Set-Public-Key:** KGC generates public key and private key for every node using their node ID and send them via a secure channel. Public key (Pub) is calculated using H0 ( ) hash function,

$$Pub = H_0(ID)$$

- **Set-Private-Key:** After computing public key for node$_i$ KGC generates Private Key for that node as follows:

$$Pr = Pub \times p * d$$

- **Sign Generation Phase:** For a given message M and public parameters provided by KGC, a sensor node with node id (ID) computes following steps using its private key $Pr$.
1) Randomly choose 2 numbers r$_1$ and r$_2$.
2) Compute $R_1 = [(H_0(ID))^e]^{r_1}$ mod n and $R_2 = [H_0(ID)]^{r_2}$ mod n.
3) Compute $h = H(R_1, R_2, ID, Pub, M)$.
4) Now set $u_1 = (H_0(ID))^{r_1 - h}$ mod $n$ and $u_2 = r_2 - Pr * h$.
5) Finally, generated resultant certificate-less signature for message M will be

$$\delta = (u_1, u_2, h).$$

- **Verification Phase:** With a signature $\delta = (u_1, u_2, h)$ on message M, verifier execute the followings:
1) Compute $R'_1 = u_1{}^e H_0(ID)^{eh}$ mod n and $R'_2 = H_0(ID)^{u_2} Pub^{Pr*h}$ mod n.
2) Accept the message with signature if and only if following equation holds

$$h = H(u_1{}^e H_0(ID)^{eh} \bmod n, H_0(ID)^{u_2} Pub^{Pr*h} \bmod n,$$
$$ID, Pub, M)$$

**Correctness:** Proof of correctness of certificate-less signature is very easy as shown in following steps:

$$H(u_1{}^e H_0(ID)^{eh} \bmod n, H_0(ID)^{u_2} Pub^{Pr*h} \bmod n, ID, Pub, M)$$
$$= H \begin{pmatrix} (H_0(ID)^{r_1 - h})^e H_0(ID)^{eh} \bmod n, \\ H_0(ID)^{r_2 - Pr*h} H_0(ID)^{Pr*h} \bmod n, ID, Pub, M \end{pmatrix}$$
$$= H \begin{pmatrix} H_0(ID)^{r_1 e - eh} H_0(ID)^{eh} \bmod n, \ H_0(ID)^{r_2} \bmod n, \\ ID, Pub, M \end{pmatrix}$$
$$= H(H_0(ID)^{r_1 e} \bmod n, R_2, ID, Pub, M)$$
$$= H((H_0(ID)^e)^{r_1} \bmod n, R_2, ID, Pub, M)$$
$$= H(R_1, R_2, ID, Pub, M)$$
$$= h.$$

## 4    Security Analysis

As security model concerns, certificate-less signature scheme is different from ordinary security schemes. There are two types of adversaries (Type I and Type II) to who certificate-less schemes are vulnerable. The adversary $\mathcal{A}I$ in Type I represents malicious outsider sensor node which attacks against the CLS scheme. So $\mathcal{A}I$ is allowed to get access to the public key and the values by using that key, but adversary cannot get access of master secret key. So $\mathcal{A}I$ can change the public keys of node in order to challenge the CLS scheme. The adversary $\mathcal{A}II$ in Type II shows a corrupt key generation center. So adversary $\mathcal{A}II$ can generate private key for a user but it cannot change the public key of a node.

In the following we will give security analysis of our proposed scheme under Random Oracle Model and show that our scheme is capable of preventing the network from Type I and Type II attacks.

### 4.1    Analysis of Type I Attack

Suppose there is an adversary $\mathcal{A}1$ whom can change the public key of a sensor node with id ID and generate a malicious signature for the verifier. To generate signature $\mathcal{A}1$ will perform following steps:

1) $\mathcal{A}1$ Generates an illegal public key $Pub'$ and change this in place of $Pub$.
2) $\mathcal{A}1$ Generates a random number $r_1$ and computes $R_1 = [(H_0(ID))^e]^{r_1} \ mod \ n$.
3) $\mathcal{A}1$ Now generates random no. $r_2$ and calculates $R_2 = [H_0(ID)]^{r_2} \ mod \ n$.
4) $\mathcal{A}1$ computes $h = H(R_1, R_2, ID, Pub', M)$ and checks weather $r_1 - h$ is divisible by $e$ . if $r_1 - h$ is not divisible by $e$, $\mathcal{A}1$ repeats step 2) and 3).
5) $\mathcal{A}1$ computes $r_1 - h = e.b$ and $u_1 = H_0(ID)^b \ mod \ n$.
6) $\mathcal{A}1$ output the signature $\delta = (u_1, u_2, h)$

Since $h = H(R_1, R_2, ID, Pub', M)$, so during the verification phase the signature will be verified by proving $R_1 = R_1'$ by the verifier (BS)

$R_1' = u_1{}^e H_0(ID)^{h.e}$
$R_1' = H_0(ID)^{e.b} H_0(ID)^{h.e}$
$R_1' = H_0(ID)^{r_1-h} H_0(ID)^{h.e} \neq R_1$

So during the verification phase verifier can identify that signature is not legal and discards that illegal message. So by this we can show that our scheme is secure against type I adversary.

### 4.2    Analysis of Type II Attack

To prove the security of our scheme under type II attack we will use following theorem under random oracles model.

**Theorem:** For a random oracles model, if there is existence of a type II adversary $\mathcal{A}_{II}$ which can request at most $q_H, q_{H_0}$ queries to hash functions $H$ and $H_0$, respectively, and $q_s$ queries to certificateless signing oracle, which can break security of our proposed certificateless signature scheme with probability $\rho$ and in a time bound $t$, then there is also the existence of another algorithm $\mathcal{B}$ which can use $\mathcal{A}_{II}$ in random oracle to solve the discrete logarithm problem.

**Proof:** Let if there is the existence of a type II adversary $\mathcal{A}_{II}$ which can break our proposed scheme. We will develop an adversary $\mathcal{B}$ which will make use of adversary $\mathcal{A}_{II}$ to solve NP-Hard discrete logarithm problem. To solve this problem in polynomial time, $\mathcal{B}$ has to simulate a challenge on random oracle model and the oracles (hash oracle, key generation oracle and signing oracle) for $\mathcal{A}_{II}$ Thereby, $\mathcal{B}$ performs this in the following ways:

**Setup:** $\mathcal{B}$ maintains three lists $H$-list, $H_0$-list and Key list which are all initially empty. Let $(e, n)$ are the system parameters provided to adversary. $d$ Is the master secret key which satisfies $ed = 1 \ mod \ \varphi(n)$, and $\mathcal{B}$ knows the values of $d \ and \ (p, q)$. Select two hash functions $H$ and $H_0$ as random oracles. Suppose $pub_{ID^*} = Y$ is public key of a user $U^*$ challenging the scheme and ID being the identity of $U^*$. Finally, $\mathcal{B}$ sends a binary string of public parameters $(e, d, n, g, H, H_0)$ to adversary $\mathcal{A}_{II}$ .

**Queries:** Adversary $\mathcal{A}_{II}$ is allowed to access hash functions as oracles given below in polynomial times. All simulations of these oracles is done by $\mathcal{B}$.

**$H$ − Hash Queries:** During this oracle query process $\mathcal{A}_{II}$ is able to request $q_H$ number of queries at most. For each query $(R_1, R_2, ID, Pub, M)$, $\mathcal{B}$ chooses $k_{ID} \in \{0,1\}^l$ randomly and sets $k_{ID} = H(R_1, R_2, ID, Pub, M)$. Finally return $k_{ID}$ to $\mathcal{A}_{II}$ and add value $(R_1, R_2, ID, Pub, M, k_{ID})$ to $H$ - list.

**$H_0$ −Hash Queries:** For this oracle query $\mathcal{B}$ randomly chooses $t_{ID} \in \varphi(n)$ and set $H_0(ID) = g^{t_{ID}}$ and return it to adversary $\mathcal{A}_{II}$ where $\varphi(n)$ is known as the Euler's phi function. Finally add $(ID, H_0(ID), t_{ID})$ to $H_0 -$ list.

**Public Key Request Queries:** Adversary $\mathcal{A}_{II}$ can query oracle using identity ID. If $ID \neq ID^*$, $\mathcal{B}$ computes $pub_{ID} = H_0(ID)$ and add record $(ID, pub_{ID}, t_{ID})$ to **Key list**. Otherwise $\mathcal{B}$ search for record $(ID^*, H_0(ID^*), t_{ID^*})$ and computes $pub_{ID^*} = Y^{t_{ID^*}}$ and add record $(ID^*, pub_{ID^*}, \perp)$ to **Key list**. Finally send $pub_{ID}$ to $\mathcal{A}_{II}$.

**Private Key Generate:** When $\mathcal{A}_{II}$ makes a private key generate query with identity ID, if $ID \neq ID^*$, $\mathcal{B}$ first searches for an available record $(ID, pub_{ID}, t_{ID})$ in **Key list** and computes $pr_{ID}$. Otherwise $\mathcal{B}$ aborts the oracle.

**Signing Oracle:** For every query on input $(M, ID)$ , if $ID \neq ID^*$ , $\mathcal{B}$ first obtains the private key using $ID$ ,then generates a signature using that private key. Otherwise $\mathcal{B}$ computes following:

1) Randomly choose $u_1 \in Z_n$ and $h \in \{0,1\}^l$, $u_2 \in Z_{\varphi(n)}$.
2) Compute values $R_1 = u_1{}^e H_0(ID)^{eh}$ and $R_2 = H_0(ID)^{u_2} Pub^{Pr*h}$ .

3) Search if there exists a record $(R_1, R_2, ID, Pub, M)$ in H-list. If record exists in $H$ - list, then abort it else set $H(R_1, R_2, ID, Pub, M) = h$ and add it to H-list.

4) Signature $\delta = (u_1, u_2, h)$ is generated and returned to $\mathcal{A}_{II}$.

**Output**: After simulating all the queries $\mathcal{A}_{II}$ outputs a forgery $(ID^*, pub_{ID^*}, M^*, \delta^* = (u_1^*, u_2^*, h^*))$ and win this analysis. For this forgery must satisfy these conditions:

1) If $\delta^*$ is a valid forgery, then $h^* = (R_1^*, R_2^*, pub_{ID^*}, ID^*, M^*)$ which is available on H-list, where $R_1^* = u_1^{*e} H_0(ID^*)^{h^* e}$ and $R_2^* = H_0(ID^*)^{u_2^*} pub_{ID^*}^{pr*h^*}$.
2) $ID^*$ is identity of challenges user and $H_0()$ is oracle queried by $ID^*$.

When $\mathcal{B}$ again plays the same random oracle tape but with different choices of oracle H. According to the forking lemma (Pointcheval 1996) , $\mathcal{B}$ can generate another valid signature which is $(ID^*, pub_{ID^*}, M^*, \delta'^* = (u_1'^*, u_2'^*, h'^*))$, then these should hold true:

$$R_2^* = H_0(ID^*)^{u_2^*} pub_{ID^*}^{pr*h^*} \text{ and } R_2^* = H_0(ID^*)^{u_2'^*} pub_{ID^*}^{pr*h'^*}.$$

So following holds true from above two relations.

$$H_0(ID^*)^{u_2^*} pub_{ID^*}^{pr*h^*} = H_0(ID^*)^{u_2'^*} pub_{ID^*}^{pr*h'^*}$$
$$\Updownarrow$$
$$(H_0(ID^*))^{u_2^* - u_2'^*} = pub_{ID^*}^{pr*h'^* - pr*h^*}$$
$$\Updownarrow$$
$$(g)^{t_{ID}*(u_2^* - u_2'^*)} = Y^{pr(h'^* - h^*)}$$
$$\Updownarrow$$
$$(g)^{t_{ID}*(u_2^* - u_2'^*)/pr(h'^* - h^*)} = Y$$

So here we got the discrete logarithm of Y to base $g$ is $t_{ID} * (u_2^* - u_2'^*)/pr(h'^* - h^*)$. So we can see that discrete logarithm problem can be solved by $\mathcal{B}$. Obviously, it is the contradiction to solve the difficulty of discrete logarithm problem.

Other than these attacks our scheme also keeps the original security properties of the signature, which are integrity, authentication and non-repudiation. In the following we will show that our scheme ensures these security essentials.

**Integrity**

In our proposed scheme base station checks the integrity of message $M$ by verifying the signature $\delta = (u_1, u_2, h)$, where $h = H(R_1, R_2, ID, Pub, M)$. The signature contains parameters $h$ which has the message $M$ , user public key and $R_1, R_2$. In order to verify the message BS uses signer's public key and KGC's master public key to compute values $R'_1$ and $R'_2$ first. Then verifier uses these values, signer's public key and message $M$ to generate $h' = H(R'_1, R'_2, ID, Pub, M)$. When verifier passes the equation $h'? = h$ and the verification of signature $\delta$, then he or she can believe that received message $M$ is equal to the value $M$ in signature $\delta$. Hence our proposed scheme provides a mechanism to convince that transmitted message and signature are correct and complete. The details of signature verification and correctness of equation $h'? = h$ are described in section 3 (verifying phase).

## Non-Repudiation

Non-Repudiation is a case where a malicious signer, who generates a signature with message $M$, but then denies the signature. In our proposed scheme when a signer generates a signature, he or she uses its $ID$ to generate the correct signature. During the sign phase node calculates the values of $R_1, R_2$, where it uses its identity value $ID$ provided by KGC. When signature $\delta = (u_1, u_2, h)$ is generated, we can see that $h = H(R_1, R_2, ID, Pub, M)$ which contains the identity. So during the verification phase verifier can easily find out that this signature is generated by node with identity $ID$. Signer has to apply one way hash function $H_0()$ on $ID$ to generate the signature, so without right ID he or she cannot produce correct signature. Hence proposed scheme can prevent the signer from repudiating their signature.

## Forgery Attack

An attacker can apply forgery in 2 ways: 1) forgery of the message, and 2) forgery of message and signature both.

### Forgery of Message

Assume that there is an attacker who is able to intercept the signature $\delta = (u_1, u_2, h)$ and message $M$ and modifies the message to $M'$. When he sends the signature $\delta = (u_1, u_2, h)$ and message $M'$ to verifier, for message and signature verification verifier uses signer's public key $Pub$ and its own master public key e to compute $R'_1 = u_1{}^e H_0(ID)^{eh} \bmod n$ and $R'_2 = H_0(ID)^{u_2} Pub^{Pr*h} \bmod n$. then verifier uses these values to generate $h' = H(R'_1, R'_2, ID, Pub, M')$ and verifies whether $h$ is equal to $h'$. In this case this will not be possible because there is an alteration in message, so verifier can easily detect that something is wrong with message and he or she will discard the message.

### Forgery of Message and Signature Both

Now assume the case where attacker intercepts the signature $\delta = (u_1, u_2, h)$ and message $M$ and modifies both message and signature as $\delta_{modify} = (u'_1, u'_2, h')$ and $M''$. Attacker will send this modified signature $\delta_{modify} = (u'_1, u'_2, h')$ and $M''$ to verifier. But as we can see that $h'$ contains $R_1, R_2, ID, Pub$ $and$ $M$ and values of $u_1$ and $u_2$ cannot be calculated without correct $Pub$ and master secret key $d$. Therefore without correct knowledge of these values attacker cannot pass the verification.

As we can see by demonstration given above, that our scheme can withstand the forgery attack in wireless sensor network.

## 4.3  Complexity and Performance Analysis

In this section we will calculate the complexity our proposed scheme and will compare with other related schemes. We will do this by finding out how many scalar multiplications [SM], exponential operations [E], hash operations [H], addition [A] and subtraction [S] operations are used in signature signing and verifying phase.

Table 1 Complexity of scheme During Signing phase

| Signing Message Phase | SM | E | A | S | H |
|---|---|---|---|---|---|
| Zang et al. [5] | 1 | 4 | 0 | 2 | 4 |
| Sharma et al. [10] | 3 | 4 | 0 | 2 | 4 |
| Our proposed Scheme | 2 | 3 | 0 | 2 | 4 |

Table 2 Complexity of scheme During Verification phase

| Verifying Message Phase | SM | E | A | S | H |
|---|---|---|---|---|---|
| Zang et al. [5] | 2 | 4 | 0 | 0 | 3 |
| Sharma et al. [10] | 2 | 4 | 0 | 0 | 3 |
| Our proposed Scheme | 4 | 4 | 0 | 0 | 3 |

As we can see that although in Table 2 during message verification phase our scheme takes more SM operations than other schemes, but as we recall that message verification is performed at base station only so if the cost for verification in our scheme is slightly more, it may be ignored because base station has unlimited energy and lifetime. So our main concern is to minimize the complexity of message signing phase which we can surely see in table 1.

Table 3 Performance analysis

| Schemes | No. of Phases in Algorithm | Security against Type I Attack | Security against Type II Attack | Authentication | Non-repudiation | Integrity |
|---|---|---|---|---|---|---|
| Zang et al. [5] | 7 | No | Yes | Yes | Yes | Yes |
| Sharma et al. [10] | 7 | Yes | Yes | Yes | Yes | Yes |
| Our proposed Scheme | 5 | Yes | Yes | Yes | Yes | Yes |

As we can see that our proposed scheme contains only 5 phases of algorithms and provides security against all mentioned security threats. So we cans argue that our algorithm perform better that other schemes in comparison in case of wireless sensor networks.

## 5    Conclusion and Future Scope

When we discuss about certificate-less signature scheme, we see that most of the existing certificate-less signature schemes were produced based on bilinear pairing technique which is very hard to build in real life environment. So in this paper we proposed a RSA based certificate-less signature scheme for wireless sensor networks which can be implemented in an efficient way. The computational complexity of Discrete Logarithm Problem is used to prove the security of our scheme in random oracles model. Our scheme uses only 5 polynomial time algorithms for signature generation. As we can see in Table 3 that our scheme has better performance than Zang et al [5] and Sharma et al. [10] in a wireless sensor network and provide almost all essential security goals.

As future work of this paper we will simulate our scheme in network security simulator tool AVISPA [19]. We will also improve the performance of our scheme and compare it with some other ID based and certificate- less signature schemes.

## References

1. Akyildiz, I.F., Su, W., Sankarasubramaniam, Y., Cayirci, E.: Wireless sensor networks: A survey. Computer Networks **38**, 393–422 (2002)
2. Al-Riyami, S., Paterson, K.: Certificateless public key cryptography. In: Advances in Cryptology (ASIACRYPT 2003). Lecture Notes in Computer Science, pp. 452–473 (2003)
3. Amin, F., Jahangir, H., Rasifard, H.: Analysis of public-key cryptography for wireless sensor networks security **2**(5), 403–408 (2008)
4. Chen, X., Makki, K., Yen, K., Pissinou, N.: Sensor network security: A survey. IEEE Communications Surveys Tutorials **11**(2), 52–73 (2009)
5. Zhang, J., Mao, J.: An efficient rsa-based certificateless signature scheme. Journal of Systems and Software **85**(3), 638–642 (2012)
6. Wood, A.D., Stankovic, J.A.: Denial of service in sensor networks. IEEE Computer **35**(10), 54–62 (2002)
7. Karlof, C., Wagner, D.: Secure routing in wireless sensor networks: Attacks and countermeasures. Elsevier's AdHoc Networks Journal, Special Issue on Sensor Network Applications and Protocols **1**(2–3), 293–315 (2003)
8. Cheng, C.-C., Sun, C.-y., Chang, S.: A strong RSA-based and certificateless-based Signature scheme. International Journal of Network Security **0**(0), 1–7 (2014)
9. Sharma, G., Verma, A.: Breaking the rsa-based certificateless signature scheme. Information-An International Interdisciplinary Journal **16**(11), 7831–7836 (2013)

10. Sharma, G., Bala, S., Verma, A.K.: An Improved RSA-based Certificateless Signature Scheme for Wireless Sensor Networks. International Journal of Network Security **0**(0), 1–8 (2014)
11. Tsai, J., Lo, N., Wu, T.: Weaknesses and improvements of an efficient certificateless signature scheme without using bilinear pairings. International Journal of Communication Systems **27**(7), 1083–1090 (2014)
12. Tso, R., Huang, X., Susilo, W.: Strongly secure certificateless short signatures. Journal of Systems and Software **85**(6), 1409–1417 (2012)
13. Tso, R., Yi, X., Huang, X.: Efficient and short certificateless signatures secure against realistic adversaries. The Journal of Supercomputing **55**(2), 173–191 (2011)
14. Gong, P., Li, P.: Further improvement of a certificateless signature scheme without pairing. International Journal of Communication Systems **27**(10), 2083–2091 (2014)
15. Zhang, X., Heys, H., Cheng, L.: Energy efficiency of symmetric key cryptographic algorithms in wireless sensor networks. In: 25th Biennial Symposium on Communications (QBSC 2010), pp. 168–172 (2010)
16. Zhang, Z., Wong, D., Xu, J., Feng, D.: Certificateless public-key signature: security model and efficient construction. In: Applied Cryptography and Network Security. Lecture Notes in Computer Science, vol. 3989, pp. 293–308 (2006)
17. He, D., Chen, J., Zhang, R.: An efficient and provably-secure certificateless signature scheme without bilinear pairings. International Journal of Communication Systems **25**(11), 1432–1442 (2012)
18. He, D., Khan, M., Wu, S.: On the security of a rsa-based certificateless signature scheme. International Journal of Network Security **15**(6), 408–410 (2013)
19. Glouche, Y., Genet, T., Houssay, E.: SPAN A Security Protocol Animator for AVISPA. IRISA, September 2008
20. Wang, C., Long, D., Tang, Y.: An efficient certificateless signature from pairings. International Journal of Network Security **8**(1), 96–100 (2009)
21. Bellare, M., Rogaway, P.: Random oracles are practical: a paradigm for designing efficient protocols. In: ACM Conference on Computer and Communications, pp. 62–73 (1993)
22. Attar, A., Tang, H., Vasilakos, A.V., Yu, F.R., Leung, V.C.M.: A Survey of Security Challenges in Cognitive Radio Networks: Solutions and Future Research Directions. IEEE Proceedings **100**(12), 3172–3186 (2012)
23. Han, K., Luo, J., Liu, Y., Vasilakos, A.V.: Algorithm Design for Data Communications in Duty-Cycled Wireless Sensor Networks: A Survey. Proceedings of IEEE Communications Magazine **51**(7) (2013)
24. Zhang, Z., Wang, H., Vasilakos, A.V., Fang, H.: ECG-Cryptography and Authentication in Body Area Networks. IEEE Transactions on Information Technology in Biomedicine **16**(6), 1070–1078 (2012)
25. Jing, Q., Vasilakos, A.V., Wan, J., Lu, J., Qiu, D.: Security of the Internet of Things: perspectives and challenges. Wireless Networks **20**(8), 2481–2501 (2014)
26. Yan, Z., et al.: A survey on trust management for Internet of Things. Journal of Network and Computer Applications **42**, 120–134 (2014)
27. Zhang, Y., et al.: A real-time dynamic key management for hierarchical wireless multimedia sensor network. Multimedia Tools and Applications **67**(1), 97–117 (2013)
28. He, D., et al.: ReTrust: Attack-Resistant and Lightweight Trust Management for Medical Sensor Networks. IEEE Transactions on Information Technology in Biomedicine **16**(4), 623–632 (2012)

29. Wang, T., et al.: Survey on channel reciprocity based key establishment techniques for wireless systems. Wireless Networks **21**(6), 1835–1846 (2015)
30. Zhou, J., et al.: 4S: A secure and privacy-preserving key management scheme for cloud-assisted wireless body area network in m-healthcare social networks. Inf. Sci. **314**, 255–276 (2015)
31. Fadlullah, Z.M., et al.: DTRAB: Combating Against Attacks on Encrypted Protocols Through Traffic-Feature Analysis. IEEE/ACM Trans. Network **18**(4), 1234–1247 (2010)
32. Chen, X., Makki, K., Yen, K., Pissinou, N.: Sensor network security: A survey. IEEE Communications Surveys Tutorials **11**(2), 52–73 (2009)

# SeaMoX: A Seamless Mobility Management Scheme for Real-Time Multimedia Traffic Over Cellular Networks

D. Kumaresh, N. Suhas, S. Garge Gopi Krishna, S.V.R. Anand
and Malati Hegde

**Abstract** Real-time multimedia applications are often deployed to provide critical information in situations such as news coverage of an event or an incident or an ambulance rushing to provide emergency care. Mobile cellular coverage and performance of a provider network varies both spatially and temporally and challenges the ability of the network to support uninterrupted network access and application continuity. We propose that inter-provider handoffs could alleviate this problem, given the recent proliferation of Dual SIM Dual Active (DSDA) devices and the possibility of Multi SIM Multi Active devices by configuring laptops with multiple wireless broadband connections. We propose an application QoS aware mobility management approach. The software implementation is termed as SeaMoX and based on SeaMo+, an earlier implementation. We use live video streaming as an example application and demonstrate the impact of network selection and handoff by examining the playback at the receiver, which uses an adaptive jitter buffer algorithm.

## 1 Introduction

Real-time multimedia services, on the move, are an increasing need for mobile users. Such services include routine requirements such as video calls, professional requirements such as video conferencing, media reporting – live, as well as critical services such as ambulatory tele-service (ATS). ATS is the example service for our study. Such a service on a mobile network infrastructure is a fairly complex

D. Kumaresh · N. Suhas · S.G.G. Krishna · S.V.R. Anand · M. Hegde(✉)
Department of Electrical Communication Engineering,
Indian Institute of Science, Bangalore 560 012, India
e-mail: gopi@serc.iisc.in, {anand,malati}@ece.iisc.ernet.in

© Springer International Publishing Switzerland 2016
S.M. Thampi et al. (eds.), *Advances in Signal Processing and Intelligent Recognition Systems*,
Advances in Intelligent Systems and Computing 425,
DOI: 10.1007/978-3-319-28658-7_59

problem in terms of managing to keep the end-to-end delay and jitter, under bounds to provide the QoS required by the application. Mobile devices evolved to have WLAN interfaces and WiFi offloading helped improve the average QoS. WiFi Coverage is limited to a few providers and dense urban environments. So, the scope to deploy multimedia services for those on the move is very limited.

In the recent past, mobile device hardware has evolved to provide a dual cellular network interface and a WLAN interface. Our interest is in devices that are termed Dual SIM Dual Active (DSDA). DSDA devices can have both the cellular network interface active, concurrently. Similarly, Multiple SIM Multiple Active (MSMA) mobile devices can be configured as in the case of a laptop with three or more wireless broadband connections. This provides a possibility to explore inter-provider handoffs, since it can potentially address the limitation of coverage and perhaps even bandwidth aggregation across the provider links whose network characteristics are somewhat similar.

Existing literature addresses WiFi offloading – a 3G to WiFi handover. In this case, the choice of hand over to the WiFi access network is a trivial one unless there are multiple such segments to hand over to. In contrast, hand overs between 3G network access segments are much more involved. While the networks are homogeneous, their characteristics are non-deterministic with temporal and spatial variations. Cellular networks exhibit larger minimum RTTs and higher RTT variability. [2] reports a fairly high variation of up to 500 ms with a loss rate of up to 2% in 3G networks. Therefore, unlike in the case of a WiFi access network, the 3G candidate networks require a prolonged sampling of the candidate networks to observe the network behavior trends to assess the suitability of the network for a hand over. Such long-term observations require constant connectivity to the candidate networks. MSMA devices are best suited for such use.

An added advantage in the context of handovers is that there is no break in the physical connectivity since all networks are already connected and the handover is effectively at the network layer, quite simply achieved by modifying the default outgoing interface. This is sufficient for simple data transfers but requires the adaptation of the upper layers – transport and other application sub layers (such as SSL) – to adapt to the network hand over.

WiFi offloading does not scale in terms of coverage since WiFi access segments are largely deployed in urban environments. Availability of WiFi access networks in suburban or non-urban environments is very low or almost nil. It is in such scenarios, there is a need for MSMA devices.

In this paper, we illustrate the use of an MSMA device in a suburban environment supporting a critical service such as ATS. WiFi does not support high speed mobile access unlike the cellular networks and therefore, our scenario extends to an urban environment as well. The application consists of an ambulance, with a patient requiring emergency care, rushing to a hospital. The patient is being monitored and data consisting of a few vital parameters and a video of the patient is sent to the hospital in real time.

A short review of related literature and the results of the evaluating the network coverage of a few providers are in section 2. The solution design and the implementation details are in section 3. Section 4 illustrates the benefits of the proposed solution with measurements on a live network and the paper concludes with a mention of the contributions made, in section 5.

## 2    Related Work

Utilizing two access links to aggregate the access bandwidth across them has been in use for a long time. Similarly, using two different paths between a source and a destination to aggregate the available bandwidth across the paths has been explored in the recent past. Multipath-TCP (MPTCP) and SCTP are two means of addressing this context. In the specific case of the DSDA devices, the two paths from the device will have different source addresses for the two paths that originate from the device and MPTCP specifically handles such contexts. The protocol specification has been released as RFC 6824. [1], [2] illustrate the operations of MPTCP and provide measurements on live networks. [2] observes that the MPTCP latencies are nearly comparable to the smallest latencies produced by WiFi or cellular network and effectively reduces the variability in download latencies.

[3] provides measurements of cellular networks and primarily observes that cellular networks deploy large buffers for reasons such as link layer retransmissions that compensate for lossy links and channel rate variation causing the traffic to be bursty. They demonstrate a RTT variation between 150 ms to 10 seconds on 3G networks. The buffering increases the network delay and jitter. [2] characterizes cellular networks as *not so fast but loss free* compared to WiFi networks. They report a 300 ms to 800 ms RTT variation on the 3G network and packet loss rate with a fairly consistent variation between 0.37 to 1.64 %. These studies give us a fair idea of what to expect. In the context of medical applications, the study of the effect of bottleneck links on the transfer of tomography and radiography images is the closest [8]. The variation in image upload times is simulated for different bottleneck link speeds and different number of concurrent uploads.

There are no specific studies in literature that study the use of multiple cellular network links, which are concurrently active, for services. Likewise there are no studies that consider the use of such an infrastructure for real time data and video services. This is our primary contribution. Our focus is on managing real time multimedia traffic which uses UDP, over multiple cellular access links available from a device, and perform the necessary adaptation before the application receives the traffic. We begin with using our earlier implementations named SeaMo [5] and SeaMo+ [4].

[4] presents a framework, VRMS, and an implementation, SeaMo+ to make the application transparent to the changes below the application layer in the context of mobility and handovers so that the application experiences a consistent network performance. The solution doesn't specifically address real time multimedia although it addresses real time traffic. The solution in [4] uses the fuzzy logic based network handoff decision making from [5]. We extend the framework in [4] to handle real time multimedia traffic, across multiple paths on the cellular network

and include the necessary adaptation techniques into the implementation, which we term as SeaMoX.

# 3    Cellular Network Performance and Radio Coverage

## 3.1    Network Coverage

In order to make the case for the need for MSMA devices and the need to use multiple paths across multiple provider networks beneficially, we illustrate the uneven coverage of the providers with measurements across a campus. The network coverage evaluation was done within a 4 Km radius within a campus with a fairly high degree of tree cover, with measurements made every 50-70 metres along the roads. A total of 350 locations were sampled. At each location, a laptop equipped with concurrent access to three providers and a GPS measure three parameters namely:

1.  The average RSSI value over a fixed duration of time
2.  Throughputs for bulk upload and downloads to/from a remote server
3.  ICMP Echo to measure the RTTs and packet loss

A throughput of 600 Kbps was labelled as good coverage considering that with MPEG-4 encoded video, 1KB I-Frames typically cause 600 Kbps bursts when compared to the 128/256B P and B frames that require about 384Kbps. Coincidentally, the network bandwidth varied up to a peak of 1.6 Mbps with an average of 600 Kbps.

In summary, there were significant null coverage areas, which were encountered where the packet losses were very high. Provider coverage varied across the area, although one provider did seem to stand out in terms of the maximum coverage. However, even the better provider had null coverage areas.

## 3.2    Network Delay, Jitter and Bandwidth

We measure the delay and jitter on the provider networks that cover a given campus. A server on the campus network is chosen as the remote server to perform measurements with. We ensure that this server's access link is not a bottleneck link. The access link is a 1 Gbps link with about 56.5% peak occupancy. Then, we originate 1024 byte ICMP ECHO requests to the remote server as well as some well known locally hosted servers via the different providers' wireless links and observe fairly similar average RTT values and mean deviations.

Next, a mobile node, with an average speed of 50 Kmph is time synchronized with a static remote server using ntp. CBR traffic at 150 Kbps is sent over UDP to the remote server. The packets are 1024 bytes; time stamped and has sequence numbers. The remote server receives the packets and measures the drops, delay and jitter. This measurement is done on three provider networks and the results are in Tables 2 and 1. Note that these measurements provide the delay and jitter faced by traffic on the uplink of the 3G wireless accesses. Based on these three parameters, the service provider (SP) networks can be ordered as SP2, SP1 and SP3 in terms of better performance.

**Table 1** Network delay and jitter on service provider networks

|  | Avg Delay in Secs | SD for Delay in Secs | Avg Jitter in Secs | SD for Jitter in Secs |
|---|---|---|---|---|
| SP1 | 1.25 | 1.92 | 35 | 56.6 |
| SP2 | 1.70 | 2.94 | 23.53 | 27.28 |
| SP3 | 1.78 | 7.12 | 62.36 | 293.19 |

**Table 2** Packet loss on service provider networks

|  | Packet loss %age | | | |
|---|---|---|---|---|
|  | Min | Max | Avg | SD |
| SP1 | 4.24 | 29.44 | 14.65 | 10.40 |
| SP2 | 0.92 | 10.56 | 4.80 | 8.54 |
| SP3 | 0.06 | 73.02 | 32.21 | 30.17 |

The average one-way delay in the cellular service provider's network varies from 1.25 to 1.78 seconds. Jitter in the provider networks is of the order of tens of seconds with a large standard deviation. These service levels are way below those required for video conferencing or VoIP. These services require latency less than 150ms, less than 30ms jitter and less than 1% packet loss.

The jitter measurements give an idea of the size of the play out buffer required for the video play back at the receiver side (hospital). Dimensioning this buffer is the key to the QoE for the video. An improper size can upset the receiver causing buffer starvation and poor rendering of multimedia streams.

# 4    SeaMoX for Ambulatory Services

## 4.1    The Application Scenario

The application considered for the study is termed as Ambulatory Tele-Services (ATS). The scenario is the described as follows: A patient requiring emergency care is being driven to the hospital in an ambulance. There is a doctor on board and there is a need for the expert at the hospital to monitor the condition of the patient being rushed to the hospital, as well as arrange for other resources like specialists or those required for a surgery. The data sent from the ambulance to the hospital, are the real time EEG/ECG data of the patient and a real time visual of the patient or specific parts of the patient's anatomy. The EEG/ECG data aggregates to a constant bit rate of 150 Kbps and the real time video (mpeg-4) aggregates to 150 Kbps with bursts of 600 Kbps. Both streams are sent but the EEG/ECG data stream is mandatory and takes priority. Our solution is specifically discussed in this context and the measurements made are relevant to this context.

The solution to handling the real time traffic is an extension of [4], at the core of which is the Virtual Real Time Multimedia Service (VRMS). The VRMS is a software agent at the application layer to which all applications direct their requests. VRMS in turn makes a copy of all the signaling information of the application and passes through all the application packets to the remote application. It uses the stored signaling information after a handover to re-initiate a session on behalf of the application. The application is transparent to the re-initiation and experiences an uninterrupted access. The VRMS can be programmed to signal the application to perform any necessary adaptation, given that there may be packet delays.

VRMS has two limitations that make it unsuitable for use with real time video – it has no support for RTP/RTSP and it is primarily designed for use on a mobile host that uses the services from/to a static remote server. It cannot handle a mobile server. In our context, such an example would be two mobile users in a videoconference or the application scenario we have considered. VRMS is enhanced to overcome these limitations.

## 4.2   No Handoffs, Switch Paths

The mobility and handoff requirements are over-ridden by the fact that the device uses multiple active cellular interfaces. However, the resource estimation on the networks requires to be done continuously to be able to send/receive traffic on the best network/path possible. VRMS switches the traffic dynamically across the network paths accessible via the multiple network interfaces.

The network path quality estimation has to be periodic and the feedback provided to the VRMS. Downloading a 100 KB file from a remote server every 10s via each of the links does the throughput estimates. The Network Selection Module (NSM) uses this data for the network selection. The frequency of the quality estimation is judiciously done to ensure that the quality estimation does not cause a significant overhead.

## 4.3   Managing RTP/RTSP

A session consists of establishing a session with RTSP signalling followed by data transfer over RTP with RTCP packets, which are used for controlling the session [6]. VRMS is aware of the protocol and saves the initial RTSP signalling packets, when the application connects using network N1. Once VRMS is instructed by the NSM module to use network N2 it resends the signalling packets establishing a new connection to the remote host, using N2. Once the new connection is established the data sent over RTP has a different set of timestamps and synchronization source (SSRC) IDs. Once the new session is established it then sends a

TEARDOWN message to stop the previous stream. VRMS modifies this new set of timestamps and SSRCs to continue from the previous ones so that the session at the client application continues seamlessly.

## 4.4  Implementation Architecture – VRMS

Fig. 1 illustrates the architecture of VRMS, which comprises of a Link Quality Estimator (LQE), which provides inputs to a Network Selection Module (NSM). The NSM receives additional inputs for decision making from the user preferences module and the battery life monitor. The VRMS receives inputs from the NSM and triggers the appropriate session re-connects to ensure application continuity.

The software implementation of VRMS is what is termed as SeaMo+ [4]. Likewise, the implementation of the LQE and the handoff algorithm in the NSM is SeaMo [5]. SeaMoX is the name for SeaMo+ (VRMS) enhanced with the RTSP support and with path switching using multiple active cellular interfaces.

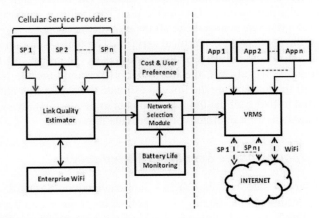

**Fig. 1** Architecture of VRMS

## 4.5  Initial Observations

A mobile node with SeaMoX was tested with a real time video stream from a static source, an IP camera. The video transmitted from the camera was 360x240, MPEG-4 encoded with a bit rate of 80 Kbps (P and B frames) peaking to 150 Kbps (I frames). VLC [7] player was used as a client on the receiving side. With the low bandwidth requirement for the live stream from the camera, a good quality stream was expected. On the network the stream was started, the reception was good. Upon switching paths, from one provider to another, the stream paused often. This occurred when the delays on the current network were larger than that of the previous network. VLC's play out buffer was not responsive to the delays and unable to adapt. It dropped the packet arriving over the new network treating them as delayed packets.

On the receiving side, VRMS used buffers to receive the packets, sequence them and hand them over to the application. Packets in sequence are handed over as and when they arrived into the VRMS buffer. The buffer was configured for 150 packets. When there is/are a packet/s that is out-of-sequence, VRMS starts buffering all incoming packets until the packet arrives and then hands over the packets to the application. The time out for the missing packet/s to arrive is 100 ms; its value is set based on size of the play out buffer of the application to ensure that it is not starved. The RTSP support in SeaMoX functioned appropriately when the traffic was transferred from one link to another. The implementation was then field-tested using a higher resolution video source.

## 5    Multiple Cellular Access Links

The application scenario (ref. § 2.4) has a mobile data source sending data to a remote static server over a cellular network access. The data is sent over the up-link of the cellular access link. The device used has multiple concurrently active cellular links and remains connected to all the provider networks.

Recall that the application is a critical service. The focus is not on optimal use of resources, but on the delivery of the critical data. Therefore, all the cellular links remain connected always, the high energy consumption due to remaining connected to all the links, using the links to estimate the access link characteristics periodically and utilizing all the links to send data to the destination are all acceptable. VRMS uses all of these effectively to provide the solution.

### 5.1    The Mobile Source Scenario (Ambulance)

The mobile sender side has the VRMS implementation, SeaMoX application loaded on it, configured and running. The application programs on the sender side connect to SeaMoX for their services. SeaMoX connects to all available cellular access links on the device and begins to estimate the link quality. The application starts up and connects to SeaMoX. SeaMoX in turn connects to the remote host/service and stores the control information depending upon the protocol used.

The ECG/EEG data is prioritized over video. The bit rate requirement for each stream is 150 Kbps, totaling to 300 Kbps. The ECG/EEG data is tolerant to delays, compared to the video, but not to packet loss or any such interruptions, whatsoever. From an application perspective, this data is critical and has to be delivered to the host/server at the hospital. The data is viewed in real time as well as archived as part of the patient's medical records. Video is relatively less critical and used for visual monitoring and assessment.

The ECG/EEG data packets are sent on all the links available, without any exception. This is done to ensure every packet sent is received at the remote host. The link with the best performance estimate is chosen to send the video. The ECG/EEG data is prioritized over video when both data use the same link.

The video traffic is switched to use the better link based on the inputs from the NSM. We term this switching of the link/path as a *soft handoff*.

## 5.2   The Receiving Host (Hospital)

The receiving host runs SeaMoX. The data destined for the applications arrive into the SeaMoX buffers. Each application has a separate buffer to ensure that the buffer size and the packet time out for each application are kept separate and dimensioned according to the application's requirements.

For the 150 Kbps ECG/EEG stream the packets arrive on all the links. The incoming packets are received into the buffer to remove duplicates. When a packet, out-of-sequence arrives, the packets are buffered until the rest of the packets arrive. When a packet is delayed, the packets arriving are delivered to the application. The delayed packet is sent to the application when it arrives. The data buffer is serviced as a FIFO. A 100-packet buffer is dimensioned with a 100 ms initial timeout value. The time out is adapted based on the average delay of the last ten packets.

## 5.3   Observations

Field trials were conducted on the same campus (ref. § 3.1). We found that the packet losses reduced significantly from what was observed earlier (ref. Table 2). A total of 23,939 packets with an average size of 800 bytes (no IP fragmentation) were sent on each provider link, during the session and a total of 8050 packets were not received. The aggregate packet loss was 11%. The average packet loss measured at the SeaMoX buffer (the application packets that never reached) was 0.82% with a SD of 0.127. These values are significantly low compared to the averages and the SD values measured on the three provider networks. The mean jitter is considerably reduced and so is the maximum value (ref. Fig. 2).

The share of packets that arrived on each link that were delivered are listed in Table 3 lists those values. The values in Table 2 and Table 3 correlate and the providers rank in the order SP2, SP1 and SP3.

## 5.4   Handling Video Playback

Handling video playback and testing video from the mobile source is work in progress. One of the playback problems was that the video would automatically pause between play times upon switching links. VLC does not grow its play out buffer to accommodate seconds worth of video. If the video lags behind by around a second or more, VLC pauses the video and resynchronizes with the stream. There are then two ways of tackling this – create a separate adaptive play out buffer to feed into the VLC buffer at the cost of including a complex algorithm into

**Table 3** Provider-wise share of the total number of packets delivered to the application

| | %age packets delivered to the application | | | |
|---|---|---|---|---|
| | Min | Max | Avg | SD |
| SP1 | 15.90 | 37.89 | 22.10 | 9.77 |
| SP2 | 38.29 | 76.68 | 66.48 | 16.52 |
| SP3 | 2.20 | 24.53 | 11.39 | 8.18 |

SeaMoX or review the link selection strategy in NSM to make it service dependent – somewhat like how OSPF has multiple network views based on different parameters such as hop count, delay, bandwidth, etc. However, in order to understand the behavior of the network and the impact on video traffic, more measurements are being made.

In terms of the video requirement, we believe there will be a need for higher resolution video, which implies a higher bit rate. There are limits on the available bit rate on 3G uplinks and an optimistic estimate is about 700 Kbps. Video conferencing (VC) quality requires 600 Kbps with MPEG-4 encoding. MJPEG encoding was an option considered because of its low computation overheads but the total bandwidth requirement for a 320x240 at 15fps is 1 Mbps and does not match the bit rate of the 3G uplink. Choosing to use VC quality for the video feed from the mobile source will necessitate considering two or more available cellular links to aggregate the bandwidth for video. These trials are under way. The ratio and the basis on which the packets are assigned and scheduled for transmission on each of the aggregated links, requires being decided. There exist numerous schemes used on wired networks. They need to be analyzed and adapted for this aggregation scenario to observe the extent of jitter variation in the mix of packets that have traversed two different links.

On the receiver side, the buffer remains a tricky question. The SeaMoX buffer feeds into the video player application's buffer. VLC [7] was used as the video

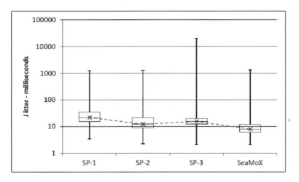

**Fig. 2** Jitter - Individual networks vs aggregation on SeaMoX

player. VLC has a play out buffer of 300 ms and adapts based on the network condition. The buffer for the video is currently set at 100 ms so that it is transparent to VLC's adaptation. It is serviced using FIFO.

When packets traverse they experience different jitter. If the SeaMoX buffer FIFO'es it to VLC, VLC will see a large variation in jitter values and may converge to the smaller one, rather than the larger one or vice versa. There is a need to analyze this further both with measurements as well as theoretically.

## 6 Conclusion

In this paper we have presented some of the performance issues that need to be addressed for supporting real-time multimedia applications in mobility scenarios across cellular networks. Our extensive field experiments in an urban setting illustrated several radio coverage voids and unpredictable network performance in terms of packet loss and delay on cellular networks. Such resource shortcomings require mobility to be addressed differently when compared to the approaches such as WiFi offloading. We proposed a seamless mobility management scheme, SeaMoX, which has been developed to meet the QoS requirements of mobile real-time multimedia applications over cellular networks. Our real-world experiments emulated ambulatory tele-services, transmitting vital medical data in real time and have shown significant improvement in terms of data reliability and end-to-end delays. The experiences gained from our work has provided insights into the effect of sending video data over multiple networks on the adaptive play out jitter buffer algorithm that runs within the media player at the receiver end. This aspect is what we believe is the take away from the paper. Our further work on delving into other aspects of the delays and adaptive play out buffering strategies is underway.

## References

1. Barré, S., Bonaventure, O., Raiciu, C., Handley, M.: Experimenting with multipath TCP. ACM SIGCOMM Computer Communication Review **41**(4), 443–444 (2011)
2. Chen, Y.C., Lim, Y.S., Gibbens, R.J., Nahum, E.M., Khalili, R., Towsley, D.: A Measurement-based study of multipath TCP performance over wireless networks. In: Proceedings of the 2013 Conference on Internet Measurement Conference, pp. 455–468. ACM, October 2013
3. Jiang, H., Wang, Y., Lee, K., Rhee, I.: Tackling bufferbloat in 3G/4G networks. In: Proceedings of the 2012 ACM SIGCOMM Conference on Internet Measurement Conference (IMC) (2012)
4. Prasad, G.B., Seema, K., Shrikant, U.H., Garge, G.K., Anand, S.V.R., Hegde, M.: SeaMo+: A virtual real-time multimedia service framework on handhelds to enable remote real-time patient monitoring for mobile doctors. In: 2013 Fifth International Conference on Communication Systems and Networks (COMSNETS), pp. 1–6, January 7–10, 2013

5. Rafiq, M., Kumar, S., Kammar, N., Prasad, G., Garge, G.K.S., Anand, S., Hegde, M.: A Vertical Handoff decision scheme for end-to-end QoS in heterogeneous networks: An implementation on a mobile IP testbed. In: 2011 National Conference on Communications (NCC), pp. 1–5, January 28–30, 2011
6. Schulzrinne, H., Casner, S., Jacobson, V., Frederick, R.: RTP: A Transport Protocol for Real-Time Applications. RFC 3550, July 2003
7. VideoLAN Organization. VideoLAN. http://www.videolan.org/videolan/
8. Mlynek, P., Misurec, J., Koutny, M., Dostal, O.: Medical traffic modeling for delay measurement in bottleneck network. In: 2011 34th International Conference on Telecommunications and Signal Processing (TSP), pp. 208–212. IEEE, August 2011

# Author Index

Printed in the United States
By Bookmasters